Atmospheric Science for Environmental Scientists

Edited by

C.N. Hewitt
Lancaster University
Lancaster, UK

and

Andrea V. Jackson
University of Leeds
Leeds, UK

Second Edition

Registered Offices
John Wiley & Sons, Inc., 111 River Street, Hoboken, NJ 07030, USA
John Wiley & Sons Ltd, The Atrium, Southern Gate, Chichester, West Sussex, PO19 8SQ, UK

Editorial Office
9600 Garsington Road, Oxford, OX4 2DQ, UK

For details of our global editorial offices, customer services, and more information about Wiley products visit us at www.wiley.com.

Wiley also publishes its books in a variety of electronic formats and by print-on-demand. Some content that appears in standard print versions of this book may not be available in other formats.

Library of Congress Cataloging-in-Publication data has been applied for

9781119515227 (paperback)

Cover Design: Wiley
Cover Image: © DANNY HU/Getty Images

Set in 9.5/12.5pt STIXTwoText by SPi Global, Pondicherry, India
Printed and bound in Singapore by Markono Print Media Pte Ltd

10 9 8 7 6 5 4 3 2 1

Contents

List of Contributors

Janet Barlow
Department of Meteorology
University of Reading
Reading, UK

Peter Brimblecombe
School of Energy and Environment
City University of Hong Kong
Hong Kong

Martyn P. Chipperfield
School of Earth and Environment
University of Leeds
Leeds, UK

Hugh Coe
School of Earth, Atmospheric, and
Environmental Sciences
The University of Manchester
Manchester, UK

Nick Hewitt
Lancaster Environment Centre
Lancaster University
Lancaster, UK

Atul Jain
Department of Atmospheric Sciences
University of Illinois
Urbana, IL, USA

Anwar Khan
School of Chemistry
University of Bristol
Bristol, UK

John Lockwood
Formerly University of Leeds
Leeds, UK

A. Rob MacKenzie
School of Geography
Earth and Environmental Sciences
University of Birmingham
Birmingham, UK

Paul Monks
Department of Chemistry
University of Leicester
Leicester, UK

Dudley Shallcross
School of Chemistry
University of Bristol
Bristol, UK

Zongbo Shi
School of Geography, Earth and
Environmental Science
The University of Birmingham
Birmingham, UK

Natalie Theeuwes
Department of Meteorology
University of Reading
Reading, UK

Joshua Vande Hey
Department of Chemistry
University of Leicester
Leicester, UK

Richard Wayne
Physical and Theoretical Chemistry Laboratory
Department of Chemistry
University of Oxford
Oxford, UK

Paul I. Williams
School of Earth and Environmental Sciences &
National Centre for Atmospheric Science
The University of Manchester
Manchester, UK

Xiaoming Xu
Department of Atmospheric Sciences
University of Illinois
Urbana, IL, USA

Preface

When we wrote the Preface to the first edition of 'Atmospheric Science for Environmental Scientists' in 2008, we noted that never before had the teaching, learning, and researching of atmospheric science been so important. We said that society must face up to the realities of global atmospheric change, including global warming and poor air quality, and that the education of students and provision of accessible information to policy makers and the public were priorities.

More than a decade later, we can only reiterate these sentiments. In 2018, the Intergovernmental Panel on Climate Change warned that the planet will reach the crucial threshold of 1.5 °C above pre-industrial levels by as early as 2030, precipitating the risks of extreme drought, wildfires, floods, and food shortages for hundreds of millions of people. And in 2018, the World Health Organization reported that 90% of the world's population lived in places where air quality exceeded WHO guideline limits, and that more than 4 million people a year died prematurely from outdoor air pollution and a further 3 million a year from indoor air pollution.

What further warnings are needed? To help society cope with the unprecedented changes that humankind is causing to our fragile atmosphere, education must be key and policy makers must act. We hope this book helps both causes.

In putting this book together, we have drawn on some of the best experts and educators in the field of atmospheric science. We hope their knowledge and enthusiasm shines through in these chapters. Our aim is to provide succinct but detailed information on all the important aspects of atmospheric science for students of environmental science and to others who are interested in learning how the atmosphere works, how humankind is changing its composition, and what effects these changes might lead to.

We are grateful to all the experts who have contributed to this book, for all reviewers' comments, and to all our students over the years who have demonstrated the need for this volume.

October 2019

Nick Hewitt
Andrea V. Jackson

Abbreviations, Constants, and Nomenclature

ADMS	Atmospheric Dispersion Modelling System
CEE	Central and eastern Europe
CCN	cloud condensation nuclei
CFC	chlorofluorocarbons
CO_2	carbon dioxide
DMS	dimethyl sulphide
DNA	deoxyribonucleic acid
EC	elemental carbon
EM	electromagnetic
ENSO	El Niño–Southern Oscillation
EPA	Environmental Protection Agency
EU	European Union
GDP	global domestic product
GEMS/AIR	Global Environmental Monitoring System/Air
GHG	greenhouse gas
HAP	hazardous air pollutant
HCFC	hydrochlorofluorocarbons
HFC	hydrofluorocarbons
IAM	integrated assessment models
IBL	internal boundary layer
IPCC	Intergovernmental Panel on Climate Change
IR	infrared
ISAM	integrated science assessment model
ITCZ	intertropical convergence zone
LAI	leaf-area index
LW	longwave
NMHC	non-methane hydrocarbons
MAP	major air pollutant
MTBE	methyl-*tert*-butyl ether
NDVI	normalized difference vegetation index
OCS	carbonyl sulphide

OECD	Organization for Economic Cooperation and Development
PAH	polycyclic aromatic hydrocarbons
PAN	peroxyacetyl nitrate
PAR	photosynthetically active radiation
PCB	polychlorinated biphenyls
PFC	perfluorogenated carbon
PM	particulate matter
PM_{10}	particles with aerodynamic diameter less than 10 μm
ppm	parts per million
ppmv	part per million by volume (1×10^{-6})
ppbv	part per billion by volume (1×10^{-9})
pptv	part per trillion by volume (1×10^{-12})
PSS	photostationary state
SAFARI	South African Regional Science Initiative
SW	shortwave
TSP	total suspended particulates
UNEP	United Nations Environmental Programme
UV	ultraviolet
VSLS	very short-lived substances
VOC	volatile organic compounds
WHO	World Health Organization
WMO	World Meteorological Organization

Constants

c	speed of light in vacuum $2.998 \times 10^8 \, \mathrm{m\,s^{-1}}$
g	acceleration due to gravity $9.8 \, \mathrm{m\,s^{-2}}$
h	Planck's constant $6.626 \times 10^{34} \, \mathrm{J\,s}$
k	Boltzmann constant $1.381 \times 10^{34} \, \mathrm{J\,K^{-1}}$
R	gas constant $8.314 \, \mathrm{J\,K^{-1}\,mol^{-1}}$ $(1.3 \times 10^5 \, \mathrm{l\,atm\,mol^{-1}\,K^{-1}})$
Γ_d	dry adiabatic lapse rate $9.81 \, \mathrm{K\,km^{-1}}$
π	3.14159
σ	Stefan–Boltzmann constant $5.67 \times 10^{-8} \, \mathrm{W\,m^{-2}\,K^{-4}}$

Nomenclature

a	radius of a particle
A	albedo
A_s	surface albedo
B	radiative intensity of a blackbody
c_p	specific heat capacity of dry air at constant pressure $(1004 \, \mathrm{J\,kg^{-1}\,K^{-1}})$

c_v	specific heat at constant volume
C	concentration (ppm or $kg\ m^{-3}$)
d	preface to variable indicating incremental quantity
d	displacement height (m)
dQ	incremental change in heat
du	incremental change in internal energy
dv	incremental change in volume
dw	incremental change in work
e	turbulent kinetic energy per unit mass ($J\ kg^{-1}$)
F_B	the total flux from a black body radiator
F_s	incoming solar radiation absorbed at the surface
F	net flux
$F\uparrow$	upwelling radiative flux
$F\downarrow$	downwelling radiative flux
$\bar{F}$	net flux leaving an element or layer
F_r	total upward reflected shortwave flux
F_0	incident solar flux
G	ground heat flux ($W\ m^{-2}$)
h	mean height of roughness elements (m)
h_b	depth of internal boundary layer (m)
H	sensible heat flux ($W\ m^{-2}$) or mean building height (m)
H_s	scale height
H/W	aspect ratio (-)
I	intensity of light
I_0	initial intensity of light
k	von Kármán's constant ≈ 0.4 (-)
k_a	absorption coefficient
k_e	extinction coefficient
k_s	scattering coefficient
l	distance through a gas interacting with light
L	latent heat of vaporization
L	Obukhov length (m)
L_w	Liquid water content
L_x	integral (or decorrelation) lengthscale (m)
m	refractive index
M	molar mass (of air unless otherwise specified)
n	number concentration, typically of absorbers or scatterers
$p(u)$	probability of windspeed u (-)
P	plant area index (-)
P, p	pressure
q	specific humidity ($kg\ kg^{-1}$, sometimes $g\ kg^{-1}$) or source emission rate ($kg\ s^{-1}$)
q_*	scaling parameter for specific humidity profile ($kg\ kg^{-1}$, sometimes $g\ kg^{-1}$)
r	radial distance

r_v mass mixing ratio of water vapour

r_w saturation mixing ratio

r_r scattered fraction reflected upwards

R radius of the Earth

$R(\lambda)$ action spectrum

R_n net radiation (W m^{-2})

S   Solar constant

Sk_u skewness statistic for downstream component of the wind (-)

Sk_w skewness statistic for vertical component of the wind (-)

t time (s)

t_r scattered fraction transmitted downwards

t_t total fraction of radiation transmitted downwards

T temperature

T_a transmittance of the atmosphere

T_e effective blackbody temperature of the Earth

T_L integral timescale (s)

T_* scaling parameter for temperature profile (°C)

u downstream velocity component (m s^{-1})

$\bar{u}$ mean component (m s^{-1})

u' fluctuation around mean component (m s^{-1})

u_* friction velocity (m s^{-1})

$\boldsymbol{U}$ velocity vector (m s^{-1})

$\bar{U}(z)$ vertical mean wind profile (m s^{-1})

$\bar{U}_h$ mean windspeed at the top of a canopy (m s^{-1})

v lateral velocity component (m s^{-1})

w vertical velocity component (m s^{-1})

W along wind building spacing (m)

x horizontal distance in downstream direction (m)

y horizontal distance in lateral direction (m)

z distance, usually altitude (m)

z_h roughness length for heat transfer (m)

z_m height of maximum plant area index (m)

z_0 roughness length for momentum transfer (m)

z_{0r} roughness length for rural surface (m)

z_{0u} roughness length for urban surface (m)

z_q roughness length for moisture transfer (m)

z^* roughness sublayer depth (m)

z/L Monin-Obukhov stability parameter (-)

σ standard deviation (depends on quantity)

σ_x downstream plume spread (m)

σ_y lateral plume spread (m)

σ_z vertical plume spread (m)

$\sigma(\lambda)$ absorption cross section as a function of wavelength

β	upscatter or backscatter fraction
γ_s	saturated adiabatic lapse rate
ΔS	storage term (W m^{-2})
ΔT	temperature difference between rural and urban areas (or heat island intensity) (°C)
∂	preface to variable indicating incremental quantity
θ	scattering angle, or potential temperature (°C or K)
θ_0	potential temperature at the surface (°C or K)
ε	emissivity
λ	wavelength of light
λE	latent heat flux (W m^{-2})
λ_F	frontal area index (-)
ρ	air density (1.2 kg m^{-3})
ρ_v	density of water vapour
ρ_a	density of dry air
ρ_{NIR}	ratio of emitted to incident near infra-red radiation (-)
ρ_{VIS}	ratio of emitted to incident visible radiation (-)
χ	optical depth
τ	momentum flux, or shear stress (kg m^{-1} s^{-2})
υ_e	extinction coefficient (-)
ω_0	single scattering albedo
ψ_m	stability function for momentum (-)

1

The Climate of the Earth

John Lockwood

Formerly University of Leeds, Leeds, United Kingdom

The causes, history, and distributions of the Earth's climates are introduced in this chapter. The combination of the distribution of incoming solar radiation across the Earth's surface and the Earth's rotation both drive and shape the observed atmosphere–ocean circulation. Important factors determining changes in climate include palaeogeography, greenhouse gas concentrations, changing orbital parameters, and varying ocean heat transport. One of the major controls of climatic changes is the greenhouse gas concentration of the atmosphere, in particular that of carbon dioxide. Before the Eocene–Oligocene boundary (≈34 Myr ago) the atmosphere–ocean circulation supported a warm atmosphere and ocean, with both poles free of permanent ice. At the Eocene–Oligocene boundary, the atmosphere–ocean circulation changed to a form similar to the present, and the first evidence of an Antarctic ice sheet is found. Falling atmospheric carbon dioxide levels probably caused this change. The waxing and waning of massive temperate latitude continental ice sheets characterize the climate of the past million years. This chapter discusses recent climate changes and evidence that they are largely driven by anthropogenic generated atmospheric carbon dioxide. In particular, recent climate changes are causing the expansion of the tropical zone and a retreat of the polar zones.

The major climate zones of the world are described, with particular attention to interannual variability, and the causes of droughts and heavy rainfalls. This includes discussions of the climatic effects of the North Atlantic Oscillation (NAO) and El Niño-Southern Oscillation (ENSO).

For more specific information on global warming and climate change science, the reader is referred to Chapter 11 in this book and to the latest reports of the Intergovernmental Panel on Climate Change, available at www.ipcc.ch.

1.1 Basic Climatology

The climate of a particular place is the average state of the atmosphere observed as the weather over a finite period (e.g. a season) for a number of different years. The so-called *climate system*, which determines the weather, is a composite system consisting of five major interactive adjoint components: the

Atmospheric Science for Environmental Scientists, Second Edition. Edited by C.N. Hewitt and Andrea V. Jackson.

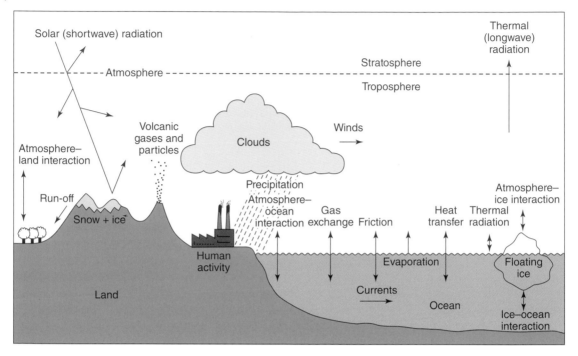

Figure 1.1 The climate system (Houghton 2005).

atmosphere, the hydrosphere, including the oceans, the cryosphere, the lithosphere, and the biosphere (Figure 1.1). All the subsystems are open and not isolated, as the atmosphere, hydrosphere, cryosphere, and biosphere act as cascading systems linked by complex feedback processes. The climate system is subject to two primary external forcings that condition its behaviour: solar radiation and the Earth's rotation. Solar radiation must be regarded as the primary forcing mechanism, as it provides almost all the energy that drives the climate system.

The distribution of climates across the Earth's surface is determined by its spherical shape, its rotation, the tilt of the Earth's axis of rotation in relation to a perpendicular line through the plane of the Earth's orbit around the Sun, the eccentricity of the Earth's orbit, the greenhouse gas content of the atmosphere, and the nature of the underlying surface. The spherical shape creates sharp north–south temperature differences, whilst the tilt is responsible for month-by-month changes in the amount of solar radiation reaching each part of the planet, and hence the variations in the length of daylight throughout the year at different latitudes and the resulting seasonal weather cycle.

The present orbit of the Earth is slightly elliptical with the Sun at one focus of the ellipse, and as a consequence the strength of the solar beam reaching the Earth varies about its mean value. At present, the Earth is nearest to the Sun in January and farthest from the Sun in July. This makes the solar beam near the Earth about 3.5% stronger than the average mean value in January, and 3.5% weaker than average in July. The gravity of the Sun, the Moon, and the other planets causes the Earth to vary its orbit around the Sun (over many thousands of years). Three different cycles are present, and when combined, produce the rather complex changes observed. These cycles affect only the seasonal and geographical distribution of solar radiation on the Earth's surface, yearly global totals remaining constant. Surplus in one season is compensated by a deficit during the opposite one; surplus in one geographical area is compensated by simultaneous deficit in some other zone. Nevertheless, these Earth orbital variations can have a significant effect on climate and are responsible for some major long-term variations.

Firstly, there are variations in the *orbital eccentricity*. The Earth's orbit varies from almost a complete circle to a marked ellipse, when it will be nearer to the Sun at one particular season. A complete cycle from near circular through a marked ellipse back to near circular takes between 90 000 and 100 000 years. When the orbit is at its most elliptical, the intensity of the solar beam reaching the Earth must undergo a seasonal range of about 30%. Second, there is a wobble in the Earth's axis of rotation causing a phenomenon known as the *precession of the equinoxes*. That is to say, within the elliptical orbits just described, the distance between Earth and Sun varies so that the season of the closest approach to the Sun also varies. The complete cycle takes about 21 000 years. Lastly, the *tilt of the Earth's axis of rotation* relative to the plane of its orbit varies at least between 21.8° and 24.4° over a regular period of about 40 000 years. At present, it is almost 23.44° and is decreasing. The greater the tilt of the Earth's axis, the more pronounced the difference between winter and summer. Technically, these three mechanisms are known as the *Milankovitch* mechanism.

If the Earth did not rotate relative to the Sun – that is, it always kept the same side towards the Sun – the most likely atmospheric circulation would consist of rising air over an extremely hot, daylight face and sinking air over an extremely cold, night face. The diurnal cycle of heating and cooling obviously would not exist, since it depends on the Earth's rotation. Surface winds everywhere would blow from the cold night face towards the hot daylight face, whilst upper flow patterns would be the reverse of those at lower levels. Whatever the exact nature of the atmospheric flow patterns, the climatic zones on a nonrotating Earth would be totally different from anything observed today. Theoretical studies suggest that if this stationary Earth started to rotate, then as the rate of rotation increased, the atmospheric circulation patterns would be progressively modified until they resembled those observed today. In very general terms, these circulation patterns take the form of a number of meridional overturning cells in the atmosphere, with separate zones of rising air motion at low and middle latitudes, and corresponding sinking motions in subtropical and polar latitudes.

1.2 General Atmospheric Circulation

A schematic representation of the mean meridional circulation between Equator and pole is shown in Figure 1.2. A simple direct circulation cell, known as the *Hadley cell*, is clearly seen equatorward from 30° latitude (Lockwood 2003). Eastward angular momentum is transported from the equatorial latitudes to the middle latitudes by nearly horizontal eddies, 1000 km or more across, moving in the upper troposphere and lower stratosphere. This transport, together with the dynamics of the middle latitude atmosphere, leads to an accumulation of eastward momentum between 30° and 40° latitude, where a strong meandering current of air, generally known as the subtropical westerly jet stream, develops (Figure 1.3). The cores of the subtropical westerly jet streams in both hemispheres and throughout the year occur at an altitude of about 12 km. The air subsiding from the jet streams forms the belts of subtropical anticyclones at about 30° to 40° N and S (Figure 1.4). The widespread subsidence in the descending limb of the Hadley cell should be contrasted with the rising limb, where ascent is restricted to a few local areas of

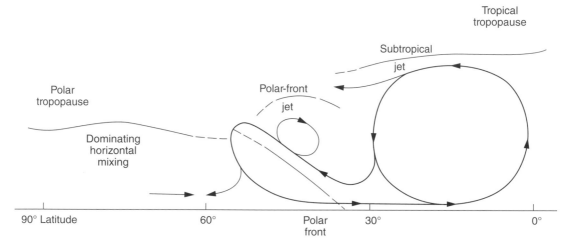

Figure 1.2 Schematic representation of the meridional circulation and associated jet-stream cores in winter. The tropical Hadley cell and middle latitude Ferrel cell are clearly visible. *Source:* from Palmen 1951.

(a)

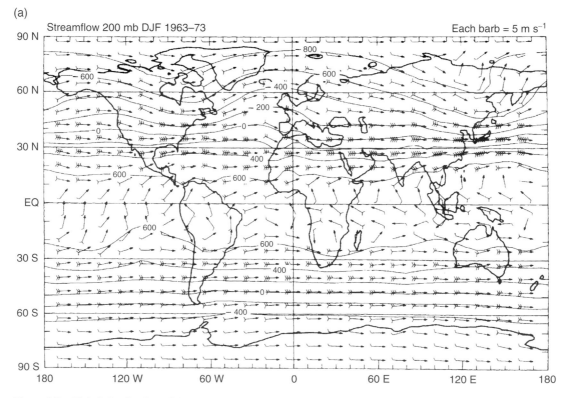

Figure 1.3 Global distribution of the mean height (1963–1973) of the 200 hPa pressure field represented as mean height minus 11 784 gpm for (a) December, January, February; (b) June, July, August. Wind speed and direction shown by arrows. Each barb on the tail of an arrow represents a wind speed of 5 m s^{-1} *Source:* from Peixoto and Oort (1992).

(b)

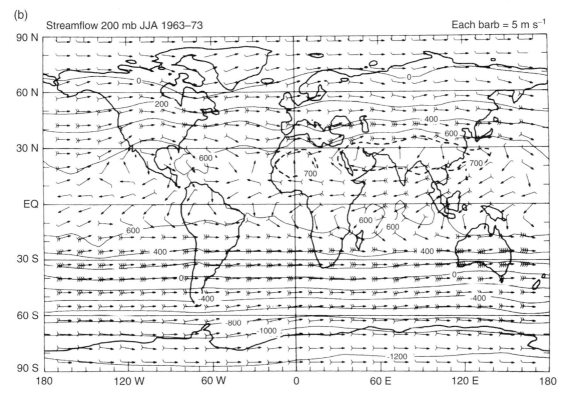

Streamflow 200 mb JJA 1963–73   Each barb = 5 m s⁻¹

Figure 1.3   (Continued)

Figure 1.4   Mean sea-level pressure (hPa)
(1963–1973) averaged for (a) December, January,
February; and (b) June, July, August. *Source:* from
Henderson-Sellers and Robinson (1986) and Oort
(1983).

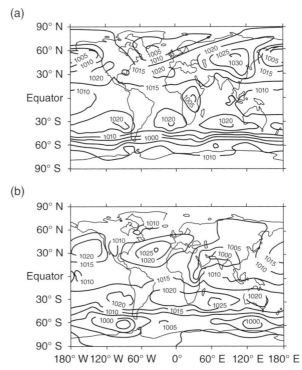

intense convection. More momentum than is necessary to maintain the subtropical jet streams against dissipation through internal friction is transported to these zones of upper strong winds. The excess is transported downwards and polewards to maintain the eastward-flowing surface winds (temperate latitude westerlies) against ground friction. The middle latitude westerly winds are part of an indirect circulation cell known as the *Ferrel cell*. The supply of eastward momentum to the Earth's surface in middle latitudes tends to speed up the Earth's rotation. Counteracting such potential speeding up of the Earth's rotation, air flows from the subtropical anticyclones towards the equatorial regions, forming the so-called *trade winds*. The trade winds, with a strong flow component directed towards the west (easterly winds), tend to retard the Earth's rotation, and in turn gain eastward momentum.

The greatest atmospheric variability occurs in middle latitudes, from approximately 40° to 70° N and S, where large areas of the Earth's surface are affected by a succession of eastward-moving cyclones (frontal depressions) and anticyclones or ridges. This is a region of strong north–south thermal gradients with vigorous *westerlies* in the upper air at about 10 km, culminating in the polar-front jet streams along the polar edges of the Ferrel cells (Figure 1.2). The zone of westerlies is permanently unstable and gives rise to a continuous stream of large-scale eddies near the surface, the cyclonic eddies moving eastward and poleward and the anticyclonic ones moving eastward and equatorward. In contrast, at about 10 km, in the upper westerlies, smooth wave-shaped patterns are the general rule. Normally, there are four or five major waves around the hemisphere, and superimposed on these are smaller waves that travel through the slowly moving train of larger waves. The major waves are often called *Rossby waves*, after Rossby who first investigated their principal properties. Compared with the Hadley cells, the middle latitude atmosphere is highly disturbed and the suggested meridional circulation shown in Figure 1.2 is largely schematic.

The extension into very high latitudes and the northward narrowing of the northern North Atlantic have consequences on the Atlantic Ocean circulation, which in turn has a series of unique effects on the climate system. This is in complete contrast to the much more benign North Pacific Ocean. Warm, saline surface water flows into the northern North Atlantic, after travelling from the Caribbean Sea, via the Gulf Stream and the North Atlantic Drift. This inflowing water, which is more saline than anywhere else in the high-latitude oceans, is finally advected to sites in the Greenland and Norwegian Seas, where extreme cooling to the atmosphere occurs and surface water sinks to the ocean depths. When cooled, water with the salinity normal in the world's oceans becomes denser, but unlike fresh water does not reach its maximum density until near its freezing point, at about −2 °C. Thus, the saltwater of the deep oceans, when cooled at the surface, goes into convective patterns, the coldest and densest portions gradually sinking from the surface to the ocean depths. Low-density surface layers in the oceans can arise either because of surface heating, or the addition of relatively fresh continental runoff or precipitation onto the ocean surface. In the cold oceans, sea-ice will form only when a layer of the ocean close to the surface has a relatively low salinity. The existence of this layer allows the temperature of the surface water to fall to freezing point, and ice to form, despite the lower levels of the ocean having a higher temperature.

1.3 Palaeoclimates

The major controls of very long-term climatic change include palaeogeography, greenhouse gas concentrations, changing orbital parameters, and varying ocean heat transport. One of the major controls on long-term climatic changes is the greenhouse gas concentration of the atmosphere and in particular that of carbon dioxide. Atmospheric carbon dioxide concentration has varied markedly during the Earth's history. Atmospheric CO_2 concentrations are controlled by the carbon cycle, and the net effect of slight imbalances in the carbon cycle over tens to hundreds of millions of years has been to reduce atmospheric CO_2. Atmospheric CO_2 concentrations remained relatively high up to about 60 Myr ago when there was a very marked fall. Atmospheric concentrations continued to fall after about 60 Myr ago, and there is geochemical evidence that concentrations were less than 300 ppm by about 20 Myr ago.

Available evidence is that during the Mesozoic Era, temperatures ranged from 10° to 20 °C at the poles to 25–30 °C at the Equator – that is, the poles were free of permanent ice fields, and the atmosphere-ocean circulation was different in some important aspects from that observed today. Slight cooling took place at the start of the Jurassic Period and marked high-latitude warming during the first half of the Cretaceous Period. Global cooling again took place towards the end of Cretaceous time, and a long-term cooling trend commenced at the start of the Eocene Epoch, some 55 Myr ago.

The sudden, widespread glaciations of Antarctica and the associated shift towards colder temperatures at the Eocene–Oligocene boundary (approximately 34 Myr ago) is one of the most fundamental reorganizations of global climate and ocean circulation known in the geological record. Prior to the Eocene–Oligocene boundary, there is little evidence of the deep cold water in the world ocean that is so common today. Indeed, before the boundary, atmospheric, and particularly oceanic circulation conditions were probably different from those observed today. After the boundary they are probably rather similar to present-day conditions. Oceanic bottom water is formed in small regions by convective buoyancy plumes that transfer relatively dense ocean water from near the surface to the ocean depths. Deep-ocean temperatures are therefore closely related to ocean surface temperatures in key regions. The surface density and salinity is usually increased by evaporation and heat transfer to the atmosphere; therefore, virtually all deep-water formation seems to be over continental shelves in low latitudes or at high latitudes. During the Cretaceous Period and up to the end of the Eocene Epoch, the ocean bottom-waters were warm, saline, and formed in shallow subtropical marginal seas. At the Eocene–Oligocene boundary, ocean bottom-water temperatures decreased rapidly to approximately present-day levels. Deep-sea cores suggest that this change occurred within 100 000 years, which is remarkably abrupt for preglacial Tertiary times, and is considered to represent the time when large-scale freezing conditions developed at sea-level around Antarctica, forming the first significant sea-ice. At this time, cold-water plumes forming off Antarctica started to dominate ocean bottom-water formation, and together with Arctic Ocean plumes they have dominated until the present day. Thus, from early Oligocene times onwards it may be considered that world climates were in the present cold or semi-glacial state.

The initial growth of the East Antarctic Ice Sheet near the Eocene–Oligocene boundary is often attributed to the opening by continental drift of ocean gateways between Antarctica and Australia (Tasmanian Passage) and Antarctica and South America (Drake Passage), leading to the organization of the Antarctic Circumpolar current and the 'thermal isolation' of Antarctica. This notion has been challenged because although most tectonic reconstructions place the opening of the Tasmanian Passage close to the Eocene–Oligocene boundary, the Drake Passage may not have provided a significant deep-water passage until several million years later. Recent model simulations (DeConto and Pollard 2003) of the glacial inception and early growth of the East Antarctic Ice Sheet suggest that declining Cenozoic carbon dioxide first leads to the formation of small, highly dynamic ice caps on high Antarctic plateaux. At a later time, a carbon dioxide threshold is crossed, initiating various feedbacks that cause the ice caps to expand rapidly with large orbital variations, eventually coalescing into a continental-scale East Antarctic Ice sheet.

According to this simulation, the opening of the two Southern Ocean gateways plays a secondary role in this transition, relative to changing carbon dioxide concentration.

1.3.1 Quaternary Glaciations

Continental ice-sheets probably appeared in the Northern Hemisphere about 3 Myr ago, occupying lands adjacent to the North Atlantic Ocean. The time of the formation of the Greenland ice-sheet is not well known from terrestrial evidence, but the presence of glacial marine sediments in North Atlantic marine cores first appeared around 3 Myr ago. The oldest glacial moraines in Iceland are dated to approximately 2.6 Myr ago. For at least the past million years, the Earth's climate has been characterized by an alternation of glacial and interglacial episodes, marked in the Northern Hemisphere by the waxing and waning of continental ice-sheets and in both hemispheres by rising and falling temperatures (Figure 1.5). The present dominant cycle is one of about 100 kyr and is seen in the growth and decay of the continental ice-sheets. The last major glacial episode started about 110 kyr ago and finished only about 10 kyr ago. These fluctuations or cycles are found in a large number of *proxy* data records, analysis of which suggests that Antarctic air temperature, atmospheric CO_2 content, and deep-ocean temperatures are dominated by variance with a 100 kyr period and vary in phase with orbital eccentricity. In contrast, global ice volume lags changes in orbital eccentricity (Shackleton 2000). Hence, the 100-kyr ice-sheet cycle does not arise from ice-sheet dynamics; instead, it is probably the response of the global carbon cycle to changes in orbital eccentricity that generates the eccentricity signal in the climate record by causing changes in atmospheric carbon dioxide concentrations.

Proxy data records can be grouped into two climatic regimes with the transitional zone about 430 kyr ago (Figure 1.5). The earlier shows higher-frequency cycles (dominance of 40-kyr cycles), with less coherence amongst the various proxy climatic records than the later one (dominance of 100-kyr cycles). This may be due to a decrease in average atmospheric CO_2 levels over the past two million years (Brook 2005).

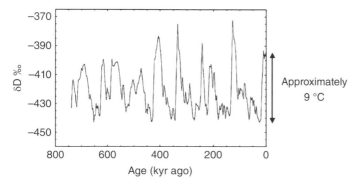

Figure 1.5   The figure shows measurements deduced from ice-cores drilled from the Antarctic ice-sheet, and analysed by the British Antarctic Survey and others as part of the European programme EPICA. The actual measurement is of the concentration of deuterium in air bubbles, and this can be related to local temperatures. The figure shows that temperature rise between the depth of the last ice age 20 000 years ago and the current interglacial is about 9 °C. *Source:* Hadley Centre for Climate Prediction and Research. From EPICA Community Members (2004).

There are differences in the amplitudes of deuterium and CO_2 oscillations before and after 430 kyr ago. The atmospheric concentration of CO_2 did not exceed 300 ppmv for the 650 000 years before the beginning (around 1750) of the industrial era. Since the Industrial Revolution, atmospheric carbon dioxide concentrations have increased by 33%. Before 430 kyr ago concentrations of CO_2 did not exceed 260 ppmv.

The transition from glacial to interglacial conditions about 430 kyr ago resembles the transition into the present interglacial period at about 10 kyr ago in terms of the magnitude of changes in temperature and greenhouse gases. As commented above, the transition 430 kyr ago also delimits the frontier between two different patterns of climate, and has been identified by recent investigations as a unique and exceptionally long interglacial. Some workers (see Brook 2005) suggest that because the orbital parameters (low eccentricity and consequently weak precessional forcing) are similar to those of the present and next tens of thousands of years, the interglacial 430 kyr ago may be the best analogue available for present and future climate without human intervention. Long interglacials with stable conditions are not, therefore, exceptional, short interglacials such as the past three are not the rule and hence cannot serve as analogues of the present Holocene interglacial.

Sudden and short-lived climate oscillations giving rise to warm events occurred many times during the generally colder conditions that prevailed during the last glacial period between 110 000 and 10 000 years ago (Lockwood 2001). They are often known as *interstadials* to distinguish them from the cold phases or *stadials*. Between 115 000 and 14 000 years ago there were 24 of these oscillations, as recognized in the Greenland ice-core records where they are called *Dansgaard–Oeschger oscillations*. These can be viewed as oscillations of the climate system about an extremely ill-defined mean state. Each oscillation contains a warm interstadial, which is linked to and followed by a cold stadial. Ice-core and ocean data suggest that the oscillations began and ended suddenly, although in general the 'jump' in climate at the start of an oscillation was followed by a more gradual decline that returned conditions to the colder 'glacial' state. Warming into each oscillation occurred over a few decades or less, and the overall duration of some of these warm phases may have been just a few decades, whereas others vary in length from a few centuries to nearly 2000 years.

Of totally different nature to Dansgaard–Oeschger oscillations are extreme and short-lived cold events, known as *Heinrich events*. These events occurred against the general background of the glacial climate and represent the climatic effects of massive surges of fresh water and icebergs from melting ice sheets into the North Atlantic, causing substantial changes in the thermohaline circulation (Lockwood 2001). Several massive ice-rafting events show up in the Greenland ice-cores as a further 3–6 °C drop in temperature from already cold glacial conditions. Many of these events have also been picked up as particularly cold and arid intervals in European and North American pollen records. The most recent Heinrich event is known as the Younger Dryas and appears as a time of glacial re-advance in Europe after the end of the main ice-age.

1.3.2 The Recent Climate Record

Extensive instrumental temperature records exist only for the period after about 1860, but recently multi-proxy data networks (e.g. Mann et al. 1999; Intergovernmental Panel on Climate Change Fifth Assessment Report, 2014) have been used to reconstruct Northern Hemisphere temperatures back to AD 1000 or earlier (Figure 1.6). These reconstructions and simulations show a long-term cooling trend in the Northern Hemisphere prior to industrialization of −0.02 °C per century, possibly related to orbital forcing, which is

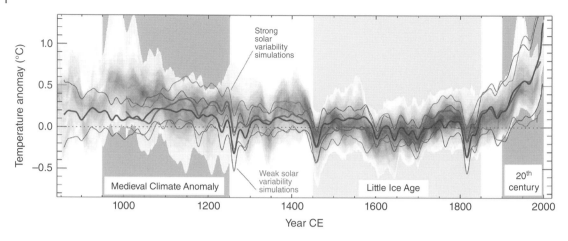

Figure 1.6 Simulated (lines) and reconstructed (grey shading) Northern Hemisphere average surface temperature changes since the Middle Ages. All data are expressed as anomalies from their 1500–1850 mean and smoothed with a 30-year filter. *Source:* data from, and more details available at, http://www.ipcc.ch/report/ar5/syr. See colour plate section for the colour representation of this figure.

thought to have driven long-term temperatures downward since the mid-Holocene at a rate within the range from −0.01 to −0.04 °C per century. The temperature reconstruction also shows that the late eleventh, twelfth and fourteenth centuries rival mean twentieth century temperature levels, whilst cooling following the fourteenth century can be viewed as the initial onset of the cold period known as the Little Ice Age. There is general agreement that the Little Ice Age came to an abrupt end around 1850, whilst studies in Switzerland indicate that overall the coldest conditions of the past 500 years were in the late seventeenth and early nineteenth centuries. The early nineteenth century was especially cold and can be considered as the 'climatic pessimism' of the past 1000 years.

Global mean surface temperature has increased dramatically during the last one hundred years or so, but not in a uniform manner (Figure 1.7). The global increase in temperature since about 1880 occurred during two sustained periods, one beginning around 1910 and another beginning in the 1970s and continuing to the present day. Best available estimates (Jones and Moberg 2003) give global temperature trends from 1910 to 1945 of 0.11 °C per decade, −0.01 °C per decade from 1946 to 1975 and 0.22 °C per decade from 1976 to 2000. In the period 2001 to 2010 warming was again ~0.11 °C per decade. Nine of the 10 warmest years observed globally since reliable observations were begun over a century and a half ago have occurred since the year 2000, and all 10 warmest years have occurred since 1998. The 20 warmest years on record have all occurred since 1995, with the five warmest years occurring since 2010. The warmest year of all (prior to 2018) was 2016.

The largest recent warming is in the winter extratropical Northern Hemisphere, with a faster rate of warming over land compared with the ocean. Using satellite-borne microwave sounding units, Qiang Fu et al. (2006) have examined atmospheric temperature trends for 1979–2005. They found that relative to the global-mean trends of the respective layers, both hemispheres have experienced enhanced tropospheric warming and stratospheric cooling in the 15–45° latitude belt. This suggests a widening of the tropical circulation zone and a poleward shift of the subtropical jet streams and their associated subtropical dry zones.

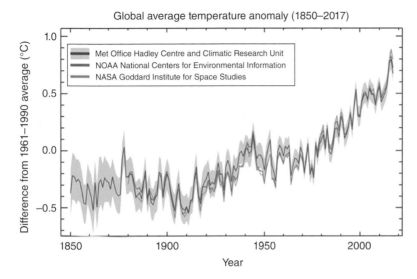

Global average temperature anomaly (1850–2017)

Figure 1.7 Global average surface temperature anomaly 1850–2017, relative to the 1981–2010 average. The grey shading shows the 95% uncertainty for the Met Office Hadley Centre/Climatic Research Unit HadCRUT4 data set and the solid lines show two other data sets. The figure is reproduced from the Hadley Centre for Climate Prediction and Research and more details are available at www.metoffice.gov.uk/research/news/2018/global-surface-temperatures-in-2017. See colour plate section for the colour representation of this figure.

Minimum temperatures for both hemispheres increased abruptly in the late 1970s, coincident with an apparent change in the character of the ENSO phenomenon, giving persistently warmer sea temperatures in the tropical central and east Pacific. The more reliable data sets show that it is likely that total atmospheric water vapour has increased several percent per decade over many regions of the Northern Hemisphere since the early 1970s. Similarly, it is likely that there has been an increase in total cloud cover of about 2% over many mid- to high-latitude land areas since the beginning of the twentieth century. The increases in total cloud amount are positively correlated with decreases in the diurnal temperature range.

The observed temperature record since the early 1900s can be partly explained in terms of natural factors such as the chaotic variability of climate due largely to interactions between atmosphere and ocean; changes in the output of the Sun; and changes in the optical depth of the atmosphere from volcanic emissions. Some of the computations made on this basis are shown in Figure 1.8, which clearly do not agree with the observations, particularly in the period since about 1970 when observed temperatures have risen by about 0.5 °C, but those simulated from natural factors have hardly changed at all.

Sulphate aerosol particles scatter some sunlight, which would otherwise reach the surface of the Earth and heat it, back to space. They therefore have a cooling influence on climate. They are formed in the lower atmosphere when sulphur dioxide, emitted by human activities such as power generation and transport, is oxidized. Other sources of sulphur dioxide are 'background' nonexplosive volcanoes, and natural di-methyl sulphate (DMS) from ocean plankton. If the climate model used in Figure 1.8 is now driven by changes in human-made factors, such as sulphate aerosols and changes in greenhouse gas

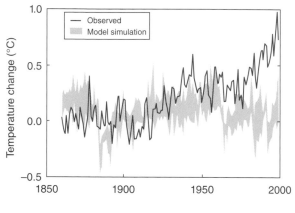

Figure 1.8 Natural factors cannot explain recent warming. Shaded area shows result of the Hadley Centre climate model driven by natural factors, but excluding anthropogenic carbon dioxide and sulphate particles. Continuous line shows actual observations. *Source:* Hadley Centre for Climate Prediction and Research.

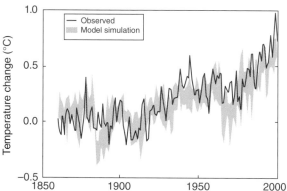

Figure 1.9 The Hadley Centre climate model is now driven by changes in anthropogenic factors – changes in greenhouse gas concentrations and sulphate particles – in addition to natural factors; observations (line) and model simulations (shaded area) are in much better agreement. *Source:* Hadley Centre for Climate Prediction and Research.

concentrations, in addition to natural factors, observations and model simulation become in much better agreement (Figure 1.9). In particular, the warming since about 1970 is clearly simulated (Stott et al. 2000). Further work has recently demonstrated that warming over individual continental areas in the past few decades can be explained only if human activities are included, and the IPCC therefore concludes that the majority of the observed warming is due to anthropogenic activity.

1.4 Polar Climates

The most distinctive climatic features in both North and South Polar Regions are the presence of ice, snow, deeply frozen ground and a long winter period of continuous darkness (Orvig 1970). The radiation budget of the polar surfaces is nearly always negative because of the absence in winter of any solar radiation, the low angle of solar incidence in the summer and the high albedo of the ice fields. Indeed, the polar regions serve as global sinks for the energy that is transported poleward from the tropics by warm ocean currents and by atmospheric circulation systems, particularly travelling cyclones and blocking anticyclones. However, there are major differences between the Arctic and Antarctic regions in terms of the distribution of land, sea, and ice. The Arctic is a largely ice-covered and landlocked ocean, whereas Antarctica consists of a continental ice-sheet over 2000 m thick surrounded by ocean.

Surrounding the Antarctic continent is a zone of floating sea-ice, about 1.5 m, thick which undergoes a marked annual cycle from a minimum in February–March to a maximum in September–October. In the summer, the Antarctic sea-ice melts almost back to the coastline of the continent, so only slightly more than 10% survives the summer. The winter advance of the sea-ice is not restricted by land at its equatorward boundary and it crosses the Antarctic circle.

In contrast, in winter the southern limits of the Arctic sea ice are constrained by the northern coastlines of Asia and North America, but pack-ice extends into middle latitudes off eastern Canada and eastern Asia where northerly winds and cold currents transport ice far southward. West of Norway there is open water to about 78° N because of the warm waters of the North Atlantic Drift and thermohaline circulation. The North Atlantic thermohaline circulation is particularly vigorous, and its climatic effect is often illustrated by comparing the surface temperature of the northern Atlantic with comparable latitudes of the Pacific, since the former is 4–5 °C warmer. During summer, the sea-ice melts back towards the pole, but much of the Arctic ice survives at least one summer melt season, and as a consequence its mean thickness is 3–4 m. The thickness is highly variable locally, with narrow openings (leads) throughout the pack-ice even in winter, and larger openings (polynyas) adjacent to coastal areas where winds blow offshore. Whilst there appears to have been no significant long-term trend in Antarctic sea-ice extent, Arctic sea-ice extent has decreased by about 2.5% per decade since 1970. More than 80% of Greenland is covered by an extensive ice sheet that rises above 3000 m in elevation.

1.4.1 Arctic

In a climatological sense, the Arctic corresponds approximately to the land areas beyond the northern limits of forests and ocean areas with at least a seasonal sea-ice cover. The boundary between Boreal forest and tundra vegetation corresponds closely to the 10° C mean isotherm of the warmest month and to the average July location of the Arctic Front separating Arctic and temperate air masses. At high latitudes, the total daily solar radiation depends largely on day-length, which in turn varies widely with the season of the year. The radiation climate produced with continuous darkness in winter and continuous daylight in summer is distinctly different from that found at lower latitudes (Serreze and Barry 2005).

There are two main areas of semi-permanent low level inversions (atmospheric temperature increases with height) in the world: the subtropical belt and the polar regions. The main cause of the polar inversions is the negative energy balance at the surface, but the inversion is also maintained by warm air advection aloft with associated subsidence. The strongest vertical temperature gradients are therefore often found near the surface and over the Central Polar Ocean the temperature inversion persists for at least 60% of the days in all months, reaching a maximum in late winter of 100%. Because of this, the surface air temperature is primarily dependent on the nature of the surface conditions. The prevailing summer melting of snow and ice holds the surface air temperature close to 0 °C, but slightly positive temperatures are usually observed near the pole in the second half of June (Figure 1.10). Relatively warm water below the sea-ice keeps winter temperatures around −30 to −35 °C, since sensible heat is conducted upwards through the sea-ice to warm the lowest air layers. Indeed, the lowest winter surface temperatures are not found over the central Polar Ocean, but rather over the land areas of northeast Siberia and the Yukon (Figure 1.10a). Here, under the influence of the continental anticyclones, calm, clear, subsiding air becomes trapped in hollows and the temperature fall below −50 °C on occasion and winter monthly means below −40 °C are common. Rapid loss of heat during the long winter nights in the

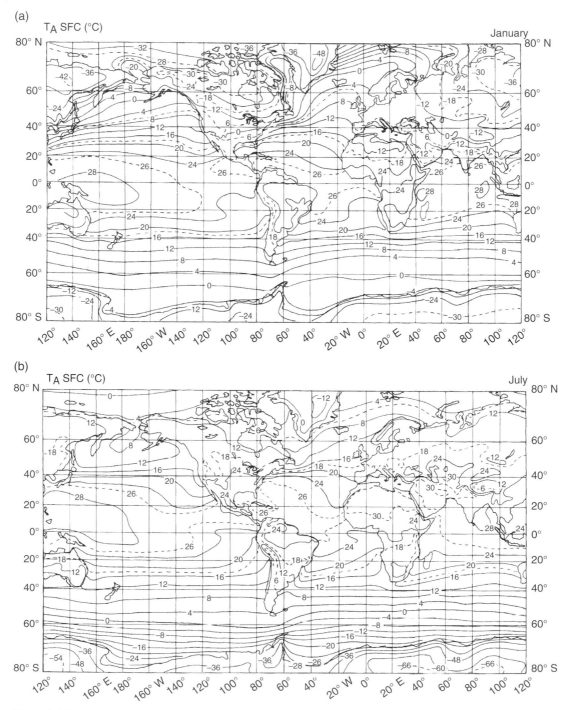

Figure 1.10 Mean surface air temperature for (a) January and (b) July. *Source:* from Natural Climate Data Center (1987) and Peixoto and Oort (1992).

form of longwave radiation cools the ground surface but in contrast to the sea-ice, this heat cannot be replaced. Similar low temperatures are found on the high plateau of the Greenland ice-sheet. Solar forcing of the diurnal temperature cycle is at a minimum in polar regions, making the polar diurnal temperature cycle extremely weak and difficult to compare directly with cycles at locations nearer to the Equator with a stronger diurnal solar radiation signal. During the winter months of December to February, there is almost continuous darkness in the Arctic. Therefore, the conventional concept of solar radiative forced maximum and minimum temperatures does not apply and diurnal temperature range is governed by the atmospheric circulation.

1.4.2 Antarctic Climates

The Antarctic continent consists of two domed ice sheets covering about 97.6% of its area at an average elevation of about 2200 m. The first evidence of the formation of Antarctic ice-sheets is at the Eocene–Oligocene boundary ($\approx$34 Myr ago). The high plateau of East Antarctica has extensive areas above 3000 m. This desolate world of ice is the coldest and windiest of the world's great land masses. On 24 August 1960, Vostock (78.5° S, 106.9° E; 3488 m) recorded a temperature of −88.3 °C, which is probably the lowest temperature recorded on the Earth's surface. Antarctica and its environs play an important role in the workings of climate on a global scale because the region acts as a large sink in the global energy cycle (King and Turner 1997; Simmons 1998).

Mean surface temperatures at coastal stations typically range, depending on location and other factors, between 0 and −5 °C in January and −15 and −25 °C in July. The seasonal cycle of surface air temperature at South Pole Station (Table 1.1) shows a number of interesting features that are peculiar to Antarctica.

Table 1.1 Mean temperature data (°C) for South Pole (1957–2010): latitude 90° S, elevation 2800 m.

Month	Daily mean
January	−28.4
February	−40.9
March	−53.7
April	−57.8
May	−58.0
June	−58.9
July	−59.8
August	−58.7
September	−59.1
October	−51.6
November	−38.2
December	−28.0

Source: After Lazzara et al. (2012).

The temperature rises rapidly in the southern spring (October and November), then averages −28 °C for the summer months of December and January. In February and March, the temperature falls rapidly by nearly half a degree Celsius per day, reaching winter levels by the March equinox.

The shortness of the Antarctic summer is caused by the variations of global radiation and the surface albedo. The maximum global radiation is on average around the time of the summer solstice. In the weeks preceding, the solstice the surface albedo decreases, due to the increasing solar elevation above the horizon and a slight metamorphosis of the snow cover. It is normal for the albedo of many surfaces to decrease with increasing solar elevation; thus, more solar radiation can be absorbed at the snow surface. After the solstice, the global radiation decreases, and the first light snowfall or influx of drifting snow restores surface conditions to those favouring higher albedo values. Hence there are good reasons to expect absorbed solar radiation to decrease rapidly in January and the surface temperature to fall quickly. During the Antarctic winter, the intense radiational cooling and the highly transmissive atmosphere generate a persistent low-level temperature inversion that reaches its greatest depth over the higher elevations of the ice-sheet, where it is present almost the entire year. The inversion is no stronger at the end than at the beginning of the polar night. This is because equilibrium is reached between the surface longwave radiation loss, which decreases as temperature falls, and the downward radiation from the inversion layer, which changes relatively little with time. The absence of a clearly defined time for the occurrence of the surface minimum temperature is a feature of the Antarctic, comprising the so-called *coreless winter.*

Katabatic winds occur when cooling causes a shallow blanket of air adjacent to the surface to become colder and therefore heavier than the atmosphere above; the air then drains downslope under the influence of gravity. These winds are most persistent where the ground is covered by ice and snow and they dominate the surface wind regime of Antarctica. Nearly everywhere on the ice-sheet the surface winds are directed downslope from the interior towards the surrounding ocean. The surface katabatic winds are related to the energy balance of the low-level inversion over Antarctica. Radiosonde measurements at the South Pole (Schwerdtfeger 1984) indicate an average radiative cooling rate of 4 °C per day at the inversion level. This is the cooling rate that appears to be occurring from radiative balance considerations alone, the energy is lost by longwave radiation to space. The real cooling rate in the atmosphere obtained from temperature observations is 1 °C or less per month. Thus, other atmospheric processes must be compensating for the longwave radiative loss to space. These additional atmospheric processes are identified as horizontal advection and adiabatic sinking. Air converges over Antarctica at middle levels, sinks, and warms due to adiabatic compression, causing the low-level inversion over the ice surface. The sinking air cools rapidly by infrared emission in the inversion layer, as explained earlier. The radiatively cooled air sinks through the inversion layer, is replaced by sinking air from above and flows away as the surface katabatic wind.

1.5 Temperate Latitude Climates

Daily weather charts for any extensive region in middle latitudes reveal well-defined *synoptic systems* that normally move from west to east with a speed that is considerably smaller than their constituent air currents. At the surface the predominant features are closed cyclonic and anticyclonic systems of irregular shape, whilst at higher altitudes in the atmosphere smooth wave-shaped patterns are the general rule.

These upper Rossby waves are important because surface synoptic systems tend to move in the direction of the broad upper flow with a velocity that is proportional to the intensity of the upper flow. So an intense upper flow pattern is associated with the rapid passage eastward of frequent surface frontal depressions and anticyclones. Surface frontal depressions also tend to form slightly downwind of upper troughs and similarly surface anticyclones tend to form slightly downwind of upper ridges. Major temperate, climatological cyclone development regions are therefore located just downwind of upper climatological troughs.

Northern temperate latitude Rossby waves tend to be locked in preferred locations, which are shown in Figure 1.3. These preferred locations may arise because the atmospheric circulation is influenced not only by the thermal properties of land and sea, but also by high mountain ranges and highlands in general. When a westerly air stream crosses a north–south aligned mountain range such as the Rocky Mountains, anticyclonic deformation of flow is found over the mountain range and cyclonic deformation to leeward. These orographically generated ridges and troughs propagate downstream and can reinforce similar features generated by the direct thermal effects of land and sea.

1.5.1 Europe

European climate is particularly interesting because it is likely that much of the southern and central parts of the continent could experience over the twenty-first century marked decreases in annual precipitation, and corresponding increases in temperature. In contrast, Scandinavia could experience increasing annual precipitation. Of the elements governing the distribution of climate, the most basic is radiation from the Sun, and its balance with outgoing longwave radiation from the Earth. The southernmost rim of Europe lies at about 35° N, almost within the subtropics, and its northernmost island extensions reach 78° N, almost in the Arctic. One important consequence of this large latitudinal spread is that the length of daylight at the solstices varies significantly over the region. Except along the southern rim, the net radiation balance of the continental surface in winter is negative, but the energy lost by the continual cooling of the land is replaced by vigorous warm air flow from the Atlantic Ocean to the west. Variations in the vigour of this westerly flow because of the NAO are of great significance in relation to the level of European winter temperatures and are discussed in detail later. Because the warm air masses from the Atlantic Ocean cool as they pass eastward over the winter continental surface, the mean isotherms in January run north–south, with temperatures falling both eastward away from the Atlantic and northward towards the Arctic (Figure 1.10a). In contrast, in summer the net radiation balance over the continental surface is positive: values fall towards the north, so isotherms are aligned east–west (Figure 1.10b).

During the summer months, the prevailing westerlies retreat northward from southern Europe, leaving it under the influence of the subsiding air of the subtropical anticyclones and creating an intense dry season with only occasional rainfall from convective showers. In winter the westerlies return to southern Europe, bringing frequent depressions that cause a precipitation maximum in the winter half of the year.

A major source of interannual variability in the atmospheric circulation over the North Atlantic and western Europe is the NAO, which is associated with changes in the strength of the oceanic surface westerlies. Its influence extends across much of the North Atlantic and well into Europe and it is usually defined through the regional sea-level pressure field, although it is readily apparent in mid-height fields in the troposphere. The NAO's amplitude and phase vary over a range of time scales from intraseasonal to interdecadal; the NAO is present throughout the year but the largest amplitudes typically occur in winter.

The NAO is often indexed by the standardized difference of December to February sea-level atmospheric pressure between Ponta Delgado, Azores (37.8° N, 25.5° W), or Lisbon, Portugal (38.8° N, 9.1° W), and Stkkisholmur, Iceland (65.18° N, 22.7° W).

Statistical analysis reveals that the NAO is the dominant mode of variability of the surface atmospheric circulation in the Atlantic and accounts for more than 36% of the variance of the mean December to March sea-level pressure field over the region from 20° to 80° N and 90° W to 40° E, during 1899 through to 1994. Marked differences are observed between winters, with high and low values of the NAO. Typically, when the index is high the Icelandic low is strong, which increases the influence of cold Arctic air masses on the northeastern seaboard of North America and enhances westerlies carrying warmer, moister air masses into western Europe (Hurrell 1995). During high NAO winters, the westerlies directed onto northern Britain and southern Scandinavia are over $8\,\mathrm{m\,s^{-1}}$ stronger than during low NAO winters, with higher than normal pressures south of 55° N and a broad region of anomalously low pressure across the Arctic. In winter, western Europe has a negative radiation balance and mild temperature levels are maintained by the advection of warm air from the Atlantic. Thus, strong westerlies are associated with anomalously warm winters, weak westerlies with anomalously cold winters, and NAO anomalies are related to downstream wintertime temperature and precipitation anomalies across Europe, Russia and Siberia.

Overlying the interannual variability, there have been four main phases of the NAO index during the historical record: prior to the 1900s the index was close to zero; between 1900 and 1930, strong positive anomalies were evident; between the 1930s and 1960s, the index was low; and since the 1980s, the index has been strongly positive. The recent persistent high positive phase of the NAO index, extending from about 1973 to 1995, is the most persistent and high of the historical record. Preliminary reconstructions from tree ring data of a 1000-year-long record suggest that the recent extended period of high values in the NAO index series may not be unique. During each positive phase, higher than normal winter temperatures prevail over much of Europe, culminating in the unprecedented strongly positive NAO index values and mild winters of 1989 and 1990. During high NAO index winters, drier conditions also prevail over much of central and southern Europe and the Mediterranean, whilst enhanced rainfall occurs over the northwestern European seaboard (Hurrell 1995). The recent high positive phase of the NAO index has resulted in an extended dry period in Morocco, with as much as 35% reduction in runoff from major river catchments and serious reduction in cereal crop production.

1.5.2   Interior North America and Asia

The interiors of both continents are far removed from oceanic influences, and because they are located in middle latitudes they experience extreme continentality of climate. The main effect of extreme continentality is to produce large seasonal temperature variations. Thus, Bergen (60° 24′ N, 5° 19′ E, 44 m) on the Atlantic coast of Norway has a mean annual temperature of 7.8 °C and an annual range of 13.7 °C, whilst at approximately the same latitude the central Asian city of Omsk (54° 56′ N, 73° 24′ E, 105 m) has a mean annual temperature of 0.4 °C and a range of 38.4 °C. Since the oceans are the main atmospheric moisture source, the continental interiors also tend to be dry, allowing the subtropical deserts to extend into temperate latitudes (see Figure 1.11).

In the United States, the region between the Rockies and 100° W has annual precipitation means between 300 and 500 mm, whereas vast areas of central Asia receive less than 500 mm, but in both cases

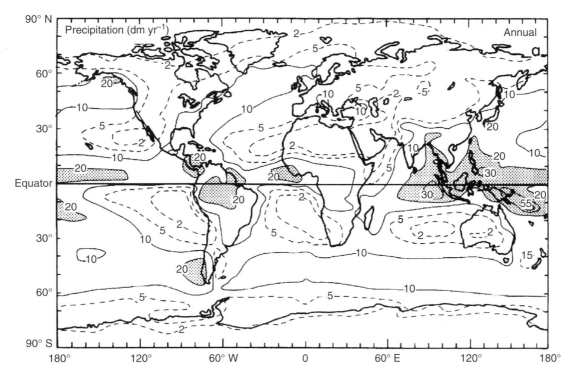

Figure 1.11   Global distribution of precipitation in decimetres per year. *Source:* from Peixoto and Oort (1992).

amounts increase towards the east coast. In North America, the mountain ranges along the west coast curtail the penetration of moist Pacific air-masses inland and have the same drying effect as the whole of lowland western Europe has on air-masses entering Asia. The interiors of both continents are open to invasions by air currents of Arctic origin. North America is also open to invasion by air currents of sub-tropical origin that contrast strongly with the cold northern air-masses, generating at times extremely active cold fronts, convective activity and tornadoes. The US midwestern and western plains' states are visited by more severe tornadoes than any other area in the world; they are most frequent during spring and summer months. In contrast, moist tropical air is blocked from penetrating into central Asia by the Tibetan Plateau and its associated mountain ranges.

Whilst the upper winter ridge over western North America directly influences the weather over the interior of that continent, a similar relationship does not exist between the ridge over western Europe and central Asia (Figure 1.3). This is because a further winter trough–ridge pair of small amplitude are located between the main ridge over western Europe and the east Asian trough, with the secondary ridge located about 85° E. In January, almost the whole of Canada (the north interior and northeast) have snow covers; similarly, all Asia north of about 40° N is snow covered. In winter, the interiors of both continents are dominated by large anticyclones (Figure 1.4a), the result of cold air ponding over the cold continental surfaces. The most intense is over Asia and is located on average to the east of the 85° E meridian, that is, just downwind of the upper climatological ridge. The anticyclone over North America is a weak and unstable feature that can be interpreted as a statistical average rather than as a *semi-permanent* and *quasi-stationary system.* Both systems act as the sources of continental polar air, which is intensely cold and dry. In summer, the continental anticyclones vanish and are replaced by shallow heat lows.

1.6 Tropical Climates

The tropical world may be considered as bounded at the surface by the two belts of subtropical high pressure, at about 30° N and 30° S, and in the upper air by the corresponding subtropical westerly jet streams (Figures 1.3 and 1.4). Under this definition, tropical climates are found in those areas dominated by the tropical Hadley cell circulations or tropical monsoon circulations.

The tropical atmosphere differs meteorologically from the middle latitude atmosphere in a number of important aspects (McGregor and Nieuwolt 1998). In the tropics, temperatures tend to be uniform in the horizontal over vast areas, so contrasting air currents are rare. In middle latitudes there are marked north–south temperature gradients, and air currents with differing origins therefore have contrasting temperatures and humidities. That is, the middle latitude atmosphere is strongly *baroclinic* with large horizontal temperature gradients and therefore strong thermal winds (measure of wind shear with height). The strongly baroclinic nature of the middle latitude atmosphere explains the formation of frontal depressions and jet streams, with the latent heat released by the condensation of water vapour only exerting a modifying influence. In the tropics, where atmospheric moisture values are often considerable, rainfall and therefore condensation rates can be high, resulting in the release of large amounts of latent heat, which can be of importance in the development of tropical circulation systems. Because the *Coriolis* parameter is small in the tropics, small pressure gradients can generate air currents as intense as those found in middle latitudes.

In the tropical world, the Sun is nearly always almost overhead at midday, with little seasonal variation in day length from about 12 hours. Seasonal temperature variations in the humid tropics are therefore generally small, particularly near the Equator (Figure 1.10). The annual temperature range (mean January minus mean July) is very small at 3 °C or less over the oceans in the equatorial zone, and is only 2 or 3 °C greater at 30° N and 30° S. Whilst values of annual temperature range are equally small over the equatorial continents, values increase rapidly towards the subtropics to reach, for example, 20 °C in the Sahara Desert and 15 °C in central Australia. Since horizontal temperature gradients in the tropical atmosphere are small, heating of the lower atmosphere often causes convective overturning because it cannot be compensated by horizontal cold air advection from elsewhere. Therefore, tropical circulation patterns are particularly influenced by heat inputs from such sources as warm ocean surfaces acting through latent heat released in deep cumulus convection. Other heat sources, such as high tropical plateaus and equatorial rain forest, are also important. These heat sources show marked latitudinal and longitudinal variations in their distributions, and also a marked tendency to vary on both annual and, in the case of the tropical oceans, on greater than annual scales. One consequence of this is that tropical rainfall patterns show both annual and greater-than-annual variations and also marked teleconnections with distant locations.

Mean winds in the tropical world reflect the mean pressure patterns (Figure 1.4). In the centres of the subtropical anticyclones and in the equatorial trough, winds are normally light and variable. Between the equatorial trough and the subtropical anticyclones there is a region of easterly winds, with a small deflection towards the Equator, usually known as the *trade winds*. A seasonal reversal of wind direction takes place over southern Asia and the northern Indian Ocean; these are the Asian summer monsoon circulations.

Table 1.2 Tropical storm[a] development regions.

Area/location	Average number of tropical storms per year
Northeast Pacific	10
Northwest Pacific	22
South Pacific	7
Northwest Atlantic (including western Caribbean and Gulf of Mexico)	7
Bay of Bengal	6
South Indian Ocean	6
Arabian Sea	2
Off northwest Australian coast	2

Source: After Lockwood (1974).
[a] Tropical storms are defined by World Meteorological Organization as a warm-core vortex circulation with sustained maximum winds of at least $20 \, \mathrm{m \, s^{-1}}$.

1.6.1 Tropical Storms

Atmospheric disturbances such as easterly waves and tropical storms (cyclones, hurricanes, and typhoons) are frequent over the western parts of the subtropical oceans (see Table 1.2). Tropical storms are driven largely by latent heat released by condensing water vapour associated with intense rainfall. Rainfall rates near the centre of a tropical cyclone can exceed 500 mm per day, and this continuous inflow of water vapour is supplied by warm ocean surfaces with temperatures above about 26.5 °C, which are normally found in the tropical oceans in the late summer and autumn. The storms decay rapidly when the supply of water vapour is cut off by the passage across cooler water surfaces or land areas. In summer and autumn, they frequently bring high winds, storm surges, and intense rainfall to the eastern coasts of the USA and Central America, subtropical southeast Asia, the Indian subcontinent, and northern Australia. As of 2018, the UK Met Office's position is that there is no clear consensus on whether global warming is currently having any measurable impact on tropical cyclones, although climate models suggest there may be an increase in tropical cyclone intensity in the future, under continued global warming. However, the models also indicate that tropical cyclone frequency will either remain unchanged or decrease, but that the frequency of the most intense storms and their associated precipitation rates will increase substantially.

1.6.2 Walker Circulations, the Southern Oscillation, and El Niño

Satellite imagery and rainfall data clearly show (Figure 1.11) three equatorial regions of maximum cloudiness and rainfall: the so-called 'Maritime Continent' of the Indonesian Archipelago, the Amazon river basin in South America, and the Zaire river basin in Africa. The rest of the equatorial region is comparatively dry, and some, such as the coasts of Peru, even desert. These longitudinal variations in rainfall are associated with east–west regional circulations along the Equator, the most important being

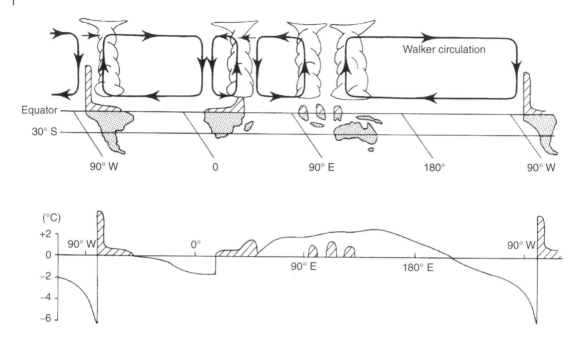

Figure 1.12 Schematic representation of the Walker circulation along the Equator during non-ENSO (El Niño–Southern Oscillation) conditions. The sea-surface temperature departures from the zonal-mean along the Equator are shown in the lower part of the figure. *Source:* from Peixoto and Oort (1992) and Wyrtki (1982).

the *Walker circulation*, which involves rising air motion over the Indonesian Archipelago and sinking over the eastern Pacific (Figure 1.12). The rising air motion takes place mostly in deep convective clouds and is associated with intense convective rainfall and therefore the wet humid climates of the Indonesian Archipelago. The subsiding air suppresses cloud formation and rainfall; therefore, it is associated with the coastal deserts of Peru. The Hadley cell circulation refers to the north–south component of these circulations: equatorward motion at low levels, rising in the convective regions near the Equator and poleward flow aloft. The Walker circulation refers to the east–west component, which is particularly prominent in the equatorial plane. Both circulations are driven by the release of latent heat in deep convective shower clouds.

The Walker circulation is closely coupled with the sea-surface temperature distribution over the Pacific, with relatively cool water in the east and warm in the west. When the Pacific Ocean off the coast of South America is particularly cold, the air above is too stable to take part in the ascending motion of the Hadley cell circulation. Instead, the equatorial air flows westward between the Hadley cell circulations of the two hemispheres to the warm West Pacific where, having been heated and supplied with moisture from the warmer waters, the equatorial air can take part in large-scale ascent (Figure 1.12). The easterly winds that blow along the Equator and the northeasterly winds that blow along the Peru and Ecuador coasts both tend to drag the surface water along with them. The Earth's rotation then deflects the resulting surface currents towards the right (northward) in the Northern Hemisphere and to the left (southward) in the Southern Hemisphere. The surface waters are therefore deflected away from the Equator in both hemispheres and also away from the coastline. Where the surface water moves away under the influence of the trade winds, colder, nutrient-rich water upwells from below to replace it. Since the newly upwelled water is colder than its surroundings, its signature in infrared satellite images takes the form of a distinctive 'cold tongue' extending westward along the Equator from the South American coast. The winds that blow along the Equator also affect the ocean *thermocline*, the boundary between the warm surface water and the deep cold water. In the absence of the wind the thermocline would be nearly horizontal; but the trade winds drag the surface water westward, raising the thermocline nearly to the ocean surface in the east and depressing it in the west. The situation in the equatorial Atlantic is analogous to the equatorial Pacific in that the warmest part is in the west, at the coast of Brazil, but west–east contrasts of water temperature are much smaller than in the Pacific. However, in January a thermally driven Walker circulation may operate from the Gulf of Guinea to the Andes, with the axis of the circulation near the mouth of the Amazon.

Associated with the Walker circulation is the *El Niño phenomenon*, when every few years the tropical Pacific Ocean off the coasts of Peru and Ecuador occasionally becomes much warmer than average for periods of several months. Under El Niño conditions, the Walker circulations become reversed, resulting in heavy rain in the normally arid areas of Peru, and drought in the western Pacific. The Southern Oscillation (SO) is dominated by an exchange of air between the Southeast Pacific subtropical high and the Indonesian equatorial low, with a period that varies between roughly one and five years. During one phase of this oscillation, the trade winds are intense and converge into the warm western tropical Pacific, where rainfall is plentiful and sea-level pressures low. At such times the atmosphere over the eastern tropical Pacific is cold and dry. During the complementary phase, the trade winds relax, the zone of warm surface-waters and heavy precipitation shifts eastwards, and sea-level pressure rises in the west whilst it falls in the east. The latter phase is the more unusual and in the eastern Pacific has become known as El Niño, whereas vigorous episodes of the former are often termed La Niña. The combined atmospheric–oceanic conditions that give rise to these changes in rainfall across the Pacific and neighbouring areas are referred to as ENSO events. The ENSO is important climatologically for two main reasons. First, it is one of the most striking examples of interannual climatic variability on an almost global scale. Second, in the Pacific it is associated with considerable fluctuations in rainfall and sea-surface temperature, and also with extreme weather events around the world (Glantz 1996).

The Southern Oscillation may be defined in terms of the difference in sea-level pressure between Darwin, Australia and Tahiti. Records are available, with the exception of a few years and occasional months, from the late 1800s. The El Niño phenomenon is associated with extreme negative SO values, but for much of the time the series exhibits continuous transitions from high to low values, with most values being positive. Of great importance to the development of the Southern Oscillation is the difference in the ways the ocean and atmosphere respond to changes in the winds and sea-surface temperature patterns, respectively. During La Niña, intense trade winds drive warm surface waters of the equatorial Pacific westward whilst exposing cold water at the surface in the east. When the winds relax during El Niño, the warm waters move back eastward, overflowing the cold water. The response of the ocean to changing winds involves not only changes in currents but also the excitation of waves that travel back and forth across the Pacific. These waves have a large signature, not at the surface, but at the thermocline, which separates the warm waters of the upper ocean from the colder water at depth. These waves eventually bring the ocean to a new equilibrium after a change in the winds, and the time it takes them to propagate across the Pacific is important in determining the time scale of the Southern Oscillation.

1.6.3 Monsoon Circulations of Southern Asia and Eastern Africa

The characteristics of the monsoon climate are to be found mainly in the Indian subcontinent, where over much of the region the annual changes may conveniently be divided into the northeast (dry) and southwest (wet) seasons (Pant and Rupa Kumar 1997). Over many tropical oceans, the atmospheric circulation undergoes very little seasonal variation. In contrast, over tropical continents the atmospheric circulation displays distinct seasonal rhythms and variations. Over the oceans, evaporation consumes a high proportion of the incident radiation, which is also rapidly absorbed by the water surface and spread over great depths by waves and turbulence, and dissipated to other latitudes by ocean currents. Over the tropical continents, particularly if they are dry because of low rainfall, solar radiation is used mostly to warm the Earth's surface, so that surface temperatures in these regions reach very high values owing to the much lower heat capacity of the soil compared with water. This warming over the tropical continents in early summer produces thermal lows which gradually take on some functions of the equatorial trough, thus forming a new region of convergence. The trade winds from the winter hemisphere cross the Equator, slow down, and create a minor area of convergence owing to their change in direction, caused by the reversal of the Coriolis force. In winter, the tropical continents experience relatively low temperatures and high surface pressure; winds are re-established flowing towards the Equator, which are similar to the trade winds. Thus, the tropical continents and adjacent oceans can experience a semi-annual reversal in wind direction characteristic of monsoons.

The southwest monsoon current in the lower 5000 m near India consists of two main branches: the Bay of Bengal branch, influencing the weather over the northeast part of India and Burma; and the Arabian Sea branch, dominating the weather over the west, central and northwest parts of India. The low-level flow across the Equator during the southwest monsoon is not evenly distributed between the longitudes 40° E and 80° E, but flows intermittently during the southwest monsoon from the vicinity of Mauritius through Madagascar, Kenya, eastern Ethiopia, Somalia, and then across the Indian Ocean towards India. A particularly important feature of this flow is the strong southerly current, with a mean wind speed of about $14 \, \mathrm{m \, s^{-1}}$ observed at the Equator over eastern Africa from April to October. The strongest flow occurs near the 1000–5000 m level, but it often increases to more than $25 \, \mathrm{m \, s^{-1}}$ and occasionally to more than $45 \, \mathrm{m \, s^{-1}}$ at heights between 1000 and 2000 and 2000–5000 m.

During the northern winter season, the subtropical westerly jet stream crosses southern Asia, with its core located at about 12 000 m altitude. It divides in the region of the Tibetan Plateau, with one branch flowing to the north of the plateau and the other to the south. The two branches merge to the east of the plateau and form an immense upper convergence zone over China. In May and June, the subtropical jet stream over northern India slowly weakens and disintegrates, causing the main westerly flow to move north into central Asia. Whilst this is occurring, an easterly jet stream, mainly at about 14 000 m, builds up over the equatorial Indian Ocean and expands westward into Africa. The formation of the equatorial easterly jet stream is connected with the formation of an upper-level high-pressure system over Tibet. In October the reverse process occurs; the equatorial easterly jet stream and the Tibetan high disintegrate, whilst the subtropical westerly jet stream reforms over northern India. The Himalayan–Tibetan plateau is of importance because it appears to accelerate the onset of the Asian monsoon and to increase its ultimate intensity. Central and southeastern parts of Tibet remain free of snow throughout most of the year, so the plateau must heat rapidly during the northern spring. This direct warming of the middle troposphere creates an upper-level anticyclone, which is readily observed on synoptic charts, with upper-level divergence and low-level convergence. Thus, suitable conditions are produced for the Asian monsoon in the northern spring. Latent heat released in intense tropical storms over India keeps the system functioning during the northern summer. Since a complex feedback system produces the southwest monsoon, failures in the system are common and produce extensive breaks in the monsoon rains when the whole system shows signs of collapse. Variations in winter snow cover over Tibet will influence the start and intensity of the southwest monsoon. General cooling over Southern Asia at the end of the northern summer causes its collapse.

Nearly half the Indian subcontinent is arid or semi-arid (Figure 1.11), and by far the larger part of this dry area is located in northwest India and southwestern Pakistan. The Arabian Sea summer monsoon circulation is linked with a heat low over Arabia, Pakistan, and northwest India. This heat low, which develops during May, establishes the low-level westerly monsoon wind regime a full month before heavy monsoon rains start over western India. In mid-July the heat low is deepest, the southwest monsoon is strongest, and the west Indian rains are heaviest.

The question then arises as to why the heat low remains cloud-free whilst rain falls to the south. The unique summer aridity of the desert belt from the western Sahara to Pakistan is strongly correlated with the forced descending motion on the northern side of the exit region of the equatorial easterly jet stream. During the northern summer, the equatorial easterly jet stream extends at about 14 000 m in the latitude belt 5–20° N from the Philippines across southern Asia and northern Africa to almost the western Atlantic. Over the whole exit region from India westwards the very gradual deceleration of the jet stream core results in widespread sinking motions on the northern side and rising motions on the southern. Since the Hadley cell circulations already generate desert conditions over North Africa and the Middle East, the deceleration of the equatorial easterly jet stream intensifies the aridity of these deserts. Similarly, numerical simulations suggest that large areas of northeast Africa would have considerably more rainfall if the Tibetan Plateau and the equatorial easterly jet stream were absent. The reverse flow is observed in the entrance region over Southeast Asia, where air sinks to the south and rises to the north, with a rainy area over southern Asia. Radiosonde ascents at Karachi and Jodhpur indicate the frequent presence of a low-level inversion, with moist air originating over the Arabian Sea below the inversion. Subsidence limits the height to which the surface air from the Arabian Sea can ascend, restricting cloud development, and thus favouring strong solar heating.

1.6.4 Australia

The chief determinant of the climate and wind field over Australia is the subtropical anticyclonic belt (Sturman and Tapper 1996). In general, the continent is affected by mid-latitude westerlies on the southern fringe, tropical convergence on the northern fringe in summer, and stable subsiding air under the subtropical anticyclones over the interior. The result is that most of the continent is covered by arid or semi-arid climates. The average altitude of Australia is only 300 m, with 87% of the continent less than 500 m and 99.5% less than 1000 m. In general, the low relief of Australia does not significantly obstruct the movement of the atmospheric systems that control the climate.

Most of Australia is warm to hot (Figure 1.10), with the exception of the alpine area in the southeast, where there is seasonal snow. The month with the highest temperature varies from November in the far north to February in the south. In the north, the build-up of monsoon clouds cools the latter part of the summer, whilst in the south the time taken to warm the ocean delays the peak temperature until late summer. July is the coldest month throughout the country. Australia is the driest continent, excluding Antarctica, and no continent has less runoff from its rivers. More than a third of the country receives, on average, less than 250 mm of rainfall annually, and only 9% receives more than 1000 mm. Most of the area south of 35° S receives mainly winter rains, whilst north of 25° S; most rain falls in summer, associated with the summer monsoon and tropical cyclones. Australian rainfall averages disguise an extremely variable rainfall, with droughts and flooding being very common.

Australian rainfall is more variable than could be expected from similar climates elsewhere in the world, mainly due to the impact of ENSO. Australian rainfall fluctuations, as well as being more severe because of ENSO's influence, also operate on very large spatial scale. High rainfall totals in Australia occur when the Southern Oscillation Index (SOI) is large and positive (La Niña events). In contrast, when the SOI is strongly negative (El Niño years), drought occurs over much of the continent. Thus, the continental scale of the 1982–1983 drought is typical of many years, although it was more severe than most. Extended periods of drought or extensive rains in Australia do not occur randomly in time, in relation to the annual cycle. The ENSO phenomenon, and Australian rainfall fluctuations associated with it, are phase-locked with the annual cycle. Thus, the heavy rainfall of an anti-ENSO event tends to start early in the calendar year and finish early in the following year. The dry periods associated with ENSO events tend to occupy a similar time period. For example, the 1982–1983 drought started about April 1982 and broke over much of the country in March and April 1983.

Rainfall is not the only aspect of Australian climate affected by ENSO. Frosts tend to be more common in inland Australia during ENSO events, because low rainfall is associated with decreased cloud cover, allowing increased radiative cooling at night. The decreased cloud cover also causes higher maximum temperatures during ENSO events. It is also reported that north of about 25° S, low-level winds in the winter during ENSO events tend to be about three to four times stronger than during anti-ENSO events. In the southeast, the winds in summer tend to be two to three times stronger in ENSO events. The higher maximum temperatures and stronger winds associated with ENSO-related droughts increase the likelihood of wildfires. The latitudinal variations in the season in which the ENSO-amplified winds occur (winter in the north, summer in the south) mean that the stronger winds occur at the time of year when fires are most likely.

1.6.5 South America

Of particular climatological significance in South America are the Andes Mountains along the west coast and the tapered shape of the continent, with much of its area lying in tropical latitudes. The latter means that the oceans are of importance in the climates over large areas of the continent.

Three circulation regimes dominate the climate of the continent (Cerveny 1998):

1) The prevailing westerly winds of the extreme southern continental latitudes
2) The semi-permanent subtropical high-pressure cells positioned over the South Atlantic and South Pacific oceans
3) The location of the intertropical convergence zone (ITCZ), a migrating band of maximum convergence, convection cloudiness and rainfall

The movements of the ITCZ strongly influence the climates of tropical South America. The ITCZ reaches its northernmost location in June and September/October, causing the season of greatest rainfall for northern South America and the Caribbean. At the height of the northern winter, the ITCZ extents southward into the central Amazon Basin. Consequently, the months of January and February mark the dry season for much of northern South America and the tropical Atlantic, whilst the Amazon Basin receives much of its annual rainfall in this season. In general, the onset of the rainy season occurs first in southeast Amazonia, with onset dates occurring progressively later towards the northwest. The demise of the rainy season occurs first in the southeast and progressively later to the northwest. Associations with the extremes of the SOI are most marked in central Amazonia and near the mouth of the Amazon. Years with the low/high SO phase are consistent with anomalously dry/wet rainy seasons in these two regions, due mainly to a late/early onset of the rainy season. During the El Niño, the near-surface Atlantic trade winds are weaker, consistent with an anomalously northward displaced ITCZ, thus the moisture input from the Atlantic into the Amazon basin is weak, the moisture convergence and convection is also weak over Amazonia, and lower rainfall is observed especially in the central and mouth of Amazon River regions. The sea-surface temperature dipole in the tropical Atlantic indicates anomalously warmer surface water to the north of the Equator. During La Niña, these patterns are reversed, with a southward displacement of the ITCZ, strong Atlantic trades winds and moisture transport into Amazonia, and an inverted sea-surface temperature dipole (anomalously warmer surface water south of the Equator). Over the western part of the basin there is a descending branch of the Walker cell during El Niño, with subsidence affecting a zonal band from the Andes to the Atlantic. There is also strong upward flow, convection, and rainfall over northern Peru to the west of the Andes, associated with anomalously warm Pacific sea-surface temperatures (SST), implying compensatory subsidence and reduced rainfall over western Amazonia. This flow pattern is inverted during La Niña. The upper-level Bolivian tropospheric high is weaker during El Niño, with the subtropical westerly jet stronger and located anomalously northward of its average position, whereas during La Niña the jet is weaker and anomalously southward of its average position.

Northeast Brazil is an exceptional area in which rainfall diminishes from in excess of 1000 mm along the coast to less than 400 mm inland (Figure 1.11). The relative dryness of northeast Brazil is caused primarily by the flow patterns of the general atmospheric circulation, particularly by thermally driven circulations of the Hadley–Walker types. To the immediate west of northeast-Brazil is the Amazon basin, where the high rainfall is associated with vigorous upward convection in cumulonimbus clouds. The water vapour condensing in these clouds releases latent heat, which warms the air and thereby maintains the ascending motion. The air rising over Amazonia descends in a Walker type circulation over the eastern subtropical Atlantic Ocean, including northeast Brazil and possibly the western coast of Africa. The circulation patterns are reflected in the rainfall and cloud distributions. Similarly, the north–south circulation patterns of the Hadley cells, with ascending motion in the convective clouds of the ITCZ and descending motion over the subtropical Atlantic of both hemispheres, also contributes to the aridity of northeast Brazil. The interannual variability, both in strength and geographical position, of the Hadley–Walker circulations results in the high interannual variability of precipitation in northeast Brazil.

In oceanic regions, the ITCZ generally lies over or near the highest SST. Therefore, a relationship should be expected to exist between the general sea-surface temperature distribution in the tropical Atlantic and the rainfall over northeast Brazil. Warmer (colder) SST in the tropical South Atlantic and colder (warmer) ones in the tropical North Atlantic are associated with wet (dry) years in northeast Brazil. Thus, in wet years (e.g. 1964, 1967, 1984, and 1985) the sea-surface temperature anomalies were positive in the tropical South Atlantic and negative or near zero in the tropical North Atlantic. In contrast, in dry years (e.g. 1951, 1953, 1958, 1970, and 1979) the sea-surface anomalies were negative in the tropical South Atlantic and positive in the tropical North Atlantic.

1.6.6 Africa

Of all the continents, Africa is the most symmetrically located with regard to the Equator, and this is reflected in the climatic zonation. The continent may be regarded as a giant plateau, for there is a relative absence of very pronounced topography, although some high mountains exist, especially in the East African region. The latitudinal position of Africa means that the continent is influenced by tropical, subtropical, and mid-latitude pressure and wind systems. As the near-equatorial trough migrates with the seasons, the ITCZ lies south of the Equator in the northern winter and north of the Equator in the summer. Connections between the equatorial easterly jet stream and North African rainfall were discussed in the section on monsoons.

Of particular interest in Africa is the Subsaharan region (Sahel), because it shows a dramatic decrease in rainfall since the late 1960s and continued severe drought conditions from the early 1970s up to at least the late 1980s. As in the case of northeast Brazil, this is probably connected with regional circulation changes and North Atlantic SST. Indeed, colder than average tropical North Atlantic SST appear to be strongly associated with drought in the African Sahel (Folland et al. 1986; Hastenrath 1995).

Questions

1   Sulphate aerosols originating from industrial areas cool the atmosphere but cause both ill health and acid rain. Attempts to remove such aerosols from the atmosphere will accelerate global warming. What therefore should be the policy on industrial pollution?

2   Summarize the main changes in the climate of your home area over the past 100 000 years.

3   How does ENSO affect local climate and human activity across the world?

4   How do major mountain ranges and plateaus affect large-scale climate?

5   What is meant by a 'proxy' indicator of climate? Give examples of how these can be used to reconstruct records of past climate.

References

Brook, E.J. (2005). Tiny bubbles tell all. *Science* 310: 1285–1287.

Cerveny, R.S. (1998). Present climates of South America. In: *Climates of the Southern Continents* (eds. J.E. Hobbs, J.A. Lindesay and H.A. Bridgman), 107–135. Chichester: Wiley.

DeConto, R.M. and Pollard, D. (2003). Rapid Cenozoic glaciation of Antarctica induced by declining atmospheric CO_2. *Nature* 421: 245–249.

EPICA Community Members (2004). Eight glacial cycles from an Antarctic ice core. *Nature* 429: 623–628.

Folland, C.K., Palmer, T.N., and Parker, D.E. (1986). Sahel rainfall and worldwide sea temperatures 1901–85. *Nature* 320: 602–607.

Fu, Q., Johanson, C.M., Wallace, J.M., and Reicher, T. (2006). Enhanced mid-latitude tropospheric warming in satellite measurements. *Science* 312: 1179.

Glantz, M.H. (1996). *Currents of Change: El Niño's Impact on Climate and Society*. Cambridge, UK: Cambridge University Press.

Hastenrath, S. (1995). Recent advances in tropical climate prediction. *Journal of Climate* 8: 1519–1532.

Henderson-Sellers, A. and Robinson, P.J. (1986). *Contemporary Climatology*. Harlow, Essex: Longman.

Houghton, J. (2005). Global warming. *Reports on Progress in Physics* 68: 1343–1403.

Hurrell, J.W. (1995). Decadal trends in the North Atlantic Oscillation: regional temperature and precipitation. *Science* 269: 676–679.

IPCC (2014). The Intergovernmental Panel on Climate Change 5th Assessment Report (AR5). https://www.ipcc.ch/report/ar5/syr/.

Jones, P.D. and Moberg, A. (2003). Hemispheric and large scale surface air temperature variations: an extensive revision and an update to 2001. *Journal of Climate* 16: 206–223.

King, J.C. and Turner, J. (1997). *Antarctic Meteorology and Climatology*. Cambridge, UK: Cambridge University Press.

Lazzara, M.A., Keller, L.M., Markle, T., and Gallagher, J. (2012). Fifty-year Amundsen–Scott South Pole station surface climatology. *Atmospheric Research* 118: 240–259.

Lockwood, J.G. (1974). *World Climatology: An Environmental Approach*. London, UK: Edward Arnold.

Lockwood, J.G. (2001). Abrupt and sudden climatic transitions and fluctuations: a review. *International Journal of Climatology* 21: 1153–1179.

Lockwood, J.G. (2003). The Earth's climates. In: *Handbook of Atmospheric Science: Principles and Applications* (eds. C.N. Hewitt and A.V. Jackson), 59–89. Oxford, UK: Blackwell.

Mann, M.E., Bradley, R.S., and Hughes, M.K. (1999). Northern Hemisphere temperatures during the past millennium: inferences, uncertainties and limitations. *Geophysical Research Letters* 27: 1519–1522.

McGregor, G.R. and Nieuwolt, S. (1998). *Tropical Climatology: An Introduction to the Climates of the Low Latitudes*. Chichester, UK: Wiley.

National Climatic Data Center (1987). *Monthly Climatic Data for the World*, vol. 40, Nos 1 and 7. Asheville, NC: National Oceanic and Atmospheric Administration.

Oort, A.H. (1983). *Global Atmospheric Circulation Statistics, 1958–1973*. Washington, DC: National Oceanic and Atmospheric Administration.

Orvig, S. (ed.) (1970). *Climate of the Polar Regions*, World Survey of Climatology, vol. 14. Amsterdam: Elsevier.

Palmen, E. (1951). The role of atmospheric disturbances in the general circulation. *Quarterly Journal of the Royal Meteorological Society* 77: 337–354.

Pant, G.B. and Rupa Kumar, K. (1997). *Climates of South Asia*. Chichester, UK: Wiley.

Peixoto, J.P. and Oort, A.H. (1992). *Physics of Climate*. New York, NY: American Institute of Physics.

Schwerdtfeger, W. (1984). *Weather and Climate of the Antarctic*. Amsterdam: Elsevier.

Serreze, M.C. and Barry, R.G. (2005). *The Arctic Climate System*. Cambridge, UK: Cambridge University Press.

Shackleton, N.J. (2000). The 100 000-year ice-age cycle identified and found to lag temperature, carbon dioxide, and orbital eccentricity. *Science* 289: 1897–1902.

Simmons, I. (1998). The climate of the Antarctic region. In: *Climates of the Southern Continents: Present, Past and Future* (eds. J.E. Hobbs, J.A. Lindesay and H.A. Bridgman), 137–160. Chichester, UK: Wiley.

Stott, P.A., Tett, S.F.B., Jones, G.S. et al. (2000). External control of twentieth century temperature variations by natural and anthropogenic forcings. *Science* 15: 2133–2137.

Sturman, A.P. and Tapper, N.J. (1996). *The Weather and Climate of Australia and New Zealand*. Melbourne, Australia: Oxford University Press.

Wyrtki, K. (1982). The Southern Oscillation, ocean–atmosphere interaction, and El Niño. *Marine Technology Society Journal* 16: 3–10.

Further Reading

Houghton, J.T. (2004). *Global Warming: The Complete Briefing*, 3e. Cambridge, UK: Cambridge University Press.

IPCC (2001). *Climate Change 2001: The Scientific Basis. Contribution of Working Group 1 to the Third Assessment Report of the Intergovernmental Panel on Climate Change* (eds. J.T. Houghton, Y. Ding, D.J. Griggs, et al.). Cambridge, UK: Cambridge University Press.

IPCC (2007). *Climate Change 2007: The Physical Science Basis. Contribution of Working Group 1 to the Fourth Assessment Report of the Intergovernmental Panel on Climate Change* (eds. S. Solomon, D. Qin, M. Manning, et al.). Cambridge, UK: Cambridge University Press.

IPCC (2007). *Climate Change 2007: Climate Change Impacts, Adaptation and Vulnerability. Contribution of Working Group 2 to the Fourth Assessment Report of the Intergovernmental Panel on Climate Change* (eds. N. Adger, P. Aggarwal, S. Agrawala, et al.). Cambridge, UK: Cambridge University Press.

Munn, T. (ed.) (2002). *Encyclopedia of Global Environmental Change*, 5 vols. Chichester, UK: John Wiley.

Oliver, J.E. (ed.) (2005). *Encyclopedia of World Climatology*. Dordrecht: Springer-Verlag.

2

Chemical Evolution of the Atmosphere

Richard Wayne

Physical and Theoretical Chemistry Laboratory, Department of Chemistry, University of Oxford, Oxford, United Kingdom

This chapter is concerned with how the planet Earth came to have an atmosphere, and how that atmosphere has been modified by chemical, physical, and biological processes to move towards its present-day composition. The story begins with the *Big Bang* in which the universe was created, and we leave it some hundreds of millions of years before present. Other chapters of this book will discuss more recent changes in composition, especially in connection with climate. As we approach our own era within a million years or so, the record of atmospheric composition and climate becomes richer and more detailed. Particularly fruitful sources of information have proved to be the examination of cores of rock drilled deep into the ocean floors, and cores obtained from ice-sheets and glaciers.

Ice cores from Greenland and Antarctica are mainstays of modern climate science. Traditionally, investigators drill in places where ice layers accumulate year after year, undisturbed by glacial flows. The long layer cake records from deep sites in the centre of Antarctica reveal how greenhouse gases have surged and ebbed across hundreds of thousands of years. But because heat from the bedrock below can melt the deepest, oldest ice, the approach has not yielded ice any older than 800 000 years, from a core drilled at Antarctica's Dome C in 2004 (Lüthi et al. 2008).

The oldest conventional ice-core is that from East Antarctica: by 2007 its depth had reached 3.2 km, corresponding to an age of 800 000 years. The core shows that there have been eight cycles of atmospheric change over this period, with peaks in the concentrations of carbon dioxide and methane being linked to a warming of the climate. Jouzel (2013) provides a valuable survey of key results obtained from layered ('cake') cores, as well as the limitations of the studies.

An alternative approach to drilling cores ever deeper investigates ancient ice sitting far closer to the surface, in the Allan Hills, a wind-swept region of East Antarctica 200 km from McMurdo Station that is famous for preserving ancient meteorites. In such 'blue ice' areas (making up just 1% of the continent's surface), the ice flows across rocky ridges, tipping the record on its side. Deep, old layers are driven up, whilst wind strips away snow and younger ice, revealing the lustrous blue of compressed ice below. But these contortions also confound the neat ordering of the annual layers – making it impossible to date the

Atmospheric Science for Environmental Scientists, Second Edition. Edited by C.N. Hewitt and Andrea V. Jackson.
© 2020 John Wiley & Sons Ltd. Published 2020 by John Wiley & Sons Ltd.

ice by counting the layers. However, the lumps of ice may be dated directly from trace amounts of argon and potassium that they contain (radioactive ^{40}K decays to ^{40}Ar with a known half-life). Although not as precise as other dating methods, the technique can date ice to within 100 000 years or so.

In 2017, it was announced (Voosen 2017) that a core drilled from blue ice in Antarctica had yielded 2.7-million-year-old ice, an astonishing find at least 1.7 million years older than the previous record-holder. Bubbles in the ice contain greenhouse gases from Earth's atmosphere from a time at which the planet's cycles of glacial advance and retreat were just beginning, potentially offering clues to what triggered the ice ages.

Part of the importance of this ancient record is associated with concerns about CO_2-driven climate change. Analysis of the new ice core reveals that CO_2 did not exceed 300 ppm in the past 2.7 million years, well below the current (October 2019) CO_2 level of 408.2 ppm (Earth's CO_2 home page: https://www.co2.earth). However, the new Allan Hills ice core provides one of the most comprehensive views of Earth's past climate, and the researchers involved are hoping to take the record back to five million years. For the purposes of our enquiry here, the significance of being able to sample trapped bubbles is that we are enabled to reach a better understanding of atmospheric composition (especially of O_2 and CO_2) in the distant past, and thus to be guided in our exploration of how the atmosphere evolved.

Earth and its neighbours Venus and Mars must have lost at an early stage any primordial atmosphere with which they might have been born. Instead, a secondary atmosphere was formed from volatile materials trapped within the solid body when it was formed, or brought in later by impacting solar-system debris (comets and meteors). Life on Earth has had an enormous effect in bringing about subsequent changes to the composition of our own atmosphere, especially in terms of the relative abundances of CO_2, N_2, and O_2. Carbon dioxide, which is present at less than 0.042% in our atmosphere, makes up more than 95% of the atmospheres of Venus and Mars. Conversely, the N_2 and O_2 that make up the bulk of our atmosphere are only minor components of the other two atmospheres. Yet it is likely that all three planets acquired secondary atmospheres that were initially similar: biological or biologically mediated processes have modified our atmosphere. A link must therefore be sought between the evolution of life and the evolution of Earth's atmosphere. What is more, we shall see later that oxygen has a critical role in protecting organisms on land from ultraviolet radiation from the Sun. Molecular oxygen and its atmospheric product ozone (O_3) are the only known absorbers of such radiation in the contemporary atmosphere. There is thus a further link between life and the atmosphere.

Figure 2.1 provides a clear illustration of some of the statements just made. The pie charts show the fractional amounts of CO_2, N_2, O_2, and Ar in the atmospheres of Earth, Venus and Mars. Mars, being a small planet with relatively low gravitational attraction, has lost much of its atmosphere as a result of erosion by impacts and by escape. The right-hand diagrams of each pair are 'reconstructions' of what the atmospheric compositions might have been like without life on Earth or losses from Mars. Venus has probably not lost substantial amounts of any of the gases under discussion; the pie chart for the present day is thus repeated on the right. There is a remarkable similarity between the three charts, indicating that the assumptions about the effects of life on Earth and of loss from Mars really can explain an evolution of the atmospheres from the same starting compositions to those found at the present day. Furthermore, although the total pressures on the three planets are now very different, as indicated on the figure, the reconstructed pressures are much more similar. The remaining differences might well come about because of the differences in sizes of the planets and in their positions in the solar system.

Venus: present day
P = 93 atm

CO$_2$ 0.965

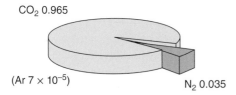

(Ar 7 × 10^{-5})

N$_2$ 0.035

Venus: present day, for comparison
P = 93 atm

CO$_2$ 0.965

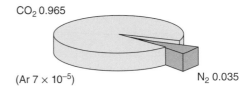

(Ar 7 × 10^{-5})

N$_2$ 0.035

Earth: present day
P = 1 atm

N$_2$ 0.781

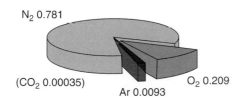

(CO$_2$ 0.00035)

Ar 0.0093

O$_2$ 0.209

Earth: effects of life removed
P = 70 atm

CO$_2$ 0.98

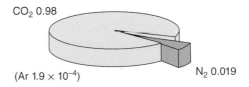

(Ar 1.9 × 10^{-4})

N$_2$ 0.019

Mars: present day
P = 0.006 atm

CO$_2$ 0.953

(O$_2$ 0.0013)

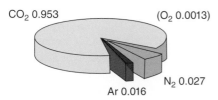

Ar 0.016

N$_2$ 0.027

Mars: adjusted for erosion and escape
P = 7.5 atm

CO$_2$ 0.98

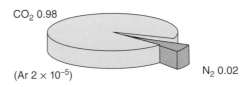

(Ar 2 × 10^{-5})

N$_2$ 0.02

Figure 2.1 Abundances of gases in the atmospheres of Venus, Earth, and Mars. The pie charts show the fractional abundances of the dominant gases, and those of other key components are shown in parentheses. The diagrams on the right for Earth and Mars show what would happen on Earth if the effects of life were removed, and on Mars if the atmosphere were adjusted for loss by erosion and escape. *Source:* data for contemporary atmospheres are taken from Wayne (2000), who cites the original references, whilst the reconstructions are reported by Morrison and Owen (1996, p. 347).

2.1   Creation of the Planets and Their Earliest Atmospheres

Hydrogen and helium were present in the universe almost immediately after the 'Big Bang', some 10–20 Gyr ago (current best estimates, based on Hubble space-telescope data, are 12 Gyr ago). Nuclear fusion transformed these elements into heavier ones such as carbon, nitrogen, oxygen, magnesium, silicon, and iron. Elements of atomic number higher than that of iron are formed in supernova explosions that scatter material through the galaxies as tiny dust grains 1–1000 nm in diameter; the grains are probably composed mainly of graphite, H_2O ice, and iron and magnesium silicates. Dust and gas in the universe are concentrated in the arms of spiral galaxies in which new stars, such as our Sun, are formed. Radiometric dating of meteorites and lunar samples suggests that the solar system was formed from the accretion of the primitive gases and dusts 4.6×10^9 years ago. According to what has now become the standard model (Wetherill 1990), planets grow by agglomeration of rocky 'planetesimals', with diameters of up to a few kilometres, that form in the solar nebula. 'Planetary embryos', larger bodies of the size of the Moon or Mercury, with masses of 10^{22} to 10^{23} kg, grow where the planetesimals are sufficiently closely packed to allow collisions between them. The largest of these bodies could form in $\approx 10^5$ years. The growth of planets the size of Earth or Venus would require the merger of about 100 of these bodies over 10^7 to 10^8 years.

Four possible mechanisms could account for the existence of atmospheres on Venus, Earth, and Mars (Cameron 1988). These are: (i) capture of gases from the primitive solar nebula; (ii) capture of some of the solar wind; (iii) collision with volatile-rich comets and asteroids; and (iv) release (outgassing) of volatile materials trapped together with the dust grains and planetesimals as the planets accreted (Wetherill 1990). One pointer to the origins of the atmospheres of the inner planets comes from the noble gases present, because chemical inertness prevents loss to surface rocks, and, except in the case of helium, they cannot readily escape to space. Two types of noble gas can be distinguished: primordial and radiogenic. Primordial isotopes such as ^{20}Ne, ^{36}Ar, ^{38}Ar, ^{84}Kr, and ^{132}Xe were present in the solar system from the time of its creation. Radiogenic isotopes, however, have built up from the decay of radioactive nuclides: ^{40}Ar from the decay of ^{40}K, and ^{4}He from the decay of ^{232}Th, ^{235}U, and ^{238}U.

In comparison with solar abundances, on a mass per unit mass basis, the Earth's atmosphere is depleted of ^{36}Ar and of ^{20}Ne by large factors. On Earth, ^{20}Ne and ^{36}Ar are depleted by a factor of nearly 10^{11} and 10^9, respectively (relative to solar abundances in units of atoms per 10^6 Si atoms (Pepin 2006). However, on this measure for carbon, much of which is bound up in involatile compounds, the depletion by a factor of 10^5 is far smaller than that for the noble gases. Evidence of this kind is taken as clear proof that Earth (and Mars and Venus) has lost almost all its primordial atmosphere, if such an atmosphere existed at all, and that the present atmosphere has been acquired later (Avice et al. 2017).

Hypotheses (i) and (ii) above argue for gravitational capture and retention by the planets after their formation either of gases of the primordial solar nebula or of the solar wind that has flowed over them during their lifetimes. The differences in ratios of noble gases between the Sun and the planets are evidence against these mechanisms. Nevertheless, these hypotheses have not been abandoned entirely (Denlinger 2005). Substantial numbers of small bodies have impacted with the planets of the inner solar system over the planetary lifetimes, and the comet–asteroid hypothesis (iii) proposes that atmospheres were brought to the planets as a result of such impacts. However, Venus and Earth have a roughly equal chance of encountering comets and asteroids, and yet Earth has nearly two orders of magnitude less ^{36}Ar on a mass for mass basis than Venus, thus suggesting that the comet–asteroid hypothesis cannot account for a substantial proportion of the present-day atmospheres (but see later). The remaining hypothesis (iv), which is that volatile materials were incorporated into the planet as it accreted, thus seems the most probable. If the planetesimals that formed the planets contained small amounts (perhaps a fraction 10^{-4} by mass) of volatile materials, then gases could be released from within the planet as it heated up (as a result of the accretion process itself, of impacts of infalling bodies on the unprotected surface, and of decay of short-lived radioactive elements). Minerals containing bound H_2O would dissociate, and physically trapped components would become degassed. One explanation for the large excess of nonradiogenic noble gases on Venus compared with Earth could be that the planetesimals that formed Venus were exposed to an intense solar wind that was absorbed before it reached the part of the solar system where Earth (or Mars) formed.

An objection to the accretion hypothesis (iv) arises because one of the most satisfactory theories on the origin of the Earth–Moon system postulates that a body of nearly the size of Mars collided with the proto-Earth to melt both the impactor and most of the Earth's mantle. This very early catastrophic event would necessarily mean that almost all the water and other moderately volatile compounds that accreted with the Earth would have been lost to space. In these circumstances, a variant of the comet–asteroid hypothesis has received renewed support. The gaseous components of present-day meteorites are of interest because they may reflect the composition of the primitive materials out of which the planets accreted, as well as providing an indicator of the materials present in the solar system that are available for impact degassing (as required by the comet–asteroid hypothesis). The striking similarity in the $^{20}Ne/^{36}Ar$ ratios for the CI group (named after the Ivuna meteorite) of carbonaceous chondrites and the planetary atmospheres has prompted speculation that there may have been a single type of parent gas reservoir, with the present small spread of abundances determined by evolutionary processes.

The way in which the inner planets heated up after they had accreted has a bearing on the composition of the earliest atmospheres if the atmospheres were primarily outgassed from the planets themselves (hypothesis iv). If the rate of accretion was sufficient that melting occurred as the planet was forming, then, at least on Earth, iron could migrate to the core, leaving an iron-free silicate mantle in an inhomogeneous process (Walker 1976). The alternative, homogeneous accretion, model proposes that the heating and differentiation of Earth took place after the accretion itself. At present, the inhomogeneous model is more widely accepted than the homogeneous model. The importance for the atmosphere is that the presence or absence of iron in the mantle would determine the oxidation state and composition of the volatiles that outgassed (Levine 1985). With iron present, reduced compounds such as CH_4, NH_3, and H_2 would be expected, whereas with outgassing through a mantle that was already differentiated, or from a veneer, the probable species would be H_2O, CO_2, and N_2, together with small amounts of the reducing gases H_2 and CO. Figure 2.2 illustrates the differences between the homogeneous and heterogeneous models.

Regardless of which gases were released initially, it does not seem likely that the reduced gases could survive for long in the early atmosphere. Molecular hydrogen, like helium, is able to escape the gravitational attraction of the Earth and this process could account for loss of almost all H_2 as long as there were adequate heat sources in the upper atmosphere. Figure 2.3 illustrates a highly simplified prebiotic hydrogen scheme in which H_2 is released from the interior of the Earth and escapes to space (Catling and Claire 2005). In fact, photolysis of molecular H_2 to atomic H by short-wavelength ultraviolet radiation allows an important additional escape process. Since H atoms possess one-half the mass of H_2, for identical translational energies their velocities are four times greater, and are thus more likely to exceed the limiting velocity of escape from Earth ($11.2\,km\,s^{-1}$).

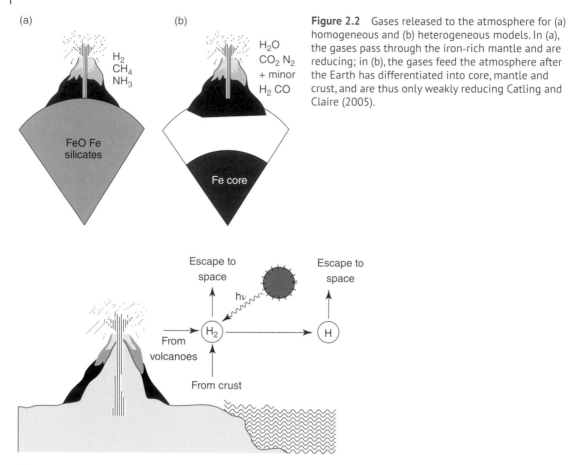

Figure 2.2 Gases released to the atmosphere for (a) homogeneous and (b) heterogeneous models. In (a), the gases pass through the iron-rich mantle and are reducing; in (b), the gases feed the atmosphere after the Earth has differentiated into core, mantle and crust, and are thus only weakly reducing Catling and Claire (2005).

Figure 2.3 Hydrogen in the atmosphere of the prebiotic Earth. Release and escape are in balance for an atmospheric mixing ratio of approximately 1000 ppmv Catling and Claire (2005).

Both NH_3 and CH_4 were readily removed from the prebiotic atmosphere. They are rapidly photolysed by solar ultraviolet radiation, and would also be attacked by OH radicals that would be formed photo-chemically if any H_2O were present in the atmosphere. Thus, in the absence of a continuous source of the gases, the abundances of these reduced species must have been low. Rocks from as long ago as 3.8×10^9 years (from Isua, in west Greenland) consist of highly metamorphosed sediments that show that abundant CH_4 was not present when they were formed. Sagan and Chyba (1997), on the other hand, argue that atmospheric methane could have been kept at relatively high concentrations, either as a result of a recycling process or, after life had been established, by the formation of the gas by methanogenetic microorganisms. They suggest that CH_4 photochemistry could have led to the formation of a high-altitude absorbing layer of organic solid aerosol that, in turn, could have protected NH_3 from rapid photolysis. Although this view is not widely accepted, it must be considered as a possibility, especially as it has implications for both greenhouse heating and the intensities of ultraviolet radiation reaching the surface of the Earth.

2.2   Earth's Atmosphere before Life Began

Section 2.1 outlined the principles that are thought to have applied to the acquisition of primitive second-ary atmospheres by Venus, Earth, and Mars. For Earth, the most likely scenario (Kasting 1993) is that it formed relatively rapidly (over a period of 10^7 to 10^8 years), that its interior was initially hot as a result of a large number of impact events, and that the core was probably formed as the planet accreted. Metallic iron could have been removed from the upper mantle, and volcanic gases could have been relatively oxidized as early as 4.5 Gyr ago. Impact probably led to the release of many of Earth's volatile materials, and thus to a transient atmosphere very heavily laden with water vapour. Infalling iron-rich planetesi-mals in this phase of planetary accretion would have led to the reduction of water to form copious amounts of H_2. After the end of the main accretionary phase, the water-vapour atmosphere would have condensed out to form the oceans, thus making possible the conversion of CO_2 in the atmosphere to carbonate rocks such as limestone and dolomite, to be discussed later.

Most current models predict that the early atmosphere consisted mostly of CO_2, N_2, and H_2O, along with traces of H_2 and CO. Such models are based on the assumption that the redox state of the upper mantle has not changed, so that the composition of volcanic gases has remained approximately constant with time. Kasting et al. (1993) argue that this assumption is probably incorrect. They believe that the upper mantle was originally more reduced than today, and has become progressively more oxidized as a consequence of the diminished release of reduced volcanic gases and the subduction of hydrated, oxi-dized seafloor. The redox state of sulphide and chromite inclusions in diamonds implies that the process of mantle oxidation was slow, so that reduced conditions could have prevailed for as much as half of Earth's history. Other oxybarometers of ancient rocks give different results, so that the question of when or if the mantle redox state changed remains unresolved. Mantle redox evolution is intimately linked to the oxidation state of the primitive atmosphere: a reduced Archaean atmosphere would have had a high hydrogen escape rate and should correspond to a changing mantle redox state, whilst an oxidized Archaean atmosphere should be associated with a constant mantle redox state.

The estimated abundance of carbon that accumulated in the crust of Earth would be sufficient to pro-duce a partial pressure of 60–80 atm if it were all in the atmosphere in the form of CO_2. Speculative reconstructions of the climates of the past have been used to infer what the partial pressure might really have been. The 'reconstruction' in Figure 2.1, reported by Morrison and Owen (1996), puts the entire burden into the atmosphere, whereas Durham and Chamberlain (1989) suggest a partial pressure of 14 atm. Whatever the original load, there is only about 4.1×10^{-4} atm in the contemporary atmosphere. The remainder is now incorporated mainly in the carbonate rocks, with a little dissolved in the oceans: the partitioning between atmosphere, hydrosphere and lithosphere is now roughly $1:50:10^5$. A most important aspect of atmospheric evolution on Earth is thus how atmospheric CO_2 might be converted to solid carbonates. Inorganic weathering reactions (Wayne 2000) can convert silicate rocks such as diop-side ($CaMgSi_2O_6$) to carbonate

$$CaMgSi_2O_6 + CO_2 \rightleftharpoons MgSiO_3 + CaCO_3 + SiO_2 \tag{2.1}$$

We should note here that even apparently abiological changes such as this weathering can be modu-lated by biological influences. Partial pressures of CO_2 in the soil where weathering occurs are 10–40 times higher than the atmospheric pressure, and these high partial pressures are maintained by soil bacteria. Thus, when appropriate organisms had developed, conversion rates could be greatly enhanced. In addition, of course, a further most important source of carbonate minerals is secretion by animals and plants, and the deposition of calcite ($CaCO_3$) shells. Life thus exerts an extremely significant influence on the removal of CO_2 from the atmosphere.

Formation of carbonates and the evolution of life itself have in common the need for liquid water, so that the nonatmospheric reservoirs for CO_2 on Earth seem linked to the presence of the liquid phase over geological time.

2.3 Comparison of Venus, Earth, and Mars

The composition and the chemical and physical behaviour of the atmospheres of Venus and Mars provide further evidence about the evolution of our own atmosphere. In particular, comparisons of the three atmospheres (e.g. Hunten 1993) show how the emergence of life on Earth modified our atmosphere dramatically from the atmospheres of our two planetary neighbours.

We have already seen how the compositions and surface pressures of the atmospheres of the three planets differ markedly. The surface temperatures are also very different: 732, 288, and 223 K, for Venus, Earth, and Mars, respectively. These temperatures can be interpreted in terms of atmospheric compositions and pressures. Carbon dioxide and water vapour are both greenhouse gases; that is, they trap outgoing infrared radiation that would otherwise escape to space, and thus raise the temperature at the surface of the planet. An estimate can be made of the temperatures that would be experienced on the planets without any atmosphere at all. A simple radiative equilibrium calculation, taking into account differing reflectivities of the planets, would suggest values of 227, 256, and 217 K, some 505, 32, and 6 K less than found for Venus, Earth, and Mars, respectively.

Because of its relative closeness to the Sun, Venus may have suffered from a 'runaway greenhouse effect' in which a positive feedback process ultimately led to vaporization of all surface water. Water vapour makes a large contribution to atmospheric heating. Since vapour pressures rapidly increase with increasing temperature, thus further increasing trapping, there exists a positive feedback in the greenhouse effect. Evaporation from a planetary surface will proceed either until the atmosphere is saturated with water or until all the available water has evaporated. What happens on any particular planet will depend on the starting temperature in the absence of radiation trapping, since that will decide whether the vapour ever becomes saturated at the temperatures reached. On Mars and Earth, the additional heating due to liberation of water vapour is not sufficient to prevent the vapour reaching saturation as ice or liquid. However, on Venus there comes a critical vapour pressure ($\approx$10 mbar) when the rate of heating begins to increase dramatically: that vapour pressure is never reached at the lower temperatures on Earth or Mars. As a result, the P–T curve for the atmospheric water vapour increases more slowly than that for the vapour–liquid phase-equilibrium curve. Condensation never occurs on Venus, and additional burdens of H_2O serve to increase the temperature even further. Certainly, this positive-feedback mechanism would explain why there is no surface water on Venus at the present day. Large amounts of water vapour could have been the dominant species in the early Venusian atmosphere, but photodissociation and escape of hydrogen to space would have removed most of the H_2O to leave the rather dry atmosphere now found (Hunten 1993). Venus contains quantities of carbon and nitrogen similar to those on Earth, but is deficient in hydrogen. Water abundance on Venus is about 42 kg m^{-2} compared with 2.7×10^6 kg m^{-2}

on Earth. There is certainly no liquid water on the surface of Venus today, and the mixing ratio for water in the atmosphere is probably not more than 2×10^{-4}.

Most mechanisms identified as potentially important for escape of hydrogen from Venus discriminate strongly against loss of deuterium, because of the large escape velocity ($10.3\,\mathrm{km\,s^{-1}}$) from that planet. Enrichment of deuterium might therefore be expected if Venus had originally possessed a water-rich atmosphere. Several pieces of evidence support deuterium enrichment, although they are not unequivocal. The ion mass spectrometer on Pioneer–Venus detected a signal at $m/e = 2$ from the upper atmosphere that can be attributed to D^+, and interpretation of the intensity data would require D/H in the bulk atmosphere of $\approx 10^{-2}$. Mass peaks at $m/e = 18.01$ and 19.01 obtained in the lower atmosphere (below 63 km) with the large-probe neutral mass spectrometer may be caused by H_2O and HDO (although the $m/e = 19$ ion could be H_3O^+). If HDO is the source of the heavier ion, then D/H on Venus is $(1.6 \pm 0.2) \times 10^{-2}$, in agreement with the ion data from the upper atmosphere. On Earth, D/H $\approx 1.6 \times 10^{-4}$ overall (and perhaps twice that value in the upper atmosphere, according to Spacelab 1 observations), so that the deuterium enrichment on Venus is 50 to 100, implying large quantities of water in the early history of the planet. This enrichment factor is the maximum that could arise, as we shall show shortly. The onset of the runaway greenhouse effect depends strongly on the surface water distribution (Kodama et al. 2018). The runaway threshold increases as the surface water distribution retreats towards higher latitude outside the Hadley circulation. The lower the water amount on a terrestrial planet, the longer the planet remains in a habitable condition.

Outgassing might be expected to release materials with oxidation states similar to those for terrestrial volcanic gases ($[CO]/[CO_2] \approx 10^{-2}$; $[H_2]/[H_2O] \approx 10^{-2}$). However, at high Venusian temperatures, the gas-phase equilibrium

$$CO + H_2O \rightleftharpoons CO_2 + H_2 \tag{2.2}$$

and reactions such as

$$2FeO + H_2O \rightarrow Fe_2O_3 + H_2 \tag{2.3}$$

at the surface would have increased the H_2 content relative to H_2O. Molecular hydrogen would thus have been the dominant gas in the early upper atmosphere of Venus.

Supersonic hydrodynamic outflow, powered by solar ultraviolet heating, would have resulted in the loss of H_2 to space. Interestingly, this flow would have entrained HD, thus sweeping deuterium away, until the mixing ratio of H_2 dropped below $\approx 2 \times 10^{-2}$. Only after this limit was passed would deuterium enrichment begin, regardless of how much water was present originally. Hydrogen, in the form of water, is now present at a mixing ratio of $\approx 2 \times 10^{-4}$, according to the Venera spectrophotometer data for 54 km altitude. Deuterium enrichment is thus limited to a factor of ≈ 100, in accordance with the apparent measured value. The escape rate calculated for loss of H_2 would have exhausted the equivalent of Earth's oceans in about 280 million years.

As the Venusian atmosphere progressed towards its contemporary water vapour content, additional hydrogen loss processes probably began to operate. Translationally 'hot' hydrogen atoms can escape if their velocities exceed $10.3\,\mathrm{km\,s^{-1}}$, and can be generated on Venus by elastic collision between 'hot' O* and ambient H

$$O* + H \rightarrow O + H* \tag{2.4}$$

A source of O* on Venus is dissociative recombination of molecular oxygen ions, which can then participate in reaction (2.4); about 15% of the resultant H atoms will have a high enough velocity to escape. Mariner 5 Lyman-α (H-resonance) airglow observations showed that there is an H-atom component with an effective temperature of 1000 K in addition to the atoms that are thermally equilibrated at 300 K. Escape via this collisional mechanism could have reduced the hydrogen content from 2% to the contemporary 0.02% in about 4.2 Gyr, which allows for the possibility that large quantities of water might once have been present. However, massive loss of hydrogen from water brings with it the problem of disposal of the oxygen. It may be that the oxygen escaped to space along with the hydrogen; alternatively, oxidation of surface material would provide a plausible sink if the surface were molten. Another problem concerns the present-day escape of hydrogen from the atmosphere of Venus. Calculations put the time taken to exhaust hydrogen from the atmosphere at the contemporary escape rate at between 500 and 1500 Myr.

The longer time is perhaps compatible with a gradual depletion of water over the life of the planet, but if the shorter time is correct, then the implication is that water is being replenished as fast as it escapes. Such replenishment could be provided by outgassing from the planetary interior (and possibly by cometary impacts). Mixing ratios for water vapour drop by a factor of about five between 10 km altitude and the surface, suggesting that there is a large flux of water from the atmosphere into the surface, which could nearly balance a relatively large flux of juvenile water from the interior. Substantial oxidation of the surface would be expected with large water fluxes through it, and some results (e.g. from Venera 13) indicate the presence of Fe[III] minerals that are consistent with a relatively highly oxidized surface.

Mars presents a sharp contrast. Surface channelling features suggest strongly that there was once liquid water on the surface. There seems to be no liquid water now, but there is water vapour in the atmosphere and clouds are observed. The polar caps (and much of the winter hemisphere) are covered with water ice, and the winter polar cap probably contains solid CO_2 as well. The escape velocity for relatively small Mars is less than half (and the energy required for escape thus less than a quarter) that for either of the two larger planets. A large proportion of the outgassed species can therefore have been lost from the Martian atmosphere.

On Mars, escape of the ^{14}N isotope is slightly faster than that of ^{15}N because of the lower mass. With the atmosphere as the reservoir of nitrogen, the N_2 remaining will have become slowly richer in ^{15}N over the life of the planet. In comparison with nitrogen on Earth, where neither isotope escapes, the Martian ^{15}N is enriched by a factor of 1.6. It follows that Mars once had 10 or more times as much nitrogen in its atmosphere as it has now. Continuous degassing of nitrogen from the planet's interior would tend to sustain the original isotopic ratio, so that the observed enrichment favours an evolutionary model in which Mars acquired its nitrogen atmosphere early in its history, with relatively little degassing later. By way of contrast to the nitrogen isotopes, the Martian ^{16}O/^{18}O ratio is almost exactly the same as that for Earth, and ^{18}O has been enriched by less than 5%. Yet Mars is losing O atoms at present at the rate of 6×10^7 atoms s^{-1} for every square centimetre of surface. Lack of ^{18}O enrichment implies a source of 'new' oxygen, in a reservoir holding at least 4.5×10^{25} atoms cm^{-2}, presumably in the form of H_2O, since the escape of hydrogen and oxygen from Mars is constrained to have a 2 : 1 stoicheiometry. There is, however, apparently an enrichment of D over H by a factor of about five over the terrestrial value. If D/H for juvenile water on Mars is the same as that for Earth, the observed enhancement must be explained by a divergent history of atmospheric evolution on the two planets. Several steps in the escape of hydrogen favour loss of H over loss of D, but, even so, the observed enrichments can be attained only if some of the (D-enriched) atmospheric water can exchange back with the condensed phase.

Model calculations, based on the assumption that D/H in primordial Martian H_2O is the same as the terrestrial value, imply an initial reservoir of hydrogen equivalent to a layer of water about 3.6 m thick, most of which has escaped, to leave a present-day residue that is 0.2 m thick. The calculations assume, amongst other things, that the escape rate has remained constant over geological time. Nevertheless, the required exchangeable surface layer thickness is almost two orders of magnitude less than geological inventories of subsurface water, so that the postulated loss is not unreasonable. Presumably, the much weaker fractionation between ^{18}O and ^{16}O compared with that between D and H has prevented a measurable enhancement of the heavier oxygen isotope even in the presence of exchange with a modest surface reservoir.

Water on Earth is particularly interesting. Geological evidence suggests that the waters of the ocean have neither completely frozen nor completely evaporated for at least the past 3.5×10^9 years. Yet the infrared luminosity of the Sun has increased substantially over that period (Sagan and Chyba 1997), leading to the 'faint young Sun paradox' to which we shall return later when we examine CO_2 concentrations in more detail. Thermostatic control seems to have been exerted by a reduction of the concentration of greenhouse gases in Earth's atmosphere, probably effected directly or indirectly by the biota. Such control lies at the heart of the Gaia hypothesis, which is described briefly in the next section. In the absence of water on Venus, carbonate rocks have probably never been formed, and most of the carbon dioxide remains in the atmosphere. The total burden of CO_2 is not much different on Earth from that on Venus; it is just the distribution between reservoirs that differs. The total amounts of atmospheric nitrogen in the atmospheres of Venus and Earth are not dissimilar, either, if the total pressures of the atmospheres are taken into account. The much greater solubility of nitrates compared with carbonates means that much less nitrogen than CO_2 has been deposited as solid insoluble minerals on Earth, and the atmospheric amounts are closer to those released by outgassing. Some CO_2 on Mars may be stored as solid at the winter poles, but the atmosphere also seems to have undergone much more physical evolution over geological time than the atmospheres of Venus or Earth as a result of the relatively small escape velocity.

2.4 Life and Earth's Atmosphere

Earth's atmosphere appears not to obey the laws of physics and chemistry: it is a disequilibrium mixture of chemical species (Lovelock 1979; Wayne 2000). Indeed, the atmosphere is like a low-temperature combustion system, and it contains a variety of easily oxidizable substances, such as H_2, CH_4, CO, H_2S, and hundreds of other minor constituents in the presence of molecular oxygen. Even the nitrogen and oxygen, the major components of the atmosphere, are not in thermodynamic equilibrium.

The key to the apparently anomalous composition of the atmosphere is the existence of living organisms: the biota bring about the entropy reduction by utilizing solar energy. Photosynthesis is the only mechanism that can account for the relative abundance of oxygen in our atmosphere. Other biological processes, mostly microbiological, release many fully or partially reduced compounds to the atmosphere. Within the atmosphere itself, photochemical transformations convert the biologically released substances, so that photochemistry plays a central part in maintaining the disequilibrium of the atmospheric gases. It is highly instructive to examine what the probable composition of our atmosphere would be like if life were not present. The contemporary and 'reconstructed' atmospheres for Earth displayed in Figure 2.1 provides the comparison for CO_2, N_2, and O_2. Figure 2.4 illustrates the differences in more detail: it gives the mixing ratios for several gases actually found in our atmosphere, and those that might

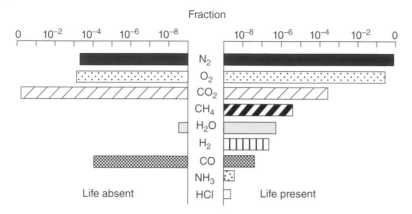

Figure 2.4 Life's influence on Earth's atmosphere. The diagram shows, on a logarithmic scale, the mixing ratios for the major gases and some significant trace species found in our atmosphere in the presence of life and those expected in its absence. (*Source:* diagram devised by Professor Peter Liss, and reproduced from Wayne (2000).)

be expected if life were absent. Bearing in mind that the scale of this diagram is logarithmic, the enormous influence of the biota is immediately evident. Oxygen concentrations in the absence of life are at least 1000 times smaller than they are in the contemporary atmosphere. The reduced compounds such as CH_4 and H_2 are virtually absent, and the only readily oxidizable species is CO, which on this view is more abundant in the absence of life (low O_2) than in its presence.

Earth's atmosphere thus has a composition that is influenced by the biota. Conversely, atmospheric composition is of evident importance for the biota. Composition, temperature, and pressure of the atmosphere can each be modulated in response to biological activity, and these atmospheric parameters themselves have consequences for that activity. Temperature, for example, is determined in part by the greenhouse heating afforded by atmospheric CO_2; the partial pressure of CO_2 in the atmosphere is strongly dependent on photosynthetic activity (short term) and chemical weathering of silicate rocks (long term) that are themselves sensitive to temperature.

It is observations of these feedbacks that have led Lovelock and co-workers (Lovelock and Margulis 1974; Lovelock 1988; Volk 1998) to their Gaia hypothesis. They see the interaction between life and the atmosphere as so intense that the atmosphere can be regarded as an extension of the biosphere: although the atmosphere is not living, it is a construction maintained by the biosphere. The Gaia hypothesis postulates that the climate and chemical composition of the Earth's surface and its atmosphere are kept at an optimum by and for the biosphere. Naturally, this idea does not receive universal acceptance, and many workers have argued (Kirchner 1989) that the close links between atmosphere, oceans, and biosphere do not necessarily imply the existence of an adaptive control system. Nevertheless, the concept of Gaia forms an interesting framework for the discussion of the evolution of the atmosphere, the evolution of life, and the possible relationship between them.

Life is unlikely to have originated, or at least persisted, on Earth immediately after the main accretionary phase had ended (Kasting 1993). Significant numbers of impactors as large as 100 km in diameter continued to hit Earth during the period of *heavy bombardment,* which lasted from 4.5 to 3.8 Gyr ago. Until 3.8 Gyr ago, the uppermost layers of the ocean would probably have been vaporized repeatedly by the impacts. Sterilization of the planet would therefore have precluded the survival of life on Earth, even if it had appeared any earlier. Nevertheless, by 3.5 Gyr ago, life was almost certainly extant. Several workers believe that there is ample morphological evidence to document the existence of microbiota in sediments that have not been metamorphosed, although the tests both for biogenicity and for the absence of rock modification continue to excite controversy. According to Schopf (2006), the oldest fossils meeting the criteria for dating authenticity are from 3.5 Gyr ago, and were found in the Apex chert of northwestern Australia. They consist of 11 types of prokaryotic thread-like organisms, and include several types of cyanobacteria (blue-green algae). Cyanobacteria not only both produce and consume oxygen, but they are also the main organisms responsible for fixing nitrogen from the present-day atmosphere. Nitrogenase, the enzyme that reduces N_2 to NH_3 or amines, is poisoned by O_2, and cyanobacteria have evolved complex mechanisms to protect their nitrogenase. Some cyanobacteria have developed specialized cells in which the N_2 fixation occurs, a topic that we touch on again in Section 2.6 when considering fossil evidence for the rise of oxygen concentrations. Other cyanobacteria photosynthesize O_2 by day, and reduce N_2 by night, or remove N_2 during the morning and release O_2 during the afternoon. The extraordinary conclusion is that these most ancient of identified fossils are of organisms that were already surprisingly advanced, and certainly suggest that life was established well before they were formed.

Circumstantial support for this conclusion is given by several discoveries of *stromatolites* (Greek for 'layered rock') from the same period: these organo-sedimentary structures are similar to mats produced by present-day cyanobacteria, and microbial communities seem to have played an active role in their formation (Bosak et al. 2013). There is even indirect evidence that life existed 3.8 Gyr ago, although supposed microfossils from the Isua sedimentary deposits of this date may have nonbiological origins, and carbon-isotope indicators of biological activity in the same deposit may have been affected by metamorphic processes. Notwithstanding these reservations, Mojzsis et al. (1996) and others argue that their ion-microprobe measurements of carbon isotopes are indeed consistent with graphitic microdomains of bio-organic origin. The oldest samples come again from Isua in West Greenland (3.8 Gyr old) and from the neighbouring island of Akilia (3.85 Gyr old: the oldest sediments yet documented), and it does seem possible that the record of life on Earth may extend back even more than 3.85 Gyr.

The evolutionary history of life on Earth traces the processes by which both living organisms and fossil organisms evolved since life emerged on the planet, until the present. Earth formed about 4.5 billion years (Ga) ago, and evidence suggests life emerged prior to 3.7 Ga (Pearce et al. 2018; Rosing 1999; Ohtomo et al. 2014) (*Note*: in this chapter, we use the abbreviations Ga and Ma to represent ages: that is, billions or millions of years *ago*). Although there is some evidence to suggest that life appeared as early as 4.1 to 4.28 Ga, this evidence remains controversial because of the nonbiological mechanisms that may have generated these potential signatures of past life (Pearce et al. 2018; Papineau et al. 2011; Bell et al. 2015; Nemchin et al. 2008). The similarities amongst all present-day organisms indicate the presence of a common ancestor from which all known species have diverged through the process of evolution (Futuyma and Kirkpatrick 2017). The oldest traces of life on Earth may lurk in Canadian rocks (Tashiro et al. 2017). Ancient rocks in northeastern Canada could contain chemical traces of life from more than 3.95 billion years ago. If confirmed, the finding would be amongst the earliest known signs of life on Earth.

It is most unlikely that the earliest organisms to appear on Earth produced oxygen. Rather, they were more probably methane-producing ('methanogens') (Maltby et al. 2018). Methanogenetic (often called methanogenic) bacteria are ancient in evolutionary terms; they are strictly anaerobic and prosper at high temperatures. The modern descendants of these organisms live in the intestines of cattle and the soils beneath rice paddies. They convert the byproducts of fermentation into methane in a process that can be represented overall by the equation

$$2CH_2O \rightarrow CO_2 + CH_4 \tag{2.5}$$

where the empirical formula CH_2O stands for the fermentable organic material. On the primitive Earth, however, the methanogens would more significantly have brought about the energetically favourable combination of the relatively abundant reduced gas H_2 and the oxidized gas CO_2:

$$4H_2 + CO_2 \rightarrow CH_4 + 2H_2O \tag{2.6}$$

Methane is a powerful greenhouse gas, and it seems reasonable to suppose that at this very early stage of Earth's history, temperature regulation was effected more by modulation of atmospheric concentrations of CH_4 than of CO_2 (Sheldon 2006).

The methanogens were probably capable of decreasing atmospheric H_2 concentrations from 0.1% down to about 0.01%. At these lower concentrations, certain photochemical steps are still capable of converting H_2 and the S-containing reduced volcanic gas H_2S first to CH_2O and then, via reaction (2.5), to CH_4

$$2H_2 + CO_2 + h\nu \rightarrow CH_2O + H_2O \tag{2.7}$$

$$2H_2S + CO_2 + h\nu \rightarrow CH_2O + H_2O + 2S \tag{2.8}$$

These processes are analogous to conventional photosynthesis, but they generate H_2O (and S compounds) rather than O_2: they constitute anoxygenic photosynthesis (Martin et al. 2018). It is believed that anoxygenic photosynthesis may have evolved as long ago as 3.5 Gyr before present, even though it is a process that superseded, or at least ran in parallel with, the even more ancient methanogenesis. There is evidence (Pace et al. 2018) that anoxygenic photosynthesis can lead to the formation of laminar stromatolites like those discussed earlier in connection with oxygenic photosynthesis.

Once released into the atmosphere, much of the CH_4 is oxidized to $CO_2 + H_2O$, but in the early anoxic atmosphere where short-wavelength ultraviolet radiation was not filtered out, the photochemical dissociation of CH_4 at high altitudes could yield H atoms or H_2 molecules that could escape to space. Figure 2.5 now adds the new steps that depend on early living organisms to the prebiotic situation that was described by Figure 2.3. Note that the wavelengths needed to photolyse H_2 directly are even shorter and substantially less intense in the solar spectrum than those that can lead to photodissociation of CH_4. As a result, methanogenesis and anoxygenic photosynthesis can both increase the rate of loss of H_2 from the atmosphere, which is why we suggested that CH_4 can be regarded as an unusual catalyst for loss of reductant. Without the escape of hydrogen from Earth, the amount of oxidant in the crust ought to be exactly equivalent to the amount of reductant on a mole-for-mole basis. That is definitely not the case, as illustrated by Figure 2.6. The oxidized material, mostly in the solid crust is more than twice as abundant as the reduced carbon; the *missing reductant* is, in reality, the enormous amount of hydrogen that has escaped to space since the Earth was formed.

Because the origin of life must have depended on the prior existence of organic species, there has long been intense interest in the possible conversion of atmospheric gases to simple organic molecules. The early experiments of Miller (1953) aroused much excitement, because it was shown that simulated

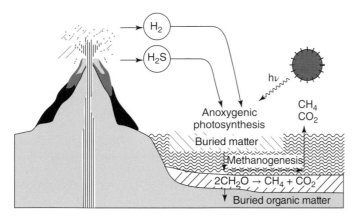

Figure 2.5 The effects on the early atmosphere of methanogenesis and anoxygenic photosynthesis brought about by living organisms. The diagram refers to a time before the origin of oxygenic photosynthesis Catling and Claire (2005).

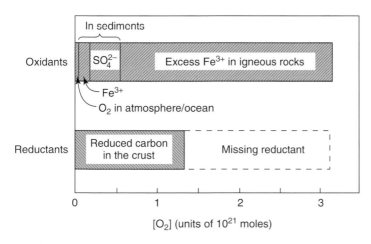

Figure 2.6 Inventory of oxidizing and reducing species in the Earth's crust, represented as their equivalents in moles of molecular oxygen. An excess of the oxidants over the reductants by a factor of more than two has been brought about by escape to space of hydrogen since the Earth was formed Catling and Claire (2005).

lightning discharges passed through mixtures of methane and ammonia produced a wide range of organic compounds that included amino acids. The less-reducing early atmosphere now thought probable (N_2, CO_2 and H_2O, together with traces of volcanic H_2 and CO: see first paragraph of Section 2.2) has more recently been shown to yield a variety of organic compounds when subject to ultraviolet irradiation, although the more reducing atmospheres generate a greater variety of compounds in larger yield. Bar-Nun and Chang (1983) found that continuous irradiation at $\lambda = 184.9$ nm of mixtures of CO and H_2O gave CO_2 and H_2 as the major products, and smaller quantities of CH_3OH, HCHO, and CH_4. Some C_2 molecules (C_2H_5OH, CH_3CHO, and C_2H_6) were also observed. Wen et al. (1989) have analysed these experiments in terms of a photochemical kinetic scheme, and the simulated abundances of most of the products are in surprisingly good accord with the experimental findings. The main steps in the main pathway involve photolysis of water

$$H_2O + h\nu \rightarrow H + OH \tag{2.9}$$

followed by termolecular addition of atomic hydrogen to CO to yield the formyl radical (HCO), which is then the precursor of the more complex organic molecules. Photolysis of carbon dioxide

$$CO_2 + h\nu \rightarrow CO + O \tag{2.10}$$

is a continuous source of CO. Wen et al. (1989) have employed their scheme to predict photochemical production rates of the organic molecules in the prebiotic atmosphere to show that substantial quantities of organic material could be formed in this way.

The simple chemical species present in the early atmosphere must have been able to produce much more complex organic compounds that ultimately became the basis of life. Methane seems an unavoidable requirement in any atmospheric source of prebiotic compounds. The CH_4 formed in the photolysis of the CO–CO_2–H_2O system could have supplemented that outgassed from the Earth. As explained at the end of Section 2.1, CH_4 is unstable against photolysis and, especially, attack by OH radicals. However, the radical products of photolysis and attack (CH_2 and CH_3) can themselves participate in processes that yield more complex organic species. The CH_2 radical offers an interesting route for the formation of nitrogen compounds before life was present. Atomic nitrogen must have been abundant in the anaerobic middle atmosphere as a result of the occurrence of several processes involving photodissociation and photoionization. Nitrogen atoms react rapidly with (triplet) CH_2 to form hydrogen cyanide

$$N + {}^3CH_2 \rightarrow HCN + H \tag{2.11}$$

thus opening up new vistas in organic chemistry (Menor-Salván 2018). The chemistry of HCN in the contemporary atmosphere is not yet completely understood, but it seems that the molecule is relatively unreactive. Addition of OH and reaction of the adduct with O_2 presumably leads to complete oxidation in an oxygen-rich atmosphere. Photolysis of HCN

$$HCN + h\nu \rightarrow H + CN \tag{2.12}$$

is followed in an oxygen atmosphere by reaction of CN with atomic or molecular oxygen and loss of the C—N bond. However, the CN radical is partially protected in a more reducing atmosphere because hydrogen abstraction regenerates HCN. Under these circumstances, reactions such as

$$CN + HCN \rightarrow C_2N_2 + H \tag{2.13}$$

$$CN + C_2H_2 \rightarrow HCCCN + H \tag{2.14}$$

offer routes to other interesting nitrogen-containing organic species. One of the most fascinating aspects of these speculations is the way in which they parallel our understanding of the chemistry of Titan's atmosphere (Wayne 2000), where organic photochemical aerosols and hazes are formed in an atmosphere consisting mainly of molecular nitrogen.

The possibility that atmospheric CH_4 fluxes were sufficient to permit occurrence of the chemistry described has recently received encouragement from suggestions that there was a significant extraterrestrial source of CO in the period of the Earth's history earlier than 3.8×10^9 years ago. The rate of impact of incoming comets and carbonaceous asteroids would have been large during this period. These bolides bring in CO ice and/or organic carbon that can be oxidized to CO in the impact plume. The elemental iron in ordinary chondritic impactors could further enhance CO by reducing CO_2. Nitric oxide (NO) is likely to be formed in a high-temperature reaction between N_2 and CO_2, and this gas also indirectly increases the ratio of [CO] to [CO_2]. A photochemical model shows that, for a total atmospheric pressure of roughly 2 atm, [CO]/[CO_2] might even have exceeded unity at times more than 4×10^9 years ago.

Comets, invoked here as a source of CO, and other extraterrestrial sources have, of course, long been looked on as potential carriers of organic molecules to Earth (e.g. Chyba et al. 1990), and even of life (Whittet 1997). This alternative view finds its most extreme expression in the proposal that interstellar molecules accumulated within the heads of comets (Hoyle 1982; Wickramasinghe and Hoyle 1998). Chemical evolution occurring a few hundred metres below the cometary surface is seen as progressing as far as biopolymers and micro-organisms. Hoyle and Wickramasinghe even contend that some past and present epidemics were initiated by the viruses and bacteria falling to Earth in cometary dust (Hoyle and Wickramasinghe 1977, 1983). However, there seem to be perfectly plausible ways in which organic molecules can be formed from the precursors already present in the atmosphere, and there are doubts about the survivability of amino acids and nucleobases subject to the pyrolytic conditions to which they would be exposed.

The question of what chemical and physical processes combined to produce the first living systems is perhaps impossible to answer with any certainty, but research continues to provide clues that may help us understand our primordial past. One important question concerns the origins of RNA and its role in contemporary biological systems. Several imaginative examinations are presented in a special issue of *Nature Chemistry* (Cantrill 2017), exemplified by the work of Tagami et al. (2017), and in the book edited by Menor-Salván (2018).

The stages between the appearance of the organic molecules on Earth and the appearance of life cannot concern us here, however fascinating those steps might be. An excellent review of the facts and speculations about the origin of life is provided by Orgel (1998).

2.5 Carbon Dioxide in Earth's Atmosphere

Both the geochemical record and the persistence of life itself indicate that the oceans can never have either frozen or boiled in their entirety. Mean surface temperatures have probably never departed from the range of 5–50 °C and may have been highest at very early periods – but this conclusion leads to a riddle! Standard stellar evolutionary models predict that at 4.5 Gyr before present the Sun's luminosity was lower than it is today by 25–30%. The standard model of evolution of solar intensity has created the 'faint young Sun paradox', mentioned earlier (Section 2.3), because it translates into a decrease of Earth's effective temperature by 8%, low enough to keep seawater frozen for ≈ 2 Gyr, if the atmosphere possessed its present-day composition and structure. The paradox may be only illusory if atmospheric behaviour 3–4 Gyr ago was different from what it is today. Explanations proposed include changes in albedo or increases in the greenhouse efficiency. Sagan and Chyba (1997), for example, argue that NH_3 concentrations might have been much higher than usually believed and that this gas could have contributed to radiation trapping. Alterations in clouds could exert a negative-feedback, stabilizing, effect, since lower temperatures would mean decreased cloud cover and reduced reflection away of solar radiation. Water vapour makes the largest contribution to the greenhouse effect in the contemporary atmosphere, but it is unlikely to be the agent of long-term temperature control. Its relatively high freezing and boiling points render its blanketing effect prone to destabilizing positive feedbacks by increasing ice and snow albedo at low temperatures (further reducing temperature), but by increasing water vapour content at high temperatures (and yet further increasing the greenhouse effect). The most straightforward explanation would be a massive atmospheric greenhouse effect, from either carbon dioxide or methane, or both. There have been speculations that volcanic outgassing or impact-vaporized materials could have released greenhouse gases. Marchi et al. (2016) propose a novel, more efficient mechanism. As the planet was pummelled by primordial asteroids – some larger than 100 km in diameter – impacts would have melted large volumes of rock, creating temporary lakes of lava. These pools of lava could have released large quantities of carbon dioxide into the atmosphere. Depending on the timescale for the drawdown of atmospheric CO_2, the Earth's surface could have been subjected to prolonged clement surface conditions or multiple freeze–thaw cycles. The bombardment is argued also to have delivered and redistributed to the surface large quantities of sulphur, one of the most important elements for life. The stochastic occurrence of large collisions could provide insights into why Earth and Venus, considered Earth's twin planet, possess radically different atmospheres. Alternative suggestions (Ozaki et al. 2018) have included the effects of primitive forms of photosynthesis involving methane-generating microbes (one transforming oceanic iron into rust and another converting hydrogen to formaldehyde [methanal, HCHO]). Another suggestion (Airapetian et al. 2016) invokes frequent and powerful coronal mass ejection (CME) events from the young Sun – so-called *superflares*. Solar flares occur when the sun emits an unusually large amount of electromagnetic radiation from its surface. These flares are often followed by CMEs, which consist of fast-moving, high-energy plasma from the Sun. Although rare now (one event in decades), Airapetian et al. estimate that the early Earth probably received one CME per day. Resultant shock waves can initiate reactions that convert molecular nitrogen, carbon dioxide, and methane to the potent greenhouse gas nitrous oxide (N_2O), as well as to hydrogen cyanide, an essential compound for life. This hypothesis has the virtue that it could explain not only additional heating through radiation trapping by N_2O, but also the prebiological *fixation* of nitrogen to form important N-containing compounds.

Whatever greenhouse-gas or other mechanism kept the Earth warm, it must have been smoothly reduced to avoid exceeding the high-temperature limit for life. Carbon dioxide seems the most likely greenhouse gas to have exerted thermostatic control of our climate (Kasting 1993), although in the earliest stages after the appearance of life, methane might also have been a critical contributor. Negative-feedback mechanisms can be identified for both CO_2 and CH_4. In the case of CO_2, nonbiological control might include acceleration of the weathering of silicate minerals to carbonate deposits in response to increased temperatures. However, as discussed previously, present-day weathering is biologically determined, and the biota both sense and amplify temperature changes. This feedback regulation of climate is seen by its proponents as evidence in support of the Gaia hypothesis. The biota both increase the partial pressure of CO_2 in soils and generate humic acids. Each of these effects increases the rate of weathering, and thus of CO_2 loss.

Temperatures were likely to have been higher in the very early history of the Earth than at present, perhaps by as much as 60 °C as a global mean (Kasting 1993). The climate and ocean pH of the early Earth are important for understanding the origin and early evolution of life. However, estimates of early climate range from below freezing to over 70 °C, and ocean pH estimates span from strongly acidic to alkaline. To better constrain environmental conditions, Krissansen-Totton et al. (2018) applied a self-consistent geological model of the carbon cycle to the last 4 billion years. The model predicts a temperate (0–50 °C) climate and circumneutral ocean pH throughout the Precambrian due to stabilizing feedbacks from continental and seafloor weathering. These environmental conditions under which life emerged and diversified were akin to those of the modern Earth. Similar stabilizing feedbacks on climate and ocean pH may operate on earthlike exoplanets, implying life elsewhere could emerge in comparable environments.

Nevertheless, firm evidence exists for glaciations around 2.5–2.3 Gyr ago, and again at about 0.8 Gyr ago: according to some reports, the earliest glaciation may even be dated back to ≈2.9 Gyr ago. Figure 2.7 shows one back-projection of the history of atmospheric CO_2 concentrations using this kind of information (Kasting 1993). A radiative–convective climate model was employed in the calculations, in which radiation trapping was brought about by CO_2 and by H_2O (with a temperature-dependent vapour pressure). The constraint on the model was that it was required to maintain the Earth's mean surface temperature at 5–20 °C at all times, and the entire surface was kept at >0 °C (ice free) always, except during the glaciations at 2.5 and 0.8 Gyr ago. Unfortunately, the probability zone (indicated by the shading) is rather wide, and encompasses concentration ranges of more than a factor of 100 at worst, and the zone narrows only as we approach the present day within a few hundred million years. The concentration soon after the formation of the Earth must have been >0.1 atm, or 300 PAL (present atmospheric level)

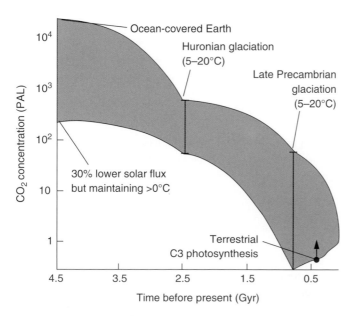

Figure 2.7 Evolution of atmospheric carbon dioxide concentrations, given relative to the PAL. The concentrations are estimated using a radiative–convective climate model to keep the Earth's temperatures within the limits marked on the diagram, and explained in the text. More [CO_2] than at present is needed early on to compensate for a Sun with an infrared intensity that had not yet built up to its full level. The shading represents the range of concentrations permitted by the various indicators discussed. *Source:* based on a figure presented by Kasting (1993).

to maintain a mean surface temperature in excess of freezing point, but may have been as high as the 70 atm (2×10^5 PAL) suggested by Figure 2.1. Such concentrations seem reasonable if we accept that in Earth's prebiological atmosphere CO_2 was a major component, as it is on Venus or Mars today.

In the prebiotic atmosphere, it is likely that CO_2 and H_2O were the only important greenhouse gases present. After life appeared on the planet, gases such as NH_3 and CH_4 could have been generated by several microbiological processes, and would have trapped additional radiation. If substantial quantities of these gases were present, then CO_2 concentrations might have been lower than those represented in Figure 2.7 (Kasting and Siefert 2002). However, this effect would be limited to the period before 2 Gyr ago, because after that concentrations of NH_3 and CH_4 would be expected to drop dramatically in response to the oxygen that then became abundant in the atmosphere. Such oxygen changes are the subject of the next section.

2.6 The Rise of Oxygen Concentrations

One of the most fascinating aspects of Earth's history concerns the build-up of oxygen from extremely low levels in the early (prebiological) atmosphere of perhaps only 0.001%, or 10^{-5} of the PAL, and the impact of that rise of free oxygen on the origin and evolution of life (Lyons et al. 2014). This low-oxygen situation seems to have persisted throughout the first half of Earth's 4.5 Gyr history. Evidence has accumulated to show that there was a permanent rise to more significant concentrations at some period between 2.4 and 2.1 Gyr before present (Ga). What is now regarded as the classical view is that the increase was a step change that has been called the *great oxidation event* or GOE (Holland 2002, 2006). We start here by presenting evidence for the increase in atmospheric oxygen concentrations somewhat before 2 Ga. In Section 2.8, some refinements to the step (and monotonic) scenario are examined.

The essential features of the evidence for a step GOE are the disappearance of easily oxidized minerals (such as pyrite, FeS_2) from ancient stream beds, and the first appearance of rusty red soils (containing Fe(III)) on land. The simplified interpretation is that, in the absence of life, and thus of photosynthetic organisms producing oxygen, inorganic chemistry provided only a weak source of oxygen. Meanwhile, there were substantial sinks (reduced species) for O_2 that consumed most of any newly generated O_2. At some point, coinciding with the supposed GOE, the original sinks of O_2 were used up, and O_2-producing photosynthetic organisms developed, leading to the conditions for increases in O_2 levels by factors of 10^4 or more.

Oxygen concentrations can only have increased from their prebiotic levels to their present-day levels through photosynthetic generation of the molecule; in turn, the increases had a critical influence on the possibility of complex life forms (Raymond and Segre 2006). The growth in atmospheric $[O_2]$ is thus clearly linked to the evolution of life on the planet.

Inorganic photochemistry can only generate a limited amount of O_2. Photolysis of water vapour to form OH and H (reaction (2.9)) and of CO_2 to form CO and O (reaction (2.10)) are the main primary steps, and secondary chemistry can lead to the formation of free O_2:

$$OH + OH \rightarrow O + H_2O \tag{2.15}$$

$$O + OH \rightarrow O_2 + H \tag{2.16}$$

$$O + O + M \rightarrow O_2 + M \tag{2.17}$$

In each case, the final stage is the combination of oxygen atoms in reaction (2.17). Two important limitations are placed on the amount of oxygen that can be produced. It has long been recognized that H_2O and CO_2 are photolysed by ultraviolet radiation in a spectral region that is absorbed by O_2 (say, at $\lambda \leq 240$ nm for H_2O and $\lambda \leq 230$ nm for CO_2). *Shadowing* by the O_2 formed thus self-regulates photolysis at some concentration. Secondly, photolysis of H_2O vapour, or of CO_2, does not, on its own, constitute a net source of O_2. Reactions (2.9) and (2.15–2.17) have the effect of converting two H_2O molecules to one O_2 molecule and four H atoms. Only if atomic hydrogen is lost by exospheric escape is there a gain in O_2, because otherwise H_2O is reformed. Addition of CO_2 photolysis (reaction 2.10) to the scheme does not alter this conclusion, since the CO product interacts rapidly with OH:

$$CO + OH \rightarrow CO_2 + H \tag{2.18}$$

The overall reaction for water-vapour photolysis now looks like:

$$2H_2O + 3h\nu_{H_2O} + h\nu_{CO_2} \rightarrow 2O + 4H \tag{2.19}$$

However, the outcome is still that for every O_2 molecule formed, four H atoms must be lost. Escape is thus the crucial event, and the rate is determined by the transport of all hydrogen species through lower levels of the atmosphere to the exosphere. Loss of O_2, for example by reaction with crustal or oceanic Fe^{2+}, or with volcanic H_2, competes with production, and so further limits the amount of free O_2 that can build up without the help of photosynthesis.

Considerable difficulties arise in giving quantitative expression to the prebiological formation of oxygen because of uncertainties in the concentrations of precursor molecules (H_2O and CO_2), temperatures and solar ultraviolet intensities. Concentrations of CO_2 might have been much greater before the gas was converted to carbonate deposits, and water-vapour levels would have been elevated had surface and atmospheric temperatures been higher than they are now. Young (*T-Tauri*) stars, which resemble the Sun at the age of a few million years, emit 10^3 to 10^4 times as much ultraviolet radiation as the present Sun. If enhanced solar-ultraviolet intensity was available during the prebiological evolutionary period of our atmosphere, then the rates of photolysis of H_2O and CO_2 would have been greatly enhanced, and become a significant source of O_2, especially if [CO_2] is high. Photochemical models developed for interpretation of the modern atmosphere can be adapted for the palaeoatmosphere by incorporating appropriate source terms, temperature profiles and boundary conditions. Such models suggest that prebiological O_2 at the surface would have been limited to about 2.5×10^{-14} PAL had both [CO_2] and solar ultraviolet intensities been at their current values. With 100 times more CO_2, and 300 times more ultraviolet radiation from the young Sun, the surface [O_2] calculated is $\approx 5 \times 10^{-9}$ PAL. The geological record provides some further information about oxygen concentrations. The simultaneous existence of oxidized iron and reduced uranium deposits in early rocks (>2.2 Gyr old) requires [O_2] to be more than 5×10^{-12} PAL, but less than 10^{-3} PAL; the values accommodated both by the model and by geochemistry thus seem to be roughly in the range 5×10^{-12} to 5×10^{-9} PAL (Frei et al. 2016).

Prebiological oxygen concentrations in the palaeoatmosphere are of importance in two ways connected with the emergence of life. Organic molecules are susceptible to thermal oxidation and photo-oxidation, and are unlikely to have accumulated in large quantities in an oxidizing atmosphere. Living organisms are photochemically sensitive to radiation at $\lambda \leq 290$ nm. Lifeforms known to us depend on an ultraviolet screen provided by atmospheric oxygen and its photochemical derivative, ozone, because deoxyribonucleic acid (DNA) and nucleic acids are readily destroyed. Biological evolution therefore seems to have proceeded in parallel with the changes in our atmosphere from an oxygen-deficient to an oxygen-rich one.

It is worth noting that, although the tiny prebiological concentrations of oxygen preclude the existence of a useful oxygen and ozone shield against ultraviolet radiation, the paucity of oxygen was probably essential in the early stages of the synthesis of complex organic molecules that became the basis of life. Organic molecules are susceptible to thermal oxidation and photo-oxidation, and could not have accumulated in large quantities in a strongly oxidizing atmosphere. Living organisms are known to develop mechanisms and structures that protect against oxidative degradation, and so are able to survive in atmospheres containing large amounts of oxygen. At about 1% of PAL, organisms can derive energy from glucose by respiration rather than by anaerobic fermentation, and they gain an energy advantage of a factor of 16. However, the fact remains that oxygen is toxic, and organisms have to trade off the energy advantage against the need to protect themselves from the oxidant.

Photosynthesis is now the dominant source of O_2 in the atmosphere. For the purposes of the present discussion, this complex and fascinating piece of chemistry can be represented by the simplified equation

$$n\text{CO}_2 + n\text{H}_2\text{O} + mh\nu \rightarrow \left(\text{CH}_2\text{O}\right)_n + n\text{O}_2 \tag{2.20}$$

The essence of the process is the use of photochemical energy to split water and thus to reduce CO_2 to carbohydrate, shown here as $[\text{CH}_2\text{O}]_n$. Isotope experiments show that photosynthetically produced O_2 comes exclusively from the H_2O and not from the CO_2. Photosynthesis releases 400×10^{12} kg of oxygen annually at present (Wayne 2000). Since the atmosphere contains about 1.2×10^{18} kg of O_2, the oxygen must cycle through the biosphere in roughly 3000 years.

It is important to realize that substantial concentrations of oxygen can build up in the atmosphere only if the carbohydrate formed in the photosynthetic process is removed from contact with the atmospheric oxygen by some form of burial. Without such burial, spontaneous oxidation would rapidly reverse the changes brought about by photosynthesis. In the contemporary atmosphere, marine organic sediment deposition buries about 0.12×10^{12} kg year^{-1} of carbon, and so releases about 0.32×10^{12} kg year^{-1} of O_2. At that rate, atmospheric oxygen could therefore double in concentration in about 4×10^6 years. There are balancing processes, including geological weathering (e.g. of elemental carbon to CO_2, sulphide rocks to sulphate, and iron(II) rocks to iron(III)) and oxidation of reduced volcanic gases (e.g. H_2 and CO). Nevertheless, marked variations in atmospheric O_2 are likely to have occurred over geological time, and such changes are a central theme of this survey.

The rise in $[O_2]$ from its prebiotic levels has generally been linked to the geological time scale either on the basis of the stratigraphical record (e.g. Frimmel 2005) of oxidized and reduced mineral deposits (including information about isotope ratios, to be explained after the next paragraph) or from fossil evidence combined with estimates of the oxygen requirements of ancient organisms. Evolution of the climate system (Walker 1990; Young 2013) is likely to have influenced the composition and mineralogy of sedimentary rocks: it is increasingly apparent that changes in atmospheric $[O_2]$ and $[CO_2]$ must be studied alongside each other, and that it is necessary to have a reliable estimate of past $[CO_2]$ in order to correctly deduce past values of $[O_2]$ (Olson et al. 2018).

We have seen that living organisms have been present on our planet from a very early stage, and perhaps from as long ago as 3.85 Gyr before present. From that time on, there has thus been a potential photosynthetic source of O_2, which could lead to an increase of atmospheric concentrations from their very low prebiotic levels. Red beds, which contain some iron in the higher oxidation state, III, were absent before roughly 2 Gyr ago, and reduced minerals, such as uraninite, were generally formed before this date. Banded-iron formations (BIFs), which contain iron (II) rather than iron (III), were also formed only up until about 1.85 Gyr ago. An anoxic deep ocean is apparently required for the deposition of BIFs (Yang et al. 2018) in order that iron can be transported over large distances as iron (II). Much evidence has now accumulated to confirm that there was a large change in atmospheric O_2 level just before 2 Gyr ago (Holland 2006). For example, the studies of carbon isotopes to be described in the next paragraph suggest that the rate of photosynthesis was enhanced markedly at this point. There is thus clear evidence that atmospheric O_2 concentrations increased very significantly over this time period. A number of investigators have used chemical profiles of palaeosols to reconstruct the evolution of atmospheric oxygen levels in Earth's early history. Rye and Holland (1998) have provided a critical review of the data. Part of the problem lies in the identification of authentic palaeosol material, but where the evidence seems firm, the profiles suggest a dramatic change in atmospheric oxygen concentrations during the period 2.2–2.0 Gyr ago. Every true palaeosol older than 2.44 Gyr suffered significant loss of iron during weathering, indicating that the atmospheric pressure of O_2 was at $\approx 5 \times 10^{-4}$ atm (2.5×10^{-3} PAL) before that date. Indeed, iron loss from palaeosols of age 2.245–2.203 Gyr is consistent with a partial pressure of $<4 \times 10^{-3}$ PAL. Nevertheless, the presence of red beds, containing some Fe(III), immediately overlying these palaeosols, suggests that by about 2.2 Gyr ago there was a substantial (but unquantified) amount of O_2 present in the atmosphere. Iron loss is negligible in palaeosols aged 2.2–2.0 Gyr and in all younger samples. The deep ocean was probably anoxic until 800 Ma, with estimates of the occurrence of deep-ocean oxygenation and the linked increase in the partial pressure of atmospheric oxygen to levels sufficient for this oxygenation ranging from about 800 to 400 Ma. Deep-ocean dissolved oxygen concentrations in this time span are typically estimated using geochemical signatures preserved in ancient continental shelf or slope sediments, which only indirectly reflect the geochemical state of the deep ocean. Stolper and Keller (2018) present a record that more directly reflects deep-ocean oxygen concentrations, based on the ratio of Fe(III) to total Fe in hydrothermally altered basalts formed in ocean basins. Their data suggest that deep-ocean oxygenation occurred between 541 Ma and less than 420 Ma, at the later end of the age range estimated previously.

Isotope ratios in mineral deposits have been noted several times as sources of information about the rise in oxygen levels from their prebiotic levels. Measurements of the abundances of the different isotopes of various elements have proved fruitful in understanding many aspects of the Earth's geological and biological history. Simple physical chemistry predicts that both the positions of chemical equilibria and the rates of chemical change may depend on the isotopic masses of the participating species. There is thus the possibility that chemical reactions could enrich the products of the process in the heavier isotope (and consequently deplete the reactants) or vice versa. Of course, the global isotopic ratios cannot change, since elements are neither created nor destroyed (except in the case of radioactive decay or radiogenic formation). However, if an ancient deposit is laid down as a result of chemistry (including biological chemistry) that is isotope dependent, and that deposit is resistant to isotope-exchange processes, then it may retain a signature of the way it was formed in its isotopic composition, which will be different from the global-mean composition.

These differences are usually quoted in terms of *delta notation,* where the symbol δ is used to represent the deviation of the isotope in question from the global mean (or an agreed standard) ratio. Since the fractional deviations are always rather small, they are multiplied by 1000 to give δ in units of 'pro mil' (symbol ‰). For example, carbon on Earth is mostly present (98.93%) as the mass 12 isotope, ^{12}C, but 1.07% is present as the heavier isotope, ^{13}C. Photosynthesis preferentially utilizes $^{12}CO_2$ rather than $^{13}CO_2$, so that plant remains are slightly depleted in ^{13}C compared with the global average, with $\delta^{13}C$ typically −20 to −30‰. We shall see below how $\delta^{13}C$ for terrestrial organic matter might provide a proxy signal for atmospheric composition in the period since about 425 Myr ago, when rich plant life provided a source of the organic carbon deposits.

A recent development in isotope studies (Thiemens 2006; Thiemens et al. 2012) has provided information of a different kind about atmospheric composition. The equilibrium and kinetic isotope effects described so far owe their origin to the slightly lowered molecular vibration frequencies that arise from the attachment of heavier mass atoms to chemical bonds: the effects are referred to as 'mass-dependent' for this reason. Some elements possess three or more stable isotopes; of immediate interest here are O (^{16}O, ^{17}O, ^{18}O) and S (^{32}S, ^{33}S, ^{34}S, ^{36}S). The extent of mass-dependent fractionation for the individual minor isotopes for an element is expected to be roughly proportional to the mass difference between the isotopes. Thus, the fractionation of $^{17}O/^{16}O$ should be half that of $^{18}O/^{16}O$: that is, $\delta^{17}O \approx 0.5 \times \delta^{18}O$. Similarly, $\delta^{33}S$ should be $\approx 0.5 \times \delta^{34}S$, ^{32}S being the most abundant isotope, and $\delta^{33}S \approx 0.25 \times \delta^{36}S$. Examination of large numbers of isotopically fractionated terrestrial samples shows that these expectations are generally matched by reality, with the three equations being $\delta^{17}O = 0.52 \times \delta^{18}O$, $\delta^{33}S = 0.52 \times \delta^{34}S$, and $\delta^{33}S = 0.27 \times \delta^{36}S$. However, a small number of samples do not follow this general rule, and these samples appear to have been formed as a consequence of non-mass-dependent (NMD) fractionation (sometimes rather confusingly and incorrectly called mass-independent fractionation). Such fractionation is now understood to arise from subtle quantum-mechanical effects dependent on symmetry. The extent to which it has occurred for any sample is represented by the difference, Δ, between the measured value of δ for the isotope in question and the value expected on the basis of normal mass dependent fractionation. Thus, using O as an example, $\Delta^{17}O = \delta^{17}O - 0.52 \times \delta^{18}O$.

Anomalous fractionation of O isotopes was first found in meteorites in 1973 (Thiemens 2006). Even for these extraterrestrial samples, finding $\Delta^{17}O \neq 0$ was a surprise, but it has subsequently turned out that several terrestrial minerals also exhibit NMD fractionation. In particular, some sulphates (gypsum and anhydrite, $\Delta^{17}O$ up to $\approx +6$‰) and nitrates (nitratine, $\Delta^{17}O \approx +20$‰) show large excesses of ^{17}O (Rumble 2005). The enrichment is believed to arise from the involvement of atmospheric ozone, O_3, in the formation of the minerals. Thiemens (2006) reviews the studies of O_3 in the atmosphere and in the laboratory that demonstrate the large relative enrichments of isotopomers such as $^{16}O^{17}O^{18}O$, which can reach 10–15% (%, not ‰!). Ivanov and Babikov (2013) are amongst those who have examined mechanisms for the NMD fractionation of oxygen isotopes in the formation of ozone. For our present enquiry, the important point is that the solid minerals provide information about the atmospheric composition at the time they were formed. The O_3 in the atmosphere is generated photochemically from O_2, so that the existence of the ^{17}O-rich sulphates and nitrates implies relatively high atmospheric oxygen concentrations. In this particular case, the results are consistent with what we know already, because the oldest minerals showing a nonzero $\Delta^{17}O$ were laid down in 'modern' times (not older than 24 Myr), in a period when it is certain that O_2 was more or less at its present level. However, the example serves to introduce investigations

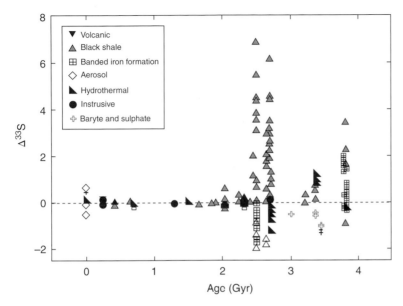

Figure 2.8 Anomalous fractionation of sulphur isotopes shown as a function of the age of the samples examined. Note the sharp difference in behaviour between samples older and younger than about 2.5 Gyr (Rumble 2005).

(Rumble 2005; Thiemens 2006; Eickmann et al. 2018) of the isotopes of S, where the mineral record extends back to nearly 4 Gyr before present.

Figure 2.8 displays data for $\Delta^{33}S$ for a variety of mineral samples of different ages. Plotting $\Delta^{33}S$ as a function of the age of the mineral makes it evident that there is a striking change in behaviour at about 2.5 Gyr ago. Minerals more recent in origin than 2.5 Gyr show $\Delta^{33}S$ varying only slightly from zero, with maximum deviations in the range -0.5 to $+0.9‰$, whereas values of $\Delta^{33}S$ for older samples cover a much wider range of -2.5 to $+8.1‰$. The phenomenon seems to exist worldwide, and the transitional date between the two types of behaviour is quite sharp. One interpretation of the observations is that NMD isotopic anomalies in atmospheric sulphur compounds could arise photochemically in the same kind of way as they do for the ozone formed photochemically from O_2. Volcanic emissions of SO_2 and H_2S can be photolysed by short-wavelength ultraviolet radiation, and the photolysis could lead to the development of NMD isotope fractionation. This signature could be preserved, for example by the formation of an aerosol that reaches the Earth's surface, thus separating products from reactants. There are two key issues. First, the short wavelengths needed to split the volcanic gases photochemically would be filtered out from the Sun's radiation by ozone in the atmosphere. Second, S compounds exist in a wider range of oxidation (valence) states in the absence of O_2 than in its presence, and these oxidation states are likely to be those involved in the formation of compounds that separate the isotopically enriched from the isotopically depleted components. Both these factors, therefore, suggest strongly that before 2.5 Gyr ago the atmosphere was largely anoxic, but that subsequently the conditions for formation and separation of the anomalous fractionation no longer existed. Support for the hypothesis is found in measurements of isotope ratios of the more recent atmosphere, as obtained from measurements made on sulphuric acid aerosols of volcanic origin that are preserved in Antarctic ice cores.

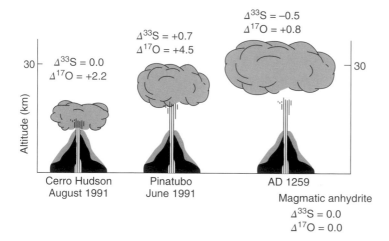

Figure 2.9 Isotopic composition of volcanic O and S in Antarctic ice-core samples. These samples are of the ash emitted and of sulphate aerosol formed from gaseous sulphur compounds (SO_2, H_2SO_4) in three different eruptions. The ash composition is not anomalous for either O or S (Δ values are zero) in any case. The $\Delta^{17}O$ value is positive for all three volcanoes, probably because of the involvement of isotopically enriched O_3 in the formation of the sulphate aerosols. Only for the two volcanoes where the gas plumes penetrated the stratosphere is $\Delta^{33}S$ nonzero. This behaviour is consistent with ^{33}S enrichment in short-wavelength photochemical processes that can occur only where solar radiation is not filtered out by stratospheric ozone (Rumble 2005).

Figure 2.9 brings together the results for three volcanic eruptions. The clouds of gas from two of the eruptions, those of AD 1259 and Mount Pinatubo penetrated the stratosphere, whereas that of Cerro Hudson did not. The magmatic anhydrite directly emitted by the volcanoes has, in none of the three cases, any isotopic anomaly ($\Delta^{17}O = 0.0$; $\Delta^{33}S = 0.0$), so this material seems to have the average isotopic makeup of the Earth's O and S. The sulphate (sulphuric acid) aerosol possesses a positive $\Delta^{17}O$, just as seen for the sulphate and nitrate aerosols found worldwide and discussed in the preceding paragraph. For $\Delta^{33}S$, however, the value is zero for the tropospheric eruption of Cerro Hudson, but nonzero (anomalous) for both the stratosphere-penetrating eruptions. Furthermore, the layers of ice in the core for the years just before and just after the stratospheric eruptions show no anomalies for the S isotopes. The observations are obviously consistent with the proposed mechanism for S-isotope anomalies at >2.5 Gyr ago. In the two eruptions that penetrated the stratosphere, the gases released were exposed to short-wavelength ultraviolet radiation, because much of the absorbing ozone lay below them, and photolysis of SO_2 could begin the chain of events leading to nonzero $\Delta^{33}S$. The SO_2 from Cerro Hudson, on the other hand, always lay below the filtering ozone, and no photochemical isotopic fractionation could arise.

The evidence just presented does suggest strongly that O_2 increased rather rapidly at around 2 Gyr ago. Rye and Holland (1998) even suggest that the concentration had reached as much as 0.15 PAL sometime between 2.2 and 2.0 Gyr before present, although other studies would put the level nearer 0.01 PAL. Kasting (1993) explains the arguments in terms of his *three-stage–three-box* model, outlined in Figure 2.10. The three boxes are the atmosphere, the surface ocean, and the deep ocean, and the separate stages of the model correspond to one after the other of these compartments becoming oxidized or oxidizing, rather than being reducing. Stage I has all three compartments reducing (anoxic), with the possible exception of

Figure 2.10 The 'three-stage–three-box' model (Kasting 1993) used to represent the increase in oxygen concentrations in the Earth's atmosphere. The three boxes are for the atmosphere, the surface ocean, and the deep ocean: R shows that the conditions were reducing, and O that they were oxidizing. There may have been oxygen-rich 'oases' in the surface ocean during stage I.

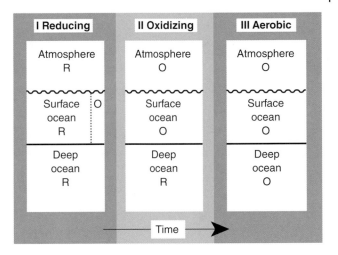

some regions of the surface ocean where oxygen production might be especially favourable. By stage II, the upper two boxes were oxidizing, but the deep ocean remained anoxic. Finally, in stage III, all parts of the oceans and atmosphere contained abundant free oxygen, as at present. Mass-balance considerations show that O_2 concentrations during stage II were ca. ≥ 0.03 PAL, because otherwise O_2 conveyed to the deep ocean would have been lost more rapidly than it was produced in the upper levels. An oxic deep ocean in stage III implies an atmospheric O_2 concentration $\geq 2 \times 10^{-3}$ PAL in order to compensate for the influx of reductants from hydrothermal vents. Anoxic bottom waters may have persisted until well after the deposition of BIFs ceased 1.85 Gyr ago, if sulphide rather than oxygen removed iron from deep-ocean water, and the aerobic deep-ocean may not have developed until 1.0–0.54 Gyr ago. Another line of evidence involves geologically stable biomarker molecules that can act as *molecular fossils*. Biomarker lipids discovered in northwestern Australia have been shown to be almost certainly contemporaneous with the 2.7-Gyr-old shales in which they are found, and thus provide fairly secure evidence for the presence of life. Some of the compounds are formed only by cyanobacteria. Photosynthetic bacterial cells of this kind initially increased the atmospheric oxygen content from $<10^{-8}$ PAL to much higher concentrations. The earliest cells lacked a nucleus, and are classified as prokaryotic: bacteria fall into this category, and can operate anaerobically. Larger cells, which are almost certainly eukaryotic, or possessing a nucleus (Cole 2016), are also found in the fossil record. It was once thought that eukaryotic organisms first appeared about 1.4 Gyr ago, but the discovery of the corkscrew-shaped organism *Grypania* puts the fossil record of eukaryotic cells back to 2.1 Gyr before present. Furthermore, the Australian biomarkers include compounds most plausibly formed from eukaryotic membranes, so that these more complex cells may already have been present 2.7 Gyr ago. The importance of this finding for the interpretation of atmospheric evolution is that almost all eukaryotic cells require large quantities of oxygen (~0.01 PAL) to function. Cell division is preceded by a clustering and splitting of the chromosomes within the nucleus (mitosis), a process dependent on the protein actomyosin that cannot form in the absence of oxygen. According to Kasting (1993), $[O_2] \approx 0.01$ PAL is below the upper limit for stage II of his model. Such concentrations could conceivably have arisen during stage I, because the possibility exists of localized oxygen oases containing as much as

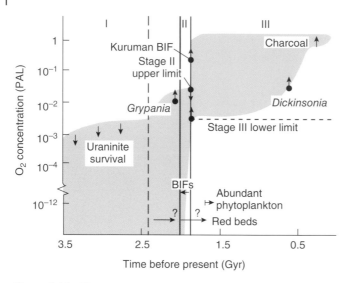

Figure 2.11 The proposed timing for the three stages of Figure 2.10 (Kasting 1993). Various indicators of concentration are marked on the figure and discussed in the text. The dashed vertical line indicates a possible earlier (2.4 Gyr before present) onset for stage II than is usually accepted.

0.08 PAL. The presence of the very early eukaryotic cells thus does not necessarily imply that the atmosphere contained abundant (>0.01 PAL) free oxygen much before 2 Gyr ago.

The data admittedly permit alternative interpretations, but accepting the general approach adopted by Kasting (1993) leads to an evolutionary picture for O_2 such as that shown in Figure 2.11, which maps out the several mineralogical and biological markers that place constraints on the oxygen concentrations, and the three stages of Kasting's model. One of the outstanding questions about the interpretation that we have adopted is the reason for the apparently sudden change in atmospheric O_2 concentrations at about 2 Gyr ago (Luo et al. 2016). Kasting (1993) reviews the evidence and cites the original literature. Perhaps the transition from stage I to stage II in the model came about when the reduced and lower-oxidation-state mineral sinks on the surface had finally been almost all consumed. Or perhaps the transition in O_2 concentrations reflects a time when the net rate of photosynthetic production of O_2 became greater than the rate of oxygen consumption in reaction with volcanic H_2 and other reduced gases. Reductions in volcanic outgassing or increased rates of carbon burial as the continental shelf area expanded are both possible explanations. These two scenarios would each be accompanied by a changing $^{13}C/^{12}C$ isotopic ratio. The observations and their interpretation may relate to a 'great oxidation event' that will be the subject of more detailed discussion in Section 2.8.

One important biological indicator at the 'transition stage' for O_2 at 2 Gyr before present is the appearance of enlarged thick-walled cells on filaments of cyanobacteria found as fossils. The significance of this observation is that cell structures were developing in response to the need to protect the organisms against the quantities of O_2 by then appearing in the atmosphere. As the atmospheric oxygen reached about 10^{-2} PAL, a revolution occurred, because eukaryotic cells, and then animal and plant life, emerged. Respiration and large-scale photosynthesis became of importance, enough free oxygen became available for the fibrous protein collagen to be formed, and the scene was set for the appearance of metazoans, or multicelled species. About 550 million years ago, the Cambrian Period opened. According to earlier ideas, this period heralded an 'evolutionary explosion' (Morris 2006; Briggs 2015). In many ways, the real significance of the Cambrian Period is that the first animals with clear external skeletons are preserved as fossils, with identities that had been recognized for centuries; remains of earlier lifeforms had not yet been discovered or understood. Metazoan fossils from the preceding 120 Myr, the Ediacaran Period, are now known. Many of these are from species resembling jellyfish. Such organisms can absorb their oxygen through the external surfaces at concentrations of about 7% of PAL. A reasonable estimate for when this level of oxygen was reached in the atmosphere can thus be set at about 670 Myr ago. The relatively impervious surface coverings of the Cambrian metazoans suggest that 120 Myr later the oxygen concentration was approaching 10^{-1} PAL. Following the opening of the Cambrian, the complexity of life is known to have multiplied rapidly and the foundations for all modern phyla were laid. 'Advanced' lifeforms (i.e. nonmicroscopic) were found ashore by the Silurian Period (420 Myr ago), and by Early Devonian times, only 30 Myr later, great forests had appeared. Soon afterwards, amphibian vertebrates ventured onto dry land.

It should be pointed out at this stage that recent research casts doubt on the idea of an 'explosion'. All the major groups of animals appear in the fossil record for the first time around 540–500 MYr ago. However, a detailed examination of the fossil record (Daley et al. 2018) indicates that the Cambrian explosion, rather than being a sudden event, unfolded gradually over the ~40 million years of the lower to middle Cambrian.

Earlier in the present section, it was pointed out in connection with the discussion of isotope ratios that $\delta^{13}C$ for terrestrial organic matter might provide a proxy signal for atmospheric composition in the period since rich plant life provided a source of organic carbon deposits. Figure 2.12 (Berner 2003) shows the results of one study that has employed the technique. The open squares show the percentage of O_2 in the atmosphere deduced from the fractionation of C and S isotopes in seawater; the results seem entirely consistent with O_2 levels inferred from the abundance of organic C and pyrite S in

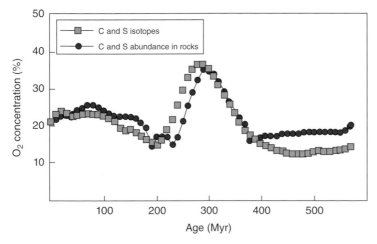

Figure 2.12 Percentage of oxygen in the atmosphere over the past 550 Myr inferred from oceanic C and S isotope abundances and abundances of organic carbon and pyrite sulphur in sedimentary rocks (Berner 2003).

sedimentary rocks (filled circles). The figure shows a clear overshoot of atmospheric O_2 levels beyond 1 PAL, peaking at roughly 300 Myr before present, and the presence of giant insects (especially dragonflies) only around that time has been explained in terms of the elevated oxygen. Several arguments link the rise of O_2 to the rise of trees and other vascular plants from about 390 Myr ago, as just mentioned. For example, these plants synthesize lignin as their '$(CH_2O)_n$', and this material is relatively resistant to biodegradation. As a result, more organic carbon becomes buried and separated from the liberated O_2, with the result that the concentrations of the O_2 can rise. Large plants can exert another influence on atmospheric composition. They possess deep and extensive root systems that greatly enhance the rate of chemical weathering (Eq. 2.3). The increased rates of weathering and of photosynthesis will both lead to a decrease in atmospheric $[CO_2]$, and thus to reduced greenhouse warming. The lowered $[CO_2]$ is believed in this way to have helped initiate the glaciation at about 290 Myr before present, which was the longest and most extensive of the past 550 Myr. It may therefore be that the rise of large land plants during the Devonian was nearly as important to the evolution of the atmosphere (and life) as was the development of microbial photosynthesis some 2 Gyr earlier (Berner 2003, 2005; Morris et al. 2018).

According to evidence from $\delta^{13}C$ and $\delta^{34}S$, another large rise in atmospheric $[O_2]$ appears to have occurred since about 205 Myr ago (Falkowski et al. 2005) at the boundary between the Triassic and Jurassic Periods. This rise can be seen clearly in Figure 2.12: after the peak at 300 Myr ago, the O_2 levels returned to ≈12%, but then possibly doubled to reach roughly 1 PAL by about 50 Myr before present. Extensive margins that developed along the Atlantic Ocean during this period may have provided a long-term storehouse for organic matter, once again separating the fuel ($(CH_2O)_n$) from the oxidant (O_2). The rise of O_2 over 150 Myr since the beginning of the Triassic almost certainly contributed to the evolution of large animals (Falkowski et al. 2005).

There is a further aspect of the rise of oxygen concentrations that will serve as an introduction to the next section. Ozone formed from oxygen in the atmosphere acts, together with the O_2 itself, as a screen for biologically damaging ultraviolet radiation. Until the atmospheric attenuation of ultraviolet intensity became sufficient, photosynthetic organisms such as the eukaryotic phytoplankton would not have been present in the oceans, and the rates both of carbon burial and of oxygen generation would have been less than they are at present. The atmospheric shield would have become effective for oceanic organisms with $[O_2]$ in the range 0.01–0.1 PAL (Kasting 1987; Reinhard et al. 2016).

2.7 Protection of Life from Ultraviolet Radiation

Complex eukaryotic life on Earth emerged and expanded in response to the evolution with time of surface oxygen levels. However, the role that planetary redox evolution has played in controlling the timing of metazoan (animal) emergence and diversification, if any, has been vigorously debated. One important issue concerns the threshold levels of free oxygen necessary in the environment for the survival of animals that evolved early. However, defining such thresholds in practice is not straightforward, and environmental O_2 levels can potentially constrain animal life in ways distinct from threshold O_2 tolerance. Reinhard et al. (2016) explore quantitatively one aspect of the evolutionary coupling between animal life and Earth's oxygen cycle – the influence of spatial and temporal variability in O_2 levels in ocean surfaces on the ecology of early metazoan organisms. Through the application of a series of quantitative biogeochemical models, they find that large spatiotemporal variations in surface ocean O_2 levels and pervasive benthic (ocean floor) anoxia are expected in a world with much lower atmospheric partial O_2 pressures than exist at present, resulting in severe ecological constraints and a challenging evolutionary landscape for early metazoan life. They argue that these effects, when considered in the light of synergistic interactions with other environmental parameters and variable O_2 demand throughout an organism's life history, would have resulted in long-term evolutionary and ecological inhibition of animal life on Earth for much of the middle of the Proterozoic eon (about 1.8–0.8 billion years ago).

The macromolecules, such as proteins and nucleic acids, that are characteristic of living cells, are damaged by radiation of wavelength shorter than about 290 nm (McLaren and Shugar 1964; Chandra et al. 2018). Ozone in the atmosphere has been seen by many as the key species in reducing mid-ultraviolet intensities to a level at which life can survive on the surface of the planet. One outstanding feature of the properties of ozone is the relationship between its absorption spectrum and the protection of living systems from the full intensity of solar ultraviolet radiation. Major components of the atmosphere, especially O_2, filter out solar ultraviolet with wavelengths <230 nm; at that wavelength, only about one part in 10^{16} of the intensity of an overhead sun would be transmitted through the molecular oxygen. At wavelengths longer than $\approx$230 nm, however, the only species in the present-day atmosphere capable of attenuating the Sun's radiation is ozone. Ozone has an unusually strong absorption just at the critical wavelengths (230–290 nm), so that it is an effective filter despite its relatively small concentration. For example, at $\lambda = 250$ nm less than one part in 10^{30} of the incident (overhead) solar radiation penetrates the ozone layer.

Ozone is formed in the atmosphere from molecular oxygen, the necessary energy being supplied by the absorption of solar ultraviolet radiation. The oxygen in the Earth's contemporary atmosphere is largely biological in origin, and ozone, needed as an ultraviolet filter to protect life, is itself dependent on the atmospheric oxygen. These links further emphasize the special nature of Earth's atmosphere. Actually, the interactions are even more subtle than we have suggested. Absolute concentrations and, indeed, the height distribution of ozone depend on a competition between production and loss. Loss of ozone is regulated by chemistry involving some of the other trace gases of the atmosphere, such as the oxides of nitrogen, which are themselves at least partly of biological origin. Biological processes thus influence both the generation and the destruction of ozone.

The starting point for many discussions of ozone concentrations in the palaeoatmosphere is the work of Berkner and Marshall (1965), who suggested that life was able to evolve in response to increasing protection from solar ultraviolet radiation as the atmospheric ozone shield developed. In its original form, the thesis propounded that life would initially develop in stagnant pools where liquid water of 10 m depth or more would be able to filter out the damaging radiation. At this stage, life in the oceans would be unlikely because organisms would be brought too close to the surface to survive. As the atmospheric content of oxygen, and thus of protective ozone, increased, life could migrate from the safety of the pools and lakes to the oceans and, finally, to dry land. Accompanying these changes would be a greatly enhanced photosynthetic and, indeed, evolutionary activity. Marine biota certainly seem to have paved the way, in terms of modification of the atmosphere, for the evolution of the terrestrial biota (Raven 1997). At one time, the opening of the Cambrian Period was thought to be characterized by an 'explosion' of evolution (see Section 2.6 and Briggs 2015; Daley et al. 2018), and it was an attractive idea that the dawn of the Cambrian was linked to the attainment in the atmosphere of an adequate ozone shield. However, as the discussion of Sections 2.4 and 2.6 will have made clear, life was abundant long before the Cambrian, and much of the reasoning for the evolutionary explosion was a result of earlier fossils not

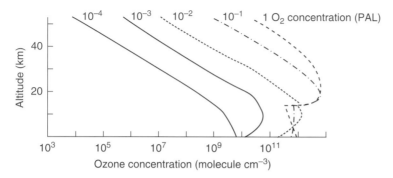

Figure 2.13 Vertical distribution of ozone for different total atmospheric oxygen contents ranging from 10^{-4} PAL to 1 PAL. The model used to obtain these results includes nitrogen, hydrogen, carbon, and chlorine chemistry; and allowance is made for ozone loss at the planetary surface. *Source:* The diagram is reproduced from Wayne (2000), and is based on data from Levine (1982).

being recognized for what they were. All the same, complexity and diversity did increase rapidly after the opening of the Cambrian, and the possibility cannot be discounted that increased mobility following sufficient protection might have been responsible.

A major question attached to the interpretation just presented is whether the biological evolutionary events were linked causally to the atmospheric changes that undoubtedly occurred. If they were, then some kind of feedback mechanism may have been in operation, since the atmospheric evolution was certainly mediated by the biota. Resolution of this question will require further information: in the first place, it is necessary to put the time history of growth of O_2 and O_3 in the atmosphere on a firmer footing than it seems to be at present. What we can do is to use atmospheric photochemical models to calculate the ozone concentrations that accompanied smaller O_2 levels in the early atmosphere. Figure 2.13 shows the results of one such calculation, in which the full chemistry of catalytic cycles (Levine 1985) was incorporated. One interesting feature of the evolution of ozone concentrations in our atmosphere is immediately apparent. At low [O_2], maximum ozone concentrations were found near the surface, but as oxygen concentrations increased, an ozone layer developed with its peak at successively higher altitudes.

Whatever shield is needed by the developing organisms, it is still difficult to specify what flux of radiation is dangerous to micro-organisms (Ranjan and Sasselov 2017). Experiments performed in space provide a way of examining biological response to the full spectrum of solar ultraviolet radiation. Kasting et al. (1989) report a maximum allowable dose for stable heredity and clone survival that would be acquired in approximately 0.3 second on an unprotected Earth. Absorption of radiation by DNA falls off at wavelengths longer than the broad maximum at $\lambda = 240$–280 nm, but most genetic damage (Gill et al. 2015) may be caused at the longest wavelengths absorbed (about 302 nm). Damage would occur primarily at the few hours near midday, when the Sun is most nearly overhead. For organisms with a generation span of one day or less, the maximum exposure is thus about four hours of high-intensity light. Experiments on genetic damage to corn pollen suggest that the maximum acceptable dose over the period of four hours would be 0.1 J m^{-2} s^{-1} for $\lambda \leq 302$ nm. An adequate ultraviolet screen would be provided by an ozone column density of 7×10^{18} molecule cm^{-2}. Other calculations of François and Gérard (1988) indicate that, at these densities, roughly 10% of a colony of cyanobacteria would survive. If a

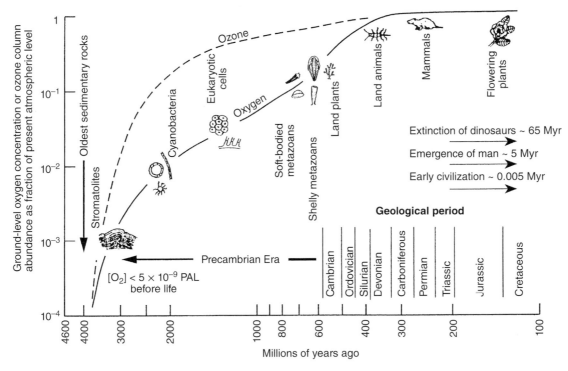

Figure 2.14 Evolution of ozone (dashed line) on Earth in response to the changes in oxygen levels. The oxygen concentration (solid line) follows trends similar to those shown in Figure 2.11, with a rather less abrupt increase at ~2 Gyr before present. Various geological and biological indicators of O_2 concentration shown on the diagram are discussed in the text. *Source:* from Wayne (2000).

survival fraction of 10^{-3} is acceptable, then the critical column density of ozone is 4.5×10^{18} molecule cm^{-2} for the most sensitive organisms. According to the relation between $[O_3]$ and $[O_2]$ given by François and Gérard, this limit is set at $[O_2]$ rather below 0.01 PAL. The calculations of Levine (1982), on the other hand, would require $[O_2]$ near 0.1 PAL. Whichever number is closer to the truth, it is abundantly clear that there would be insufficient ozone in the atmosphere to afford adequate protection for land-based life during the initial stages of oxygen growth. The largest prebiotic levels of ozone, based on the most extreme scenario, come nowhere near the lowest quantities needed. Enough ozone would not be present until at least approximately 2 Gyr ago, and possibly not until the opening of the Cambrian (say, 550 Myr before present). Figure 2.14 summarizes the material presented so far in this section. The growth in surface $[O_2]$ is identified on this plot by markers denoting the geochemical and fossil evidence; it corresponds roughly to the higher limits represented in Figure 2.11, but with a less abrupt rise at the 2 Gyr transition. Column ozone abundances calculated using the model just described are also shown in the figure. Levine's estimates of the flux of ultraviolet radiation that is tolerable to life, and hence the column density of ozone that furnishes an adequate screen, can be combined with his O_3–O_2 relation to suggest that protection would be sufficient at $[O_2] \approx 0.1$ PAL.

It is also interesting to examine the emergence of life onto dry land in relation to the growth in protection by atmospheric ozone. For $[O_2] \leq 0.01$ PAL, a layer of water of thickness > 4–5 m is needed for additional protection; by $[O_2] \approx 0.1$ PAL, water is probably no longer required. Does the transition to $[O_2] \geq 0.1$ PAL then really explain the appearance of life on dry land? Shelled organisms require dissolved oxygen that would be in equilibrium with >0.1 PAL in the atmosphere, so that the critical level of O_3 for biological protection would have already been passed when the organisms appeared abundantly in the Cambrian Period (550 Myr ago). According to the arguments presented in this chapter, even in the worst case there must have been an adequate atmospheric filter by 550 Myr ago at the latest to protect life on dry land from solar ultraviolet radiation. Yet life was apparently not firmly established on land until some 170 Myr later, towards the end of the Silurian. Thus, the possibility exists that the development of the ozone shield was not directly linked to the spread of life onto land. A positive linkage would certainly be regarded favourably by the supporters of the Gaia hypothesis, and the connection between the emergence of life out of water and the filtering of ultraviolet radiation by the atmosphere remains tantalizing.

Some researchers (Rambler and Margulis 1980) have argued that an early atmospheric ultraviolet screen was not needed at all. Many present-day bacteria seem rather resistant to ultraviolet sterilization techniques, and it is known that other organisms can protect themselves by producing coatings that absorb the ultraviolet. However, such sophisticated defence mechanisms were unlikely to be available to the simplest early cells. Perhaps we must look to the cyanobacteria that can protect themselves by forming covering mats, or to protection by prebiotic organic polymers and inorganic absorbers in the oceans. Lovelock (1988) has proposed that an atmospheric screen could have been afforded by a hydrocarbon-particle smog, initiated by photolysis of CH_4, well before 2.5 Gyr ago, and far earlier than the ozone screen is generally thought to have developed. This idea of a hydrocarbon aerosol screen has subsequently been taken up again by Sagan and Chyba (1997), as discussed at the end of Section 2.1. Yet another suggestion originated with Kasting et al. (1989), who proposed that SO_2 and H_2S, released as volcanic gases, could be photolysed to yield oligomers of sulphur (especially S_8) within the atmosphere, and that these particles could absorb and scatter solar ultraviolet radiation to an extent adequate for the survival of life on the planetary surface (Catling and Kasting 2017).

2.8 The Great Oxidation Event and Related Issues

Holland (2002) drew attention to the balance between oxygen production and its loss through reaction with reduced compounds in determining the redox state of the Earth and the amount of free oxygen in the atmosphere. He called the transition (at around 2.4–2.1 Ga) from fundamentally reducing to oxidizing the Great Oxidation Event (GOE) (Section 2.6). In Section 2.6, it was shown that studies of NMD fractionation of sulphur isotopes could provide an indicator of atmospheric oxygen levels. Such studies narrowed the window for the transition to 2.4–2.3 Ga on the basis of a presumed disappearance of the NMD sulphur-isotope signals (Bekker et al. 2004). It is now clear that a significant shift in oxygen levels did occur over this time period, but it has also become apparent that the rise in oxygen did not take place as a unidirectional step, as earlier envisaged (Kump 2008). Lyons et al. (2014) review more recent evidence in some detail, and show that the probable course of events was a sharp (but not step) rise in oxygen levels from 10^{-5} PAL or lower before 2.5 Ga to overshoot to as high as 1 PAL by 2.1 Ga, before falling back to 10^{-2} PAL by 2.0 Ga, and remaining low for the next 1.5 Gyr. Recovery of oxygen to something approaching 1 PAL took place around 600 Ma (Lyons et al. 2014). This second increase oxygenated most of the deep ocean (see later) and surface waters, providing a welcoming environment for the first animals. The picture is further complicated by evidence for 'whiffs' of oxygen before the GOE (Anbar et al. 2007). From studies of ages for U and Pb, Gumsley et al. (2017) suggest the onset of the GOE to have occurred between 2.5 Ga and 2.43 Ga, roughly 100 Myr earlier than previous estimates. The early fluctuations in oxygen levels is supported by evidence (Koehler et al. 2018) obtained by a study of isotope ratios for nitrogen and selenium, and selenium abundances. Once again, these investigations provide evidence of transient surface ocean oxygenation ~260 My before the GOE, as well as of a possible muted pulse of oxidative continental weathering; they provide the oldest firm evidence for nitrification and denitrification metabolisms.

The revised view of the GOE is thus oscillatory rising and falling of oxygen levels in both atmosphere and oceans, superposed on a rising trend that started perhaps as early as 3.0 Ga, and that lasted for up to 2.5 Gyr. Lyons et al. (2014) suggest that the 'event' of the GOE should be regarded more as a *transition,* and that the adjective *great* could be open to argument. As for the timing, the GOE has been defined as the time interval when sufficient atmospheric oxygen accumulated to prevent the generation and preservation of NMD fractionation of sulphur isotopes in sedimentary rocks. Cui et al. (2018) have attempted to establish an authentic signature of the GOE from sulphur-isotope studies of sedimentary rocks of the Huronia Supergroup. This supergroup consists mostly of metamorphosed sandstones and mudstones found in Canada (Ontario and Quebec) whose deposition has been dated to between 2.480 Ga and 2.219 Ga. Cui et al. (2018) investigated abundances of all four stable isotopes of sulphur (^{32}S, ^{33}S, ^{34}S, and ^{36}S) in the rocks for the first time, using sophisticated techniques for isotope analysis. Small but analytically resolvable nonzero values of Δ^{33}S (from −0.07‰ to +0.38‰) and Δ^{36}S (from −4.1‰ to +1.0‰) persist throughout the lower Huronian Supergroup. Neither pronounced NMD-S signals nor a transition from NMD to mass-dependent fractionation are seen in this study. Four scenarios are proposed for the genesis of small nonzero Δ^{33}S and Δ^{36}S values, and the study suggests that different analytical methods and sample history (primary vs. metamorphic) may have caused inconsistent interpretations of S-isotope profiles of the GOE successions on a global scale.

One problem in reconciling the observations lies with the disappearance of the NMD isotope signals with other markers for the rise of oxygen (Fakhraee et al. 2018). Trace-metal and sulphur-isotope data point to oxygen production long before the disappearance of NMD sulphur isotope fractionation (Reinhard et al. 2009). One suggestion (Reinhard et al. 2009, 2013; Philippot et al. 2018) is that NMD signals that formed in an oxygen-poor atmosphere were captured in pyrite (FeS_2) and other minerals in sedimentary rocks that were later uplifted as mountain ranges, oxidized, and entered rivers that delivered material with the NMD signals to oceans, from which sediments were deposited long after atmospheric oxygen rose. Philippot et al. (2018) therefore suggest that globally asynchronous sulphur isotope signals require re-definition of the Great Oxidation Event.

Several other issues arise from the concept of the GOE; for example, what was the impact of rising atmospheric oxygen on the ocean? Canfield (1998) promoted the idea that the ocean remained anoxic and probably 'euxinic' (no oxygen and a raised level of free H_2S, as in the contemporary Black Sea) for a billion years around 1.5 Ga. That is, oxygenation in the deep ocean lagged behind the atmosphere by as much as almost 2 Gyr (Lyons et al. 2014; Stolper and Keller 2018),

Since conventional 'oxygenic' photosynthesis (Section 2.6, Eq. (2.20)) is the source of virtually all the O_2 in the present-day atmosphere, it is obviously highly pertinent to examine how the development of photosynthesis fits in with the timing of the GOE. Another form of photosynthesis, *anoxygenic* ('not oxygen-producing') photosynthesis (Hanada 2016) seems to have been active long before the GOE. Shales rich in organic chemical species are common in the Archaean rock record (>2.5 Ga), and the organic material appears to have been formed by anoxygenic photosynthesis that would convert atmospheric CO_2 to deposited carbohydrate (see Eqs. (2.7) and (2.8), for examples). Photosynthesis requires light and an electron donor (reducing agent): the widespread water molecule is the electron donor for oxygenic photosynthesis. Alternative electron donors that could have delivered significant amounts of organic carbon to marine sediments include H_2S, H_2, and Fe(II). The oxidation products are elemental sulphur, water, and Fe(III), but *not* O_2. Anoxygenic phototrophic bacteria that could mediate such photosynthesis were discovered more than 100 years ago (Hanada 2016). For many years, it has been believed that the GOE was, at least in part, brought about by the evolution of oxygenic photosynthesis, and that the organisms responsible for the photochemical process were cyanobacteria whose activity blossomed at about the time of the GOE (Sections 2.4 and 2.6). However, evidence is now appearing (Cardona 2016; 2018) to suggest that oxygenic photosynthesis originated a billion years earlier (before 3.4 Ga) than was thought previously, and that cyanobacteria were not the micro-organisms responsible, but rather that simpler bacteria produced oxygen first. Of course, this discovery does not preclude the evolution of cyanobacteria at around the time of the GOE, but it does place some form of oxygenic photosynthesis at a very early stage of evolutionary history.

Fossils of macroscopic eukaryotes are rarely older than the Ediacaran Period (635–541 Ma), and their interpretation remains controversial. Zhu et al. (2016) report the discovery of macroscopic fossils from the 1560 Ma Gaoyuzhuang Formation, Yanshan area, North China, that exhibit both large size and regular morphology. Preserved as carbonaceous compressions, the Gaoyuzhuang fossils are up to 30 cm long and nearly 8 cm wide, suggesting that the Gaoyuzhuang fossils are a record of multicellular eukaryotes from the ocean floor of unprecedentedly large size. The Edicaran organisms described earlier date from roughly 631 Ma, and mark the beginning of the Phanerozoic Eon. The Cambrian Period followed the Edicaran at about 541 Ma. Thus, the first organisms showing some similarity to present-day life seem to not to have evolved until more than 1.5 GYr after the GOE (or oxygen transition). Yet it seems clear that rising oxygen levels were closely related to the evolution of life. More complex cells could function, and the ocean became oxygenated; organisms were protected from solar ultraviolet radiation by atmospheric O_2 and O_3; and so on. It is worth pointing out a further consequence of the oceans becoming oxic rather than remaining euxinic. Essential micronutrients in seawater are needed by relatively complex organisms (Moore et al. 2013): for example, molybdenum, Mo, is needed by the enzymes that fix and use nitrogen from the atmosphere (Lyons et al. 2014). However, many such micronutrients are readily scavenged in seawater in the presence of H_2S (Anbar and Knoll 2002), and the lack of the necessary nutrients would have inhibited the early diversity, distribution, and abundance of eukaryotes (Dupont et al. 2010). In that case, there would be an interesting parallel between the delay in the transition in oxygen levels despite the existence of photosynthetic organisms long before, and the burst in evolutionary activity hundreds of millions of years after oxygen levels reached 10^{-2} PAL. Favoured explanations in the first case include consumption of O_2 by reducing sinks until the sinks were used up, and in the second case a deficiency of essential nutrients until the oceans ceased to be euxinic. An interesting side issue is that in each case H_2S is closely implicated, so that there may be an evolutionary involvement of changing patterns of volcanism. However, Li et al. (2015) conducted a Fe–S–C–Mo geochemical study of the ~1.65 Ga Chuanlinggou Formation in North China, and their results suggest the presence of anoxic but non-euxinic (ferruginous: containing iron) conditions that persisted below the surface mixed layer for the deepest portion of the continental rifting basin and indicate that this pattern is apparently independent of the local organic carbon content. See Gasol and Kirchman (2018, further reading) for additional insight.

The period 1.6–1.0 Ma is becoming recognized as a key interval in the history of biological evolution on Earth. Atmospheric oxygen levels were thought to have been $<10^{-2}$ PAL, low enough to have inhibited the evolution of animal life, whereas after 550 Ma, levels were probably >0.2 PAL, large enough to sustain respiration in large animals. However, Zhang et al. (2016) have studied the patterns of enrichment of redox-sensitive trace metals in 1.4 Ga sediments in North China. These investigations are used to argue for the presence of oxygenated bottom waters during the deposition of the sediments, whilst there is also evidence for the presence of green sulphur bacteria in the water column. A model indicates atmospheric oxygen levels >0.4 PAL. This level of oxygen was more than adequate to fuel animal respiration, but at a time long before the evolution of the animals themselves. Such an unexpected result is naturally likely to provoke dispute, and Planavsky et al. (2016) were quick to claim that the evidence was faulty! These latter workers argue that the trace metal data presented by Zhang et al. lack the resolving power to justify the conclusions about elevated oxygenation of bottom waters. Further, Planavsky et al. are concerned about the validity of diagnostic biomarkers used in the work of Zhang et al., suggesting that, for example, there could be contamination of the samples by exogenous hydrocarbons. Planavsky et al. state clearly that 'there remains no robust evidence for high atmospheric oxygen levels that could have fostered life 1400 million years ago'. What the present author concludes from such controversies is that this field of research is a highly active and exciting one that requires very careful, precise measurements and interpretations that cannot be invalidated by adventitious contamination. Then we may learn more about the possible links between changing atmospheric oxygen levels and animal evolution. Beyond the investigation of the oxygen levels themselves, several researchers pursue the parallel line of how much oxygen was needed for the rise of animals (Planavsky et al. 2014), for the evolution of complex eukaryotes (Zhang et al. 2018), and for the Cambrian explosion (Zhang and Cui 2016; Wei et al. 2018). Wei et al. examined limestones that had formed in shallow Edicaran and Cambrian seas to determine the concentrations of the uranium isotopes ^{235}U and ^{238}U. Uranium compounds precipitated in oxygenated water are enriched (as a result of kinetic isotope effects) in the heavier isotope. Interestingly, the rock samples of Wei et al. show two incidents where $^{238}U/^{235}U$ ratios dropped sharply (at 542–541 Ma and at 534–523 Ma). These dates match what the fossil record tells us correspond to two pulses of enhanced evolution. This seems to be remarkable timing, and is unlikely to be a coincidence. Wei et al. interpret the observations as meaning an environmental catastrophe (the drop in oxygen wiped out many earlier organisms), and new groups emerged to fill the vacant ecological niches. This situation is analogous to the asteroid impact(s) that wiped out the dinosaurs, and opened the way to massive biological diversion. Curiously, then, this view combines the necessity for rising oxygen to permit animals to thrive, with the opposite in which a decrease in oxygen was a trigger for an increase in diversity of the biota. Meanwhile, the work of Zhang et al. (2018) suggests the early birth of complex life (Klein 2018). Once again it may be that a shift in oxygen levels (previously unknown) was the cause of the evolutionary events.

The connections, interactions, and feedbacks in the evolution of our atmosphere and of life are proving ever more fascinating, and the unravelling of the complex story will stimulate scientists of many disciplines for years to come.

2.9   The Future

Despite several imaginative and speculative ideas noted in earlier sections, it is widely accepted that ozone is at present the key atmospheric filter for ultraviolet radiation in the region $\lambda \approx 200$–300 nm. Indeed, much of the research described in this chapter owes its existence to the perceived threat of destruction of the ozone layer as a consequence of anthropogenic release of various species that can act as catalysts for removal of stratospheric ozone (Wayne 2000). Lovelock and Whitfield (1982) looked far into the future to raise the question of how much longer the biosphere could survive. They argued that the Sun's luminosity is likely to have increased by 25–30% over the past 4.5 Gyr, and that there is no reason to suppose that the trend has ceased. Whether by active or passive control, changes in CO_2 concentration appear to have compensated for the increasing solar intensity in such a way as to keep the planetary temperature very nearly constant. But the capacity for control might now be nearly exhausted, because $[CO_2]$ is approaching the lower limit tolerable for photosynthesis. If that limit is taken to be 150 ppm, then the CO_2 control of a climate favourable for life can continue for another 30–300 Myr. Some adaptation to lower CO_2 concentrations and to higher temperature is possible, but according to Lovelock and Whitfield it would not buy much time. Caldeira and Kasting (1992) are more encouraging: they employed a more elaborate model that treats greenhouse heating better, includes biologically mediated weathering, and allows for some plant photosynthesis to persist to $[CO_2] < 10$ ppm This treatment gives the biosphere another 0.9–1.5 Gyr of life, after which Earth might lose all its water to space within a further 1 Gyr. The outcome will then be that Earth's atmosphere will perhaps once again come to resemble that of its sister, Venus, as it did before the evolution of life modified it. See also Beerling (2007) in *Further Reading*: Beerling describes how plants changed Earth's history.

Questions

1   How has photosynthesis altered the composition of the Earth's atmosphere?

2   How did the appearance of ozone in the atmosphere allow life to develop on Earth?

3   What is meant by the term *Great Oxidation Event*?

4   What are stromatolites, and what is their significance?

5   How can different isotopes of the same element take part in chemical reactions at different rates?

6   How can the study of isotopic abundances provide information about oxygen levels in the ancient atmosphere?

References

Airapetian, V.S., Glocer, A., Gronoff, G. et al. (2016). Prebiotic chemistry and atmospheric warming of early Earth by an active young Sun. *Nature Geoscience* 9: 452–455.

Anbar, A.D. and Knoll, A.H. (2002). Proterozoic Ocean chemistry and evolution: a bioinorganic bridge? *Science* 297: 1137–1142.

Anbar, A.D., Duan, Y., Lyons, T.W. et al. (2007). A whiff of oxygen before the great oxidation event? *Science* 317: 1903–1906.

Avice, G., Marty, B., and Burgess, R. (2017). The origin and degassing history of the Earth's atmosphere revealed by Archean xenon. *Nature Communications* 8: 15455.

Bar-Nun, A. and Chang, S. (1983). Photochemical reactions of water and CO in Earth's primitive atmosphere. *Journal of Geophysical Research* 88: 6662–6672.

Bekker, A., Holland, H.D., Wang, P.L. et al. (2004). Dating the rise of atmospheric oxygen. *Nature* 427: 117–120.

Bell, E.A., Boehnke, P., Harrison, T.M., and Mao, W.L. (2015). Potentially biogenic carbon preserved in a 4.1 billion-year-old zircon. *Proceedings of the National Academy of Sciences* 112: 14518–14521.

Berkner, L.V. and Marshall, L.C. (1965). On the origin and rise of oxygen concentration in the Earth's atmosphere. *Journal of the Atmospheric Sciences* 22: 225–261.

Berner, R.A. (2003). The rise of trees and their effects on Paleozoic atmospheric CO_2 and O_2. *Comptes Rendus Geoscience* 335: 1173–1177.

Berner, R.A. (2005). A different look at biogeochemistry. *The American Journal of Science* 305: 872–873.

Bosak, T., Knoll, A.H., and Petroff, A.P. (2013). The Meaning of Stromatolites. *Annual Review of Earth and Planetary Sciences* 41: 21–44.

Briggs, D.E.G. (2015). The Cambrian explosion. *Current Biology* 25: R864–R868.

Caldeira, K. and Kasting, J.F. (1992). The life span of the biosphere revisited. *Nature* 360: 721–723.

Cameron, A.G.W. (1988). Origin of the solar system. *Annual Review of Astronomy and Astrophysics* 26: 441–472.

Canfield, D.E. (1998). A new model for Proterozoic Ocean chemistry. *Nature* 396: 450–452.

Cantrill, S. (ed.) (2017). On the origins of life. *Nature Chemistry* 9: 297–402.

Cardona, T. (2016). Origin of Bacteriochlorophyll *a* and the early diversification of photosynthesis. *PLoS One* 11: e0151250. https://doi.org/10.1371/journal.pone.0151250.

Cardona, T. (2018). Early Archaean origin of heterodimeric photosystem I. *Heliyon* 4: e00548.

Catling, D.C. and Claire, M.W. (2005). How Earth's atmosphere evolved to an oxic state: a status report. *Earth and Planetary Science Letters* 237: 1–20.

Catling, D.C. and Kasting, J.F. (2017). *Atmospheric Evolution on Inhabited and Lifeless Worlds*. Cambridge, UK: Cambridge University Press. ISBN: 9780521844123.

Chandra, A., Cogdell, R., Crespo-Hernández, C.E. et al. (2018). Light induced damage and repair in nucleic acids and proteins: general discussion. *Faraday Discussions* 207: 389–408.

Chyba, C.F., Thomas, P.J., Brookshaw, L., and Sagan, C. (1990). Cometary delivery of organic molecules to the early earth. *Science* 249: 366–373.

Cole, L.A. (2016). *Biology of Life*. San Diego, CA: Academic Press. ISBN: 978-0-12-809685-7.

Cui, H., Kitajima, K., Spicuzza, M.J. et al. (2018). Searching for the great oxidation event in North America: a reappraisal of the Huronian Supergroup by SIMS Sulfur four-isotope analysis. *Astrobiology* 18: 519–538.

Daley, A.C., Antcliffe, J.B., Drage, H.B., and Pates, S. (2018). Early fossil record of Euarthropoda and the Cambrian Explosion. *Proceedings of the National Academy of Sciences* 115: 5323–5331.

Denlinger, M.C. (2005). The origin and evolution of the atmospheres of Venus, Earth and Mars. *Earth, Moon, and Planets* 96: 59–80.

Dupont, C.L., Butcher, A., Valas, R.E. et al. (2010). History of biological metal utilization inferred through phylogenomic analysis of protein structures. *Proceedings of the National Academy of Sciences* 107: 10567–10572.

Durham, R. and Chamberlain, J.W. (1989). A comparative study of the early terrestrial atmospheres. *Icarus* 77: 59–66.

Eickmann, B., Hofmann, A., Wille, M. et al. (2018). Isotopic evidence for oxygenated Mesoarchaean shallow oceans. *Nature Geoscience* 11: 133–138.

Fakhraee, M., Crowe, S.A., and Katsev, S. (2018). Sedimentary sulfur isotopes and Neoarchean Ocean oxygenation. *Science Advances* 4: e1701835.

Falkowski, P.G., Katz, M.E., Milligan, A.J. et al. (2005). The rise of oxygen over the past 205 million years and the evolution of large placental mammals. *Science* 309: 2202–2204.

François, L.M. and Gérard, J.-C. (1988). Ozone, climate, and biospheric environment in the ancient oxygen-poor atmosphere. *Planetary and Space Science* 36: 1391–1414.

Frei, R., Crowe, S.A., Bau, M. et al. (2016). Oxidative elemental cycling under the low O_2 Eoarchean atmosphere. *Scientific Reports* 6: 21058.

Frimmel, H.E. (2005). Archaean atmospheric evolution: evidence from the Witwatersrand gold fields, South Africa. *Earth-Science Reviews* 70: 1–46.

Futuyma, D.J. and Kirkpatrick, M. (2017). *Evolution*, 4e. Oxford, UK: Oxford University Press. ISBN: 9781605356051.

Gill, S.S., Anjum, N.A., Gill, R. et al. (2015). DNA damage and repair in plants under ultraviolet and ionizing radiations. *The Scientific World Journal* 2015: 250158.

Gumsley, A.P., Chamberlain, K.R., Bleeker, W. et al. (2017). Timing and tempo of the great oxidation event. *Proceedings of the National Academy of Sciences* 114: 1811–1816.

Hanada, S. (2016). Anoxygenic photosynthesis – a photochemical reaction that does not contribute to oxygen reproduction. *Microbes and Environments* 31: 1–3.

Holland, H.D. (2002). Volcanic gases, black smokers and the great oxidation event. *Geochimica et Cosmochimica Acta* 66: 3811–3826.

Holland, H.D. (2006). The oxygenation of the atmosphere and oceans. *Philosophical Transactions of the Royal Society B: Biological Sciences* 361: 903–915.

Hoyle, F. (1982). Comets – a matter of life and death. *Vistas in Astronomy* 24: 123–139.

Hoyle, F. and Wickramasinghe, N.C. (1977). Does epidemic disease come from space? *New Scientist* 76: 402–404.

Hoyle, F. and Wickramasinghe, N.C. (1983). Bacterial life in space. *Nature* 306: 420.

Hunten, D.M. (1993). Atmospheric evolution of the terrestrial planets. *Science* 259: 915–920.

Ivanov, M.V. and Babikov, D. (2013). On molecular origin of mass-independent fractionation of oxygen isotopes in the ozone forming recombination reaction. *Proceedings of the National Academy of Sciences* 110: 17708–17713.

Jouzel, J. (2013). A brief history of ice-core science over the last 50 yr. *Climate of the Past* 9: 2525–2547.

Kasting, J.F. (1987). Theoretical constraints on oxygen and carbon dioxide concentrations in the Precambrian atmosphere. *Precambrian Research* 34: 205–239.

Kasting, J.F. (1993). Earth's early atmosphere. *Science* 259: 920–926.

Kasting, J.F. and Siefert, J.L. (2002). Life and the evolution of Earth's atmosphere. *Science* 299: 1066–1068.

Kasting, J.F., Young, A.T., Zahnle, K.J., and Pinto, J.P. (1989). Sulfur, ultraviolet radiation, and the early evolution of life. *Origins of Life* 19: 95–108.

Kasting, J.F., Eggler, D.H., and Raeburn, S.P. (1993). Mantle redox evolution and the oxidation state of the Archean atmosphere. *Journal of Geology* 101: 245–257.

Kirchner, J.W. (1989). The Gaia hypothesis – can it be tested? *Reviews of Geophysics* 27: 223–235.

Klein, A. (2018). The early birth of complex life. *New Scientist* 238: 6.

Kodama, T., Nitta, A., Genda, H. et al. (2018). Dependence of the onset of the runaway greenhouse effect on the latitudinal surface water distribution of earth-like planets. *Journal of Geophysical Research, Planets* 123: 559–574.

Koehler, M.C., Buick, R., Kipp, M.A. et al. (2018). Transient surface ocean oxygenation recorded in the ~2.66-Ga Jeerinah formation, Australia. *Proceedings of the National Academy of Sciences* 115: 7711–7716.

Krissansen-Totton, J., Arney, G.N., and Catling, D.C. (2018). Constraining the climate and ocean pH of the early Earth with a geological carbon cycle model. *Proceedings of the National Academy of Sciences* 115: 4105–4110.

Kump, L.R. (2008). The rise of atmospheric oxygen. *Nature* 451: 277–278.

Levine, J.S. (1982). The photochemistry of the palaeoatmosphere. *Journal of Molecular Evolution* 18: 161–172.

Levine, J.S. (ed.) (1985). *The Photochemistry of Atmospheres*. Orlando, FL: Academic Press.

Li, C., Planavsky, N.J., Love, G.D. et al. (2015). Marine redox conditions in the middle Proterozoic ocean and isotopic constraints on authigenic carbonate formation. Insights from the Chuanlinggou formation, Yanshan Basin, North China. *Geochimica et Cosmochimica Acta* 150: 90–105.

Lovelock, J.E. (1979). *Gaia: A New Look at Life on Earth*. Oxford, UK: Oxford University Press.

Lovelock, J.E. (1988). *The Ages of Gaia: A Biography of our Living Earth*. Oxford, UK: Oxford University Press.

Lovelock, J.E. and Margulis, L. (1974). Atmospheric homeostasis by and for the biosphere. *Tellus* 26: 2–9.

Lovelock, J.E. and Whitfield, M. (1982). Life-span of the biosphere. *Nature* 296: 561–563.

Luo, G., Ono, S., Beukes, N.J. et al. (2016). Rapid oxygenation of Earth's atmosphere 2.33 billion years ago. *Science Advances* **2**, e1600134.

Lüthi, D., Le Floch, M., Bereiter, B. et al. (2008). High-resolution carbon dioxide concentration record 650,000–800,000 years before present. *Nature* 453: 379–382.

Lyons, T.W., Reihard, C.T., and Planavsky, N.J. (2014). The rise of oxygen in Earth's early ocean and atmosphere. *Nature* 506: 307–315.

Maltby, J., Steinle, L., Löscher, C.R. et al. (2018). Microbial methanogenesis in the sulfate-reducing zone of sediments in the Eckernförde Bay, SW Baltic Sea. *Biogeosciences* 15: 137–157.

Marchi, S., Black, B.A., Elkins-Tanton, L.T., and Bottkea, W.F. (2016). Massive impact-induced release of carbon and sulfur gases in the early Earth's atmosphere. *Earth and Planetary Science Letters* 449: 96–104.

Martin, W.F., Bryant, D.A., and Beatty, J.T. (2018). A physiological perspective on the origin and evolution of photosynthesis. *FEMS Microbiology Reviews* 42: 205–231.

McLaren, A.D. and Shugar, D. (1964). *Photochemistry of Proteins and Nucleic Acids*, International Series of Monographs on Pure and Applied Biology, vol. 22, 462. Oxford, UK: Pergamon Press.

Menor-Salván, C. (ed.) (2018). *Prebiotic Chemistry and Chemical Evolution of Nucleic Acids*, . Cham, Switzerland: Springer.

Miller, S.L. (1953). A production of amino acids under possible primitive Earth conditions. *Science* 117: 528–529.

Mojzsis, S.J., Arrhenius, G., McKeegan, K.D. et al. (1996). Evidence for life on Earth before 3800 million years ago. *Nature* 384: 55–59.

Moore, C.M., Mills, M.M., Arrigo, K.R. et al. (2013). Processes and patterns of oceanic nutrient limitation. *Nature Geoscience* 6: 701–710.

Morris, S.M. (2006). Darwin's dilemma: the realities of the Cambrian 'explosion'. *Philosophical Transactions of the Royal Society B: Biological Sciences* 361: 1069–1083.

Morris, J.L., Puttick, M.N., Clark, J.W. et al. (2018). The timescale of early land plant evolution. *Proceedings of the National Academy of Sciences* 115: E2274–E2283.

Morrison, D. and Owen, T. (1996). *The Planetary System*. Reading, MA: Addison-Wesley.

Nemchin, A.A., Whitehouse, M.J., Menneken, M. et al. (2008). A light carbon reservoir recorded in zircon-hosted diamond from the Jack Hills. *Nature* 454: 92–95.

Ohtomo, Y., Kakegawa, T., Ishida, A. et al. (2014). Evidence for biogenic graphite in early Archaean Isua metasedimentary rocks. *Nature Geoscience* 7 (1): 25–28.

Olson, S.L., Schwieterman, E.W., Reinhard, C.T., and Lyons, T.W. (2018). Earth: atmospheric evolution of a habitable planet. In: *Handbook of Exoplanets*. (eds. H. Deeg and J. Belmonte). Heidelberg, Germany: Springer.

Orgel, L.E. (1998). The origin of life – a review of facts and speculations. *Trends in Biochemical Sciences* 23: 491–495.

Ozaki, K., Tajika, E., Hong, P.K. et al. (2018). Effects of primitive photosynthesis on Earth's early climate system. *Nature Geoscience* 11: 55–59.

Pace, A., Bourillot, R., Bouton, A. et al. (2018). Formation of stromatolite lamina at the interface of oxygenic–anoxygenic photosynthesis. *Geobiology* 16: 378–398.

Papineau, D., de Gregorio, B.T., Cody, G.D. et al. (2011). Young poorly crystalline graphite in the >3.8-Gyr-old Nuvvuagittuq banded iron formation. *Nature Geoscience* 4: 376–379.

Pearce, B.K.D., Tupper, A.S., Pudritz, R.E., and Higgs, P.G. (2018). Constraining the time interval for the origin of life on Earth. *Astrobiology* 18: 343–364.

Pepin, R.O. (2006). Atmospheres on the terrestrial planets: clues to origin and evolution. *Earth and Planetary Science Letters* 252: 1–14.

Philippot, P., Ávila, J.N., Killingsworth, B.A. et al. (2018). Globally asynchronous Sulphur isotope signals require re-definition of the great oxidation event. *Nature Communications* 9: 2245.

Planavsky, N.J., Reinhard, C.T., Wang, X. et al. (2014). Earth history. Low mid-Proterozoic atmospheric oxygen levels and the delayed rise of animals. *Science* 346: 635–638.

Planavsky, N.J., Cole, D.B., Reinhard, C.T. et al. (2016). No evidence for high atmospheric oxygen levels 1,400 million years ago. *Proceedings of the National Academy of Sciences of the United States of America* 113: E2550–E2551.

Rambler, M. and Margulis, L. (1980). Bacterial resistance to ultraviolet radiation under anaerobiosis: implications for pre-Phanerozoic evolution. *Science* 210: 638–640.

Ranjan, S. and Sasselov, D.D. (2017). Constraints on the early terrestrial surface UV environment relevant to prebiotic chemistry. *Astrobiology* 17: 169–204.

Raven, J.A. (1997). The role of marine biota in the evolution of terrestrial biota: gases and genes–atmospheric composition and evolution of terrestrial biota. *Biogeochemistry* 39: 139–164.

Raymond, J. and Segre, D. (2006). The effect of oxygen on biochemical networks and the evolution of complex life. *Science* 311: 1764–1767.

Reinhard, C.T., Raiswell, R., Scott, C. et al. (2009). A late Archean sulfidic sea stimulated by early oxidative weathering of the continents. *Nature* 326: 713–716.

Reinhard, C.T., Planavsky, N.J., and Lyons, T.W. (2013). Long-term sedimentary recycling of rare sulphur isotope anomalies. *Nature* 497: 100–103.

Reinhard, C.T., Planavsky, N.J., Olson, S.L. et al. (2016). Earth's oxygen cycle and the evolution of animal life. *Proceedings of the National Academy of Sciences* 113: 8933–8938.

Rosing, M.T. (1999). ^{13}C-depleted carbon microparticles in >3700-Ma sea-floor sedimentary rocks from West Greenland. *Science* 283: 674–676.

Rumble, D. (2005). A mineralogical and geochemical record of atmospheric photochemistry. *American Mineralogist* 90: 918–930.

Rye, R. and Holland, H.D. (1998). Paleosols and the evolution of atmospheric oxygen: a critical review. *American Journal of Science* 298: 621–672.

Sagan, C. and Chyba, C. (1997). The early faint Sun paradox: organic shielding of ultraviolet-labile greenhouse gases. *Science* 276: 1217–1221.

Schopf, J.W. (2006). Fossil evidence of Archaean life. *Philosophical Transactions of the Royal Society B: Biological Sciences* 361: 869–885.

Sheldon, N.D. (2006). Precambrian aerosols and atmospheric CO_2 levels. *Precambrian Research* 147: 148–155.

Stolper, D.A. and Keller, B. (2018). A record of deep-ocean dissolved O_2 from the oxidation state of iron in submarine basalts. *Nature* 553: 323–327.

Tagami, S., Attwater, J., and Holliger, P. (2017). Simple peptides derived from the ribosomal core potentiate RNA polymerase ribozyme function. *Nature Chemistry* 9: 325–332.

Tashiro, T., Ishida, A., Hori, M. et al. (2017). Early trace of life from 3.95 Ga sedimentary rocks in Labrador, Canada. *Nature* 549: 516–518.

Thiemens, M.H. (2006). History and applications of mass-independent isotope effects. *Annual Review of Earth and Planetary Sciences* 34: 217–262.

Thiemens, M.H., Chakraborty, S., and Dominguez, G. (2012). The physical chemistry of mass-independent isotope effects and their observation in nature. *Annual Review of Physical Chemistry* 63: 155–177.

Volk, T. (1998). *Gaia's Body: Toward a Physiology of Earth*. New York, NY: Springer-Verlag.

Voosen, P. (2017). Record-shattering 2.7-million-year-old ice core reveals start of the ice ages. *Science* 357: 630–631.

Walker, J.C.G. (1976). Implications for atmospheric evolution of the inhomogeneous accretion model of the origin of the earth. In: *The Earliest History of the Earth* (ed. B.F. Windley). New York, NY: Wiley.

Walker, J.C.G. (1990). Precambrian evolution of the climate system. *Palaeogeography, Palaeoclimatology, Palaeoecology* 82: 261–289.

Wayne, R.P. (2000). *Chemistry of Atmospheres*, 3e. Oxford, UK: Oxford University Press.

Wei, G.-Y., Planavsky, N.J., Tarhan, L.G. et al. (2018). Marine redox fluctuation as a potential trigger for the Cambrian explosion. *Geology* 46: 587–590.

Wen, J.-S., Pinto, J.F., and Yung, Y.L. (1989). Photochemistry of CO and H_2O: analysis of laboratory experiments and applications to the prebiotic Earth's atmosphere. *Journal of Geophysical Research* 94: 14957–14970.

Wetherill, G. (1990). Formation of the Earth. *Annual Review of Earth and Planetary Sciences* 18: 205–256.

Whittet, D.C.B. (1997). Is extraterrestrial organic matter relevant to the origin of life on earth? *Origins of Life and Evolution of the Biosphere* 27: 249–262.

Wickramasinghe, N.C. and Hoyle, F. (1998). Infrared evidence for panspermia: an update. *Astrophysics and Space Science* 259: 385–401.

Yang, X., Zhang, Z., Santosh, M. et al. (2018). Anoxic to suboxic Mesoproterozoic Ocean: evidence from iron isotope and geochemistry of siderite in the banded iron formations from north Qilian, NW China. *Precambrian Research* 307: 115–124.

Young, G.M. (2013). Evolution of Earth's climatic system: evidence from ice ages, isotopes, and impacts. *GSA Today* 23: 4–10.

Zhang, X. and Cui, L. (2016). Oxygen requirements for the Cambrian explosion. *Journal of Earth Science* 27: 187–195.

Zhang, S., Wang, X., Wang, H. et al. (2016). Sufficient oxygen for animal respiration 1,400 million years ago. *Proceedings of the National Academy of Sciences of the United States of America* 113: 1731–1736.

Zhang, K., Zhu, X., Wood, R.A. et al. (2018). Oxygenation of the Mesoproterozoic Ocean and the evolution of complex eukaryotes. *Nature Geoscience* 11: 345–350.

Zhu, S., Zhu, M., Knoll, A.H. et al. (2016). Decimetre-scale multicellular eukaryotes from the 1.56-billion-year-old Gaoyuzhuang formation in North China. *Nature Communications* 7: 11500.

Further Reading

Atreya, S.K., Pollack, J.B. and Matthews, M.S. (eds) (1989). *Origin and Evolution of Planetary and Satellite Atmospheres*. University of Arizona Press, Tucson, AZ.

Beerling, D. (2007). *The Emerald Planet: How Plants Changed Earth's History*. Oxford University Press, Oxford, UK.

Catling, D.C. and Kasting, J.F. (2017). *Atmospheric Evolution on Inhabited and Lifeless Worlds*. Cambridge University Press, Cambridge, UK. ISBN: 9780521844123.

Gasol, J.M. and Kirchman, D.L. (2018). *Microbial Ecology of the Oceans*, 3e. Wiley Blackwell, Hoboken, NJ.

Lammer, H., Zerkle, A.L., Gebauer, S., et al. (2018). Origin and evolution of the atmospheres of early Venus, Earth and Mars. *The Astronomy and Astrophysics Review* 26 (2) doi: https://doi.org/10.1007/s00159-018-0108-y.

Lenton, T.M. (2003). *Evolution on Planet Earth: The Impact of the Physical Environment*, Pp. 35–53. Academic Press, Cambridge, MA.

Lenton, T.M. and Daines, S.J. (2017). Matworld – the biogeochemical effects of early life on land. *New Phytologist*, 215, 531–537.

Lewis, J.S. (1997). *Physics and Chemistry of the Solar System* (revised edition). Academic Press, San Diego, CA.

Lewis, J.S. and Prinn, R.G. (1984). *Planets and their Atmospheres: Origin and Evolution*. Academic Press, Orlando, FL.

Rollinson, H. (2007). *Early Earth Systems*. Blackwell, Oxford, UK.

3

Atmospheric Energy and the Structure of the Atmosphere

Hugh Coe

School of Earth, Atmospheric, and Environmental Sciences, The University of Manchester, Manchester, United Kingdom

The Sun provides a massive 5×10^{24} J year^{-1} energy input to the Earth, its oceans and atmosphere; compared with the internal energy of the Earth, which generates around 10^{21} J year^{-1}. This solar input is responsible for driving atmospheric circulation, maintaining the temperature structure of the atmosphere and evaporating water into the atmosphere to initiate the hydrological cycle as well as initiating many chemical processes. It is also central to photosynthesis, the process by which the biosphere reduces carbon dioxide (CO_2) to carbohydrates. It is therefore important to understand the radiation balance of the Earth and the atmosphere because the transfer of solar radiation underpins so many of the important processes in the atmosphere, biosphere, and oceans.

The average solar flux reaching the top of the Earth's atmosphere and the average temperature of the Earth vary by only fractions of a per cent from one year to the next, which indicates that although the Earth receives a huge amount of energy each year, it does not retain it and loses the same amount to space. The system is in balance and we must be able to understand the way this balance is maintained if we are to predict the effect future changes to parts of the system will have on the whole. What is clear is that human influences are significant, and are likely to be increasingly so in the future, causing increases in global temperature with significant regional variations.

3.1 The Vertical Structure of Earth's Atmosphere

Figure 3.1 shows the variability of temperature and density with altitude and although there is considerable variability from day to day, seasonally and from one location to another, the main features in the profile are typical of the vertical structure of the atmosphere. The pressure profile of the atmosphere can be calculated from the change in pressure, dp, experienced in a small change of height, dz:

$$dp = g\rho dz \tag{3.1}$$

Atmospheric Science for Environmental Scientists, Second Edition. Edited by C.N. Hewitt and Andrea V. Jackson.
© 2020 John Wiley & Sons Ltd. Published 2020 by John Wiley & Sons Ltd.

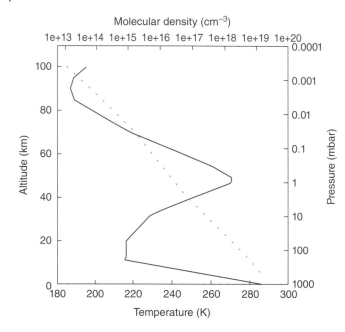

Figure 3.1 The average vertical temperature and molecular density structure of the Earth's atmosphere. The data are taken from the USA standard atmosphere and represent a time and spatial average, a local instantaneous sounding will vary considerably from this profile.

where, g is the acceleration due to gravity and ρ is the density of air. The acceleration due to gravity can be treated as constant as the atmosphere is thin with respect to the radius of the Earth, but density clearly varies with altitude. Air behaves more-or-less as an ideal gas so:

$$\rho = \frac{Mp}{RT},$$ (3.2)

where M is the molar mass of air, R is the gas constant ($8.314\,\mathrm{J\,K^{-1}\,mol^{-1}}$), and T is the temperature in Kelvin. Substituting and rearranging gives:

$$\frac{dp}{p} = -\frac{gM}{RT}dz$$ (3.3)

which can be integrated from the surface ($z = 0$, $p = p_0$):

$$p = p_0 \exp \int_0^z \frac{dz}{(RT/gM)}$$ (3.4)

This is known as the *hydrostatic equation*. The denominator, RT/gM, has units of length and is often referred to as the *scale height*, the vertical distance over which the pressure falls to $1/e$ of its initial value.

As the temperature does not remain constant throughout the atmosphere so the scale height also changes. This analysis assumes that the atmosphere is composed of a gas of single molar mass, M, but in reality the atmosphere is composed of several gases and so we might expect a different scale height for each gas at any altitude. If this was the case, then the composition of the atmosphere in the absence of sources and sinks would vary with height as the heavier molecules have smaller scale heights. This is not observed below 100 km, as the mean free path of molecules, or molecular mixing length, is much smaller than the macroscopic mixing length resulting from turbulence and convection. As a result, macroscopic mixing processes act equally on all molecules and dominate the molecules' specific diffusion processes. Above 100 km this is not the case, and molecular separation with height is observed. The mean molar mass of air in the lower part of the atmosphere is determined by the mix of molecular nitrogen and oxygen and is 28.8 g.

The atmosphere can be subdivided into layers based on its thermal structure. Closest to the surface, the temperature reduces with height to a minimum at around 10 km, known as the *tropopause*. The region closest to the surface is known as the *troposphere* (*tropos*–Greek for 'turning') and at the surface the temperature varies from minima of around −50 °C at the wintertime poles to maxima of 40 °C over the continents close to the Equator. The temperature in the troposphere decreases by, on average 6.5 K km^{-1} from the surface to the tropopause. The *stratosphere* (*stratus* – Greek for 'layered') is a region between 10 and 50 km in which the temperature profile of the atmosphere increases to a maximum at the *stratopause*. This inversion is caused, as we shall see later, by absorption of solar ultraviolet (UV) radiation by a layer of ozone (O_3). Increasing temperature with height suppresses vertical mixing through the stratosphere and so causes its layered structure.

Above 50 km, warming by UV absorption can no longer compete with the cooling processes and temperatures begin to decrease. This region is called the *mesosphere* and extends to around 90 km, where the atmospheric temperature reaches a second minimum, the *mesopause*. However, unlike the troposphere, where the rate of decrease of temperature with height, the *lapse rate*, is sufficient for convection to occur, the mesospheric lapse rate is only around 2.75 K km^{-1} and the layer remains stable. Above the mesopause, the temperature again increases through the so-called *thermosphere*. At these altitudes, the air becomes so thin that molecular collisions become infrequent. Thus, atomic species with high translational energy cannot redistribute that energy to highly excited vibrational and rotational states in molecular species. The high temperatures in the thermosphere do not reflect a large energy source but, rather, the inability of the thin atmosphere at these altitudes to lose energy via radiative transfer.

3.2 Solar and Terrestrial Radiation

3.2.1 Solar Radiation

The Sun is composed of approximately 75% hydrogen and 25% helium, and its energy is derived from the fusion of hydrogen into helium nuclei, which are then transferred to the surface via shortwave electromagnetic (EM) radiation. Although the Sun has a radius of 7.0×10^5 km, virtually all the energy received by the Earth is emitted by the outer 500 km, known as the *photosphere*, which emits light across the entire EM spectrum, from gamma rays to radio waves. Most of the radiative power incident at the top of the Earth's atmosphere is due to light of wavelength between 200 nm (ultraviolet) to 4 μm (infrared), with a peak intensity at about 490 nm (green part of the visible region of the spectrum).

The Sun has a temperature of approximately 5800 K and can be thought of as a *blackbody emitter* at this temperature (emits the maximum amount of radiation possible at each wavelength at a given temperature).

Planck related the emissive power, or intensity, of a blackbody $B(\lambda, T)$ at a given wavelength, λ, to the temperature, T, of the emitter by

$$B\left(\lambda,\, T\right) = \frac{2hc^2}{\left(\lambda^5 \exp^{hc/k\lambda T} - 1\right)} \tag{3.5}$$

where k is the Boltzmann constant $(1.381 \times 10^{-23}\, \mathrm{J\,K^{-1}})$, c is the speed of light in vacuum $(2.998 \times 10^8\, \mathrm{m\,s^{-1}})$, h is Planck's constant $(6.626 \times 10^{34}\, \mathrm{J\,s})$, and the irradiance of the Sun as a function of wavelength and the blackbody curve for an emitter at a temperature of 5800 K can be seen in Figure 3.2.

The total flux emitted by a blackbody radiator, F_B, and the total emitted intensity B, can be found by integrating the Planck blackbody function (3.5) over all wavelengths:

$$F_B = \pi B = \pi \int_0^\infty B\left(\lambda,\, T\right) d\lambda = \sigma T^4 \tag{3.6}$$

where σ is the Stefan–Boltzmann constant $(5.671 \times 10^{-8}\, \mathrm{W\,m^{-2}\,K^{-4}})$.

Solar radiation has an average intensity of approximately $1370\, \mathrm{W\,m^{-2}}$ at the distance of the Earth from the Sun, often referred to as the solar constant, S, but varies with time on a wide variety of scales. The Sun rotates with a period of 27 days and both active, brighter regions known as *faculae* and less active, darker regions known as *sunspots* face the Earth during each rotation. The output from these different regions of the Sun varies by between 0.1 and 0.3% of the total flux. The number of sunspots on the surface of the Sun varies with a cycle of 11 years, causing variations in radiative flux at the top of the Earth's atmosphere of the order of 1%. Lower frequency variations in the solar flux, again of the order of 1–2%, have also been inferred from isotopic abundance measurements.

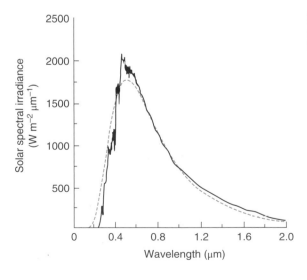

Figure 3.2 The spectral irradiance from the Sun compared with that of a blackbody at 5777 K. *Source:* taken from Iqbal 1983.

3.2.2   Terrestrial Radiation

The Earth also acts as a blackbody radiator, but as its global mean surface temperature, T_s, is 288 K, most of the irradiance from the Earth is in the infrared part of the spectrum and peaks at about 10 μm. Figure 3.3a shows a blackbody curve for an emitter of temperature 288 K compared with one at 5777 K, representing the solar spectrum. There is very little overlap between the incoming solar radiation at UV and visible wavelengths and the outgoing infrared radiation from Earth's surface. Thus, incoming solar radiation and outgoing terrestrial radiation are distinct from one another, separated by a gap at around 4 μm, and are often referred to as shortwave (SW) and longwave (LW) radiation, respectively.

Figure 3.3   (a) The blackbody curves for 5777 K and 280 K, representing the solar photosphere and the Earth's surface. The Sun emits mainly in the visible, whereas the Earth emits predominantly in the infrared. The so-called incoming shortwave and outgoing longwave radiation is separated by a gap at around 4 μm. (b and c) The fractional absorption of radiation from the top of the atmosphere to 10 km and the surface of Earth respectively. The main absorbers in each wavelength region are indicated in (c). Most absorption of longwave outgoing radiation occurs in the troposphere and is due principally to water vapour and CO_2. However, note that the strong absorption band of O_3 that occurs at 9.6 μm in the centre of the so-called atmospheric window is due mainly to stratospheric absorption.

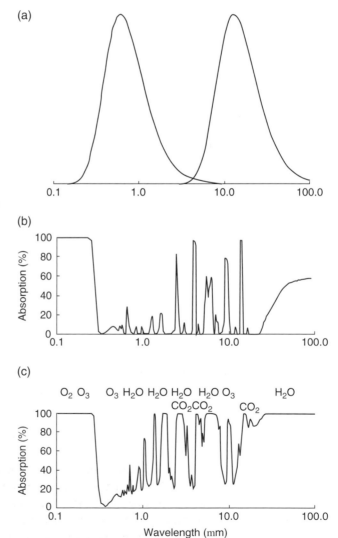

As the mean surface temperature of the Earth changes little from year to year and has varied by less than 5 °C in the past 20 000 years it is clear that the system is in equilibrium and the energy inputs must be balanced by energy losses. The effective area of the Earth receiving sunlight at any one time is given by πR^2, where R is the radius of the Earth, yet the total area of the Earth is $4\pi R^2$, so the average radiant flux over the Earth is given by $S/4$. However, not all of the incoming radiation is absorbed by the surface, some is reflected back to space by either the surface, clouds, aerosol particles, or scattering from molecules in the atmosphere. The fractional reflectance is known as the global mean planetary reflectance or *albedo*, A. The average surface albedo is around 0.15 but the high reflectivity of clouds leads to an overall planetary albedo, A, of 0.3. The incoming irradiance absorbed by the Earth's surface, F_s, is given by

$$F_s = (1 - A)\frac{S}{4} \tag{3.7}$$

and has a value of 240 W m^{-2}.

The incoming irradiance absorbed, F_s, must be balanced by the outgoing blackbody radiation of the Earth, given by σT_e^4; where T_e is the effective blackbody temperature of the Earth–atmosphere system. Equating incoming and outgoing fluxes gives an expression for T_e

$$T_e = \left[\frac{(1 - A)S}{4\sigma} \right]^{1/4} \tag{3.8}$$

that yields an equilibrium temperature of 255 K, compared with 288 K, the average surface temperature of the Earth. The fact that the Earth's surface is 33 K warmer than predicted by this simple calculation suggests that other processes act to offset the loss of heat by longwave cooling. Even if the albedo is halved to completely remove the contribution of clouds to the planetary albedo the equilibrium temperature only rises to 268 K. To understand why the atmosphere is warmer than predicted by the simple calculation above, the interaction between trace constituents in the atmosphere and incoming and outgoing radiation needs to be examined.

Worked Example 3.1

1 Calculate the percentage change in the average solar radiation incident on top of the Earth's atmosphere when the Earth is closest to the Sun (*perihelion*) and when it is furthest from the Sun (*aphelion*), compared with the solar radiation at the mean Earth–Sun distance. The Earth–Sun distances at perihelion and aphelion are 147×10^6 km and 152×10^6 km, whereas the mean distance is 149.6×10^6 km.

Answer

This can be easily calculated by appreciating that the radiation is inversely proportional to the inverse of the square of the radius. The ratios are:

Perihelion: $\left(\dfrac{149.6}{147} \right)^2$, a 3.6% increase

Aphelion: $\left(\dfrac{147}{152} \right)^2$, a 3.1% decrease

2 Calculate the change in the radiative equilibrium temperature of the Earth that would result from such changes in the incident radiation, assuming an equilibrium temperature of 256 K.

Answer

This can be solved by calculating the solar constant at each of the positions in the Earth's orbit and finding the ratios. A more elegant approach uses Eq. (3.8):

$$T_e = \left[\frac{(1-A)S}{4\sigma}\right]^{1/4}$$

and differentiating produces:

$$\frac{dT_e}{dS} = \frac{1}{4}\frac{(1-A)}{4\sigma}\left[\frac{(1-A)S}{4\sigma}\right]^{-3/4}$$

Dividing both sides by Eq. (3.8) once more:

$$\frac{1}{T_e}\frac{dT_e}{dS} = \frac{1}{4}\frac{(1-A)}{4\sigma}\left[\frac{(1-A)S}{4\sigma}\right]^{-3/4}\left[\frac{(1-A)S}{4\sigma}\right]^{-1/4} = \frac{1}{4S}$$

$$\frac{dT_e}{T_e} = \frac{1}{4}\frac{dS}{S}$$

if $T_e = 256$ K, this gives a 2.3 K increase in temperature at perihelion and a 2.0 K decrease at aphelion.

3.2.3 Absorption of Radiation by Trace Gases

It has so far been assumed that the atmosphere acts simply to scatter and reflect incoming shortwave radiation and does not absorb light. However, this is not the case. The atmosphere interacts with both incoming solar radiation and outgoing terrestrial radiation, and the strength of the interaction as a function of wavelength is responsible for the heating of the lower atmosphere.

Figure 3.3 illustrates the effect of absorption by trace gases in the atmosphere on the transmission of incoming shortwave radiation from the Sun and outgoing longwave radiation from the Earth. Figure 3.3a shows the blackbody curves for emitters at 5777 K and 280 K, respectively representing the solar and terrestrial emission spectra. Figures 3.3b and c show the fraction of light entering the top of the Earth's atmosphere that is absorbed before reaching 10 km and sea level respectively, as a function of wavelength. At 10 km, the top of the troposphere, virtually all radiation below 290 nm has been absorbed. All radiation below 100 nm is absorbed in the thermosphere above 100 km. Molecular oxygen absorbs strongly at wavelengths between 100 and 200 nm and also in a weaker band between 200 and 245 nm. The oxygen absorptions appreciably attenuate incoming UV radiation of wavelengths <200 nm above an altitude of 50 km. Light of wavelengths between 200 and 300 nm is strongly absorbed in the stratosphere by the oxygen trimer, ozone, and transmission of radiation of wavelengths <290 nm is negligible below 10 km. Between 300 and 800 nm the stratosphere is only weakly absorbing and most of the solar radiation at these wavelengths is transmitted into the troposphere. A comparison of Figure 3.3b and c shows that there is little further absorption in the troposphere at wavelengths below 600 nm, but H_2O and CO_2 absorption bands, whose abundances are dominated by their tropospheric concentrations, deplete the near infrared part of the incoming solar flux appreciably. As a result, the solar irradiance at the surface is dominated by visible wavelengths.

The interaction between the outgoing longwave radiation and the atmosphere can also be seen in Figure 3.3b and c and compared with an irradiance spectrum of a blackbody emitter of temperature 288 K, representing the radiation emitted from the surface of the Earth in Figure 3.3a. Several different molecules are efficient absorbers of infrared radiation, and many of these are most abundant in the troposphere. Consequently, much of the outgoing radiation is absorbed in the lowest 10 km of the atmosphere. Much of the outgoing radiation of wavelengths less than 7 μm is absorbed by water vapour, with some contribution from methane and nitrous oxide, N_2O. Light of wavelengths longer than 13 μm is efficiently absorbed by CO_2, whose absorption band is centred at 15 μm. This band is particularly important as it lies close to the maximum of the longwave irradiance spectrum. At longer wavelengths water vapour is excited into many rotational states that effectively form an absorption continuum beyond 25 μm. Minor absorbers between the CO_2 and water bands are mainly N_2O and CH_4. The only fraction of the outgoing radiation that is transmitted through the troposphere without undergoing appreciable absorption lies in the so-called *atmospheric window* between 7 and 13 μm.

The only significant absorptions of infrared radiation in the stratosphere are due to ozone. The 9.6 μm band of ozone happens to lie in the middle of the atmospheric window and as a result means that stratospheric ozone plays a significant role in the outgoing longwave radiation budget of the Earth.

The absorption of radiation by gases in the atmosphere is complex and mainly involves several trace gas species rather than major constituents. These interactions are key to the chemical composition, thermal structure, and radiative balance of our atmosphere.

3.3   Solar Radiation, Ozone, and the Stratospheric Temperature Profile

As discussed above, ozone is a very efficient absorber of solar radiation between 200 and 300 nm, and its formation and presence control the stratospheric temperature profile. Figure 3.4 shows an average vertical profile of ozone through the atmosphere compared to an average temperature profile. The main layer of ozone in the atmosphere is situated between 15 and 30 km and reaches a maximum concentration of around 5×10^{12} molecules cm^{-3} at 22 km; however, the maximum temperature at the top of the stratosphere occurs at around 50 km, well above the main ozone layer.

Ozone is formed from the photo-dissociation of molecular oxygen but is itself removed by *photodissociation* (the fragmentation of a molecule as a result of its absorption of a photon that is energetic enough to break its molecular bonds). Both O_2 and O_3 absorb UV light very strongly and prevent highly energetic radiation penetrating to lower altitudes (see Figure 3.5). The peak in the absorption cross section of O_2 occurs in the Schmann-Runge continuum at around 145 nm, as shown in Figure 3.6. At wavelengths less than 175 nm O_2 is dissociated into two oxygen atoms, one of which is electronically excited. This strong absorption prevents sunlight of wavelengths below 175 nm from penetrating below around 70 km (Figure 3.5). The oxygen atoms formed as a result of O_2 photolysis react with other molecules

Figure 3.4 The average vertical profile of ozone and temperature through the atmosphere. The ozone profile is represented as both a mixing ratio and a molecular concentration. Data from the US Air Force Geophysics Laboratory (AFGL) standard ozone and temperature profiles.

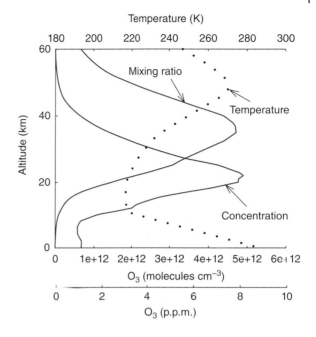

Figure 3.5 The extent that ultraviolet (UV) solar radiation penetrates through the atmosphere as a function of wavelength. The penetration altitude is the altitude at which the initial intensity at any wavelength is attenuated to e^{-1} of its original intensity. *Source:* from Salby 1996.

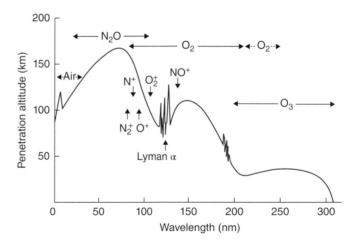

of O_2 to form ozone. Figure 3.7 shows the strong absorption cross sections of ozone occurring between 240 and 300 nm with a maximum value of 1.1×10^{-17} cm^2 at 255 nm. As a result, above 60 km ozone is photolysed very efficiently back to O_2 and atomic oxygen, reducing its concentration and favouring the existence of atomic oxygen.

The Herzberg continuum between 200 and 240 nm is responsible for photolysis of O_2 below 60 km because the shorter wavelengths have already been removed (Figure 3.5), whilst radiation of these wavelengths penetrates down to around 20 km.

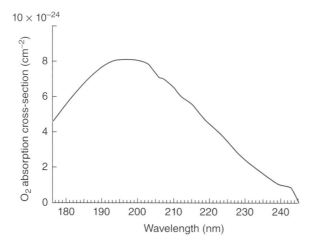

Figure 3.6 The absorption cross-section of molecular oxygen as a function of wavelength. *Source:* from Brasseur and Solomon 1986.

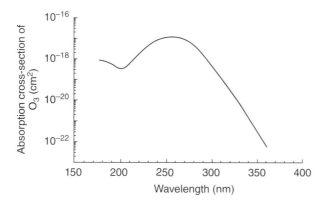

Figure 3.7 The absorption cross-section of ozone at 273 K as a function of wavelength between 170 and 360 nm. *Source:* taken from Seinfeld and Pandis 1998.

Little O_3 is produced high up in the atmosphere as the air density is low and there is little O_2 to be photolysed or to subsequently react with the atomic oxygen formed by its photolysis. Density increases with descent through the atmosphere, favouring ozone formation via the combination of atomic and molecular oxygen, and the concentration of ozone increases to a maximum at around 20 to 25 km. Lower in the stratosphere the overhead ozone column is significant and absorbs much of the radiation between 200 and 290 nm, thus limiting photolysis of oxygen and slowing the rate of ozone formation. The concentration of ozone reduces and reaches a minimum by the tropopause where radiation of less than 290 nm is almost completely removed.

The absorption of UV radiation by both oxygen and ozone leads to their photolysis and the energy involved in these sunlight-induced reactions produces local warming. The temperature at a particular altitude will then be a combination of the rates of photolysis of the two oxygen species, in particular ozone, and the air density. The rates of photolysis will depend on the local incidence of radiation and thus on the optical density of the atmosphere in the column above at a given wavelength. This, in turn, will be dependent on the overhead concentration profile of O_2 and O_3 themselves. As the air density increases, any products of photochemical processes that remain energetically excited are deactivated more rapidly via an increased chance of collisions, leading to an increase in temperature. Although the temperature profile is strongly linked to that of ozone, its maximum occurs not at the maximum ozone concentration but above it and close to the region where the photolytic formation and loss processes of ozone are most rapid.

3.4 Trapping of Longwave Radiation

Incoming visible and UV radiation from the Sun is energetic enough to excite electrons within certain optically active molecules and in the cases of ozone and molecular oxygen the photon energy is sufficient to fragment the molecule and cause its photolysis. Less energetic outgoing photons of infrared wavelengths induce vibrational and rotational excitations of molecules. These excitations do not cause chemical changes in the absorbing molecule, instead the excited molecule, below 100 km at least, is rapidly deactivated by collisions and the energy absorbed from the original photon is distributed thermally.

The effect as a result of these interactions on a layer of the atmosphere will be that some fraction of the outgoing longwave radiation entering the base is absorbed by molecules such as CO_2, H_2O, and CH_4 in the layer (see Figure 3.3c). The absorbed energy is transferred to kinetic energy by collisions between the absorbing molecules and others in the layer. The layer will itself act as a blackbody and re-radiate infrared radiation; however, the layer will radiate uniformly in all directions and so acts to increase the longwave flux through the lower layers of the atmosphere. This process raises the local temperature in the lower layers of the atmosphere above that predicted from a straightforward surface budget calculation of the kind described above.

3.5 A Simple Model of Radiation Transfer

Several gases in the atmosphere absorb strongly in the infrared part of the spectrum, and each has a complex absorption pattern made up of many different individual vibrational and rotational transitions. The way these different absorptions interact is not straightforward and should be accounted for in a detailed description of radiative transfer through the atmosphere as discussed in the next section.

However, a general picture of the processes taking place in the atmosphere can be shown by deriving a simple model of radiative transfer, known as a *Grey atmosphere*. This is based on a simplified atmosphere that it is transparent to incoming shortwave radiation, includes only one trace gas that absorbs uniformly at all infrared wavelengths, where the effect of scattering is removed, and where it is assumed that radiation is either emitted or absorbed in the vertical direction and that each level of the atmosphere is in local thermodynamic equilibrium.

The absorption of light by an absorbing species in the atmosphere must firstly be described. The intensity of light of wavelength λ, $I(\lambda)$, which passes through a depth dz of an absorber with number concentration, n, is reduced by an amount $dI(\lambda)$ given by:

$$dI(\lambda) = -I(\lambda)n\sigma(\lambda)dz = I(\lambda)d\chi \qquad (3.9)$$

where $\sigma(\lambda)$ is the absorption cross-section at wavelength λ and is constant for any given species, and χ is the *optical depth*. We can obtain the intensity of light transmitted a distance z through the absorber, $I_z(\lambda)$, by integrating (3.9):

$$I_z(\lambda) = I_0(\lambda)\exp\left\{-\int_0^z n\sigma(\lambda)dz\right\} \tag{3.10}$$

where $I_0(\lambda)$ is the initial intensity of light of wavelength λ. In cases where the concentration of the absorber is independent of the depth of the absorbing slab the above relation becomes the *Beer–Lambert law*:

$$I_z(\lambda) = I_0(\lambda)\exp\left\{-n\sigma(\lambda)z\right\} \tag{3.11}$$

This is not the case for a vertical slice through the atmosphere.

In our simplified model, the single species absorbs uniformly over all wavelengths so we can simplify (3.10) to give:

$$I_z = I_0 \exp\left\{-\int_0^z n\sigma dz\right\} = I_0 \exp\left\{-\int_{\chi_0}^0 d\chi\right\} \tag{3.12}$$

where χ_0 is by convention the optical depth at the base of the atmosphere.

Only the absorption of light has so far been considered; however, the layer will re-emit radiation as a blackbody in a similar way so the intensity of emitted radiation, B, must also be included, and assuming Kirchoff's law:

$$dI = -In\sigma dz + Bn\sigma dz = (I - B)d\chi \tag{3.13}$$

Furthermore, in any slice of the atmosphere there may be some downwelling longwave radiation arising from blackbody emission of the layers above so both the upwelling and downwelling radiative fluxes, $F^\uparrow$ and $F^\downarrow$, should be treated separately:

$$\frac{dF^\uparrow}{d\chi} = F^\uparrow - \pi B \text{ and } -\frac{dF^\downarrow}{d\chi} = F^\downarrow - \pi B \tag{3.14}$$

The net flux through a layer is given by $F = F^\uparrow - F^\downarrow$, the difference between the upwelling and downwelling radiation. As it has been assumed that the atmosphere is in local thermodynamic equilibrium the flux must not change with height and is therefore constant throughout the depth of the atmosphere:

$$\frac{dF}{d\chi} = \bar{F} - 2\pi B \text{ and } \frac{d\bar{F}}{d\chi} = F \tag{3.15}$$

where $\bar{F} = F^\uparrow + F^\downarrow$ is the total flux leaving one layer.

As F is constant, these expressions are easily integrated to give:

$$\bar{F} = 2\pi B \text{ and } \bar{F} = F\chi + \text{constant.} \tag{3.16}$$

The blackbody emission flux of the outermost layer of the Earth's atmosphere is given by πB_0, and this must be equal to half the total flux from this layer, $\overline{F}$. As there are no overlying layers to supply a contribution to $F^\downarrow$, $\overline{F} = F$ and so:

$$B = \frac{F}{2\pi}\chi + B_0 = \frac{F}{2\pi}(\chi + 1). \tag{3.17}$$

In this model, the blackbody emission decreases linearly with height from the surface to the top of the atmosphere and there is a constant difference between the up- and downwelling fluxes, i.e. $F = F^\uparrow - F^\downarrow$ is constant. Furthermore, as there is no heat gained or lost by the atmosphere, the upwelling longwave radiation leaving the top of the atmosphere must be equal to the solar radiation absorbed at the Earth's surface, F_s (Eq. 3.7), so at the top of the atmosphere:

$$F\uparrow = F = F_s. \tag{3.18}$$

Considering the boundary conditions at the surface, the upward flux of radiation emitted by the Earth at a temperature T_s, $\pi B(T_s)$, must be balanced with the downwelling short and longwave radiation:

$$\pi B(T_s) = F_s + F^\downarrow(\chi_s) \tag{3.19}$$

where χ_s is the optical depth of the lowest layer of the atmosphere. However,

$$\overline{F} - F = 2F^\downarrow = 2\pi B - F_s \tag{3.20}$$

which, when evaluated at the surface, gives

$$2\pi B(T_s) = \pi B(\chi_s) + F_s/2 \tag{3.21}$$

This expression implies that there is a temperature discontinuity between the surface and the cooler lowest layer of the atmosphere. Figure 3.8 shows schematically how the radiation fluxes vary

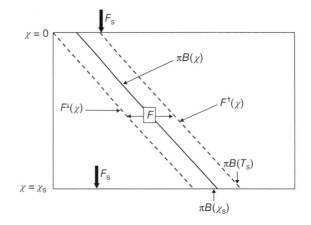

Figure 3.8 Schematic of the variation of upwelling and downwelling radiation and blackbody emission fluxes with optical depth through the atmosphere. The difference between the upwelling and downwelling radiation fluxes is constant with height and the model predicts a discontinuity in the blackbody emission flux at the surface.

with optical depth through the atmosphere and emphasizes the discontinuous blackbody emission flux at the surface.

If it is assumed that the absorber varies in concentration solely as a function of pressure, then its optical density in the atmosphere can be expressed as:

$$\chi = \chi_s \exp\{-z/H_s\} \tag{3.22}$$

where z is the height above the surface and H_s is the scale height in the surface layers, approximately 7 km. When the atmosphere is optically thin, $\chi < 1$, then radiation traverses the level with little interaction. When $\chi > 1$, radiation is absorbed efficiently within the layer and successive layers do not strongly interact. Choosing a value of $\chi_s = 1$ crudely illustrates how the model works. Optical depth, χ, can be evaluated, and using (3.17) we can derive B as a function height. However, from Eq. (3.6):

$$F_B = \pi B = \pi \int_0^\infty B(\lambda, T) d\lambda = \sigma T^4 \tag{3.23}$$

and the temperature at any level in the atmosphere can be calculated. The results of such a calculation are shown in Figure 3.9. The discontinuity in the temperature at the surface is again obvious, reducing from 282 K to 255 K above the surface. The temperature falls rapidly with height and tends to some limit, the so-called *skin temperature*, the temperature as z tends to infinity, $(F_s/2\sigma)^{1/4}$. If the atmosphere did not interact with outgoing longwave radiation ($\chi = 0$), the temperature of the atmosphere would be constant with altitude and be equal to the skin temperature.

Clearly the temperature in the troposphere does not vary in this way, however; the temperature profile in the lower stratosphere (Figure 3.1) is close to that described above. As will be discussed below, air in contact with the ground is heated by the surface, becomes buoyant, and initiates convection and the rate of decrease of temperature with height, or *lapse rate*, though most of the troposphere is determined by heat transport by convection rather than radiative transfer. The temperature is greater than that predicted by the radiative scheme to a height of 8 km, above which the radiative scheme predicts warmer

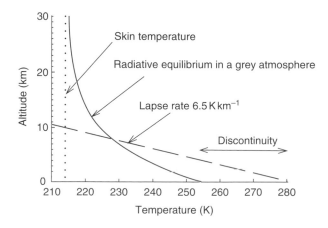

Figure 3.9 The temperature structure of the grey atmosphere. The temperature predicted by a model of radiative equilibrium in a grey atmosphere tends towards the skin temperature at high altitudes. At the surface there is a marked discontinuity. In reality the warm surface of the Earth heats the air immediately above and initiates convection, modelled by the average tropospheric lapse rate of 6.5 K km^{-1}.

temperatures than the lapse rate. So the model predicts a convective lower atmosphere that is turbulent and well mixed and a transition at around 8 km to a stable atmosphere whose temperature structure is controlled by radiative processes. The transition level in this model is a little lower than the observed tropopause but nevertheless this simple scheme predicts the broad temperature characteristics of the atmosphere below 20 km.

Worked Example 3.2

The measured effective surface temperatures (T_m) of the Earth and Venus are 250 K and 230 K, respectively, whereas the measured surface temperatures are 280 K and 750 K. Calculate the optical thickness of the atmospheres of the Earth and Venus.

Recall that $B(\chi_0) = \dfrac{F}{2\pi}(\chi_S + 1)$ (Eq. 3.17),

and $B(T_S) = B(\chi_s) + \dfrac{F_s}{2\pi}$ (Eq. 3.21), where

$B(T_s)$ and $B(\chi_0)$ are the blackbody emission fluxes at the surface (temperature T_s) and in the lowest layer of the atmosphere (with an optical depth χ_0); F is the difference between the upward and downward longwave fluxes and F_s is the incoming shortwave flux.

Answer

The net longwave flux and the incoming shortwave flux must balance:

$$F = F_s$$

So

$$B(T_S) = \frac{F_s}{2\pi}(\chi_S + 2)$$

The effective surface temperature T_m can be related to $B(\chi_s) = \sigma T^4{}_m$ and $F = \pi\sigma^4{}_m$. The blackbody flux from the surface where T_S is the measured surface temperature

$$B(T_S) = \sigma T_S^4$$

Substituting in

$$B(T_S) = \frac{F}{2\pi}(\chi_S + 2)$$

gives

$$\chi_0 = \left(\frac{T_S}{T_m}\right)^4 - 2$$

Using the values of T_S and T_m given, for Earth $\chi_S = 1.14$ and for Venus $\chi_S = 224$.

Worked Example 3.3

A type of glass is transparent to sunlight but absorbs 80% of radiation emitted from the ground. Show that in a greenhouse covered with this glass, the total radiant energy incident on the surface of the Earth is $5S/3$, where S is the incident flux of solar radiation.

Answer

Let the surface emit flux, F. Hence, the glass absorbs $0.8F$, and as the system is in equilibrium, the glass must re-emit this energy equally in the upwards and downwards direction, $0.4F$ either way. Hence, the total radiative flux downwards at the surface is $S+0.4F$. However, as we are in equilibrium, this must be balanced by the upward flux at the surface, F. Hence, $S+0.4F = F$ and so $S = 0.6F$, and the total downward flux must be $S+0.4(F/0.6) = 5S/3$.

3.6 Light Scattering

So far, absorption has been considered as the main mechanism directing light away from the direct beam. However, when light passes through a medium some of it is directed away from its direction of travel and any photons that are diverted from their direction of propagation are scattered. Whereas absorption leads to changes in local heating, scattering does not, but scattering is important as the atmospheric component of the planetary albedo. In the atmosphere, both scattering and absorption occur, and their combination is known as *extinction*.

If we have a medium containing a number of scattering elements, we can define a *scattering coefficient*, k_s, analogous to an absorption coefficient of $k_a = n\sigma$, where n is the number of absorbers per unit volume. The intensity of light of wavelength λ, after passing through a medium of thickness l, $I(\lambda)$, may be given by:

$$I(\lambda)=I_0(\lambda)\exp(-k_s l)=I_0(\lambda)\exp(-\chi) \tag{3.24}$$

where $I_0(\lambda)$ is the intensity of light entering the medium and $k_s l$ is the *optical depth*, χ.

When both scattering and absorption occur, the overall *extinction coefficient* can be defined as:

$$k_e = k_s + k_a \tag{3.25}$$

and the intensity can be found again using the Beer–Lambert law:

$$I(\lambda)=I_0(\lambda)\exp(-k_e l) \tag{3.26}$$

3.6.1 Rayleigh Scattering

When the scattering elements are small compared with the wavelength of light being scattered, as is the case when considering scattering by molecules, the treatment is simpler than in other cases. The analysis was first performed in 1871 by Lord Rayleigh to explain why the sky is blue. As the EM field of the incident light varies, the molecules and small particles are continuously redistributing their charges in response. The field causes the charges to oscillate at the frequency of the radiation, and these charges then re-radiate an EM field at the same frequency as the forcing field. However, the emitted radiation will not be propagated in the same direction as the forcing field and will not necessarily have the same polarization. A detailed treatment is beyond the scope of this chapter but can be found in Bohren and Huffman (1983). The intensity of scattered light, I, at a distance, r, and angle, θ, from a spherical particle of radius, a, and refractive index, m, is given by:

$$I = \frac{a^6}{r^2}\left(\frac{2\pi}{\lambda}\right)^4\left(\frac{m^2-1}{m^2+2}\right)\left(1+\cos^2\theta\right)I_0 \qquad (3.27)$$

where λ is the wavelength of light and I_0 is the initial intensity.

This result provides us with insights into the main physical effects of Rayleigh scattering. For example, the angular distribution of the scattered intensity varies from 2 in the forwards and backwards directions (where $\cos\theta = 0°$ and $180°$), to 1 at $90°$. The scattering intensity and energy removed from the incident beam is proportional to the sixth power of the radius (or for nearly spherical particles to the square of the volume). Lastly, the scattered intensity and energy removed from the incident beam is proportional to λ^{-4}.

These results lead to some readily observed effects. The sky is blue as blue light is scattered out of the direct solar beam much more efficiently than longer wavelengths. The horizon appears whiter than the zenith because the increased path leads to increased scattering. During sunrise and sunset, blue and yellow light are efficiently removed from the direct beam over the very long atmospheric paths at the high solar zenith angles, leading to spectacular red skies.

3.6.2 Mie Scattering

If the scatterers are large compared with the wavelength of light (e.g. raindrops), then geometric optics provide a good approximation. However, in most atmospheric situations aerosol particles and cloud droplets are neither large nor small enough to be treated simply. For the general case, the treatment is rather complex and is known as Mie scattering, after Gustav Mie who first solved the problem in 1908. In general, particles absorb as well as scatter and Mie gave solutions for the absorption, scattering, and extinction coefficients. An important parameter relating the scattering and absorption contributions to the extinction coefficients is the single scattering albedo, ω_0, which is defined as:

$$\omega_0 = \frac{k_s}{k_e} \qquad (3.28)$$

It is unitless and varies between 0 and 1, as the aerosol particles vary from entirely absorbing to entirely scattering. Marine aerosol particles, composed mostly of inorganic salts, often in solution, have values of ω_0 close to unity, whereas biomass burning and combustion aerosol particles close to source regions have considerable amounts of black carbon in them and ω_0 values as low as 0.5 have been observed. As the particles are transported in the atmosphere, nonabsorbing material can condense onto them, and their ω_0 increases to 0.9 and above with time.

3.6.3 Scattering Model of an Aerosol Particle Layer

The following model, shown diagrammatically in Figure 3.10, illustrates the importance of scattering by aerosol particles in the atmosphere and its impacts on the radiation budget in the atmospheric column.

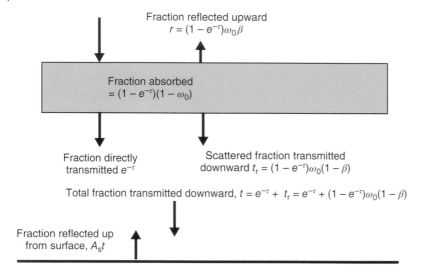

Figure 3.10 Schematic representation of an aerosol layer overlying a surface of albedo, A_s. The equations are detailed in the text.

Consider a direct solar beam impinging on an aerosol layer. Assume that the sun is directly overhead – at its zenith. The fraction of the incident beam that is transmitted directly through the layer is $e^{-\chi}$ (from Eq. 3.24). We can define an upscatter fraction, β, which is the fraction of the scattered radiation that is in the upward hemisphere relative to the local horizon. It is the upscatter fraction that is important in this context, not the backscatter relative to the incident beam, although the two are the same in our simple model. The nontransmitted fraction is therefore $1 - e^{-\chi}$, which is comprised of the scattered fraction, $\omega_0(1 - e^{-\chi})$, and the absorbed fraction $(1 - \omega_0)(1 - e^{-\chi})$. Of the scattered fraction, the fraction r_r is reflected upwards and t_r is transmitted downwards:

$$r_r = \omega_0\beta\left(1 - e^{-\chi}\right) \tag{3.29}$$

and

$$t_r = \omega_0\left(1 - \beta\right)\left(1 - e^{-\chi}\right) \tag{3.30}$$

The total fraction of radiation transmitted downwards is therefore

$$t_t = e^{-\chi} + t_t = e^{-\chi} + \omega_0\left(1 - \beta\right)\left(1 - e^{-\chi}\right) \tag{3.31}$$

If the albedo of the underlying surface is A_s, then the fraction of the radiation incident on the surface that is reflected is tA_s.

The total upward reflected shortwave flux is then:

$$\begin{aligned}F_r &= \left[r_r + A_s t_r^2\left(1 + r_r A_s + r_r^2 A_s^2 + r_r^3 A_s^3 + ...\right)F_0\right] \\ &= \left[r_r + t_r^2 A_s / \left(1 - A_s r_r\right)\right]F_0 = A_t F_0\end{aligned} \tag{3.32}$$

The series can be reduced to $1/(1-A_s r_r)$ using a Taylor expansion. With A_t the total reflectance of the aerosol–surface system, the change in the reflectance as a result of the presence of an aerosol layer is then:

$$\Delta A_a = A_t - A_s = \left[r_r + t_r^2 A_s / \left(1 - A_s r_r \right) \right] - A_s \tag{3.33}$$

We have so far assumed that the atmosphere itself does not scatter or absorb, yet we have already seen that this is not the case. To a first approximation, we can assume that the atmospheric extinction takes place above the aerosol layer; hence, the flux on top of the aerosol layer is, $T_a F_0$, not F_0. The total upward flux can now be written as:

$$T_a^2 F_r = \left[r_r + t_r^2 A_s / \left(1 - A_s r_r \right) \right] \left(T_a F_0 \right) T_a \tag{3.34}$$

The term $(T_a F_0)$ is the fractional transmittance to the top of the layer and the second T_a on the right-hand side is the fractional transmittance of the upwelling flux leaving the layer before leaving the atmosphere.

The change in the *outgoing* radiative flux as a result of an aerosol layer underlying an atmospheric layer is:

$$\Delta F = F_0 T_a^2 \left[r_r + t_r^2 A_s / \left(1 - A_s r_r \right) - A_s \right] \tag{3.35}$$

The quantity $\Delta F > 0$ if ΔA_a is positive. When interpreted in terms of climate forcing (see section below) we are concerned with the *incoming* flux so the change in forcing is $-\Delta F$, which depends on the following: the incident solar flux (F_0); the transmittance of the atmosphere (T_a); the albedo of the underlying surface (A_s); the single scattering albedo of the aerosol (ω_0); the upscatter fraction of the layer (β) and its optical depth (χ).

The problem is that ω_0 depends on the aerosol size distribution, its chemical composition, and is wavelength dependent; β depends on aerosol size (larger particles scatter more in the forward direction) and composition; and χ depends largely on the mass of the aerosol.

A striking example of the effect of an aerosol layer can be seen by applying this simple model to an important example of aerosol in a climatically important region. Biomass burning is widespread over southern Africa during the dry season. Large quantities of biomass are advected off the west coast of southern Africa over the South Atlantic Ocean. Global models, such as that described in Penner et al. (1992), have considered the aerosol as partially absorbing but essentially independent of other features in the model so, aerosol over a dark ocean increases scattering back to space. Even though the aerosol is partially absorbing, the sea surface is darker and the net effect is a cooling. Large negative numbers over the Gulf of Guinea show a predicted net cooling as more radiation is reflected to space than if the aerosol layer were not present. However, in that part of the world, a large sheet of low-lying stratocumulus cloud is semi-persistent. Such clouds have high albedos (0.8 is not uncommon) and as the biomass-burning aerosol is lofted into the atmosphere over the continent, it has been observed to exist above the cloud deck (e.g. Haywood et al. 2003; during the South African Regional Science Initiative [SAFARI] experiment). An example of this is shown in Figure 3.11. The simple model derived above can be used to demonstrate this by including typical values measured during the SAFARI campaign. The input

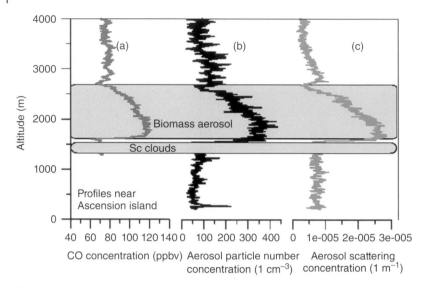

Figure 3.11 Vertical profiles of carbon monoxide, aerosol particle number concentration, and aerosol scattering coefficient measured near to Ascension Island using instruments on board the UK Meteorological Office C-130 aircraft during South African Regional Science Initiative (SAFARI) (Haywood et al. 2003). Note the presence of stratocumulus cloud underlying the biomass burning layer. There is a narrow gap between these two layers that is surprisingly clean.

Table 3.1 Typical values of key quantities measured during the South African Regional Science Initiative (SAFARI) experiment (Haywood et al. 2003) used to demonstrate the importance of the underlying surface when considering the role of partially absorbing aerosols. The data have been used to derive values of ΔA_a and ΔF from Eqs. (3.33) and (3.35).

Variable	Ocean	Land
Incident solar flux (F_0) [$\mathrm{W\,m^{-2}}$]	670	670
Transmittance of the atmosphere (T_a)	0.49	0.49
Albedo of the underlying surface (A_s)	0.1	0.8
Single scattering albedo of the aerosol (ω_0)	0.89	0.89
Upscatter fraction of the layer (β)	0.2	0.2
Optical depth (t)	0.45	0.45
ΔA_a	0.011	−0.014
$\Delta F\,[\mathrm{W\,m^{-2}}]$	7.3	−9.5

parameters for two cases where the underlying surface is sea and cloud are shown in Table 3.1. The flux for the midday sun at 60° solar zenith angle is represented by F_0, the albedo of the sea is 0.1, and that of cloud is approximately 0.8. The single scattering albedo and the optical depth used are from measurements made with the UK C130 aircraft during SAFARI (Haywood et al. 2003). The table also shows the change in the albedo and the flux resulting from the predictions in Eqs. (3.33 and 3.35).

The predictions in the absence of cloud show that there is a net loss of radiation to space from the top of the atmosphere. However, once the cloud is included, there is a net increase in radiation and a positive forcing. This is because the aerosol is now more absorbing that the underlying, highly reflective cloud and consequently more of the downwelling radiation is absorbed, rather than reflected, leading to a net warming. These simple results agree reasonably well with much more complex radiation modelling using satellite information and other constraints by Keil and Haywood (2003) and illustrate the importance of considering coupled behaviour in global climate models.

Worked Example 3.4

The optical depth and single scattering albedo of biomass burning aerosol have been measured from aircraft over Africa and show values of 0.45 and 0.87 respectively. Calculate the change in the shortwave flux at the top of the atmosphere assuming these aerosols are transported over the ocean. Use values of $\beta = 0.2$, $F_0 = 670\,\mathrm{W\,m^{-2}}$, $T_a = 0.49$, and $A_s = 0.07$. If the layer overlays an extensive low-lying cloud deck of albedo 0.85, show how this will affect your result. Explain why these results are so different.

Answer

From Eqs. (3.29) and (3.30), for an albedo of 0.07: $r_r = 0.063053$ and $t_r = 0.8898$, which gives $\Delta DF = 7.8\,\mathrm{W\,m^{-2}}$, calculated from Eq. (3.35). This is an increase in the flux upward at the top of the atmosphere and so a net cooling.

For an albedo of 0.85 r_r and t_r remain the same, but the surface albedo has changed: from Eq. (3.35) this gives $\Delta F = -12.2\,\mathrm{W\,m^{-2}}$. This is a decrease in the flux upward at the top of the atmosphere and therefore a net warming.

The reflected and transmitted fractions of radiation resulting from the aerosol layer r_r and t_r, remain unchanged after each interaction, it is the change in underlying albedo that affects the coupling between the layers. A more reflective surface allows a greater interaction with the aerosol. Over a highly absorbing surface the aerosol layer acts primarily to scatter, its absorbing properties are not as strong as the surface, producing a net cooling. However, over a highly reflective surface the aerosol absorbs significant amounts of radiation and so acts as a net warmer, compared to if it were not there at all.

Worked Example 3.5

A shallow marine stratocumulus cloud has a droplet concentration of 10 droplets $\mathrm{cm^{-3}}$. A pollution source raises the number of activated particles to 80 droplets $\mathrm{cm^{-3}}$. Assuming that the liquid water content (LWC) of the cloud remains unchanged and the clouds do not absorb radiation, calculate the fractional increase in the optical depth caused by the pollution source. The optical depth, χ, can be approximated by $\chi = 2\pi h r^2 N$, where r is the droplet radius and N is the cloud droplet number.

Answer

The mass of water in a single droplet is $4/3\pi r^3 N\rho_v$, so the LWC of the cloud, L_w, is $4/3\pi r^3 N\rho_v$, where ρ_v is the density of water.

In the background:

$$\chi_b = 2\pi h r_b^2 N_b \text{ and } L_b = 4/3\pi r_b^3 N_b \rho_v$$

In the polluted layer:

$$\chi_\pi = 2\pi h r_p^2 N_p \text{ and } L_p = 4/3\pi r_s^3 N_p \rho_v$$

The LWC of both clouds remains the same,

$$L_p = L_b$$

so

$$N_b/N_p = r_p^3/r_b^3 \text{ and } \chi_p/\chi_b = r_b/r_p$$

$$\chi_p/\chi_b = r_b/r_p = r_b/\left(r_b\left(N_b/N_p\right)^{1/3}\right) = \left(N_p/N_b\right)^{1/3} = 2$$

3.7   Conduction, Convection, and Sensible and Latent Heat

Although radiative transfer of energy is very important in the atmosphere, energy may also be transferred through *conduction* and *convection*. The process of conduction occurs by the transfer of kinetic energy from one molecule to an adjacent one, which is most efficient when the molecules are tightly constrained in solids and, especially, when there is a defined structure to the material, such as a metal. Gases, including air, have low thermal conductivities and so the atmosphere is a poor conductor of heat.

Convection occurs much more efficiently than conduction in fluids, as warmer parts of the mass can mix much more rapidly with cooler parts and transfer heat. This transfer of heat on the macroscale is far faster than transfer on the molecular scale and makes this process extremely important when considering heat transfer in the atmosphere. Heat is exchanged between the Earth's surface, which is radiatively heated, and the lowest layer of the atmosphere by conduction at the molecular level. The heating of the air causes density changes in the fluid and locally the air expands. This makes the warmed parcel more buoyant and may in itself cause the parcel to mix through the bulk of the air above, a process known as *free convection*.

However, the atmosphere is continually stirred by large-scale winds generated by pressure gradients and motion around and over mountain ranges, and as a result the air above becomes mixed. This process forces the heated air close to the ground to mix through the air and warm the whole air mass. Hence, this process is known as *forced convection*.

Convection then mixes parcels of warm and cold air together and so changes the temperature of the two parcels. The warm parcel loses heat as it cools and the colder parcel gains heat as it warms. Enthalpy, or specific heat, is transferred along this temperature gradient. The specific heat content of a parcel of air of unit mass is defined as $c_p T$, where c_p is the specific heat at constant pressure and T is the temperature of the parcel.

Energy may be transferred indirectly, without changing the temperature of the air parcel, through a change in phase of water in the atmosphere, otherwise known as *latent heat*. A large amount of heat is required to change liquid water to water vapour and the same amount of energy is released when water vapour condenses and a cloud forms. The release of energy during cloud formation has an effect on the temperature profile compared with the dry atmosphere. The *latent heat of vaporization, L*, is the energy required to convert 1 kg of liquid water to water vapour at the same temperature: at $0\,^{\circ}C\,L = 2.5 \pm 10^{6}\,J\,kg^{-1}$. The latent heat of melting is the energy required to melt 1 kg of ice to form liquid water. At $0\,^{\circ}$ C this is around $3.3 \pm 10^{5}\,J\,kg^{-1}$.

3.7.1 Sensible Heat and the Temperature Structure of the Dry Atmosphere

A rising parcel of air will expand because the pressure exerted on the parcel by the surrounding air reduces with height. The work done by the air parcel in expanding is at the expense of its own internal energy and so the temperature of the parcel falls. The effect of a small change in altitude on a dry air parcel of unit mass can be considered in terms of small perturbations to the pressure, dp, volume, dv, and temperature, dT, of that parcel using the ideal gas law ($pv = RT$):

$$(p + dp)(v + dv) = R(T + dT) \tag{3.36}$$

where R is the molar gas constant ($8.314\,J\,mol^{-1}\,K^{-1}$). Assuming the products of the increments are negligible:

$$pdv + vdp = RdT \tag{3.37}$$

and the work done by the parcel of gas at pressure p in expanding by a volume increment dv is:

$$dw = pdv \tag{3.38}$$

From the first law of thermodynamics we know that the quantity of heat, dQ, supplied to a unit mass of a gas is balanced by an increase in the internal energy of the gas, du, and the external work done by the gas, dw:

$$dQ = du + dw = du + pdv \tag{3.39}$$

We should now introduce the specific heats at constant pressure and volume, c_p and c_v, which are defined in the following way:

$$c_p = \left(\frac{dQ}{dT}\right)_p \text{ and } c_v = \left(\frac{dQ}{dT}\right)_v \tag{3.40}$$

Let us first consider the case when there is no volume change, dv = 0. In this case:

$$dQ = du \tag{3.41a}$$

and so from (3.40)

$$du = c_v dT \tag{3.41b}$$

In general, then:

$$dQ = c_v dT + pdv = c_v dT + RdT - vdp \tag{3.42}$$

We may now consider the particular case when there is no pressure change, $dp = 0$.

$$dQ = (c_v + R)dT \tag{3.43a}$$

and so from (3.40)

$$c_p = \left(\frac{dQ}{dT}\right)_p \tag{3.43b}$$

we have

$$dQ = c_p dT - v dp \tag{3.44}$$

In most situations in the troposphere, vertical motion of air is rapid enough to far outweigh any heat transfer by conduction or radiation. Under these circumstances there is no net exchange of heat and $dQ = 0$; such processes are known as *adiabatic* changes and

$$c_p dT = v dp = RT\left(\frac{dp}{p}\right) \tag{3.45a}$$

or

$$\frac{dT}{T} = \frac{R dp}{c_p p} \tag{3.45b}$$

On integrating, we can see that

$$\frac{T}{p^\kappa} = \text{constant}, \qquad \kappa = \frac{R}{c_p} = \frac{c_p - c_v}{c_p} = 0.288 \tag{3.46}$$

This sole dependence of the temperature of an air parcel on pressure during an adiabatic process means that we can find the temperature an air parcel would have if it was moved from some arbitrary pressure level to 1000 mbar adiabatically. This temperature is known as the *potential temperature*, θ, where

$$\theta = T\left(\frac{1000}{p}\right)^\kappa \tag{3.47}$$

If θ is constant with height, then the atmosphere is said to be in convective equilibrium and the fall of temperature with height, or *lapse rate*, is found by substituting the hydrostatic equation into (3.45). Remembering that we have considered a parcel of unit mass, so $v\rho = 1$:

$$c_p dT + g dz = 0 \tag{3.48}$$

or

$$\frac{dT}{dz} = -\frac{g}{c_p} = -\Gamma_d = -9.8\,\text{K km}^{-1} \tag{3.49}$$

where Γ_d is the *dry adiabatic lapse rate* and is the rate of reduction of temperature with height through the atmosphere assuming that the air is unsaturated and in convective equilibrium. That is to say, if a parcel of air is displaced vertically, its temperature at the new pressure will be the same as the surrounding air at that level. Although not true locally, on average the atmosphere would show a lapse rate close to this value as long as the air is dry and there is no contribution from phase transitions of water to the energy budget.

3.7.2   Stability of Dry Air

The measured lapse rate in the atmosphere, γ, is the observed rate of change of temperature with height, and this can be compared with the dry adiabatic lapse rate to investigate the likely extent of vertical motion of an air parcel in that layer. The two real environment temperature curves in Figure 3.12A and B have lapse rates that are greater and less than Γ_d, respectively.

First consider environment curve A. If a parcel of air at point O is displaced upwards, its temperature reduces along the dry adiabatic lapse rate and will therefore be higher than that of its surroundings. Since the pressure is the same, the density of the parcel must be less than that of its surroundings. The air parcel will then have positive buoyancy and will be accelerated upwards. Similarly, if the parcel is displaced downwards, then it becomes cooler and denser than its surroundings and sinks further. An atmosphere under these conditions is said to be *unstable*.

Conversely, when we consider a parcel of air at point O on environment curve B, an upward displacement of the air parcel along the dry adiabatic lapse rate curve will result in the temperature of the parcel displacement being cooler than that of its surroundings and have a greater density. Likewise, a downward displacement will result in the parcel being warmer and more buoyant than its surroundings. In both cases, the parcel is subjected to a restoring force that tends to return it back to point O. Vertical columns of air with temperature profiles similar to the environment curve B are said to be *stable*.

Figure 3.12 The lapse rates of the two environment curves A and B are respectively greater and less than the dry adiabatic lapse rate of 9.8 K km^{-1}. If an air parcel at O ascends adiabatically its temperature changes at a rate Γ_d. If the surrounding environment has a lapse rate greater than this, (A), then the ascending parcel is warmer than the air around it, it is buoyant and is unstable. If the lapse rate of the surrounding air is less than that of the ascending parcel, (B), then the displaced parcel is cooler than its surroundings and the air is colder.

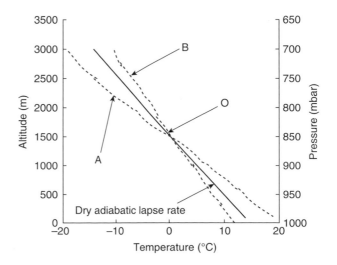

3.7.3 Latent Heat and the Effect of Clouds on the Vertical Temperature Structure

It has so far been assumed that the atmosphere is dry; however, water can change phase readily under atmospheric conditions and such changes of state produce large changes in the energy budget of the system.

The amount of water vapour in a parcel of air may be expressed as the mass mixing ratio, r_v, where $r_v = \rho_v/\rho_a$ and ρ_v and ρ_a are the densities of water vapour and dry air in the parcel. The mixing ratio is usually expressed in units of grammes of water per kilogramme of air, $g\,kg^{-1}$. The maximum amount of water vapour an air parcel can hold at any given temperature is given by the saturation mixing ratio, r_w. A further increase in water into an already saturated air parcel will lead to condensation. The saturation mixing ratio is solely a function of temperature: as the parcel warms its capacity to hold water vapour increases, and conversely at colder temperatures air can hold less moisture before condensation occurs.

The adiabatic changes experienced by a cloudy air parcel are shown schematically in Figure 3.13. Consider an unsaturated air parcel close to Earth's surface. As the air parcel rises it will cool and its saturation mixing ratio decreases. If there is no exchange between the air parcel and its environment the water vapour mixing ratio of the parcel remains constant. If the parcel continues to rise, it will eventually reach a level at which the ambient mixing ratio of the parcel is equal to the saturation mixing ratio. This point is known as the *condensation level* and is often observed by a clearly defined base to a layer of clouds. As the air parcel continues to rise the water vapour mixing ratio is at, or slightly above, r_w and so condensation occurs and heat is released into the air parcel. Some of the energy required to expand the parcel as it rises into a lower pressure environment is met by this energy release and so the parcel no longer needs to meet all of the work of expansion from its own internal energy. The effect of condensation within a cloud is therefore to offset some of the temperature reduction with height of a rising parcel under dry adiabatic conditions by releasing energy through the phase transition of water.

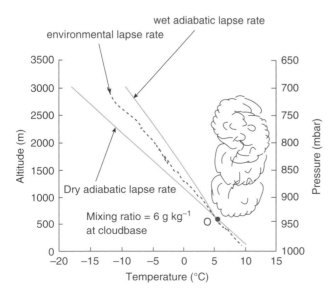

Figure 3.13 The effect of moisture on the atmospheric lapse rate. A parcel of air at O has a mixing ratio of $6\,g\,kg^{-1}$ of water vapour and so is just at saturation. As the air parcel rises adiabatically its temperature will decrease at a rate γ_s as some of the energy required to expand the parcel is supplied by the condensation of water vapour. If the environment curve is greater than γ_s and less than Γ_d the parcel is stable in dry air but becomes unstable when cloud is present, in other words conditionally unstable.

We can modify our mathematical description of the dry atmosphere to account for these changes in the following way. Unlike the dry atmosphere, where there is no change in heat of the parcel with height, heat is provided to the parcel by the mass of water condensed. If we are considering a parcel of unit mass, the latent heat released will supply a change in heat $dQ = -Ldr_v$, where dr_v is the change in mixing ratio of water vapour in the parcel. We can therefore modify Eq. (3.44) to give:

$$dQ = -Ldr_v = c_p dT - vdp \tag{3.50}$$

and by substituting the hydrostatic equation:

$$-Ldr_v = c_p dT + gdz. \tag{3.51}$$

As a result, we can see that the reduction in temperature with height of a cloudy air parcel under adiabatic conditions, the *saturated adiabatic lapse rate*, γ_s, is given by:

$$\gamma_s = -\frac{dT}{dz} = \frac{g}{c_p} + \frac{L}{c_p}\frac{dr_v}{dz} = \Gamma_d + \frac{L}{c_p}\frac{dr_v}{dz}. \tag{3.52}$$

As the cloudy air parcel ascends, the water vapour mixing ratio decreases with height as water is condensed onto cloud droplets so dr_v/dz is negative. This reduces γ_s below the dry adiabatic lapse rate, Γ_d, by an amount that is directly proportional to the mass of water vapour condensing in the rising air over a fixed height interval.

It is clear that as the amount of water vapour a parcel can hold at saturation is a strong function of temperature, the saturated adiabatic lapse rate will vary markedly, depending on the temperature of the air parcel. At high latitudes, or altitudes with very cold temperatures, an air parcel can hold very little water vapour at saturation. As a result, most of the energy for expansion of a rising cloudy air parcel must still come from the internal energy of the parcel and the lapse rate is little different from the parcel in dry conditions. However, at low latitudes air temperatures are higher and there is a considerable amount of water vapour held in the parcel at saturation. Under these circumstances, condensation may contribute significantly to the energy budget of the parcel. At high ambient temperatures it may be that γ_s may be as low as $0.35\,\Gamma_d$.

It should also be pointed out that the extremely low temperatures at the tropopause act as a cold trap. At these very low temperatures, air can sustain very little water in the vapour phase before reaching saturation. As water vapour is condensed and precipitates out of the parcel, the remaining parcel is very dry and little water vapour is transferred into the stratosphere.

Clearly, the average lapse rate of the troposphere will be less than the dry adiabatic lapse rate but will be considerably more than the $3.5\,K\,km^{-1}$ experienced in the lower levels of equatorial cumulonimbus clouds. As discussed above, the average of the lapse rate in the troposphere is around $6.5\,K\,km^{-1}$.

Worked Example 3.6

1 The average solar radiation incident at the top of Earth's atmosphere is $342\,W\,m^{-2}$. The global planetary albedo is 0.31 and $67\,W\,m^{-2}$ of the incident radiation is absorbed by the atmosphere. On average, $390\,W\,m^{-2}$ is lost from Earth's surface as infrared radiation but the Earth absorbs 83% of this amount

in the form of infrared radiation emitted downward by the atmosphere. If the global average sensible heat flux away from the surface is $24\,\mathrm{W\,m^{-2}}$, calculate the global latent heat flux.

Answer

Amount of shortwave radiation absorbed at the surface
$= 342 - (0.31 \pm 342) - 67 = 169\,\mathrm{W\,m^{-2}}$.
Net longwave loss $= 390(1-0.83) = 66.3\,\mathrm{W\,m^{-2}}$.
Latent heat flux $= 169-66.3-24 = 78.7\,\mathrm{W\,m^{-2}}$.

2 If 25% of all the surface water evaporated goes on to form ice, what is the mass of water precipitated globally in one year? The latent heat of fusion is $334\,\mathrm{kJ\,kg^{-1}}$ and the latent heat of vaporization is $2260\,\mathrm{kJ\,kg^{-1}}$ and the radius of the Earth is approximately $6400\,\mathrm{km}$.

Answer

The same amount of water must condense and be removed from the atmosphere each year through precipitation as is evaporated into it from the surfaces of the land and ocean.

Suppose M is the mass of water precipitated. Energy released by condensation of evaporated water to liquid water in one year is $2\,260\,000\,\mathrm{J}$. Energy released by freezing of 25% of this condensed water to ice in one year is $0.25\,M(334\,000)$. The sum of these two must be equal to the annual latent heat flux over the globe.

$$0.25M\left(334000\right)+M2\ 260000 = 78.7\times24\times60\times60\times365\times4\times3.14\times\left(6.4\times10^{6}\right)^{2}$$

Hence:

$$M = 5.45\times10^{17}\,\mathrm{kg}$$

3.7.4 Stability in Cloudy Air

It follows from the above that the rates of change of temperature of an ascending air parcel will be different in dry and cloudy air and that an air parcel will respond differently to small vertical perturbations in position depending on whether cloud is present in the air parcel. What is certainly true is that if an air parcel is unstable in dry air (see discussion in Section 3.8.2) then it must be unstable in cloudy conditions. Put another way, if the ambient temperature reduces with height more steeply than the dry adiabatic lapse rate, it must fall faster than the saturated lapse rate.

However, the converse clearly does not hold. Consider an air parcel at point O in Figure 3.13. The fall in ambient temperature with height is less than the dry adiabatic lapse rate and in dry air this will lead to any vertical motions of a parcel being suppressed as the air parcel is stable. However, if the temperature structure of the atmosphere is the same but the air parcel at O has a water vapour mixing ratio of $6\,\mathrm{g\,kg^{-1}}$, then the parcel at point O will be just saturated and so will be at cloudbase. The reduction in the adiabatic lapse rate caused by heat released during condensation is sufficient to reduce the saturated adiabatic lapse rate to less than the ambient temperature profile in the cloudy column above point O. Any vertical displacement of an air parcel subjected to these conditions will lead to it becoming more

Table 3.2 Stability criteria for a moist air parcel. The ambient reduction in temperature with height, γ, may be greater or less than the change in temperature with height induced by either a dry or a saturated adiabatic process, γ_d. or γ_s.

1	$\gamma < \gamma_s$	Absolutely stable
2	$\gamma = \gamma_s$	Saturated neutral
3	$\gamma_s > \gamma > \gamma_d$	Conditionally unstable
4	$\gamma = \gamma_d$	Dry neutral
5	$\gamma > \gamma_d$	Absolutely unstable

buoyant than the air around it and hence the parcel will be unstable. This is known as *conditional instability,* and a stable column of air that is forced to rise over orography or a frontal zone may cool until condensation occurs. This may release enough latent heat to make the air column positively buoyant and hence unstable.

Table 3.2 shows the five different stability criteria possible from absolutely unstable to absolutely stable. Neutral stability occurs in dry air when any vertical movement of an air parcel neither increases nor decreases its buoyancy relative to the surrounding air. The lapse rate under these conditions will be equal to Γ_d. Similarly, neutral stability can also occur in cloudy air, although of course the lapse rate under cloudy conditions will then be equal to γ_s.

3.8 Energy Budget for Earth's Atmosphere

The contributions of each of the above processes to the overall energy budget of the Earth–atmosphere system will now be considered, first in terms of the energy budget averaged over the whole globe and then the temporal variation across the globe.

3.8.1 Average Energy Budget

During the course of one year, around 5.5×10^{24} J of solar energy is received at the top of Earth's atmosphere in the form of visible and UV radiation. We will assume this input to be 100 units, as shown in Figure 3.14.

The average planetary albedo is approximately 0.3 and so 30% of the incoming shortwave radiation is reflected back to space with no interaction. Two-thirds of this reflection is from cloud tops, one-fifth is from molecular scattering in the atmosphere and the remainder is from reflection at the Earth's surface. Of the remaining 70 units of energy, 19 are absorbed by the atmosphere and 51 are absorbed by the Earth's surface.

Earth itself acts as a blackbody radiator and emits radiation from its surface, mainly at infrared wavelengths. Overall the number of units of energy lost from the surface as longwave radiation is 117, greater than the energy input from the Sun. However, of this large loss from the surface only 6 units escape to space directly whilst 111 units are absorbed by the atmosphere mainly by water vapour, CO_2, and clouds.

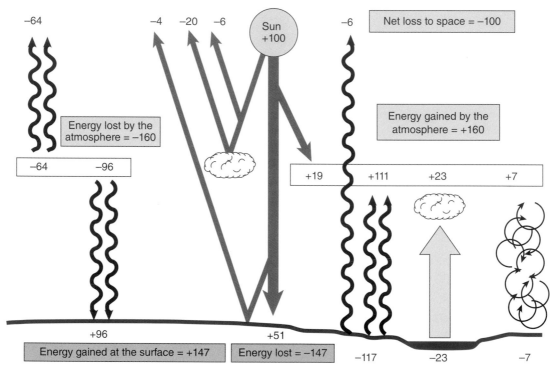

Figure 3.14 The average energy budget of the Earth–atmosphere system. The contributions from shortwave and longwave radiation to the energy budget are indicated by straight and wavy arrows respectively. One-fifth of the loss of energy from the surface is in the form of sensible and latent heat, the remainder is due to longwave radiation.

In addition to the absorption of short- and longwave radiation, the atmosphere gains a further 23 units from evaporation of surface water into the atmosphere in the hydrological cycle and also 7 units from sensible heat transfer by convective processes. In total the atmosphere gains 160 units of energy per year.

Clearly, the atmosphere is not undergoing a very large warming, and this energy gain is balanced overall by losing the same amount of energy in the form of longwave radiation. Of the 160 units lost by the atmosphere, 96 units are re-radiated back to the surface and 64 units are radiated out to space.

This longwave re-radiation by the atmosphere also balances the energy budget at Earth's surface, where in total 147 units of energy are received in the form of short- and longwave radiation and lost by longwave radiation, evaporation and convective transfer. Likewise, the incoming 100 units of solar energy are balanced by a loss of 100 units from reflection and longwave radiative emission from both Earth and its atmosphere. A review of the flow of energy through Earth's climate system can be found in Trenberth et al. (2009).

3.8.2 Variations in the Heat Budget Across the Globe

The energy budget is averaged over all latitudes and seasons. Clearly, the shortwave radiation input is not uniform over the entire globe. Low latitudes receive considerably more energy from solar radiation than higher latitudes. Furthermore, the albedo of the surface determines the fraction of sunlight reflected away from the surface and so affects the amount of shortwave radiation absorbed. For example, the albedo of fresh snow is very high (0.8–0.9), whereas that of the ocean is as low as 0.08.

Figure 3.15a shows the calculated short and longwave energy budgets of the Earth–atmosphere system averaged over time as a function of latitude. There is an excess of incoming shortwave radiation between 35° S and 40° N and a deficit at higher latitudes compared with the outgoing longwave radiation budget. If equilibrium were to be maintained at every latitude, the short- and longwave radiation should balance locally and the two curves in Figure 3.15a would be identical. The fact that they are not, and as local

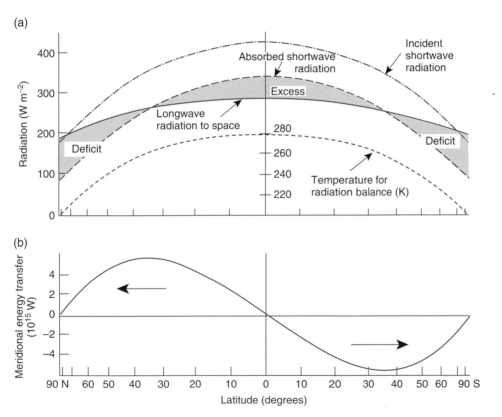

Figure 3.15 (a) The calculated short and longwave energy budgets of the Earth–atmosphere system averaged over time as a function of latitude. There is an excess of shortwave solar radiation at low latitudes relative to the long wavelength losses in the same region of the planet. This increase in heat flux is balanced by a net loss of radiation at latitudes above 40°. As the temperatures of the lower atmosphere in both the tropical and polar regions are approximately constant with time there must be a poleward transfer of energy, shown in (b). *Source:* taken from Wells 1997.

mean temperatures close to the Equator are not increasing with time and those close to the poles are not decreasing, means that heat energy must be transported from low latitudes poleward. This is achieved by circulation within both the ocean and the atmosphere, transporting heat away from the Equator towards the pole, and maintaining a higher temperature at latitudes greater than 50° than would be possible from a system in radiative equilibrium, illustrated by the thin dashed curve in Figure 3.15a.

The difference between the incoming shortwave and outgoing terrestrial longwave radiation is known as the *net radiation* and is shown as the shaded area in Figure 3.15a. Net radiation is positive close to the Equator indicating energy gain, and negative above 40°. The energy transfer from Equator to pole can be calculated for each hemisphere by integrating the net radiation from the Equator towards the pole. The result of this calculation is shown in Figure 3.15b. The energy transfer reaches a maximum around 40°, the latitude below which the Earth–atmosphere system on average gains energy. The energy is transported to higher latitudes where the system on average loses more energy than it receives. This heat pump supplies the energy required to drive the ocean circulation and global wind patterns in the atmosphere.

Detailed satellite measurements of net radiation can be used to deduce the size of the energy transfer from the Equator to the poles. However, this gives us no indication of the relative size of the contributions made to this transfer by the circulation of the ocean and atmosphere.

3.9   Aerosols, Clouds, and Climate

It is becoming increasingly apparent that the steady-state balance in the Earth's climate system has indeed changed markedly over the past 100 years, and recent improvements in global climate models are predicting such changes to become greater over the next 100 years with increasing certainty. It is not possible to provide a detailed discussion of climate change in this chapter, but several text and web references are provided for further reading (e.g. www.ipcc.ch).

Radiative forcing is a commonly used concept to link the change in a particular atmospheric component to a subsequent rise in temperature.

For example, assume a simple Earth–atmosphere system that is in equilibrium where the radiative flux in at the top of the atmosphere equals the flux leaving the atmosphere, and the net radiation, ΔI is zero. Now suppose that there is a sudden doubling of CO_2. This would reduce the longwave radiation, leaving the top of the atmosphere by an amount ΔI. To compensate for this, the top of the thermal energy within the Earth–atmosphere system must increase, increasing temperature of the surface by an amount, ΔT_s, and hence leading to greater upwelling longwave flux at the top of the atmosphere. This is the concept of radiative forcing.

In reality, radiative forcing is calculated using global climate models, which are run with and without a particular climate variable in them. The change in the radiation budget (often calculated at the tropopause) induced by the variable is linked to the temperature change. Radiative forcing is therefore a commonly used method of estimating the global mean surface temperature changes for a wide range of influences on climate.

Aerosols and clouds contribute to anthropogenic change in climate by directly acting to change the surface temperature directly via the global radiative budget – known as *forcings*; or by responding to changes induced by forcing agents that rapidly changing internal energy flows in the system, for example altering cloud cover, − so-called *rapid adjustments*; or by changing in response to changes in surface temperature and hence amplifying or dampening the change – so called *feedbacks*. Most recently, these effects and responses have been combined together and quantified as an *Effective Radiative Forcing*.

3.9.1 Aerosol Radiation Interactions

Several different types of anthropogenic aerosol are considered to be radiatively important: sulphate, nitrate (Martin et al. 2004), fossil-fuel-combustion-derived organic (e.g. Maria et al. 2004) and black carbon (Menon et al. 2002), biomass burning (Osborne et al. 2004; Thornhill et al. 2018), and mineral dust (Tegen et al. 1996; Highwood et al. 2003; Haywood 2011), although sulphate aerosol has been treated in significantly more detail than the other types (Boucher and Pham 2002). Most aerosol types scatter radiation at visible wavelengths but absorb little or no radiation at these wavelengths. However, black carbon, emitted directly from all sources of combustion such as traffic, coal-fired power generation, cookstoves, and wildfires absorbs visible radiation very efficiently and hence can locally warm the atmosphere (Bond et al. 2013). Although a considerable amount of dust occurs naturally, a significant amount results from land-use changes exposing soils.

The key parameter for determining whether an aerosol will warm or cool climate is the single scattering albedo, ω_0. Both scattering and absorbing aerosols reduce the solar radiation reaching the surface, thereby tending to cause local surface cooling; however, absorbing aerosols may warm the atmosphere itself, altering temperature and relative humidity profiles. Aerosols are far more spatially heterogeneous than the radiatively active gases, and their vertical profile is also often highly structured. This, as we have seen in Section 3.6, can greatly affect their local effect on the surface radiation budget and the local heating in the column. This is exacerbated if multiple aerosol layers are present. When absorbing aerosols coincide with clouds, local heating may evaporate the cloud. However, if black carbon overlays a low-level stratiform cloud layer, the heating aloft may strengthen the inversion at boundary layer top and reinforce the cloud. The overall impact of the aerosol on a regional scale is therefore considerably larger than the radiative forcing value suggests. At the present time, the impact of such heterogeneities on regional climate is very poorly understood.

3.9.2 Aerosol Cloud Interactions

Changes in the microphysical properties of the aerosol and their number entering clouds, induce a change in the optical properties of the cloud as they act as the sites on which the cloud droplets form. Such changes can then have an impact on the way the cloud scatters and transmits radiation. In addition, it has been postulated that changes in cloud properties will induce further changes to clouds that affect their interaction with radiation. These include suppression of precipitation due to smaller cloud drops when more aerosols are present, increasing cloud lifetime and deepening the cloud (Albrecht 1989). In mixed-phase clouds, additional effects affecting the glaciation process may also affect cloud reflectivity and lifetime but these are at present very poorly quantified (e.g. Lohmann and Neubauer 2018).

To determine the cloud droplet number in a liquid water cloud, it is important to know the effectiveness of the aerosol population in forming cloud droplets (cloud condensation nuclei, [CCN]). This property is dependent on the size, number, chemical composition, and mixing state of the particles and their geographical distribution. Much progress has been made over the past 20 years in understanding

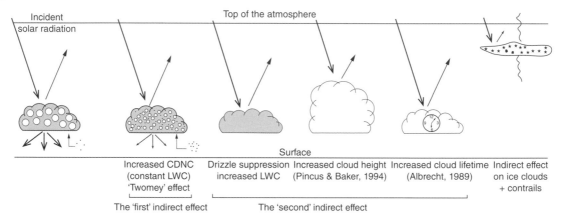

Figure 3.16 Aerosols can have an impact on the radiation budget through their control of cloud properties or behaviour. This can occur in a number of ways: increasing aerosol number concentration increases cloud droplet number concentration (CDNC) and hence the reflectivity of the cloud; this process leads to smaller cloud droplets for the same initial available water, hence drizzle is suppressed and the liquid water content (LWC) of the formed cloud is larger. This again increases cloud reflectivity; increased LWC leads to enhanced heating and hence increased cloud depths; this may also lead to increased cloud lifetimes. *Source:* taken from Haywood and Boucher 2000.

the role of inorganic particulates as CCN; however, the role of the organic fraction, often mixed in with the same particles as the inorganic material, remains uncertain, since they may form surface active layers (Lin et al. 2018) or play a role in adding mass during activation by co-condensation with water vapour (Topping et al. 2013).

To date, progress in demonstrating the climatic effects of clouds has been limited to those effects that directly affect the reflectance of the cloud through increase in cloud droplet number, the so-called Twomey effect (Twomey 1977; Figure 3.16). However, as Figure 3.16 also shows, there are several other parameters that change that cannot be treated as straightforward effects without considering detailed feedbacks.

The best estimates of the effective radiative forcing due to aerosols, including both radiation interactions and cloud interactions (IPCC, 2015), is Wm^{-2} with a range from -1.9 to 0.1 Wm^{-2}, that is to say, a net cooling effect.

3.9.3 Climate Feedbacks

Changing surface temperatures will subsequently affect many different variables in the Earth system, which may respond and either reinforce the change or counteract it. These can be described as feedback mechanisms. For example, it was shown earlier that a doubling of CO_2 led to a ΔI of $4.6\,\mathrm{W\,m^{-2}}$ and a ΔT_s of 1.37 K. However, this, like early climate models, assumed there were no feedbacks between CO_2 and other elements in the climate system. Clearly, temperature change will give rise to other changes in the Earth system, and these may well feed back positively, or negatively, onto those changes.

An example of an effect that reinforces global temperature rise (a positive feedback) is the melting of ice and snow. This lowers the average surface albedo and hence leads to more absorption of incident shortwave radiation. There is further concern here as once such a change has been invoked its reversal is made more difficult as a much cooler temperature is required to begin accumulation. A second positive feedback is that increasing water vapour in the air reduces the longwave transmission and increases longwave re-emission towards the surface.

Feedbacks can also be induced by increasing the concentration of CO_2 in the atmosphere. In the past, global climate models have calculated temperature change in two stages. A model of the carbon cycle is used to calculate the future atmospheric concentrations of carbon dioxide. Then climate change is calculated utilizing a separate global climate model using the estimates of CO_2 rise as input, with no feedback from climate change to the carbon cycle. An example of an important feedback is surface seawater and surface polar temperature rise, which may lead to reduced ocean circulation, a slowing of the thermohaline circulation (Broecker 1997), and a decrease in the CO_2 adsorption by surface waters. In addition, this may enhance the decay rates of organic materials, leading to greater emissions of CO_2 and CH_4 to the atmosphere. Changes in ocean–atmosphere coupling have been shown to impact on the storm track of the North Atlantic (Woollings et al. 2012).

On the other hand, there are many possible feedback mechanisms that act to mitigate increasing temperature. An example of a negative physical feedback is the decrease in the atmospheric adiabatic lapse rate due to increased cloudiness aloft. Biological processes may also be important. For instance, increased algal growth in warmer waters requires increased CO_2 for photosynthesis and hence reduces the atmospheric reservoir.

Interactions between land and the atmosphere are also extremely important, but scientific understanding is not as developed as the coupling between oceans and the atmosphere. Recent evidence has been provided to show that in northern latitudes, warming changes snow cover and near-surface permafrost extent (Koven et al. 2011). Soil moisture is determined by the balance between rainfall and runoff, and this plays an important role in evapotranspiration. The ability of models to capture these couplings effectively remains a challenge (Seneviratne et al. 2010). The latest generation of Earth system models are providing improved representation of climate-carbon feedbacks, and improved constraints are becoming available (e.g. Wenzel et al. 2014; however, carbon emissions resulting from changes in land use, forest regrowth, and nitrogen cycling remain challenging issues to resolve.

3.10 Solar Radiation and the Biosphere

In addition to providing the energy to our Earth–atmosphere system and providing a climatology in which life can flourish, solar radiation has many direct influences on the biosphere (the narrow band of the atmosphere–earth–ocean system where living organisms are found). There are also indirect effects and complex feedback systems between the atmosphere, biosphere, and radiative transfer, especially when human's activity is considered as part of the biosphere – for instance, anthropogenic emissions of carbon dioxide and methane and ozone depletion. This final section of the chapter will concentrate on the more direct effects of solar radiation on lifeforms.

In many respects, the behaviour of living organisms in sunlight is not an effect, in the sense of an existing organism responding to an externally imposed stimulus, but rather, an example of the evolutionary adaptation of the organism to utilize an available resource. Many of the so-called effects of radiation (usually detrimental) come from upsetting the balance between the radiation climate in which the organism evolved and that to which it is currently exposed: this change might be considered as an external stimulus.

The radiation balance of the Earth-atmosphere discussed in earlier sections is concerned only with the total energy in the complete solar waveband (0.3–4 µm). However, many of the photoreactions initiated by sunlight are wavelength dependent, and different parts of the solar spectrum have to be considered for different reactions. In addition, the simple physical energy contained in a waveband is not always a good indicator of its potential to induce the desired reaction; the action spectrum is also required. The action spectrum, or response spectrum, for a given reaction describes the wavelength-dependent sensitivity of the reaction or target body to the incident radiation, often normalized to unity at the wavelength of maximum response. Thus, if the incident solar spectral intensity is $I(\lambda)$ and the action spectrum of interest is denoted by $R(\lambda)$, then the intensity of biologically effective radiation (i.e. the physical energy weighted with its effectiveness in producing the specified reaction) will be

$$\int I(\lambda) R(\lambda) d\lambda$$

(3.53)

Many of the important biological action spectra are in the UV and visible portions of the solar spectrum, where the individual photons have most energy. Approximately half the total solar energy is in the visible part of the spectrum, with a small amount in the UV and the rest at infrared wavelengths (see Figure 3.3).

Two fundamental uses of solar radiation by inhabitants of the biosphere are vision and photosynthesis. Respectively, they allow the majority of mobile creatures to see, and plants to convert solar energy into a useable form (sugar), initially for themselves and then as plant matter for other levels of the food chain. Our optical system (the eye) responds to visible radiation (wavelengths between 400 and 700 nm), and the photopic response peaks in the middle of this range (green light). It is no coincidence that the solar spectrum, both extraterrestrially and at the surface, is a maximum in the same wavelength region. Understanding the illuminance (visually effective radiation with $R(\lambda)$ equal to the photopic response in Eq. (3.53)) is important in, for example, building design and is measured with a lux metre that has a response spectrum very similar to the photopic response of the eye.

Systems that photosynthesize also make good use of the same waveband. Photosynthesis by chlorophyll-containing plants is the process by which water and atmospheric CO_2 are converted into simple sugars (and thence more complex compounds), oxygen, and water. The absorptance (ratio of absorbed to incident radiation) of typical green leaves exceeds 90% at blue and red wavelengths, but decreases to less than 80% in the green waveband, where reflectance and transmittance increase (hence the observed green colour). Absorptance drops precipitously at the longwave end of the visible spectrum and is less than 5% between 0.7 and 1 µm, thereafter increasing again. The reflection and transmission of infrared radiation helps to prevent the plant from overheating.

Given the comparatively constant spectral composition to photosynthetically active radiation (PAR), the photosynthesis rate will depend on the intensity of the radiation, plus water availability, carbon dioxide concentration, and the presence or lack of other environmental stresses (e.g. temperature). In the absence of other limiting factors, the photosynthesis rate increases almost linearly, with increasing incident radiation up to a limiting value. At this point the plant is light saturated and further increases in radiation are not beneficial. The light saturation point occurs at different irradiance and photosynthesis rates, depending on the type of plant: both are low for shade-loving plants and increase until, for some plants, it is difficult to reach light saturation in sunlight.

At UV wavelengths photon energies become sufficient to cause damage and it is the detrimental effects of UV that are most often cited, although there are beneficial effects as well. Prominence has been given to the UVB waveband as it is radiation in this waveband that is most affected by changes in stratospheric ozone. Ozone depletion leads to increased UVB at the surface and a shift of the short wavelength limit of the solar spectrum to shorter (more damaging) wavelengths. In humans and animals, UVB radiation is necessary for skeletal health as it initiates the cutaneous synthesis of vitamin D, but it also damages DNA, produces sunburn/tanning, affects skin-mediated immunosuppression, and causes damage to the eye. Sunburn is probably the best-known detrimental effect, with an action spectrum that includes both UVB and UVA radiation. It is also associated with increased risk of skin cancer, particularly the most fatal variety, malignant melanoma. Skin cancers, such as chronic eye damage resulting in cataracts, are diseases whose risks increase with accumulated lifetime exposure to UV, moderated for skin cancers by the skin's natural sensitivity to UV radiation. Fair-skinned people are at greater risk than those with naturally high levels of pigmentation. The relation between skin colour and UV effects is a good example of an evolutionary balancing act. People originating from high latitudes (low levels of sunlight and UV) have fair, sun-sensitive skins with little melanin (a competing absorber for UV photons and responsible for colour in tanned, brown, or black skin), enabling them to take advantage of available UV for vitamin D synthesis. Movement to higher radiation environments (equatorwards) increases the risks of sunburn and skin cancer. Conversely, highly pigmented peoples from low latitudes have natural protection against high levels of UV there, but if they move polewards where UV is reduced, they can become susceptible to vitamin D deficiency.

3.11 Summary

Earth's energy system and its atmosphere are driven by solar energy. Although the atmosphere is essentially transparent to incoming shortwave radiation, several components, including water vapour, carbon dioxide, methane, nitrous oxide, and ozone, absorb infrared radiation emitted by the Earth. It is this interaction that increases the temperature of the surface and lower atmosphere substantially. Absorption of solar radiation in the UV is absorbed by ozone and leads to its photolysis. This process gives rise to the heating of the stratosphere. Unlike UV radiation, absorption of longwave radiation does not lead to molecular fragmentation but is re-radiated in the upward and downward directions. A simple treatment of radiation transfer through a vertical column was introduced based on a single absorbing species and an assumption that the atmosphere was in radiative equilibrium at every level. Although this model gives some insight into the radiative transfer through the atmosphere, it does not explain many observations. For this it is necessary to understand the complexity of absorption of longwave radiation.

Light is not only absorbed, it is also scattered. Molecular scattering has a strong wavelength dependence in the visible and explains why the sky is blue and sunsets appear red. Particles are much larger, approximately the size of the wavelength of light, and hence interact with light in a more complex way. A simple model was used to illustrate the complex behaviour of aerosol particle scattering and absorption. The surface reflectance is clearly important and can lead to the aerosols acting to either warm or cool the atmosphere, changing the sign of their effect.

The loss of heat from the surface takes place either directly, or through evaporation of water, and is important as it closes the net radiation budget. The global average budget shows that the exchanges of longwave radiation between the atmosphere and the Earth's surface greatly exceed the net fluxes at the top of the atmosphere. It is also clear that although the Earth system is in global balance, there is considerable variation across the globe. Net heat gain occurs close to the Equator, which is transported to the poles mostly by the atmosphere but with a significant contribution from the ocean, particularly in the subtropics.

It is now recognized that significant human-induced changes on climate are occurring. The largest of these arises from carbon dioxide, whose atmospheric reservoir has increased in the past 150 years through changes in land use and fossil-fuel burning. Other agents are also having an effect and lead to a current radiative forcing of around $2.5\,\mathrm{W\,m^{-2}}$. However, aerosol particles have a significant impact on the radiative budget by directly scattering and absorbing light, and through their influence on cloud properties and behaviour, and hence their reflectivity. These effects are thought to cool the lower atmosphere and are spatially heterogeneous, mainly as particles vary widely in space and time, and remain rather poorly quantified despite a considerable research effort to reduce the uncertainties.

Feedbacks and coupling between the ocean, land, and atmosphere are now recognized as very important. The new generation of Earth system models couples these interactions and is a major subject for ongoing work.

Radiation plays a key role in the biosphere, driving photosynthesis and stimulating growth. However, as UV transmission increases due to reduced ozone concentrations in the stratosphere, so harmful radiation can reduce crop yield.

Questions

1   Describe how and why solar irradiance at the Earth's surface differs from that of a blackbody radiator at 5777 K.

2   What is meant by albedo? Why is planetary albedo important in the energy budget of the Earth?

3   How is outgoing longwave radiation trapped in the Earth's atmosphere?

4   How is solar radiation scattered in the atmosphere, and what effect does this scattering have on the energy budget of the Earth?

5   What direct and indirect effects do aerosols have on the radiation budget of the Earth?

References

Albrecht, B. (1989). Aerosols, cloud microphysics and fractional cloudiness. *Science* 245: 1227–1230.

Bohren, C.F. and Huffman, D.R. (1983). *Absorption and Scattering of Light by Small Particles*. New York, NY: Wiley-Interscience.

Bond, T.C., Doherty, S.J., Fahey, D.W. et al. (2013). Bounding the role of black carbon in the climate system: a scientific assessment. *Journal of Geophysical Research – Atmospheres* 118 (11): 5380–5552. https://doi.org/10.1002/jgrd.50171.

Boucher, O. and Pham, M. (2002). History of sulfate aerosol radiative forcings. *Geophysical Research Letters* 29: 22–25.

Brasseur, G. and Solomon, S. (1986). *Aeronomy of the Middle Atmosphere*, 2e. Dordrecht, the Netherlands: Reidel.

Broecker, W.S. (1997). Thermohaline circulation, the Achilles heel of our climate system: will man-made CO2 upset the current balance? *Science* 278: 1582–1588.

Haywood, J.M. (2011). Geostationary earth radiation budget intercomparison of longwave and shortwave radiation (GERBILS). *Quarterly Journal of the Royal Meteorological Society* 137: 1105–1105. https://doi.org/10.1002/qj.884.

Haywood, J.M. and Boucher, O. (2000). Estimates of the direct and indirect radiative forcing due to tropospheric aerosols: a review. *Reviews of Geophysics* 38 (4): 513.

Haywood, J.M., Osborne, S.R., Francis, P.N. et al. (2003). The mean physical and optical properties of regional haze dominated by biomass burning aerosol measured from the C-130 aircraft during SAFARI 2000. *Journal of Geophysical Research* 108 (D13): 8473.

Highwood, E.J., Haywood, J.M., Silverstone, M.D. et al. (2003). Radiative properties and direct effect of Saharan dust measured by the C-130 aircraft during Saharan Dust Experiment (SHADE): 2. Terrestrial spectrum. *Journal of Geophysical Research* 108: SAH 5.

IPCC (2014). The Intergovernmental Panel on Climate Change 5th Assessment Report (AR5). https://www.ipcc.ch/report/ar5/syr.

Iqbal, M. (1983). *An Introduction to Solar Radiation*. Toronto, Canada: Academic Press.

Keil, A. and Haywood, J.M. (2003). Solar radiative forcing by biomass burning aerosol particles during SAFARI 2000: a case study based on measured aerosol and cloud properties. *Journal of Geophysical Research* 108 (D13): 8467.

Koven, C.J., Friedlingstein, P., Ciais, P. et al. (2011). Permafrost carbon-climate feedbacks accelerate global warming. *Proceedings of the National Academy of Sciences of the United States of America* 108: 14769–14774.

Lin, J.J., Malila, J., and Prisle, N.L. (2018). Cloud droplet activation of organic–salt mixtures predicted from two model treatments of the droplet surface. *Environmental Science: Processes & Impacts* 20: 1611.

Lohmann, U. and Neubauer, D. (2018). The importance of mixed-phase and ice clouds for climate sensitivity in the global aerosol–climate model ECHAM6-HAM2. *Atmospheric Chemistry and Physics* 18: 8807–8828. https://doi.org/10.5194/acp-18-8807-2018.

Maria, S.F., Russell, L.M., Gilles, M.K., and Myneni, S. (2004). Organic aerosol growth mechanisms and their climate-forcing implications. *Science* 306: 1921–1924.

Martin, S.T., Hung, H.M., Park, R.J. et al. (2004). Effects of the physical state of tropospheric ammonium-sulfate-nitrate particles on global aerosol direct radiative forcing. *Atmospheric Chemistry and Physics* 4: 183–214.

Menon, S., Hansen, J., Nazarenko, L., and Luo, Y. (2002). Climate effects of black carbon aerosols in China and India. *Science* 297: 2250–2253.

Osborne, S.R., Haywood, J.M., Francis, P.N., and Dubovik, O. (2004). Short-wave radiative effects of biomass burning aerosol during SAFARI 2000. *Quarterly Journal of the Royal Meteorological Society* 130: 1423–1448.

Penner, J., Dickinson, R., and O'Neill, C. (1992). Effects of aerosol from biomass burning on the global radiation budget. *Science* 256 (5062): 1432–1434.

Salby, M.L. (1996). *Fundamentals of Atmospheric Physics*, International Geophysics Series, vol. 61. San Diego, CA: Academic Press.

Seinfeld, J.H. and Pandis, S.N. (1998). *Atmospheric Chemistry and Physics: From Air Pollution to Climate Change*. New York, NY: John Wiley & Sons.

Seneviratne, S.I. et al. (2010). Investigating soil moisture-climate interactions in a changing climate: a review. *Earth-Science Reviews* 99: 125–161.

Tegen, I., Lacis, A.A., and Fung, I. (1996). The influence on climate forcing of mineral aerosols from disturbed soils. *Nature* 380: 419–421.

Thornhill, G.D., Ryder, C.L., Highwood, E.J. et al. (2018). The effect of South American biomass burning aerosol emissions on the regional climate. *Atmospheric Chemistry and Physics* 18: 5321–5342. https://doi.org/10.5194/acp-18-5321-2018.

Topping, D., Connolly, P., and Mcfiggans, G. (2013). Cloud droplet number enhanced by co-condensation of organic vapours. *Nature Geoscience* 6 (6): 443–446. https://doi.org/10.1038/ngeo1809.

Trenberth, K.E., Fasullo, J.T., and Kiehl, J. (2009). Earth's global energy budget. *Bulletin of the American Meteorological Society* 90: 311–323.

Twomey, S. (1977). Influence of pollution on shortwave albedo of clouds. *Journal of Atmospheric Science* 34 (7): 1149–1152.

Wells, N. (1997). *The Atmosphere and Ocean: A Physical Introduction*, 2e. New York, NY: John Wiley & Sons.

Wenzel, S., Cox, P.M., Eyring, V., and Friedlingstein, P. (2014). Emergent constraints on climate-carbon cycle feedbacks in the CMIP5 Earth system models. *Journal of Geophysical Research – Biogeosciences* 119: 794–807. https://doi.org/10.1002/2013JG002591.

Woollings, T., Gregory, J., Pinto, J. et al. (2012). Response of the North Atlantic storm track to climate change shaped by ocean–atmosphere coupling. *Nature Geoscience* 5: 313–317.

Further Reading

Gill, A.E. (1982). *Atmosphere–Ocean Dynamics*. London, UK: Academic.

Goody, R.M. and Yung, Y.L. (1989). *Atmospheric Radiation: Theoretical Basis*, 2e. New York, NY: Oxford University Press.

Houghton, J.T. (1986). *The Physics of Atmospheres*, 2e. Cambridge, UK: Cambridge University Press.

Wayne, R.P. (2000). *Chemistry of Atmospheres: An Introduction to the Chemistry of the Atmosphere*, 3e. Oxford, UK: Oxford University Press.

4

Biogeochemical Cycles

Dudley Shallcross and Anwar Khan

School of Chemistry, University of Bristol, Bristol, United Kingdom

The elements carbon (C), nitrogen (N), sulphur (S), hydrogen (H), oxygen (O), and the halogens (X = F, Cl, Br, I) play key roles in the maintenance of life on Earth. Quite apart from being essential components of the fundamental building blocks of life, such as proteins, these elements perform vital regulatory functions in the Earth system. For example, carbon dioxide (CO_2), methane (CH_4), and nitrous oxide (N_2O) are naturally occurring greenhouse gases, trapping heat in the form of infrared radiation that otherwise would escape to space and thus ensures that the surface temperature of the Earth is habitable for life. Elements travel or cycle through the whole Earth system (i.e. the oceans, the atmosphere, the land, etc.) and may be transformed from one compound to another as they do so. This is the basis of the study of biogeochemical cycles. In this chapter we will consider how these elements cycle through the atmosphere and how and why they are partitioned into the various compounds observed.

A fundamental concept is the budget of a compound put simply, we consider the sources (direct emissions and/or photochemical productions) of the compound and the sinks (loss processes chemically and/or physical) for the compound. If the sources and sinks balance, this will lead to a constant concentration of the compound being observed in the atmosphere (Figure 4.1a); if sources are greater than sinks, the concentration of that compound in the atmosphere will increase with time (Figure 4.1b); and if sources are less than sinks, the concentration of that compound in the atmosphere will decrease with time (Figure 4.1c). However, can we be more quantitative? If the budget is balanced, what will the constant concentration be, and what controls this parameter? It turns out (see the derivation of this equation that follows) that the concentration of a compound A at time t, $[A]_t$, is

$$[A]_t = E/R\left(1-\exp\left(-R_t\right)\right) \tag{I}$$

where E is the source or emission rate and R is the loss rate. When sufficient time has elapsed, a constant concentration of A, $[A]_{constant}$ is reached and is equal to E/R. If the emission rate E increases, a new $[A]_{constant}$ is established that is bigger than before, and if the emission rate decreases the new $[A]_{constant}$ is smaller than before. If the loss rate R increases, $[A]_{constant}$ decreases, and if the loss rate decreases, $[A]_{constant}$ increases. Hence, the concentration of any compound in the atmosphere is determined rather simply by E and R.

Atmospheric Science for Environmental Scientists, Second Edition. Edited by C.N. Hewitt and Andrea V. Jackson.
© 2020 John Wiley & Sons Ltd. Published 2020 by John Wiley & Sons Ltd.

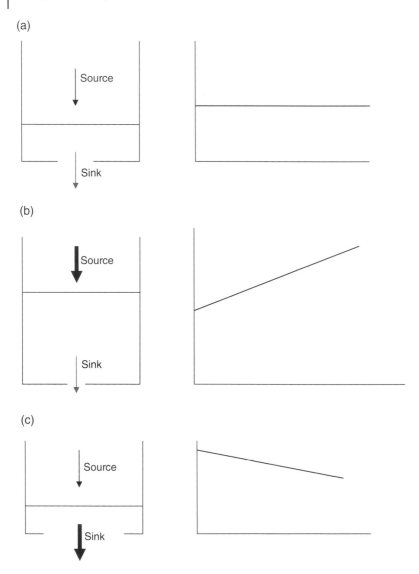

Figure 4.1 (a) Source and sink terms are equal; the concentration remains constant with time, e.g. N_2 and O_2. (b) Source is greater than the sink. The concentration increases with time, e.g. hydrochlorofluorocarbons (HCFCs) and CO_2 (currently). (c) Source is smaller than the sink. The concentration decreases with time, e.g. CH_3CCl_3 and chlorofluorocarbons (CFCs).

Derivation of Eq. (I)

Imagine compound A has a total source or total emission rate E (having units of mass emitted per unit time) and is removed from the atmosphere at a rate R (having units of per unit time). We can write the rate of change of the concentration of A, $(d[A]/dt)$ as

$$d\left[A\right]/dt = E - R\left[A\right]$$ (i)

When the emission and removal rates are equal, the concentration of A is constant with time (i.e. $d[A]/dt = 0$), hence

$$E - R\big[A\big]_{\text{constant}} = 0 \tag{ii}$$

$$\big[A\big]_{\text{constant}} = E / R \tag{iii}$$

What about the time scale required for this constant concentration to establish? Let

$$y = E - R\big[A\big] \tag{iv}$$

$$\Rightarrow dy = -R d\big[A\big] \tag{v}$$

$$\Rightarrow -dy / R = d\big[A\big] \tag{vi}$$

Substituting (iv) and (vi) into (i) we obtain

$$dy / dt \cdot \big(-1 / R\big) = y \tag{vii}$$

or

$$dy / y = -R dt \tag{viii}$$

Integrating leads to

$$\ln y = -Rt + C \tag{ix}$$

where C is a constant of integration and substituting for y we obtain

$$\ln\big\{E - R\big[A\big]\big\} = -Rt + C \tag{x}$$

Finally, assuming that $[A] = 0$ at $t = 0$

$$\big[A\big]_t = E/R\big(1 - \exp\big(-Rt\big)\big) \tag{I}$$

Thus, A will reach its steady state or constant value E/R with an e-folding time of $1/R$.

Example

Given the following data, calculate the steady-state concentration or constant concentration of A and plot the change in concentration with time in each case:

1) $E = 10^{12}\,\text{kg year}^{-1}$ $R = 10^{-6}\,\text{s}^{-1}$
2) $E = 10^{12}\,\text{kg year}^{-1}$ $R = 10^{-4}\,\text{s}^{-1}$
3) $E = 10^{10}\,\text{kg year}^{-1}$ $R = 10^{-6}\,\text{s}^{-1}$

Answer

Let t be the number of seconds in a year $= 60 \times 60 \times 24 \times 365 = 3.1536 \times 10^7 \, \mathrm{s}$

1) $E/R = 10^{12}/(10^{-6} \times t) = 3.2 \times 10^{10} \, \mathrm{kg}$
2) $E/R = 10^{12}/(10^{-4} \times t) = 3.2 \times 10^{8} \, \mathrm{kg}$
3) $E/R = 10^{10}/(10^{-6} \times t) = 3.2 \times 10^{8} \, \mathrm{kg}$

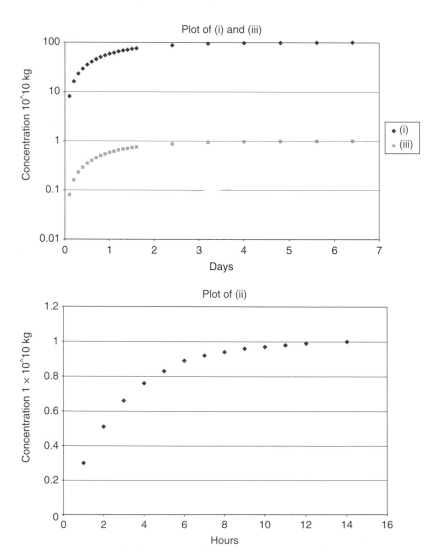

Notes

With a much shorter lifetime, scenario **2** reaches the steady-state concentration much more quickly than **1** or **3**. Scenarios **1** and **3** take the same time to reach steady state because they have the same lifetime, but because the emission rate is greater for **1** than **3** the steady-state concentration is much larger.

4.1 Sources

Sources (sometimes referred to as emissions) of compounds in the atmosphere can be divided into primary and secondary sources. A primary source is one where the compound is released directly into the atmosphere, whereas a secondary source is one where the compound is generated *in-situ* in the atmosphere. An example is carbon monoxide (CO) that has both primary and secondary sources in the atmosphere. CO is emitted directly into the atmosphere by a number of sources, such as from the exhaust of road vehicles during high-temperature incomplete combustion of fossil fuels. It is also produced in the atmosphere from the photooxidation of hydrocarbons that are themselves emitted into the atmosphere (see Chapter 5). In addition, there are two major primary sources, anthropogenic (from human activity) and biogenic (from natural sources).

Typical anthropogenic sources include mobile sources such as vehicles, railways, vessels and aircraft, and stationary sources such as manufacturing industries, power plants, waste incineration, buildings, and homes. Biogenic sources include all living systems such as plants, trees, animals, insects, and microbes. However, there are nonliving natural sources of compounds in the atmosphere. For example, dust in the atmosphere can come from wind-blown soil, and lightning is an important source of nitric oxide (NO) in the atmosphere. Biomass burning can be both anthropogenic (deliberate or accidental burning of vegetation by humans) and biogenic (e.g. natural forest fires) and are an important source of a number of compounds. In general, anthropogenic source strengths can be quite constant (e.g. power plant emissions) or at least tend to follow a regular pattern (vehicle emissions follow a regular diurnal pattern). Hence, it is possible to determine the overall anthropogenic source strength of a particular compound by constructing an emission inventory (several examples follow in this chapter), but such inventories are still an approximation. Biogenic source strengths are often determined by levels of sunlight, temperature, and nutrients, and can therefore be highly seasonal in nature. Therefore, it is much harder to estimate the biogenic source strength of a particular compound.

4.2 Sinks

There are two main sink or loss processes for compounds in the atmosphere, namely depositional (sometimes referred to as physical) loss processes and photochemical loss processes.

4.2.1 Depositional Loss Processes

There are two types of depositional loss, wet deposition and dry deposition. Wet deposition involves the uptake of soluble compounds into aqueous media (e.g. aerosols, fogs, clouds, and during rainfall). Therefore, the more soluble a species is, the faster the uptake and therefore the greater the loss rate (see Chapter 6 for more details). A measure of the solubility of a species is the Henry's law coefficient K_H

$(mol\,l^{-1}\,atm^{-1})$, and the fraction of compound C remaining in the gas-phase after dissolution can be expressed as

$$C/C_0 = 1/\left(1 + 10^{-6} K_H RT \left(L/\rho_w\right)\right) \tag{II}$$

where C/C_0 is the fraction of species C remaining in the gas-phase, R is the gas constant $(0.082051\,atm\,mol^{-1}\,K^{-1})$, T is the temperature in Kelvin, L $(g\,m^{-3})$ is the liquid water content, and ρ_w is the density of water $(1\,g\,cm^{-3})$. Solubility increases as the relative amount of oxygen in the compound increases, so methane (CH_4) is rather insoluble, whereas methanoic acid $(HCOOH)$ is moderately soluble. Compounds such as hydrogen peroxide (H_2O_2) are highly soluble and wet deposition is an appreciable loss process in this case. Strong acids such as nitric acid (HNO_3) and sulphuric acid (H_2SO_4) will not only dissolve in aqueous media but will undergo rapid dissociation once in the aqueous phase:

$$HNO_3(g) \leftrightarrow HNO_3(aq) \tag{4.1}$$

$$HNO_3(aq) \rightarrow H^+(aq) + NO_3^-(aq) \tag{4.2}$$

Such strong acids undergo near total uptake into the aqueous phase.

Dry deposition occurs when a compound can be adsorbed or react on surfaces at the ground and removed from the atmosphere. There are several factors that determine the dry deposition loss rate. The first is the characteristics of the gas. The more polar the gas the more likely it is to have a significant dry deposition rate (polarity arises when there is an uneven distribution of charge about a chemical bond and is particularly associated with compounds that contain elements such as F, O, Cl). Second, the surface type is of course important, the rate of dry deposition to an ice surface is likely to be very different to that to a surface covered with vegetation. It is usual to relate the flux F $(mol\,m^{-2}\,s^{-1})$ to the surface to the gas concentration C $(mol\,m^{-3})$ at a reference height, where the constant of proportionately is the dry deposition velocity (V_d). Hence

$$F = -V_d C \tag{III}$$

Dry deposition velocity is normally associated with a reference height of $10\,m$ and will be positive for a depositing species. The V_d will vary with gas and surface type and is typically 0.1–$1\,cm\,s^{-1}$ for gases. The simple concept of deposition velocity (V_d) is useful for incorporation into complex atmospheric models, but this hides the complexity of the deposition process. In reality, the dry deposition process consists of three distinct phases. First, turbulent transfer from the bulk atmosphere transports material to a thin layer of stagnant air adjacent to the surface. Second, molecular diffusion transports material across this layer to the surface. Third, material can now be taken up by the surface. It is usual to model each of these processes by a concentration difference and an associated resistance to transfer. The total resistance R_t can be related back to the deposition velocity via

$$R_t = R_a + R_b + R_c = V_d^{-1} \tag{IV}$$

where R_a is aerodynamic resistance, R_b is the quasi-laminar resistance and R_c is the surface resistance. Such a model of dry deposition is known as the bulk resistance model.

4.2.2 Photochemical Loss Processes

Compounds can also be converted to another compound in the atmosphere via photochemical processes. The simplest photochemical loss process is photolysis itself, where a compound absorbs a photon of sunlight, and this energy leads to dissociation of that molecule or isomerization, where the new molecule has the same composition but is structurally different from its parent compound. In the lowest 10 km of the atmosphere, the troposphere, the wavelengths of solar energy available are those longer than 300 nm. Aldehydes, ketones, and iodocarbons in particular can be dissociated by solar photons in this region. At higher altitudes in the atmosphere, in the stratosphere (around 10–50 km) higher energy photons as short as around 200 nm are available. Here, many molecules such as H_2O_2, HNO_3, methyl chloride (CH_3Cl), methyl bromide (CH_3Br), and the chlorofluorocarbons (CFCs) can be dissociated. The rate of photolysis J is given by

$$J = \int \Phi \Psi \sigma d\lambda \qquad (V)$$

where Φ is the quantum yield, i.e. the number of molecules dissociated per photon (usually Φ is 1), Ψ is the photon flux, i.e. the number of photons at a particular wavelength per unit area per second, and σ is the absorption cross-section, which is a measure of how likely the molecule will absorb a photon at a particular wavelength.

Worked Example 4.1

Calculate the photolysis rate of nitrogen dioxide (NO_2) given the following data averaged over the wavelength range 300–400 nm:

$$\Phi = 1.0; \; \Psi = 5 \times 10^{15} \, cm^{-2} s^{-1}; \; \sigma = 2 \times 10^{-19} \, cm^2$$

Answer

$$J(NO_2) = \int \Phi \Psi \sigma d\lambda = 1.0 \times 5 \times 10^{15} \times 2 \times 10^{-19}$$
$$= 1 \times 10^{-3} \, s^{-1}$$

Chemical reaction can be initiated by free radicals such as hydroxyl (OH), chlorine (Cl), and nitrate (NO_3) or by reaction with ozone (O_3) (see Chapter 5) and *stabilized Criegee Intermediates (sCIs)*. Throughout the daytime atmosphere, the OH radical is an important oxidant, whereas the NO_3 radical is important only at night. In the marine environment and in polar regions, Cl atoms can be an important oxidant in the troposphere; their role in the stratosphere is discussed in Chapter 8. Ozone can react with unsaturated species such as alkenes. The primary ozonide formed from the reaction of ozone with an alkene dissociates into a vibrationally excited Criegee intermediate that can then either undergo unimolecular decomposition or form an sCI. sCIs can be an important oxidant over the regions of high emissions of the precursors (e.g. unsaturated hydrocarbons). Table 4.1 shows how the rate coefficients of the reactions of OH with some hydrocarbons vary and their corresponding lifetimes (see the worked example of lifetime determination that follows). If one considers the alkanes, e.g. CH_4, ethane (C_2H_6) and propane (C_3H_8), the reaction with OH proceeds via abstraction of a hydrogen atom:

$$OH + RH \rightarrow R + H_2O \tag{4.3}$$

Therefore, it is possible to explain the increase in rate coefficient from CH_4 to C_3H_8 by the fact that there are progressively more H atoms available. Therefore, one would conclude that very large alkanes react very quickly and have short lifetimes; this is indeed the case. If one compares CH_4 with CH_3Cl, by our previous argument, we would assume that CH_3Cl would react more slowly than CH_4, with 3 H atoms as opposed to 4. However, this is not the case. Now we must consider the effect of the C—Cl bond. We know that Cl is electronegative and will draw electron density away from C and leave the 3 H atoms with a slight positive charge. This induced dipole increases the reaction between the H and OH. If we replace all the H atoms in CH_4 with Cl, i.e. CCl_4, we reach a situation where OH cannot react with the compound; it is an endothermic process. Hence, CCl_4 can accumulate in the troposphere and will reach the stratosphere where it is broken down by ultraviolet light, releasing chlorine (Chapter 8). Once again, if we consider the alkene, e.g. ethene (C_2H_4), by considering the number of H atoms, we would conclude that OH should react more quickly with C_2H_6 and at about the same rate as CH_4. Inspection of Table 4.1 shows that OH reacts 10 000 times faster with C_2H_4 than it does with CH_4. The reason is that a different mechanism, addition, is occurring:

$$OH + CH_2 = CH_2 + M \rightarrow CH_2(OH) - CH_2 \tag{4.4}$$

The OH radical adds to one end of the double bond rapidly, forming a radical. Isoprene (C_5H_8) contains two double bonds and is therefore even more reactive.

Table 4.1 Examples of rate coefficients for the reaction of hydrocarbons with OH, and their corresponding lifetimes assuming an [OH] $= 1 \times 10^6$ molecule cm^{-3}, kinetic data taken from Atkinson et al. (2006) and DeMore et al. (1997).

Hydrocarbon	Rate coefficient at 298 K (cm^3 molecule^{-1} s^{-1})	Lifetime
CH_4	7×10^{-15}	10 years
C_2H_6	3×10^{-13}	4 months
C_3H_8	1×10^{-12}	1 month
C_2H_4	1×10^{-11}	3 days
C_5H_8	1×10^{-10}	$^1/_2$ day
CH_3Cl	1×10^{-13}	1 year
CO	2×10^{-13}	6 months
CCl_4	a	50 years^a

^{a} Is destroyed in the upper atmosphere, and the lifetime is determined by transport.

Whereas deposition usually leads to the permanent removal of that compound from the atmosphere, it is possible in some cases for the compound to be reformed if the photochemical reaction converts the compound into a temporary reservoir, which can then be decomposed into its component parts, e.g.

$$OH + NO_2 + M \rightarrow HNO_3 + M \tag{4.5}$$

$$HNO_3 + h\nu \rightarrow OH + NO_2 \tag{4.6}$$

In reaction (4.5) OH radicals and NO_2 combine to form nitric acid (HNO_3), but in the stratosphere this can be photolysed (reaction 4.6) to release back OH and NO_2. Note that $h\nu$ represents a photon of light. Therefore, the full equation that describes the budget of a compound x can be written:

$$dX/dt = E_P + P_{PC} - V_d\left[X\right]/h - k_w\left[X\right] - L_{PC}\left[X\right] \tag{VI}$$

where E_P is the emission rate from primary sources, P_{PC} is the production rate from in-situ photochemistry, V_d/h is the dry deposition loss rate, k_w is the wet deposition loss rate, and L_{PC} is the loss rate from in-situ photochemistry (chemical reactions and photolysis).

A simple way to determine the lifetime of compound x is to consider the following example of species x, which is removed primarily by reaction with OH, photolysis, dry deposition, and wet deposition:

$$-d\left[X\right]/dt = k_r\left[OH\right]\left[X\right] + J\left[X\right] + V_d\left[X\right]/h + k_w\left[X\right] \tag{i}$$

$$\Rightarrow -d\left[X\right]/\left[X\right] = \left(k_r\left[OH\right] + J + V_d/h + k_w\right)dt$$
$$\Rightarrow \int d\left[X\right]/\left[X\right] = \int -\left(k_r\left[OH\right] + J + V_d/h + k_w\right)dt \tag{ii}$$
$$\Rightarrow \ln\left\{\left[X\right]/\left[X\right]_0\right\} = -\left(k_r\left[OH\right] + J + V_d/h + k_w\right)t$$

where $[X]_0$ is the initial concentration and $[X]$ the concentration after time t.

The time scale for the decay is given by the term $(k_r[OH] + J + V_d/h + k_w)$, and the lifetime is $1/\{k_r[OH] + J + V_d/h + k_w\}$, i.e. the lifetime is the reciprocal of the sum of the loss rates. The larger the loss rate, the shorter the lifetime.

Worked Example 4.2

Calculate the lifetime of hydrogen (H_2) at Earth's surface given the following information:

$$OH + H_2 \rightarrow products; \; k = 1 \times 10^{-14} \; cm^3 \; molecule^{-1} \; s^{-1}$$
$$H_2 + h\nu \rightarrow H + H; \; J = 1 \times 10^{-10} \; s^{-1}$$
$$k_w \; (H_2) = 1 \times 10^{-9} \; s^{-1}$$
$$V_d \; (H_2) = 1 \times 10^{-9} \; ms^{-1}$$

Assume that $[OH] = 1 \times 10^6$ molecule cm^{-3} and h = 0.1 m.

Answer

$$\text{Lifetime of } H_2 = 1 / \left\{ k_r \left[OH \right] + J + V_d / h + k_w \right\}$$
$$= 1 / \left(1 \times 10^{-14} \times 1 \times 10^6 + 1 \times 10^{-10} + 1 \times 10^{-9} + 1 \times 10^{-9} / 0.1 \right)$$
$$= 47393365 \text{ seconds}$$
$$= 1.5 \text{ years}$$

4.3 Carbon

All living matter is carbon based and the element is an essential part of life on Earth. The major reservoirs of carbon in the atmosphere are CO_2 (~400 ppmv), CH_4 (~1.8 ppmv), and CO (~0.1 ppmv), but myriad other volatile organic compounds (VOCs) are emitted into Earth's atmosphere from a variety of sources and constitute a large pool of atmospheric carbon.

4.3.1 Carbon Dioxide

Measurements of levels of CO_2 over the past 50 years or so (Figure 4.2) show a dramatic rise from around 315 ppmv in the late 1950s to the present-day value of over 400 ppmv. Analysis of ice cores has facilitated an assessment of the changes to atmospheric levels of CO_2 (Barnola et al. 1987). The analysis shows that up until 1800, levels of CO_2 were pretty constant at 280 ppmv, but have risen dramatically since the onset of the Industrial Revolution. The cause of this rapid change is a combination of the burning of fossil fuel,

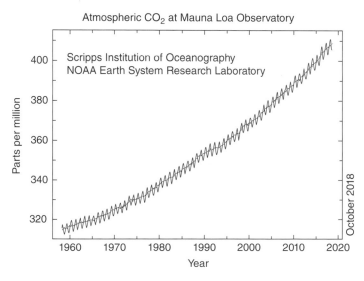

Figure 4.2 Schematic of the rise in atmospheric CO_2 over the past 50 years recorded at Mauna Lao, Hawaii, USA. Location 19°N, 155°W; sample height 3397 m a.s.l. *Source:* adapted from Earth System Research Laboratory, National Oceanic and Atmospheric Administration, https://www.esrl.noaa.gov/gmd/ccgg/trends/full.html.

deforestation, and changes in land use. Carbon dioxide is the major contributor (~60%) to the total radiative forcing of the long-lived greenhouse gases, trapping outgoing terrestrial radiation in the infrared region of the spectrum, and an increase in its atmospheric mixing ratio is expected to lead to a warming of Earth's surface. The extent of the warming predicted is currently of considerable debate, but estimates based on future global population and land use suggest that levels of CO_2 may double from preindustrial levels to around 700 ppmv, and that global surface temperatures may increase by around 2–4.5 °C over the next 100 years (Solomon et al. 2013). Current estimates for the budget of CO_2 (see Table 4.2), based on measurements over the period 2002–2011, show that for a perturbation of 8.3 GtC year^{-1}, approximately half is retained in the atmosphere, with major known sinks being the ocean (~ 2.1 GtC year^{-1}) and uptake by land (~ 1.6 GtC year^{-1}). If CO_2 levels rise as expected and the associated warming of the surface is realized, this will have potential consequences for the CO_2 sink terms. Carbon dioxide is less soluble in warmer water, and global warming will lead to a gradual rise in sea-surface temperatures, hence reducing the effectiveness of the ocean as a sink for CO_2. In addition, as the amount of dissolved inorganic carbon in the ocean increases, a decrease in ocean buffering is expected, again reducing the effectiveness of the ocean to take up CO_2. Nevertheless, biological processes in the ocean may themselves counteract these physical changes. For example, favourable changes in the external supply of biologically limiting nutrients such as iron (Fe), silicon (Si), phosphorous (P), and nitrogen (N), could increase the strength of biological production within the ocean surface and assimilate more CO_2 dissolved in the ocean.

It is readily apparent that the climate system is extremely complex. The exchange of CO_2 with terrestrial plants is another important sink. Plants assimilate CO_2 in the process of photosynthesis, releasing O_2, where the gain in carbon is known as the gross primary production (GPP). However, plants also release CO_2 through respiration processes (taking up O_2) and so the difference between assimilation and release of CO_2 gives the net primary production (NPP), i.e. the net amount of new stored carbon. Land NPP is found in tropical forests (33%), tropical savannas and grasslands (22%), cultivated land (6%), temperate forests (12%), wetlands (6%), temperate grassland and shrubland (10%), boreal forests (4%), and deserts and semi-deserts (4%) (Ajtay et al. 1979; Solomon et al. 2013).

Table 4.2 Global budget for an anthropogenic perturbation of CO_2.

	1750–2011 Cumulative PgC	1980–1989 PgC year^{-1}	1990–1999 PgC year^{-1}	2000–2009 PgC year^{-1}	2002–2011 PgC year^{-1}
Atmospheric increase	240 ± 10	3.4 ± 0.2	3.1 ± 0.2	4.0 ± 0.2	4.3 ± 0.2
Fossil-fuel combustion and cement production	375 ± 30	5.5 ± 0.4	6.4 ± 0.5	7.8 ± 0.6	8.3 ± 0.7
Ocean-to-atmosphere flux	−155 ± 30	−2.0 ± 0.7	−2.2 ± 0.7	−2.3 ± 0.7	−2.4 ± 0.7
Land-to-atmosphere flux	30 ± 45	−0.1 ± 0.8	−1.1 ± 0.9	−1.5 ± 0.9	−1.6 ± 1.0
Land-to-atmosphere flux partitioned as					
Net land-use change	180 ± 80	1.4 ± 0.8	1.5 ± 0.8	1.1 ± 0.8	0.9 ± 0.8
Residual land sink	−160 ± 90	−1.5 ± 1.1	−2.6 ± 1.2	−2.6 ± 1.2	−2.5 ± 1.3

Source: based on Solomon et al. (2013).

Assessing the response of plants to increasing CO_2 levels is far from simple. It has been shown that growth promotion occurs under elevated CO_2 levels, and studies suggest that an increase in growth of more than 30% is typical for a doubling of CO_2 concentration. However, growth temperature is very important, and it has been observed that simultaneous increases in air temperature and CO_2 concentration offset the stimulation of biomass and grain yield in rice compared with increases in CO_2 alone. Elevated CO_2 does seem to raise the maximum temperature at which plants can survive, but for any plant, exceeding their optimal growing temperature will have serious consequences on their growth. Therefore, the response of a particular species to changes in CO_2 levels will depend on how far away from its optimal growing temperature it is. Again, in studies on rice, it was found that season-long enrichments of CO_2 from 300 to 700 ppmv, resulted in a 21–27% increase in net canopy photosynthesis and a 10% reduction in total evapotranspiration. Higher levels of CO_2 lead to partial closure of stomata, reducing the stomatal conductance and leading to increasing water-use efficiency. Therefore, increasing CO_2 levels may have an overall positive effect on terrestrial plant growth, but there are caveats, which will be discussed in a later section. The overall lifetime of CO_2 is around 120 years, and it is therefore well mixed in the atmosphere.

4.3.2 Methane

Methane (CH_4) is the most abundant organic species in the atmosphere and contributes about 18% to the radiative forcing attributed to long-lived greenhouse gases. On a molecule per molecule basis, CH_4 is a far more effective greenhouse gas (factor of 20 based on a 100-year time horizon) than CO_2 and therefore understanding its budget is of great importance. In a similar way to CO_2, levels of methane have risen dramatically over the past 200 years. Pre-industrial levels were around 750 ppbv, but now stand at ~1800 ppbv in the Northern Hemisphere and ~1700 ppbv in the Southern Hemisphere. Methane increased at around 1.3% per year for most of the twentieth century until the early 1990s (Blake and Rowland 1988), when the rate of growth slowed to 0.6% per year (Steele et al. 1992) and a near zero growth rate was observed between 1999 and 2006 (Solomon et al. 2013). It is not known with certainty why there has been a slowdown: reduction in sources or an increase in the concentration of the OH radical, the major sink for CH_4, are possibilities. Since 2007, CH_4 levels have risen again by approximately 1%. The current estimated budget for CH_4 is shown in Table 4.3, where it is apparent that both natural and anthropogenic sources are equally important. The anthropogenic sources are evenly spread between fossil-fuel-related release (30% of the total anthropogenic source), waste management (22%), enteric fermentation of cattle (27%), biomass burning (10%), and rice paddies (11%). Natural sources of CH_4 are dominated by wetland emissions, particularly in the tropics. The decomposition of organic matter under oxygen-deficient conditions leads to the production of CH_4 and it is no surprise that where temperature is highest and microbial activity most intense, the largest natural source of CH_4 is found. Termites also produce a non-negligible quantity of CH_4, and other insects may well do the same. The dominant loss process for CH_4 in the atmosphere is via reaction with the hydroxyl radical, OH, $k_{298} = 6 \times 10^{-15}$ cm^3 molecule^{-1} s^{-1} (DeMore et al. 1997), with minor contributions from stratospheric removal and consumption of methane in soils by methanotrophic bacteria. The global residence time of CH_4 in the atmosphere is approximately 10 years, leading to a slight inter-hemispheric gradient.

Table 4.3 Sources and sinks for methane in the atmosphere expressed as Tg CH_4 per year.

Tg (CH_4) year^{-1}	1980–1989		1990–1999		2000–2009	
	Top-down	Bottom-up	Top-down	Bottom-up	Top-down	Bottom-up
Natural sources	193 (150–267)	355 (244–466)	182 (167–197)	336 (230–465)	218 (179–273)	347 (238–484)
Natural wetlands	157 (115–231)	225 (183–266)	150 (144–160)	206 (169–265)	175 (142–208)	217 (177–284)
Other sources	36 (35–36)	130 (61–200)	32 (23–37)	130 (61–200)	43 (37–65)	130 (61–200)
Freshwater (lakes and rivers)		40 (8–73)		40 (8–73)		40 (8–73)
Wild animals		15 (15–15)		15 (15–15)		15 (15–15)
Wildfires		3 (1–5)		3 (1–5)		3 (1–5)
Termites		11 (2–22)		11 (2–22)		11 (2–22)
Geological (incl. oceans)		54 (33–75)		54 (33–75)		54 (33–75)
Hydrates		6 (2–9)		6 (2–9)		6 (2–9)
Permafrost (excl. lakes and wetlands)		1 (0–1)		1 (0–1)		1 (0–1)
Anthropogenic sources	348 (305–383)	308 (292–323)	372 (290–453)	313 (281–347)	335 (273–409)	331 (304–368)
Agricultural and waste	208 (187–220)	185 (172–197)	239 (180–301)	187 (177–196)	209 (180–241)	200 (187–224)
Rice		45 (41–47)		35 (32–37)		36 (33–40)
Ruminants		85 (81–90)		87 (82–91)		89 (87–94)
Landfills and waste		55 (50–60)		65 (63–68)		75 (67–90)
Biomass burning (incl. biofuels)	46 (43–55)	34 (31–37)	38 (26–45)	42 (38–45)	30 (24–45)	35 (32–39)
Fossil fuels	94 (75–108)	89 (89–89)	95 (84–107)	84 (66–96)	96 (77–123)	96 (85–105)
Sinks						
Total chemical loss	490 (450–533)	539 (411–671)	515 (491–554)	571 (521–621)	518 (510–538)	604 (483–738)
Tropospheric OH		468 (383–567)		479 (457–501)		528 (454–617)
Stratospheric OH		46 (16–67)		67 (51–83)		51 (16–84)
Tropospheric Cl		25 (13–37)		25 (13–37)		25 (13–37)
Soils	21 (10–27)	28 (9–47)	27 (27–27)	28 (9–47)	32 (26–42)	28 (9–47)
Global						
Sum of sources	541 (500–592)	663 (536–789)	554 (529–596)	649 (511–812)	553 (526–569)	678 (542–852)
Sum of sinks	511 (460–559)	567 (420–718)	542 (518–579)	599 (530–668)	550 (514–560)	632 (592–785)
Imbalance (sources minus sinks)	30 (16–40)		12 (7–17)		3 (−4–19)	
Atmospheric growth rate	34		17		6	

Note: The values in brackets represents minimum and maximum values.
Source: data taken from Solomon et al. (2013).

4.3.3 Carbon Monoxide

The budget of CO is dominated by the atmospheric oxidation of methane and other VOCs (see next section), initiated by the OH radical (see Chapter 5) and from incomplete combustion associated with biomass burning and fossil fuels. Vegetation and oceans are also a non-negligible source of CO. Hence, anthropogenic sources dominate, although it should be noted that the oxidation of natural VOCs released from vegetation provides over half the source from VOC oxidation. Despite having no significant direct impact on global warming, CO is an extremely important atmospheric species, whose major loss process is reaction with OH ($k_{298} = 2.4 \times 10^{-13}$ cm^3 molecule^{-1} s^{-1}; DeMore et al. 1997) giving the atmospheric residence time of ca. two months.

$$CO + OH \rightarrow CO_2 + H \tag{4.7}$$

In fact, reaction with CO constitutes about 70% of the total sink for OH, hence the importance of CO in Earth's atmosphere. A strong gradient exists between Northern Hemisphere and Southern Hemisphere CO levels. For example, current background levels of CO are 60–70 ppbv in the Southern Hemisphere and 120–180 ppbv in the Northern Hemisphere, with extremely high levels experienced in urban and industrialized areas (as high as ppm levels). The strong gradient exists because the Northern Hemisphere has much larger anthropogenic sources than does the Southern Hemisphere. Furthermore, the abundance of the OH radicals within the tropics prevents large-scale inter-hemispheric transport. A summary of the global atmospheric budget is given in Table 4.4.

4.3.4 Volatile Organic Compounds

Myriad VOCs are emitted into the atmosphere by both biogenic and anthropogenic sources. These VOCs have varying lifetimes, depending on their structure, with the vast majority undergoing photooxidation initiated by the OH radical (Chapter 5), ultimately leading to the production of CO_2 and H_2O.

Table 4.4 Global budget for carbon monoxide in the atmosphere.

In-situ		Direct		Sinks (Tg CO per year)	
		Sources (Tg CO per year)			
Oxidation of CH$_4$	800	Biomass burning	700	Reaction with OH	2100 ± 600
Oxidation of isoprene	270	Vegetation	150	Uptake by soils	400 ± 100
Oxidation of industrial NMHC	110	Ocean	50	Flux into stratosphere	110 ± 30
Oxidation of biomass NMHC	30	Fossil and domestic fuel	650		
Oxidation of acetone	20				
Subtotal	1230		1550		
Total sources			2780 ± 1000	Total sinks	2600 ± 780

NMHC, non-methane hydrocarbons.
Sources: data taken from Graedel and Crutzen (1993) and Solomon et al. (2013).

Photooxidation may lead to the production of intermediate compounds such as carboxylic acids and nitrates, which are soluble and therefore can be removed by wet deposition. In general, the biogenic VOCs are more reactive in the atmosphere than the anthropogenic VOCs. Thus, the biogenic VOCs play a significant role in photochemical oxidation cycles even when their concentrations are lower. Natural sources of VOCs are believed to dominate (Singh and Zimmerman 1992), constituting around 90% of the carbon flux. Apart from methane and dimethyl sulphide, the major source of biogenic VOCs is terrestrial plants (Fall 1999). In some cases, the reason why plants manufacture certain VOCs is well known. For example, ethene plays a key role in growth and development and its production is greatly enhanced following plant wounding, exposure to chemicals and infection (Fall 1999). However, for other VOCs such as isoprene (2-methyl 1,3-butadiene, C_5H_8), the reason for its production is less clear.

Emission of isoprene is in fact the single largest source of carbon to the atmosphere after methane. In general, deciduous trees and most woody plants emit isoprene, although ferns, vines, and some herbaceous plants are also emitters. Emission of isoprene is light dependent, dropping to zero in the dark, and is also temperature dependent, with the flux increasing rapidly with temperature. Figure 4.3 shows the distribution of isoprene emissions derived from a land-surface vegetation model (Wang and Shallcross 2000). Singsaas et al. (1997) have suggested that isoprene acts as a thermal protectant, which is consistent with the observation that emissions are elevated as temperature increases. However, there are problems with this explanation in that many species in hot desert environments do not emit isoprene at all, whereas

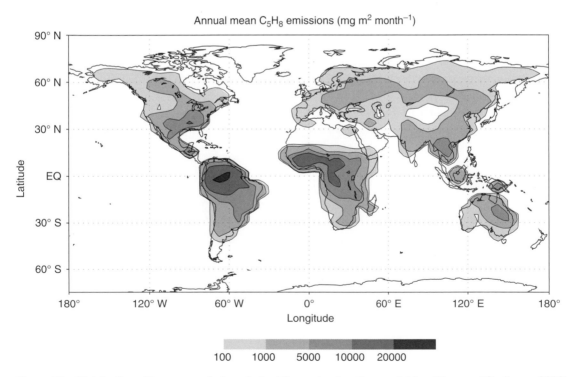

Figure 4.3 Distribution of isoprene emissions derived from a land-surface model (see Wang and Shallcross 2000) using emission algorithms from Guenther et al. (1995).

plants of the same genus in more temperate climates do (Fall 1999). Another possible explanation includes neutralization of free radicals within the plant. Isoprene is extremely short-lived, whose rate coefficient for reaction with OH is near the gas-collision limit ($k_{298} = 1.01 \times 10^{-10}$ cm^3 molecule^{-1} s^{-1}, DeMore et al. 1997), giving rise to a lifetime of a few hours. Consequently, levels of isoprene in the atmosphere are quite variable and display a strong diurnal (day–night) cycle, peaking in mid-afternoon and rapidly decaying at night, as well as a strong seasonal cycle, peaking in late spring. In forests, levels of isoprene can reach 15 ppbv, although pptv levels are more typical elsewhere.

Chapter 5 presents a detailed discussion of the role played by VOCs in the presence of NO$_x$ (NO and NO$_2$) in the production of tropospheric ozone. At Earth's surface, ozone is deleterious to plants, entering the stomata and destroying cell lining, whilst being itself a greenhouse gas. As levels of CO$_2$ increase, it is anticipated that global emissions of isoprene will also increase, driven mainly by increasing surface temperature. It has been noted that isoprene may be responsible for a significant proportion of ozone production in the troposphere, and it is possible that isoprene emissions will change in the future, changing tropospheric ozone. Once again, natural biogeochemical cycles are intricately coupled with the climate system.

Monoterpenes (C$_{10}$H$_{16}$) are another major class of VOC manufactured by plants. A selection of the most common monoterpenes is shown in Figure 4.4, whose ecological role includes herbivore defence, attraction of pollinators, and allelopathic effects on competing plants. Monoterpene emissions are dependent on temperature rather than light, so emissions continue into the night. Monoterpenes rapidly react with the OH radical, and also with the NO$_3$ radical and like isoprene have short lifetimes of a few hours to days. Plants also emit C$_6$ aldehydes and alcohols known as the hexenal family, which have antibiotic properties (Hatanaka 1993).

Anthropogenic emissions of VOCs are dominated by combustion of fossil fuel, with alkanes, aromatics, and alkenes being the major VOC classes. In urban areas, emissions of VOCs follow a typical diurnal cycle, tracking rush-hour traffic flow. Photochemical smog formation in the summer months is driven by the emission of these VOCs, in the presence of NO$_x$ and sunlight (Chapter 5). In urban areas, individual VOC levels can be of the order of some ppbv, whereas in background air levels of the more reactive species can be in the pptv range or less.

Worked Example 4.3

Carbon cycle
Given the concentration of CO$_2$ (400 ppmv), CH$_4$ (1.8 ppmv), and CO (0.1 ppmv), what is the fraction of carbon in each reservoir?

Simple Answer

$$CO_2 = 400/401.9 = 0.9953$$
$$CH_4 = 1.8/401.9 = 0.0045$$
$$CO = 0.1/401.9 = 0.0003$$

Not all the compound is carbon, though:

$$CO_2 \text{ fraction of carbon} = 12 / 44$$
$$= \text{molar mass of carbon} / \text{molar mass of } CO_2$$

Figure 4.4 A selection of the most common monoterpenes found in the atmosphere.

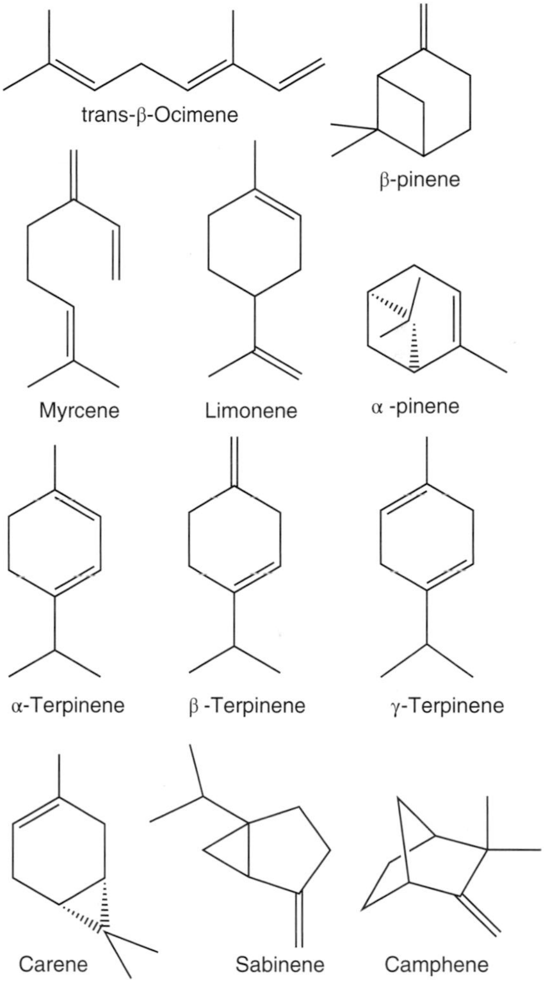

trans-β-Ocimene

β-pinene

Myrcene   Limonene   α-pinene

α-Terpinene   β-Terpinene   γ-Terpinene

Carene   Sabinene   Camphene

CH_4 fraction of carbon $= 12/16$

    $=$ molar mass of carbon / molar mass of CH_4

CO fraction of carbon

    $= 12/28$ molar mass of carbon / molar mass of CO

Revised Answer

    $CO_2 = 109.1/110.5 = 0.9874$
    $CH_4 = 1.35/110.5 = 0.01222$
    $CO = 0.04/110.5 = 0.0004$

Not a large difference, but the importance of CH_4 has doubled!

4.4 Nitrogen

Nitrogen (N_2) represents nearly 80% of Earth's atmosphere and is chemically inert throughout the lower atmosphere. Only in the upper atmosphere, where significant fluxes of very short wavelength radiation can be found, N_2 can be broken down by photolysis. Despite their small and highly variable sources, the other nitrogen-containing species present in the atmosphere at significant levels, ammonia (NH_3), NO, and NO_2 play a disproportionately large role in determining atmospheric composition. Ammonia, for example, is the only alkaline gas in the atmosphere and is important in the neutralization of acid aerosols. Nitrogen oxides, NO and NO_2, play a vital role in the production and destruction of ozone in both the troposphere (Chapter 5) and the stratosphere (Chapter 8) and therefore affect the radiative budget of Earth and the oxidizing capacity of the atmosphere.

4.4.1 Nitrous Oxide

Nitrous oxide (N_2O) is one of the most important greenhouse gases, whose current atmospheric mixing ratio is about 330 ppbv, compared with an estimated preindustrial level of around 285 ppbv with a mean growth rate of ~0.9 ppb year^{-1} over the past 10 years. N_2O has a global warming potential (GWP), which is ~300 times higher than that of CO_2 for a time horizon of 100 years. Nitrous oxide has an extremely long lifetime (~150 years) and is virtually inert in the troposphere, being destroyed in the stratosphere by direct photolysis and reaction with $O(^1D)$ atoms:

$$N_2O + h\nu \rightarrow N_2 + O \tag{4.8}$$

$$N_2O + O(^1D) \rightarrow N_2 + O_2 \qquad \sim 40\% \tag{4.9}$$

$$N_2O + O(^1D) \rightarrow NO + NO \qquad \sim 60\% \tag{4.10}$$

Nitrous oxide is released to the atmosphere from both soils and aquatic systems with soils under natural vegetation (undisturbed soils) (37%), cultivated soils (agriculture) (23%), and oceans (21%) making up the bulk of the known sources. The fossil-fuel combustion and various industrial processes can also make a non-negligible contribution to the sources of N_2O. Overall, the contribution of anthropogenic sources is about the same as natural terrestrial sources (Solomon et al. 2013). Denitrifying bacteria transform nitrate to N_2 and some N_2O under anaerobic conditions, which can then evade to the surface and enter the atmosphere. The major uses of nitrogen fertilizers in the Northern Hemisphere make a small northern to southern gradient of ~1.2 ppbv. A budget for N_2O is presented in Table 4.5 based on the Solomon et al. (2013) assessment. There are many uncertainties, including the estimation of the ocean source, since N_2O is both lost to and emitted from the oceans. Since it delivers NO_x to the stratosphere, N_2O plays an important role in controlling the abundance of stratospheric ozone.

4.4.2 Ammonia

Ammonia has a short lifetime in the atmosphere of approximately one to two days and is removed mainly by both wet and dry deposition processes, with some additional small loss processes via reaction with OH

Table 4.5 Estimates of the global sources and sinks for N_2O in Tg N per year.

Anthropogenic sources	IPCC estimate 2006–2011
Fossil-fuel combustion and industrial processes	0.7 (0.2–1.8)
Agriculture	4.1 (1.7–4.8)
Biomass and biofuel burning	0.7 (0.2–1.0)
Human excreta	0.2 (0.1–0.3)
Rivers, estuaries, coastal zones	0.6 (0.1–2.9)
Atmospheric deposition on land	0.4 (0.3–0.9)
Atmospheric deposition on ocean	0.2 (0.1–0.4)
Surface sink	−0.01 (0–0.1)
Total anthropogenic sources	6.9 (2.7–11.1)
Natural Sources	IPCC estimate 2006–2011
Soils under natural vegetation	6.6 (3.3–9.0)
Oceans	3.8 (1.8–9.4)
Lightning	–
Atmospheric chemistry	0.6 (0.3–1.2)
Total natural sources	11.0 (5.4–19.6)
Total natural + anthropogenic sources	17.9 (8.1–30.7)
Stratospheric sink	14.3 (4.3–27.2)
Observed growth rate	3.61 (3.5–3.8)

Source: data taken from Solomon et al. (2013).
Note: The values in brackets represents minimum and maximum values.

and NO_3 radicals. The main sources arise from biological activity, such as the decomposition of urea in animal urine by enzymes, the decomposition of excrement and the release from soils and the ocean following mineralization of organic material. Anthropogenic sources are roughly four-fold higher than natural sources, which centre around its use in fertilizers and as a byproduct of waste production. Since deposition processes dominate its loss and sources are diverse, levels of ammonia are highly variable, ranging from 0.1 to 10 ppbv over continental regions. It has already been noted that ammonia is the only alkaline gas in the atmosphere and neutralizes H_2SO_4, HNO_3, and HCl to form ammonium salts leading to aerosols. Deposition of ammonium salts to the soil decreases its pH, leading to a decline in growth and the aerosols from these salts can deteriorate regional air quality and atmospheric visibility and influence global radiation budgets (see Chapter 2).

4.4.3 Nitrogen Oxides

In the atmosphere, NO and NO_2 (NO_x) are extremely tightly coupled during sunlit hours and rapidly interconvert with one another in the presence of ozone (see Chapter 5):

$$NO + O_3 \rightarrow NO_2 + O_2 \tag{4.11}$$

$$NO_2 + h\nu \rightarrow NO + O \tag{4.12}$$

$$O + O_2 + M \rightarrow O_3 + M \tag{4.13}$$

Individually, NO and NO_2 have extremely short lifetimes of the order of seconds, but by considering the two compounds together as NO_x, the lifetime is lengthened to many hours. Hence, NO and NO_2 display strong diurnal cycles, and their concentrations will display a seasonal cycle. In urban areas, NO_x can reach hundreds of ppbv, and in particularly polluted environments ppmv levels, whereas clean maritime air will have levels of only 5–10 pptv. The major loss processes for NO_x are conversion to HNO_3 via reaction with OH:

$$NO_2 + OH + M \rightarrow HNO_3 + M \tag{4.5}$$

and dry deposition of NO_x. Nitric acid can be removed by wet and dry deposition, constituting a loss of NO_x from the atmosphere. In the troposphere, in the presence of VOCs, NO_x can promote the formation of ozone (Chapter 5) and can also be sequestered to form temporary nitrate reservoirs such as peroxyacetylnitrate (PAN), which is an acyl nitrate ($CH_3C(O)O_2NO_2$). PAN is an excellent indicator of the photochemical processing of an air mass, which can allow NO_x to be transported away from source regions and influence chemistry on regional and global scales.

For most sources, NO_x is emitted in the form of NO. Natural sources of NO are from soil processes and lightning discharge, whilst an ever-increasing source is that from the high-temperature combustion of fossil fuels. At the high temperatures inside an internal combustion engine

$$O + N_2 \rightarrow NO + N \tag{4.14}$$

$$N + O_2 \rightarrow NO + O \tag{4.15}$$

lead to the formation of NO. There are considerable uncertainties within the budget for NO_x, such as the soil and lightning source strengths. However, the burden from fossil fuels is reasonably well defined and set to increase globally, despite the growing use of three-way catalysts in vehicles. This is summarized in Table 4.6.

4.5 Sulphur

Sulphur is incorporated into the amino acid residue cysteine and methionine and is vital for the structural integrity of proteins; therefore, it is an essential element for living organisms on Earth. Cross-linking in proteins via sulphur–sulphur bonds is of great importance; intermolecular 'disulphide' linkages can lead to large-scale structures such as nails, whilst intramolecular linkages allow the protein to adopt specific configurations as in the case of an enzyme. On decomposition by bacteria, organic sulphur compounds usually release hydrogen sulphide (H_2S). However, H_2S is but one of many sulphur compounds released into the atmosphere, which include sulphur dioxide (SO_2), carbonyl sulphide (OCS), carbon disulphide (CS_2) and dimethyl sulphide (CH_3SCH_3, also known simply as DMS).

Since the onset of the Industrial Revolution, the sulphur burden in the atmosphere has increased dramatically due to the burning of fossil fuels that inevitably contain some sulphur. Hence, the anthropogenic

Table 4.6 Atmospheric budget for NO_x and NH_3 (TgN year^{-1}).

Anthropogenic	NO_x	NH_3
Fossil-fuel combustion and industrial processes	28.3	0.5
Agriculture	3.7	30.4
Biomass and biofuel burning	5.5	9.2
Total	**37.5**	**40.1**
Natural sources		
Soils under natural vegetation	7.3 (5–8)	2.4 (1–10)
Oceans	–	8.2 (3.6)
Lightning	4 (3–5)	–
Total	**11.3**	**10.6**
Total anthropogenic + natural	**48.8**	**50.7**
Deposition from the atmosphere		
Continents	27.1	36.1
Oceans	19.8	17.0
Total	**46.9**	**53.1**

Source: data from Solomon et al. (2013).
Note: The values in brackets represents minimum and maximum values.

contribution to the total sulphur emission budget, mainly in the form of SO_2, approaches 75%, with the bulk of these emissions (around 90%) emanating from the Northern Hemisphere. Natural emission sources, which make up the remaining 25%, are pretty evenly distributed over the two hemispheres, with a slight bias towards the Northern Hemisphere (Brasseur et al. 1999). It should be noted that a very large amount of sulphate is released into the atmosphere from the oceans in the form of sea salt. However, these very coarse particles are rapidly deposited back to the ocean and play no further role in the global sulphur cycle. In addition, sulphur-containing minerals are transported around the globe during wind-driven erosion of soils.

A summary of the most recent estimates of the global budgets for each of the five main sulphur species is presented in Table 4.7. It is apparent that each of these budgets has a large error associated with it, despite intensive research over the past 25 years. The diverse range of sources and their often inaccessibility to study means that the sulphur budget is still only known approximately and new sources and sinks for each compound may well emerge in the future.

4.5.1 Sulphur Dioxide

The burning of fossil fuels is the major source of SO_2 in the atmosphere, with volcanoes, smelting of metal sulphides and oil refineries being other significant sources. The contribution to the SO_2 budget from the oxidation of natural reduced sulphur compounds (see later) is quite minor by comparison. The sinks of SO_2 are dry and wet deposition, and oxidation by OH and sCI in the gas phase or by O_3 and H_2O_2

Table 4.7 Global annual sources and sinks of OCS, CS$_2$, H$_2$S, DMS and SO$_2$ in Tg per year.

		OCS	CS$_2$	H$_2$S	DMS	SO$_2$
Source	Open ocean	0.10±0.15[a] 0.89[b]	0.11±0.04[a,c] 0.24[b]	1.50±0.60[a]	20.70±5.20[a] 28.1 (17.6–34.4)[d]	
	Coastal ocean	0.10±0.05[a]	0.04±0.02[a,c]	0.30±0.10[a]		
	Salt marshes	0.10±0.05[a]	0.03±0.02[a]	0.50±0.35[a]	0.07±0.06[a]	
	Anoxic soils	0.02±0.01[a]	0.07±0.06[a]			
	Vegetation	0.37±0.07[a]			1.58±0.86[a]	
	Tropical forests	0.42±0.12[a]			1.60±0.50[a]	
	Soils	0.29±0.17[a]			0.29±0.17[a]	
	Wetlands	0.03±0.03[a]	0.02±0.02[a]	0.20±0.21[a]	0.12±0.07[a]	
	Volcanism	0.05±0.04[e] 0.01[b]	0.05±0.04[e] 0.02[b]	1.05±0.94[f]	20[g]	
	Precipitation	0.13±0.08[a]				
	OCS + OH	0.08±0.07[e]				
	DMS oxidation	0.17±0.04[h]				
	CS$_2$ oxidation	0.42±0.12[e]				
	Biomass burning	0.07±0.05[i] 0.04[b]	1.8[g]			
	Anthropogenic	0.12±0.06[a] 0.38[b]	0.34±0.17[e] 0.70[b]	3.30±0.33[j]	0.13±0.04[a] 1.50[b]	176[g]
Total		1.31±0.68[a]	0.66±0.37[a]	7.72±2.79[a]	24.49±5.30[a]	200
Sink	Oxic soil	0.92±0.78[a]	0.44±0.38[a]			
	Vegetation	0.56±0.10[a]				
	Reaction with OH	0.13±0.10[e]	0.57±0.25[e]	8.50±2.80[a]	25.00±1.30[a]	20[g]
	Reaction with O	0.02±0.01[e]				
	Photolysis	0.03±0.01[e]				
	Cloud scavenging	180[g]				
Total		1.66±1.00	1.01±0.63	8.50±2.80	25.00±1.30[a]	200
Total imbalance		0.35±1.68	0.35±1.00	0.78±5.59	0.51±6.60	–

Sources:
[a] Watts (2000);
[b] Lee and Brimblecombe (2016).
[c] Xie et al. (1997);
[d] Lana et al. (2011);
[e] Chin and Davis (1993);
[f] Andreae (1990);
[g] Möller (1984);
[h] Barnes et al. (1994);
[i] Nguyen et al. (1995);
[j] Watts and Roberts (1999);

in the aqueous phase, which makes its lifetime from a few hours to several days. Sulphur dioxide emissions are strongly linked with acid rain formation, which can damage plant and animal life as well as damage buildings and monuments and can lead to respiratory problems in humans and animals.

The presence of aerosols in the atmosphere has an influence on Earth's climate, resulting from both direct and indirect effects on the radiation budget (see Chapter 2). The direct effect arises because aerosols scatter and absorb incoming solar radiation, thereby reducing the energy reaching ground level. The indirect effect results from the role of aerosols in cloud formation, since clouds reflect incoming radiation. Both effects are influenced by the number, size distribution, and chemical composition of the aerosol, and are currently believed to lead to atmospheric cooling, which offsets the warming influence of radiatively active greenhouse gases such as CO_2.

The majority of aerosols in the atmosphere are generated as a result of gas-to-particle conversion processes, although there are substantial contributions from other sources, such as resuspended mineral dust and sea-salt aerosols. An essential prerequisite for gas-to-particle conversions to occur is the presence of a species in the gas phase at a partial pressure in excess of its saturation vapour pressure with respect to the condensed phase (i.e. condensable material). It is generally accepted that the most significant condensable molecule formed in the troposphere is sulphuric acid (H_2SO_4), which has also been long recognized as the most important from the point of view of the nucleation of new particles. The major source of H_2SO_4 results from the oxidation of anthropogenically derived SO_2, for which the predominant gas phase oxidation pathway is initiated by reaction with the OH radical and sCIs:

$$OH + SO_2 + M \rightarrow HOSO_2 + M \tag{4.16}$$

$$HOSO_2 + O_2 \rightarrow SO_3 + HO_2 \tag{4.17}$$

$$sCI + SO_2 \rightarrow SO_3 + carbonyl \tag{4.18}$$

$$SO_3 + H_2O + M \rightarrow H_2SO_4 + M \tag{4.19}$$

In recent years, the role of sCIs has become important in sulphur chemistry due to the rapid oxidation of SO_2, which accounts for significant production of SO_3 (and ultimately H_2SO_4). The oxidation of SO_2 by sCIs can rival the OH oxidation pathway, especially in high alkenes emission terrestrial regions (e.g. rainforest, boreal forest, heavily polluted urban environments). The typical rate coefficients for $k_{4.18}$ ~ 4.0×10^{-11} cm^3 molecule^{-1} s^{-1} and $k_{4.16}$ ~ 9.0×10^{-13} cm^3 molecule^{-1} s^{-1} and a daytime [OH] ~ 1.0×10^6 molecule cm^{-3} results in a [sCI] ~ 2.3×10^4 molecule cm^{-3} being required for the two SO_2 oxidation rates to be equal. A variety of model-measurement studies have estimated surface level of sCI in the range of 1×10^4 molecule cm^{-3} to 1×10^5 molecule cm^{-3} (Khan et al. 2018), so sCI can make a significant contribution to H_2SO_4 formation in the terrestrial boundary layer.

Once formed, sulphuric acid will either be taken up by existing aerosols or will create new ones. Sulphur dioxide is sufficiently soluble for aqueous phase oxidation to be another important route for its conversion to sulphuric acid (K_H ~1.23 mol l^{-1} atm^{-1} at 25° C). Once taken up into the aqueous phase, SO_2 establishes an equilibrium with the bisulphite ion and the sulphite ion:

$$SO_2(aq) \leftrightarrow H^+ + HSO_3^- \tag{4.20}$$

$$HSO_3^- \leftrightarrow H^+ + SO_3^{2-} \tag{4.21}$$

where K_1 is the equilibrium constant for reaction (4.20) and K_2 is the equilibrium constant for reaction (4.21). The Henry's law coefficient for SO_2, $K_H(SO_2)$, can be expressed as

$$K_H\left(SO_2\right) = \left[SO_2\left(aq\right)\right]/P\left(SO_2\right) \tag{4.22a}$$

where (SO_2 [aq]) is the aqueous concentration of SO_2 and $P(SO_2)$ is the partial pressure of SO_2 in the gas-phase, but, because of the formation of bisulphite (the dominant species in HSO_3^- in aqueous media) and sulphite ions, it is more common to use an effective *Henry's law coefficient,* termed $k_{Heff}(SO_2)$ and defined as

$$k_{Heff}\left(SO_2\right) = \left(\left[SO_2\left(aq\right)\right] + \left[HSO_3^-\right] + \left[SO_3^{2-}\right]\right)/P\left(SO_2\right) \tag{4.22b}$$

Eq. (4.22) can then be rearranged in the form

$$K_{Heff}\left(SO_2\right) = k_H\left(SO_2\right)\left(1 + K_1/\left[H^+\right] + K_1 K_2/\left[H^+\right]^2\right) \tag{4.23}$$

Inspection of Eq. (4.23) shows that solubility of SO_2 is dependent on pH, decreasing at low solution pH, where the $k_{Heff}(SO_2)$ then approaches the value of $k_H(SO_2)$. The aqueous-phase oxidation of the bisulphite ion to the sulphate ion is dominated by H_2O_2 for pH < 5, with oxidation by dissolved ozone becoming the main contributor for pH > 4.5. Other mechanisms for aqueous-phase oxidation exist (see Jacob and Hoffmann 1983) but are generally less important. Depending on the amount of moisture in the atmosphere, 20–80% of the SO_2 emitted is oxidized to sulphate, with the remainder being removed by dry deposition. The mixture of SO_2 and sulphate (i.e. sulphur tropospheric aerosol) has a lifetime of between two and six days before being lost via wet or dry deposition. Because of the short-lived nature of sulphur tropospheric aerosols, the cooling effect is very localized and short-lived compared with the warming effects of greenhouse gases.

Highest SO_2 mixing ratios are found over the major industrial regions of the world, the eastern USA, Europe, and the far east, where levels are as high as a few ppb. Reliable measurements of SO_2 in remote regions are problematic due to loss of SO_2 on moist surfaces, such as instrument inlets, but, mixing ratios drop to below 100 pptv.

4.5.2 Carbonyl Sulphide

There are a large variety of OCS sources, both natural and anthropogenic. Coastal regions of oceans and areas of high biological activity are found to have higher fluxes of OCS than open waters, yet overall the oceans are a significant source. Biomass burning accounts for almost 10% of the global OCS emissions yet there is no clear distinction as to what percentage is natural (from forest fires) or man-made (from burning agricultural waste, deforestation, firewood burning). Soils, wetlands, and volcanoes are other sources of OCS, but their impact is far less important than the other sources. The two other important sources of OCS are the oxidation of CS_2 and DMS. OCS is relatively insoluble ($K_H \sim 2.2 \times 10^{-2}$ mol l^{-1} atm^{-1} at 25° C) and the major loss processes are uptake by oxic soils (such as Aridsols) and vegetation. The reaction of OCS with the OH radical is slow ($k_{298} = 2.0 \times 10^{-15}$ cm^3 molecule^{-1} s^{-1}; Atkinson et al. 2004), and therefore gas-phase removal in the troposphere is a minor loss process.

Plants are found to be able to take up OCS by the enzyme carbonic anhydrase under dark conditions as long as the plant is able to take up OCS through the leaf stomata or the leaf cuticle. Soils have a temperature optimum for OCS uptake, probably due to the dependence on enzyme catalysed processes, where the enzymes amplify the trace gas exchange up to a certain threshold. The vegetation and soil uptake fluxes of OCS are still sparse and underestimated, which can reflect its global budget shown in Table 4.7. The lifetime of OCS in the atmosphere is very large (approximately 4 years) with respect to all loss processes, making it the most long-lived of the sulphur species considered. Since OCS is so long-lived, it can be transported up to the stratosphere, where it is photolysed by UV radiation, or reacts with O atoms to subsequently act as a source of SO_2 and ultimately sulphate particles (stratospheric sulphate aerosol known as the *Junge layer*). The Junge layer is formed of a mixture of 25% H_2O and 75% H_2SO_4, which occurs at an altitude of approximately 20 km during volcanically quiescent periods. Due to the longer residence time of the stratospheric aerosol compared with the tropospheric aerosol (which are rapidly scavenged in the atmosphere) and because of their cooling effects, stratospheric aerosols can minimize the warming due to increasing greenhouse gas levels in the atmosphere.

Typical mixing ratios of OCS in the troposphere are around 500 pptv, and do not vary dramatically with altitude. The combination of low solubility, low reactivity with OH, and somewhat diffuse sources would be concomitant with these observations. Latitudinal measurements differ somewhat, but more recent studies suggest that a gradient does exist across the hemispheres that changes with season: higher concentrations in the summer in the Southern Hemisphere perhaps reflecting the dependence of production from the ocean and from the oxidation of DMS on the availability of sunlight and lower concentrations in the summer in the Northern Hemisphere likely to follow that of vegetation, which is the dominant sink of OCS.

4.5.3 Carbon Disulphide

Carbon disulphide (CS_2) is naturally produced from rotting organic material in oceanic, coastal, and marshlands, but spatial analysis indicates large concentrations over continental masses, suggesting the main sources of CS_2 are anthropogenic (Watts 2000). Anthropogenic sources such as chemical processing dominate, being over 50% of the estimated source, most notably in the production of cellulose. Soils and wetlands are thought to be the other natural sources, but the emission fluxes of CS_2 from these sources are sparse (Watts 2000), which make them extremely hard to estimate. The significantly shorter lifetime for CS_2, compared with OCS, results in a highly nonuniform distribution in the troposphere, with very little penetration into the stratosphere expected. Observed levels range from 2 pptv under clean marine conditions up to approximately 300 pptv in areas heavily influenced by anthropogenic emissions.

The lifetime of CS_2 is of the order of a week: the major loss process is reaction with the OH radical ($k_{298} = 1.7 \times 10^{-12}$ cm^3 molecule^{-1} s^{-1}, Atkinson et al. 2004), with SO_2 and OCS as major products. These oxidation products can increase the formation of the global atmospheric sulphate aerosol. The dry deposition through vegetation is also a significant loss process of CS_2. Carbon disulphide is another relatively insoluble species ($K_H = 5.5 \times 10^{-2}$ mol l^{-1} atm^{-1} at 25° C), and therefore wet deposition is unlikely to be important.

4.5.4   Hydrogen Sulphide

The budget of H_2S is the least well characterized of all the sulphur compounds considered here. However, it is widely agreed that H_2S is the major reduced sulphur compound released from soils and vegetation and is also a significant component of the marine budget (Möller 1984). Like the other sulphur compounds considered thus far, it is quite insoluble (K_H ~9.5×10^{-2} mol l^{-1} atm^{-1} at 25° C) and in conjunction with available seawater measurements it would appear that H_2S is supersaturated in seawater.

Hydrolysis of OCS and reduction of sulphates appear to be the main source of H_2S in seawater, although there is evidence suggesting that H_2S formation is related to primary production in the oceans. Hydrolysis of OCS is a chemical process that occurs in the oxic water column to maintain H_2S concentrations. Bacteria in anoxic environments can reduce sulphate to respire, producing H_2S as a byproduct. The H_2S can then be transported up the water column towards the surface where it can escape to the atmosphere. The shallow water environments such as estuaries, mudflats, salt marshes, and swamps are all active H_2S emission regions for this reason. In soils, H_2S is produced by the degradation and mineralization of sulphur containing organic matter by bacteria. Another major natural source of H_2S is volcanoes. Like CS_2, anthropogenic sources of H_2S, such as the combustion of fossil fuel, dominate, contributing nearly half the total burden.

The main sink of H_2S is the reaction with OH radical ($k_{298} = 4.7 \times 10^{-12}$ cm^3 molecule^{-1} s^{-1}, Atkinson et al. 2004) to form the HS radical giving rise to a lifetime of just two to three days. The HS radical is, in turn, rapidly oxidized in the atmosphere to SO_2, although the precise details of the mechanism are uncertain. The short lifetime of H_2S means that its distribution in the troposphere is highly variable, and marine levels can vary between 5 and 100 pptv, rising to as high as many hundreds of pptv in wetland regions (Brasseur et al. 1999).

4.5.5   Dimethyl Sulphide

Dimethyl sulphide (CH_3SCH_3) commonly known as DMS, is the most abundant natural sulphur compound emitted into the atmosphere. Haas first discovered the release of DMS from phytoplankton in 1935, and in 1948 Challenger and Simpson showed that DMS was generated from the Zwitter ion, dimethyl-sulphone-propionate (DMSP), also known as dimethyl-β-propiothetin. It is commonly believed that DMSP is produced by phytoplankton as an osmoregulating substance, which is released by grazing from zooplankton, cell leakage, senescence, or viral infection (Watts 2000). Once in the water, DMSP can be metabolized by the enzyme DMSP-ase, found intracellularly and in DMS-producing bacteria leading to DMS production.

Dimethyl sulphide is largely insoluble (K_H ~4.74×10^{-1} mol l^{-1} atm^{-1} at 25 °C and 32.5 salinity units) and, therefore, degasses from the water column. Although other sources of DMS have been identified, such as from vegetation, soils, wetlands, and anthropogenic sources, production from the oceans is believed to dominate (Table 4.7). However, it should be noted that estimates of the oceanic are wide ranging, from around 10 Tg year^{-1} up to 110 Tg year^{-1} (Watts 2000) (1 Tg = 1×10^{12} g).

Two approaches have been used to estimate the oceanic flux, one using seawater measurements and a mass transfer coefficient from sea to air (Liss and Slater 1974; Lana et al. 2011), the other using air concentrations of DMS and its lifetime (Watts 2000). Both methods have their flaws, for the former the actual mass transfer coefficient used will depend on which tracer, such as CO_2 or Rn, the measurement is based on, for example, Liss and Merlivat (1986) and will vary with wind-speed, for which several parameterizations exist (e.g. Wanninkhof 1992). Air–sea transfer is also affected by the composition of the microlayer at the interface, which is poorly understood. For the latter approach a sure knowledge of the lifetime for DMS is required; thus, if additional loss processes do exist then this estimate will of course be a lower limit. Both methods require a compilation of many seawater and atmospheric observations over the whole globe throughout the year which can improve the estimation of DMS oceanic flux.

The residence time of DMS is assumed to be very short, of the order of a day as it is rapidly removed by reaction with both OH ($k_{298} = 6.6 \times 10^{-12}$ cm^3 molecule^{-1} s^{-1}, Atkinson et al. 2006) and the NO$_3$ radical ($k_{298} = 1.1 \times 10^{-12}$ cm^3 molecule^{-1} s^{-1}, Atkinson et al. 2006). Removal of DMS via reaction with halogen radicals such as Cl or BrO has been speculated and may well be significant additional loss processes in the marine environment (e.g. James et al. 2000; Khan et al. 2016).

The oxidation mechanism for DMS is extremely complex and has been the focus of numerous studies. Reaction of DMS with OH is thought to proceed via two channels, abstraction (reaction 4.24a) and addition (reaction 4.24b):

$$CH_3SCH_3 + OH \rightarrow CH_3SCH_2 + H_2O \tag{4.24a}$$

$$CH_3SCH_3 + OH + M \rightarrow CH_3S(OH)CH_3 + M \tag{4.24b}$$

The subsequent fate of the CH$_3$SCH$_2$ radical and the CH$_3$S(OH)CH$_3$ adduct in the atmosphere is an area of considerable debate, with new laboratory studies constantly refining the assumed mechanism, but subsequent reactions in this complex scheme lead to the formation of either SO$_2$ or CH$_3$SO(O)OH, known as methanesulphonic acid (MSA). In doing so, NO may be oxidized to NO$_2$, giving rise to the possibility that ozone may be formed.

The importance of the oxidation of SO$_2$ to H$_2$SO$_4$ has been discussed in detail, as H$_2$SO$_4$ can act as a condensation nuclei and lead to the formation of new particles. In contrast, MSA tends to stick to existing aerosol surfaces and does not lead to the formation of new particles (Brasseur et al. 1999).

The distribution of DMS reflects the fact that it is so short-lived and has a strong biological source from the oceans. Therefore, DMS concentrations are highest around high-productivity coastal and open-ocean regions, and display a strong seasonal cycle, reflecting the seasonal pattern of phytoplankton growth and decay. Since DMS is also so short-lived, it also displays a very strong diurnal cycle.

4.5.6 The CLAW Hypothesis

In 1987, a link between phytoplankton and cloud albedo was proposed. In their seminal paper, Charlson, Lovelock, Andreae, and Warren suggested that phytoplankton are able to modify Earth's climate through the generation of clouds by producing DMS, which is oxidized to give non-sea-salt sulphate (NSSS), leading to increased cloud formation and number density of cloud droplets, which in turn increases the cloud albedo counteracting global warming (negative feedback). There is evidence to suggest that the individual steps in the CLAW hypothesis are correct, but there is great uncertainty in the subsequent effect that the increased cloud albedo will have on the production of DMS. This hypothesis has been highly contentious over the years, and adversaries to the CLAW hypothesis suggest that increased CO$_2$ levels in the atmosphere decrease the pH of the oceans and could lead to reduction in production of DMS from phytoplankton, which would intensify warming (positive feedback) due to increased greenhouse gas levels in the atmosphere (Six et al. 2013). Despite the CLAW hypothesis opening the door to a new area of interest and research, the scientific community is still yet to reach a conclusive answer to its validity.

4.6 Halogens

The halogens chlorine, bromine, and iodine are present in the Earth's atmosphere in both inorganic and organic forms. Natural ecosystems use halogen-containing compounds for myriad purposes (Gribble 1994), whilst the use of halogen-containing species, such as chlorofluorocarbons (CFCs), hydrochloro-fluorocarbons (HCFCs), hydrofluorocarbons (HFCs), hydrofluroolefins (HFOs), and halons (containing bromine as well as fluorine and possibly chlorine) in a variety of industrial applications has led to a dramatic increase in the halogen burden in the atmosphere. In Chapter 8, the impact of the CFCs and their replacement compounds on stratospheric ozone is discussed in more detail and will only be briefly mentioned here. Recently the anthropogenic halogen load has become more controlled due to legislation. Thus, biogenic halogen emissions begin to play a more important role as they make up an increasing part of the total halogen load. The total atmospheric budget for the halogens is far from complete. However, in this section, the sources, sinks, and lifetimes of the major known species will be presented.

4.6.1 Organohalogens

Both monohalomethanes and polyhalomethanes may be formed by enzyme-catalysed reactions naturally. A variety of studies have begun to elucidate the potential roles of organohalogens in natural ecosystems. Nightingale et al. (1995) showed that levels of CH_3I, CH_2Br_2, $CHBr_3$, $CHBr_2Cl$, $CHBrCl_2$, and $CHCl_3$ were elevated in beds of the marine macroalgae *Laminaria digitata*. Release rates were influenced by partial desiccation, light availability, tissue age, tissue wounding, and grazing. These compounds may act as antimicrobial agents or grazing deterrents, since herbivory of marine macroalgae is intense and often the primary factor affecting their distribution and abundance. The simpler organohalogens may act as intermediates in the synthesis of more complex halogenated antigrazing compounds, or could result from the breakdown of such compounds following plant death.

In soils, very high haloperoxidase activity has been reported in the surface organic layer, where high microbial activity (e.g. fungi and bacteria) is known to be involved in the degradation of organic matter. The function of haloperoxidase activity in soil can be intra- or extracellular formation of organohalogens for the chemical defence of the producing microorganisms. Alternatively, haloperoxidases could be released by microorganisms in order to produce hypohalites. Formation of hypohalites *in situ* in the soil will cause oxidation of the refractory organic matter in the organic layer to produce more soluble and accessible organic compounds, but also organohalogens such as chloroform ($CHCl_3$).

Methyl halide formation occurs in fungi. Methyl chloride is an effective precursor to the biosynthesis of veratryl alcohol (3,4-dimethoxybenzyl alcohol). Veratryl alcohol is a secondary metabolite, which is biosynthesized by many white rot fungi and plays a central role in lignin degradation. A further possibility for the role of CH_3Cl is associated with the high methoxyl content of lignin, which can be as much as 20%. The diversion of excess one-carbon fragments from this source into volatile CH_3Cl, which will rapidly diffuse from the cell, may represent a means of overcoming the biochemical handicap likely to be imposed by a very large one-carbon pool in the cell. The sources, sinks, and residence times of a range of organohalogens will now be discussed.

4.6.2 Chlorofluorocarbons

Table 4.8 shows the uses of the major CFCs and other chlorinated halocarbons prior to controls on their use introduced by the Montreal Protocol. Originally thought to be inert in the atmosphere, it has been known for over 30 years that CFCs lead to the destruction of stratospheric ozone. Provided the implementation of the Montreal Protocol and subsequent legislation is adhered to, the stratospheric abundance

Table 4.8 Historical applications (prior to controls over use) of chlorofluorocarbons (CFCs), hydrochlorofluorocarbons (HCFCs), and current applications of hydrofluorocarbons (HFCs).

Sector	Compound	Application
Aerosols	CFC-11 ($CFCl_3$)	Used as pure fluids or mixtures for aerosol propellants
	CFC-12 (CF_2Cl_2)	
	CFC-114 ($CClF_2CClF_2$)	
	HCFC-22 (CHF_2Cl)	
	HCFC-142b ($C_2H_3ClF_2$)	
	HFC-134a (CH_2FCF_3)	
	HFC-152a (CH_3CHF_2)	
Foam blowing	CFC-11	Used as foam-blowing agent
	CFC-12	
	HCFC-22	
	HCFC-141b ($C_2H_3Cl_2F$)	
	HFC-134a	
	HFC-152a	
	HFC-245fa ($CF_3CH_2CHF_2$)	
	HFC-365mfc ($CF_3CH_2CF_2CH_3$)	
	HFC-227ea (CF_3CHFCF_3)	
	E-HFC-1234ze (E-$CHF{=}CHCF_3$)	
Solvents	CFC-113 (CCl_2FCClF_2)	Solvents used for electronics, precision cleaning and dry cleaning
	CH_3CCl_3	
	CFC-11	
	CCl_4	
Refrigeration and air-conditioning	CFC-11	Used for wide range of refrigeration and air-conditioning applications
	CFC-12	
	HCFC-22	
	HFC-134a	
	HFC-152a	
	HFC-143a (CH_3CF_3)	
	HFC-32 (CH_2F_2)	
	HFC-125	
	HFC-227ea	
	HFC-1234yf ($CH_2{=}CFCF_3$)	

of halogenated ozone-depleting substances is expected to return to the pre-1980 level of 2 ppbv chlorine equivalent by about 2050. Based on AGAGE (Advanced Global Atmospheric Gases Experiment) measurements, the three most abundant CFCs are CFC-12, CFC-11, and CFC-113 with mixing ratios of 516.2, 230.1, and 71.5 pptv in 2016 which declined by 21.8, 13.8, and 5.1 pptv from 2008, respectively.

The CFCs are destroyed by photolysis in the stratosphere or via reaction with O(^{1}D) atoms, e.g. for CFC-11

$$CFCl_3 + h\nu \rightarrow CFCl_2 + Cl \tag{4.25}$$

$$CFCl_3 + O\left(^1D\right) \rightarrow CFCl_2 + ClO \tag{4.26}$$

The effectiveness of the halogen species in destroying ozone, i.e. their ozone depletion potentials, is discussed in Chapter 8. Lifetimes for these compounds are long, 45 years for CFC-11, 100 years for CFC-12, and 85 years for CFC-113 (Solomon et al. 2013). CFCs account for about 70% of the total halocarbon radiative forcing, with CFC-11, CFC-12, and CFC-113 contributing 0.062, 0.17, and 0.022 Wm^{-2}, respectively.

4.6.3 Replacement Compounds: Hydrochlorofluorocarbons, Hydrofluorocarbons, and Hydrofluoroolefins

The phase-out of CFC production generated an immediate need to find replacements to meet the continued industrial demand. However, scope for direct substitution is limited, and some historical CFC requirements are being superseded by alternative technology, such as the use of pump-action aerosols or the use of hydrocarbons as refrigeration (McCulloch 1994; McCulloch et al. 2005). Alternative replacement compounds must not only match the properties needed to provide the specific effect desired but must also fulfil other criteria such as energy efficiency, safety (nontoxic and nonflammable), and environmental acceptability.

Hydrochlorofluorocarbons (HCFCs) and hydrofluorocarbons (HFCs) were introduced as an interim CFC replacement, since the presence of a labile C—H bond allows the compound to be oxidized via reaction with the OH in the troposphere. Tropospheric removal via reaction with OH considerably shortens the atmospheric residence time of the HCFC and HFC compared with the CFC it replaces. HFCs have effectively zero ozone-depleting potential (ODP) as they contain no chlorine. HCFCs do contain chlorine, but they deplete stratospheric ozone to a much lesser extent than CFCs. HCFCs and HFCs have GWP that is smaller than the compound they replace, CFCs, but that are larger than CO_2, thus they contribute to climate change directly through radiative forcing by absorption of infra-red radiation. The reaction with OH in the troposphere does not eliminate the possibility of the HCFCs contributing to stratospheric ozone depletion, and for HCFCs and HFCs contributing to global warming or other adverse effects as there is still the possibility the compound, or its oxidation products, will reach the stratosphere.

The most popular HCFCs used in the refrigeration sector as a replacement of CFCs are HCFC-22, HCFC-141b, HCFC-142b whose global productions have not been phased out completely under the Montreal Protocol Agreement especially in south and Southeast Asia, which is reflected in their increased AGAGE measured average mixing ratios from 2010 to 2016 by 30.1, 4.0, and 1.7 pptv, respectively. Three HCFCs (HCFC-22, HCFC-141b, and HCFC-142b) account for roughly 15% of the total halocarbon radiative forcing, whereas six HFCs (HFC-23, HFC-32, HFC-125, HFC-134a, HFC-143a, HFC-152) contribute only about 5% of the total halocarbon radiative forcing. The most commonly used is HFC-134a as a replacement for CFC-12, which has a history of use in automobile air conditioners and foam-blowing applications. The AGAGE measured mixing ratios of HFC-134a reached 60.7 ppt in 2016, with an increase of 31.8 ppt since 2010. The largest emissions of HFC-134a found in North America, Europe, and East Asia. The worldwide use of HFCs increased their concentrations significantly in the last decade, which is alarming in respect of global climate because of their high GWPs. Under an amendment to the Montreal Protocol (Kigali amendment, 2016), a global agreement was reached to limit the future use of HFCs: developed countries reduce HFCs use by 2019 and two groups of developing countries freeze HFCs use by 2024 and 2028, respectively.

As an alternative of HFCs (i.e. fourth-generation refrigerant), hydrofluoroolefins (HFOs), the unsaturated hydrofluorocarbons have emerged which have no ODP and very low GWP. HFOs can react with OH, Cl, O_3, and NO_3 making their lifetimes very short, c. a few days. One of the main oxidation products from these reactions is trifluoroacetic acid (TFA), which can be washed out of the atmosphere into the aqueous environmental compartment or can react with sCI to form highly oxygenated species with low vapour pressures that can lead to secondary organic aerosol formation.

4.6.4 Methyl Chloroform

Methyl chloroform (CH_3CCl_3) was used primarily as a cleaning solvent. Industrial sales have declined rapidly in recent years as control measures have become effective, thus the global emissions of CH_3CCl_3 amounted to ~2 Gg in 2012 and have decreased by around 80% since 2005 (WMO 2014). The only natural source of CH_3CCl_3 is biomass burning, which can be important, but Simpson et al. (2007) found it as a negligible source with the emission flux of 0.014 Gg year^{-1}. The seasonal variations of methyl chloroform are found for AGAGE measurement sites with summer minimum, which is driven by increased scavenging by OH. The residence time of CH_3CCl_3 is five years due to reaction with OH, although uptake by the oceans may also be a non-negligible loss.

4.6.5 Carbon Tetrachloride

In 1931, carbon tetrachloride (CCl_4) began to be used as a chemical intermediate in the production of CFCs and also as a solvent, with usage increasing rapidly from the 1950s to the 1980s. Since the introduction of the Montreal Protocol, which listed CCl_4 as a controlled substance, large-scale production has declined rapidly, and been accelerated by the global phase-out of CFC production, which consumed 80–90% of the total CCl_4 production. The decrease in AGAGE measured mixing ratios of CCl_4 from 2008 to 2016 is 8.9 ppt, with an atmospheric mixing ratio of 80.1 pptv in 2016. The North–South interhemispheric gradient of about 1.3 ppt since 2006 shows that there are some emissions sources of CCl_4 in the Northern Hemisphere (WMO 2014). The possible emission sources can be fugitive emissions from industries, productions during bleaching, or emissions from old landfills.

Carbon tetrachloride is removed by photodissociation in the stratosphere, where it eventually yields phosgene ($COCl_2$). The ocean is also a sizable sink for CCl_4 with approximately 20% of atmospheric CCl_4 being consumed by ocean mixing and hydrolysis. The soil can also act as a sink of CCl_4. The combined loss processes lead to a residence time for CCl_4 of 26–35 years.

4.6.6 Dichloromethane

Dichloromethane (CH_2Cl_2) is a man-made solvent that is used in a wide range of applications, such as the decaffeination of coffee, metal degreasing, drug preparation, and paint stripping. It is also used as a blowing agent in the foam plastics industries. The surface-mixing ratios of CH_2Cl_2 (2000–2012) increased at a global mean rate of ~8% per year, with the largest growth observed in the Northern Hemisphere (a factor of three greater than in the Southern Hemisphere) due to the increased anthropogenic emissions. The continuous increase in growth of CH_2Cl_2 could significantly offset a portion of the decline in anthropogenic chlorine provided by the Montreal Protocol, which can delay ozone hole recovery (Hossaini et al. 2017). Ambient mixing ratios of 40–60 pptv and 15–20 pptv have been observed in the Northern and Southern Hemispheres, respectively. There is no direct evidence for substantial natural sources, although atmospheric measurements of background concentrations in the Southern Hemisphere are double than those that have been calculated using known emission fields. Gribble (1994) notes that CH_2Cl_2 has been observed in the oceans and marine algae and may well have a seasonal maritime source. Biomass burning is also a significant source especially in the Southern Hemisphere (Lobert et al. 1999). The tropospheric lifetime for CH_2Cl_2 is around 0.4 years, with reaction with OH being the dominant loss process ($k_{298} = 1.15 \times 10^{-13}$ cm^3 molecule^{-1} s^{-1}; DeMore et al. 1997) and photolysis as a minor loss process.

4.6.7 Chloroform

Chloroform ($CHCl_3$) is known to have both anthropogenic and natural atmospheric sources, which are approximately of equal magnitude. Anthropogenic emissions arise from the use of $CHCl_3$ as a chemical intermediate in industrial processes, as a solvent, as a secondary product during chlorination processes, and from coal combustion and waste incineration. Chlorination processes that lead to the production of chloroform include sewage treatment, paper pulp bleaching, the chlorination of drinking water. However, paper manufacture is no longer an important source because of the phase out of chlorine in the paper processing industries.

Natural production of chloroform appears to be very widespread amongst different biota, and known sources include rice fields, soil ecosystems, termites, fungi and moss, although marine algae are believed to represent the dominant natural source. Biomass burning is another significant source of chloroform. Anthropogenic sources are better constrained, but natural emissions have not been fully explored or quantified and so no significant trends of AGAGE observations are apparent since the beginning of the record in 1994. Reaction with the OH radical is the dominant loss process ($k_{298} = 1.0 \times 10^{-13}$ cm^3 molecule^{-1} s^{-1}; DeMore et al. 1997) giving a lifetime for chloroform of about 0.5 years.

4.6.8 Perchloroethene and Trichloroethene

Perchloroethene (PCE) and trichloroethene (TCE) are used as industrial solvents and degreasers. Industrial emission estimates suggest that source regions are broadly similar to those of CH_2Cl_2. There has been a suggestion that certain types of macroalgae in the subtropical oceans may be capable of synthesizing chlorinated alkenes such as TCE and PCE, with the dominant source regions being located within the Northern Hemisphere.

Consumption of PCE has been declining steadily since the 1970s with the introduction of more efficient cleaning equipment, which has enabled a greater proportion of solvent to be recovered and recycled rather than be released into the environment. The primary loss process for both species in the atmosphere is reaction with the OH radical, with residence times of 96 days for PCE ($k_{298} = 1.72 \times 10^{-13}$ cm^3 molecule^{-1} s^{-1}; DeMore et al. 1997) and five days for TCE ($k_{298} = 2.2 \times 10^{-12}$ cm^3 molecule^{-1} s^{-1}; DeMore et al. 1997).

4.6.9 Chloromethane or Methyl Chloride

Chloromethane or methyl chloride is by far the most abundant organohalogen in the atmosphere, with man-made emissions being negligible in comparison with natural ones. CH_3Cl currently contributes around 17% of tropospheric chlorine which concentration (~540 pptv) has remained fairly constant over the last few years (WMO 2014). A short residence time (1–1.5 year) following reaction with OH ($k_{298} = 3.76 \times 10^{-14}$ cm^3 molecule^{-1} s^{-1}; DeMore et al. 1997) and loss to soil and oceans. The absence of an interhemispheric gradient is consistent with widespread natural sources, not concentrated within the industrialized Northern Hemisphere.

A global source strength of 0.75 Tg year^{-1} is required to balance the global budget of CH_3Cl (Table 4.9). The known natural sources of CH_3Cl include the ocean, tropical forests, fungi, salt marshes, wetlands, rice paddies, and mangroves from where the ocean (~700 Gg year^{-1}) and tropical forests (~2050 Gg year^{-1}) contribute most of its global sources. Anthropogenic sources identified include biomass burning and other combustion processes and mixing ratios as high as 2 ppbv have been observed in smoke from forest fires. Biomass burning is estimated to produce ~500 Gg year^{-1}. Current evidence would tend to indicate that the overwhelming proportion of the atmospheric CH_3Cl burden is of direct biological origin, and biosynthesis has been demonstrated in a wide variety of organisms, such as polypore fungi, macroalgae, and higher plants. Studies have concluded that higher tropospheric boundary layer mixing ratios of CH_3Cl in air masses of marine origin are consistent with a widespread natural flux from the oceans.

4.6.10 Methyl Bromide and Other Organobromo Compounds

Atmospheric methyl bromide (CH_3Br) is derived from both anthropogenic and biogenic sources. The budget of CH_3Br (Table 4.9) has been the subject of recent debate and it is still not fully resolved. The phase-out of controlled emissions of CH_3Br doesn't decrease its global burden as expected, suggesting that the balance of emissions is now overwhelmingly of natural origin. Methyl bromide is believed to be the main carrier of bromine to the stratosphere, where bromine is extremely efficient at promoting ozone destruction (see Chapter 8). The mean AGAGE measured tropospheric concentration was 7.0 pptv in 2016, which was a reduction of 2.2 pptv from peak levels measured during 1996–1998. A strong North–South interhemispheric gradient is seen as suggesting that either there are greater sources in the Northern Hemisphere or greater sinks in the Southern Hemisphere. The oceans are now believed to be a net sink for tropospheric CH_3Br, although parts may act as a source. CH_3Br is also used as a pesticide in the cultivation of high-value agricultural produce, although its use is being phased out as part of the Montreal Protocol, and about 40% of the applied compound may reach the atmosphere. Soils themselves are a sink for CH_3Br, with a total sink of around 30 Gg year^{-1}. On the other hand, vegetation (mangroves, rapeseed), fungus, saltmarsh, shrublands, rice paddies, and wetlands are the other sources of CH_3Br. Around 3 Gg

Table 4.9 Atmospheric budget for CH_3Cl and CH_3Br.

	CH_3Cl (Gg year^{-1})	CH_3Br (Gg year^{-1})
Anthropogenic Sources		
Leaded gasoline	N.Q.	0–3
Coal combustion	162 (29–295)	9.2
Fumigation	N.Q.	9.9 (8.6–11.3)
Biomass burning	468 (198–738)	23 (10–36)
Oceanic Source	700 (510–910)	32 (22–44)
Terrestrial Sources		
Tropical and subtropical plants	2040 (1430–2650)	N.Q.
Mangroves	12 (11–12)	1.3 (1.2–1.3)
Rapeseed	N.Q.	5.1 (4.0–6.1)
Fungus	145 (128–162)	2.2 (1.0–5.7)
Salt marshes	85 (1.1–170)	7.0 (0.6–14.0)
Wetland	27 (5.5–48)	0.6 (−0.1–1.3)
Rice paddies	3.7 (2.7–4.9)	0.7 (0.1–1.7)
Shrublands	15.0 (9.0–21.0)	0.7 (0.5–0.9)
Total Sources	**3658**	**84**
Sinks		
Reaction with OH	2832 (2470–3420)	56 (48–63)
Loss in soil	1058 (664–1482)	30 (19–41)
Loss in ocean	370 (296–445)	33 (20–44)
Loss in stratosphere	146	4
Total Sinks	**4406**	**123**
Net (Sources-Sinks)	**−748**	**−39**

Source: data from WMO (2014).
Note: The values in brackets represents minimum and maximum values. N.Q. means not quantified. Biomass burning includes both indoor biofuel use and open field burning.

of CH_3Br is generated from vehicle emissions. Biomass burning is another potentially large source of CH_3Br (~23 Gg year^{-1}), but current uncertainties in emission rates are large. The combined residence time of CH_3Br, allowing for loss to the oceans, soils, and by reaction with the OH radical ($k_{298} = 2.98 \times 10^{-14}$ cm^3 molecule^{-1} s^{-1}; DeMore et al. 1997), is currently estimated to be 0.8 years.

The short-lived brominated species (e.g. $CHBr_3$, CH_2Br_2, $CHBr_2Cl$, and $CHBrCl_2$) can be a significant source of stratospheric bromine with a contribution of up to 5 pptv, representing 25% of total inorganic stratospheric bromine (WMO 2014). Bromoform ($CHBr_3$) and dibromomethane (CH_2Br_2) are the most abundant short-lived brominated species accounting for ~75% of the short-lived bromocarbons in the marine environment. The emissions of $CHBr_3$, CH_2Br_2, are predominantly from marine macro-algae and phytoplankton, displaying large emission variabilities. The abiotic substitution of bromine in $CHBr_3$ with chlorine is supposed to be the main source for bromochloromethanes (e.g. $CHBr_2Cl$ and $CHBrCl_2$). The other minor sources of bromocarbons are from water treatment and disinfection; they are formed as byproducts when chlorine is added to water to kill bacteria. The long-term in-situ measurement data of $CHBr_3$ and CH_2Br_2 at Mace Head gives a seasonal trend with maximum concentrations during summer months and minimum concentrations during winter months. An excellent correlation between $CHBr_3$ and CH_2Br_2 suggests a common marine source, presumably originating from macroalgae in coastal region (Carpenter et al. 2003). The combined loss by OH and photolysis give the residence times of $CHBr_3$, CH_2Br_2, CH_2BrCl, and $CHBrCl_2$ of 24, 123, 137, and 78 days, respectively (WMO 2014).

4.6.11 Methyl Iodide

Methyl iodide (CH_3I) is believed to be the main gaseous iodine species in the troposphere usually with a concentration of 0.1–5 pptv, which plays an important role in the natural iodine cycle. The time series of atmospheric CH_3I show an increasing trend from 2003/2004 to 2009/2010 by several tens of percent in the marine environment (WMO 2014). The C—I bond is easily cleaved and CH_3I has the longest residence time of any organoiodine compound. Methyl iodide is photolysed, giving rise to a residence time of 2–10 days, depending on location and season. Methyl iodide is almost exclusively of marine origin, with some minor contribution from biomass burning, rice paddies, fungi, volcanoes, peatland ecosystems, and wetlands. The oceanic productions include photochemical degradation of organic matter in seawater and biogenic activities of phytoplankton and macroalgae. There have been no enhanced levels of CH_3I observed in urban areas, suggesting no emissions due to fossil-fuel combustion and industrial activities. The levels of CH_3I exhibit a seasonal cycle, with highest levels in summer and early autumn and has been found to be more abundant in the tropics and over oceanic regions of high biomass productivity. In the warm tropical surface waters, subject to high solar irradiance, photochemical production of CH_3I is significant. This contrasts with negative saturation anomalies (-0.65 pmol kg^{-1}) measured in cold surface waters of the open ocean around Greenland, subject to low light levels. Therefore high-latitude oceans may be a sink during autumn and winter. Photochemical production of CH_3I would go some way to explaining the lack of correlation between CH_3I and CH_3Cl, CH_3Br, CH_2Br_2, $CHBr_3$, $CHBr_2Cl$, and chlorophyll. Kelp, bacteria, phytoplankton, and macroalgae are all believed to be small but significant sources of oceanic CH_3I.

4.6.12 Other Organoiodine Compounds

There is much uncertainty about the contribution of individual organic iodine compounds to the global iodine budget and a small number of short-lived organoiodine species have been observed, including C_2H_5I, CH_2ICl, CH_2IBr, and CH_2I_2, whose residence times are approximately four days, a day, an hour, and a few minutes, respectively. The total organic iodine from CH_3I, CH_2ICl, CH_2IBr, and CH_2I_2 contributes only 20% of the reactive iodine flux in the marine boundary layer (MBL), suggesting some other contributors (e.g. inorganic iodine) may be significant for the reactive iodine flux in the MBL (WMO 2014). The atmospheric measurements of these organoiodine compounds are sparse. An extensive suite of alkyl iodides and bromides were measured at Mace Head in Ireland, Hateruma Island in the East China Sea, and Cape Ochiishi in the eastern coast of Hokkaido, Japan. Positive correlations were found between CH_2IBr/CH_2I_2, CH_2ICl/CH_2I_2, and CH_3I/C_2H_5I, which was interpreted as signifying common or linked marine sources (Carpenter et al. 1999; Yokouchi et al. 2011). The rapid photolysis of these compounds can lead to IO formation via the reaction of iodine radicals with ozone. Measurements of these organoiodine compounds in coastal and open oceans suggest that the organoiodine flux to the atmosphere may be considerably underestimated if CH_3I alone is considered. However, due to the short residence times, it is believed that very low levels of iodine precursors enter the stratosphere.

It is worth noting that several thousand naturally occurring organohalogen compounds have been identified in aquatic and terrestrial systems. The edible seaweed, *Asparagopsus taxiformis,* for example, contains over 100 such organohalogen compounds, including mixed chlorinated, brominated and iodinated ketones, and alkenes. Whether there is a significant flux of these compounds to the atmosphere is still unknown.

4.6.13   Inorganic Halogens

One of the important inorganic halogens is hydrogen chloride (HCl), which is directly emitted from volcanic eruptions, biomass burning, incineration of wastes, and industrial processes (e.g. semiconductor and petroleum manufacturing). Another important source results from the reaction of sea salt with gas-phase acidic compounds in the atmosphere:

$$NaCl(s) + HNO_3(g) \rightarrow HCl(g) + NaNO_3(s) \tag{4.27}$$

$$2NaCl(s) + H_2SO4(g) \rightarrow 2HCl(g) + Na_2SO_4(s) \tag{4.28}$$

Sea salt also contains NaBr, and the reaction of sea salt with N_2O_5 can initiate the release of bromine or/and chlorine nitrite, which is itself rapidly photolysed, yielding Br or/and Cl atoms:

$$NaCl(s) / NaBr(s) + N_2O_5(g) \rightarrow ClNO_2(g) / BrNO_2(g) + NaNO_3(s) \tag{4.29}$$

$$BrNO_2(g) / ClNO_2(g) \rightarrow Br(g) / Cl(g) + NO_2(g) \tag{4.30}$$

The emission of short-lived organohalogens can produce inorganic halogen compounds (e.g. BrCl, I_2, HOCl, HOBr) in the atmosphere, and these may, in turn, generate free halogen atoms.

HCl and HBr adsorbed on ice can participate heterogeneous reactions with hypohalous acids (HOX) or halogen nitrates (XONO$_2$) in the dark winter of the polar region and convert photochemically inactive chlorine and bromine reservoir compounds (e.g. Br_2, BrCl, and/or Cl_2). During spring, these compounds release halogen atoms through photolysis.

$$HOBr(g) / BrONO_2(g) + HBr(s) \rightarrow Br_2(g) + H_2O(s) / HNO_3(s) \tag{4.31}$$

$$HOBr(g) / BrONO_2(g) + HCl(s) \rightarrow BrCl(g) + H_2O(s) / HNO_3(s) \tag{4.32}$$

$$HOCl(g) / ClONO_2(g) + HBr(s) \rightarrow BrCl(g) + H_2O(s) / HNO_3(s) \tag{4.33}$$

$$HOCl(g) / ClONO_2(g) + HCl(s) \rightarrow Cl_2(g) + H_2O(s) / HNO_3(s) \tag{4.34}$$

$$BrCl(g) + HBr(s) \rightarrow Br_2(g) + HCl(s) \tag{4.35}$$

$$Br_2 + hv \rightarrow Br + Br \tag{4.36}$$

$$BrCl + hv \rightarrow Br + Cl \tag{4.37}$$

In general, once released, halogen atoms may react with O_3 (X is Cl, Br, and I), forming XO:

$$X + O_3 \rightarrow XO + O_2 \tag{4.38}$$

Both IO and BrO can either react with themselves or each other, in either case regenerating the free halogen that can again remove an ozone molecule or produce other halogen products.

$$XO + XO \rightarrow X + X + O_2 \tag{4.39a}$$

$$XO + XO + M \rightarrow XOOX + M \tag{4.39b}$$

$$XO + XO \rightarrow OXO + X \tag{4.39c}$$

$$XO + XO \rightarrow X_2 + O_2 \tag{4.39d}$$

The molecule OIO can also be formed by cross reaction between IO and BrO (DeMore et al. 1997). Considerable interest surrounds OIO, which has been observed in the MBL, although its chemistry is uncertain at this time. XO can also form temporary reservoir species, HOX and $XONO_2$, via reaction with HO_2 and NO_2:

$$XO + HO_2 \rightarrow HOX + O_2 \tag{4.40}$$

$$XO + NO_2 + M \rightarrow XONO_2 + M \tag{4.41}$$

For Br and particularly I, these reservoirs are quite short-lived in the troposphere, being either taken up into aerosol, or destroyed by photolysis in the case of $IONO_2$. The role of halogen reservoirs in the stratosphere is discussed in Chapter 8. XO can also react with NO to regenerate X:

$$XO + NO \rightarrow X + NO_2 \tag{4.42}$$

Chlorine atoms can also react with hydrocarbons (RH) to form HCl, which limits its effectiveness in destroying ozone in the troposphere. However, Br can only abstract hydrogen from labile species such as aldehydes and HO_2, although addition to unsaturated species is also possible. Iodine atoms are unable to react with any hydrocarbons and reaction with ozone is by far the dominant channel in the troposphere. Iodine chemistry is therefore the most likely to lead to significant ozone destruction in the troposphere, although, under special circumstances, such as in the boundary layer in the Arctic spring, bromine chemistry can initiate rapid surface ozone destruction.

It is worth noting that, once released, fluorine atoms can react with O_3, but on reaction with RH forms the very stable HF. Since release of fluorine from its major reservoir compounds, such as the CFCs, HCFCs, HFCs, and the PFCs (perfluorocarbons), occurs in the stratosphere, HF is transported back down to the troposphere where it is removed by wet deposition. The lifetime of inorganic halogens and their reservoirs is highly variable and will depend strongly on the aerosol loading present.

4.7 Hydrogen

Hydrogen (H_2) is the second most abundant oxidizable trace gas in the atmosphere which has a globally averaged tropospheric mixing ratio of around 530 ppbv with a lifetime of about 2 years. Figure 4.5 shows how the concentration of H_2 varies with time at the Earth's surface in both the Northern and Southern Hemispheres. Although the levels are very similar, the trends and depth of the seasonal cycle are not, i.e. H_2 levels are about 3% higher in the Southern Hemisphere compared with that in the Northern Hemisphere and the seasonal variation in the Northern Hemisphere is around 50 ppbv, whereas it is only 25 ppbv in the Southern Hemisphere. Hydrogen is produced by the photolysis of HCHO:

$$HCHO + h\nu \rightarrow H_2 + CO \tag{4.43}$$

$$HCHO + h\nu \rightarrow H + HCO \tag{4.44}$$

where HCHO is itself produced from the oxidation of methane and other VOCs in the atmosphere. Hydrogen is also emitted during incomplete combustion of fossil fuel and biomass burning and from nitrogen fixation in the continental and marine biosphere, therefore it is surprising that its levels are higher in the Southern Hemisphere than the Northern Hemisphere. It is destroyed by reaction with OH and by dry deposition to soils. The uptake by soil is the most dominating loss process of H_2 which is

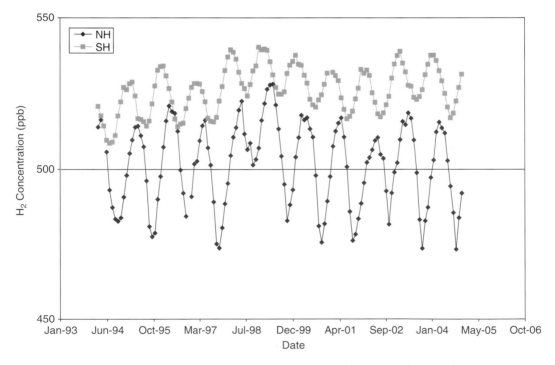

Figure 4.5 Hydrogen (H_2) measured at the surface in the Northern Hemisphere (NH) and Southern Hemisphere (SH). *Source:* data courtesy of Dr S.J. O'Doherty, School of Chemistry, University of Bristol.

associated with enzymatic and microbial activities linked to H_2 diffusivity. Thus, the soil-loss process gives rise not only to the smaller levels H_2 in the Northern Hemisphere but also the stronger seasonal cycle.

4.8 Summary

Budget. For each element and each compound containing that element, a budget can be constructed where the known sources and sinks are compiled. If the sources outweigh the sinks, then the concentration of that compound will increase with time in the atmosphere. If the sinks outweigh the sources, then the concentration will decrease with time.

Sources. There are primary sources, where the compound is released directly to the atmosphere, and secondary sources where the compound is formed in the atmosphere. Of the primary sources, there are anthropogenic sources (arising from human activity) and biogenic sources (natural).

Sinks. There are two depositional sinks, wet and dry deposition, where compounds are physically removed by incorporation into aqueous media or destruction at a surface. There are also photochemical sinks. Photolysis is where the compound is broken apart by sunlight and chemical loss via reaction with oxidants, such as the OH radical and O_3.

Carbon. The major form of carbon in the atmosphere is CO_2, with CH_4 and CO the next two most important forms. There are myriad VOCs emitted into the atmosphere that are oxidized to CO_2 and CO. CO_2 and CH_4 are very important greenhouse gases in the atmosphere.

Nitrogen. The major form of nitrogen in the atmosphere is N_2 itself, and with N_2O this constitutes 99.99% of the total N. However, NO_x and NH_3 are present in trace levels and have a profound impact on the composition of the atmosphere, particularly photochemical smog.

Sulphur. The major emission of sulphur into the atmosphere is in the form of SO_2, but the longest-lived reservoir is OCS. DMS is the largest biogenic emission of S into the atmosphere. Sulphur is important in the atmosphere because it can lead to aerosol formation and clouds and therefore cool the planet (negative greenhouse effect).

Halogens. The major natural sources of Cl, Br, and I in the atmosphere are CH_3Cl, CH_3Br, and CH_3I. Many chlorine containing compounds have been made for industrial applications and in particular the CFCs, chlorofluorocarbons, have become the dominant source of chlorine in the atmosphere. Recent legislation has banned the use of CFCs.

Hydrogen. Hydrogen has a strong soil loss process, and this gives rise to very different seasonal cycles in the Northern and Southern Hemispheres.

Questions

1   Discuss the main removal mechanisms (sinks) for pollutants in the troposphere.

2   Relative measurements were made of ethane (squares) and isoprene (triangles) at a rural site in the Northern Hemisphere, the monthly average levels are shown below: What are the major sources and sinks for each of these compounds? Explain why the seasonal cycles vary in the way that they do.

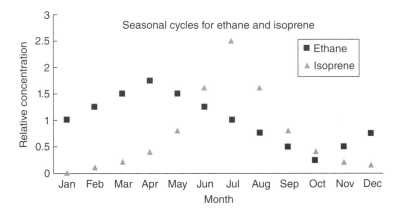

3   The concentration of NO on a weekday in Bristol, a small city in the United Kingdom, is shown in the figure below. Why are there peaks at around 9 a.m. and 5 p.m., and why are the two peaks not the same size?

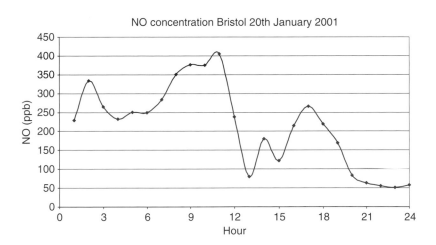

4   What is the CLAW hypothesis, and why is sulphur chemistry so important in the atmosphere?

5   The concentration of halogenated compounds, CFC11 ($CFCl_3$), CFC12 (CF_2Cl_2), CCl_4, and CH_3CCl_3 from 1994 to 2016 are shown in the figure below. All these compounds are banned and their production and subsequent emission should have ceased since the mid-1990s. Explain why the concentration-time profiles have the form they do. (Data courtesy of Dr. S.J. O'Doherty.)

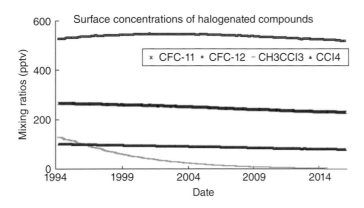

References

References

Ajtay, G.L., Ketner, P., and Duvigneaud, P. (1979). Terrestrial primary production and phytomass. In: *The Global Carbon Cycle: SCOPE*, vol. 13 (eds. B. Bolin, E.T. Degens, S. Kempe and P. Ketner), 129–181. Chichester: Wiley.

Andreae, M.O. (1990). Ocean–atmosphere interactions in the global biogeochemical Sulphur cycle. *Marine Chemistry* 30: 1–29.

Atkinson, R., Baulch, D.L., Cox, R.A. et al. (2004). Evaluated kinetic and photochemical data for atmospheric chemistry: volume I- gas phase reactions of Ox, HOx, NOx and SOx species. *Atmospheric Chemistry and Physics* 4: 1461–1738.

Atkinson, R., Baulch, D.L., Cox, R.A. et al. (2006). Evaluated kinetic and photochemical data for atmospheric chemistry: volume II gas phase reactions of organic species. *Atmospheric Chemistry and Physics* 6: 3625–4056.

Barnes, I., Becker, K.H., and Patroescu, I. (1994). The tropospheric oxidation of DMS: a new source of OCS. *Geophysical Research Letters* 21: 2389–2392.

Barnola, J.M., Raynaud, D., Korotkevich, Y.S., and Lorius, C. (1987). Vostok ice core provides 160,000 year record of atmospheric CO_2. *Nature* 329: 408–414.

Blake, D.R. and Rowland, F.S. (1988). Continuing worldwide increase in troposheric methane, 1978–1987. *Science* 239: 1129–1131.

Brasseur, G.P., Orlando, J.J., and Tyndall, G.S. (1999). *Atmospheric Chemistry and Global Change*. New York, NY: Oxford University Press.

Carpenter, L.J., Sturges, W.T., Penkett, S.A., and Liss, P.S. (1999). Short-lived alkyl iodides and bromides at Mace Head, Ireland: links to biogenic sources and halogen oxide production. *Journal of Geophysical Research* 104: 1679–1689.

Carpenter, L.J., Liss, P.S., and Penkett, S.A. (2003). Marine organohalogens in the atmosphere over the Atlantic and Southern Oceans. *Journal of Geophysical Research* 108: 4256.

Challenger, F. and Simpson, M.I. (1948). Studies on biological methylation. Part XII. A precursor of dimethyl sulphide evolved by Polysiphonia fastigiata. Dimethyl-2-carboxyethylsulphonium hydroxide and its salts. *Journal of the Chemical Society* 2: 1591–1597.

Charlson, R.J., Lovelock, J.E., Andreae, M.O., and Warren, S.G. (1987). Oceanic phytoplankton, atmospheric sulfur, cloud albedo and climate. *Nature* 326: 655–661.

Chin, M. and David, D.D. (1993). Global sources and sinks of OCS and CS_2 and their distribution. *Global Biogeochemical Cycles* 7: 321–337.

DeMore, W.B., Sander, S.P., Golden, D.M. et al. (1997). *Chemical Kinetics and Photochemical Data for Use in Stratospheric Modeling. Evaluation 12.* In: *Jet Propulsion Laboratory Publication 97–4.* Los Angeles, CA.: National Aeronautics and Space Administration.

Fall, R. (1999). Biogenic emissions of volatile organic compounds from higher plants. In: *Reactive Hydrocarbons in the Atmosphere* (ed. C.N. Hewitt), 43–91. San Diego, CA: Academic Press.

Fried, A., Henry, B., Ragazzi, R.A. et al. (1992). Measurements of OCS in automotive exhausts and an assessment of its importance to the global Sulphur cycle. *Journal of Geophysical Research* 97: 14621–14634.

Graedel, T.E. and Crutzen, P.J. (1993). *Atmospheric Change: An Earth System Perspective.* New York, NY: W.H. Freeman and Co.

Gribble, G.W. (1994). Natural organohalogens. *Chemistry in Britain* 71: 907–911.

Guenther, A., Hewitt, C.N., Erickson, D. et al. (1995). A global-model of natural volatile organic-compound emissions. *Journal of Geophysical Research-Atmospheres* 100 (D5): 8873–8892.

Haas, P. (1935). The liberation of methyl sulphide by seaweed. *Biochemical Journal* 29: 1297–1299.

Hatanaka, A. (1993). The biogeneration of odour by green leaves. *Phytochemistry* 34: 1201–1218.

Hossaini, R., Chipperfield, M.P., Montzka, S.A. et al. (2017). The increasing threat to stratospheric ozone from dicholoromethane. *Nature Communications* 8: 15962(1–9).

Jacob, D.J. and Hoffmann, M.R. (1983). A dynamic model for the production of H^+, $NO3^-$ and SO_4^{2-} in urban fog. *Journal of Geophysical Research* 88: 6611–6621.

James, J.D., Harrison, R.M., Savage, N.H. et al. (2000). Quasi-Lagrangian investigation into dimethyl sulfide oxidation in maritime air using a combination of measurements and model. *Journal of Geophysical Research-Atmospheres* 105 (D21): 26379–26392.

Khan, M.A.H., Gillespie, S.M.P., Razis, B. et al. (2016). A modelling study of the atmospheric chemistry of DMS using the global model, STOCHEM-CRI. *Atmospheric Environment* 127: 69–79.

Khan, M.A.H., Percival, C.J., Caravan, R.L. et al. (2018). Criegee intermediates and their impacts on the troposphere. *Environmental Science: Processes & Impacts* 20: 437–453.

Lana, A., Bell, T.G., Simó, R. et al. (2011). An updated climatology of surface dimethylsulfide concentrations and emission fluxes in the global ocean. *Global Biogeochemical Cycles* 25: GB1004.

Lee, C.-L. and Brimblecombe, P. (2016). Anthropogenic contributions to global carbonyl sulfide, carbon disulfide and organosulfides fluxes. *Earth-Science Reviews* 160: 1–18.

Liss, P.S. and Merlivat, L. (1986). Air–sea gas exchange rates: introduction and synthesis. In: *The Role of Air–Sea Exchange in Geochemical Cycling* (ed. P.B. Menard), 113–127. Dordrecht, the Netherlands: Reidel.

Liss, P.S. and Slater, P.G. (1974). Flux of gases across the air-sea interface. *Nature* 247: 181–184.

Lobert, J.M., Keene, W.C., Logan, J.A., and Yevich, R. (1999). Global chlorine emissions from biomass burning: reactive chlorine emission inventory. *Journal of Geophysical Research* 104: 8373–8390.

McCulloch, A. (1994). Sources of hydrochlorofluorocarbons, hydrofluorocarbons and fluorocarbons and their potential emissions during the next 25 years. *Environmental Monitoring Assessement* 31: 167–174.

McCulloch, A., Midgley, P.M., and Lindley, A.A. (2005). Recent changes in the production and global atmospheric emissions of chlorodifluoromethane (HCFC-22). *Atmospheric Environment* 40 (5): 936–942.

Möller, D. (1984). Estimation of the global man-made Sulphur emission. *Atmospheric Environment* 18 (1): 19–27.

Nguyen, B.C., Mihalopoulos, N., Putard, J.P., and Bonsang, B. (1995). OCS emissions from biomass burning in the tropics. *Journal of Atmospheric Chemistry* 22: 55–65.

Nightingale, P.D., Malin, G., and Liss, P.S. (1995). Production of chloroform and other low-molecular weight halocarbons by some species of macroalgae. *Limnology and Oceanography* 40: 680–689.

Simpson, I.J., Blake, N.J., Blake, D.R. et al. (2007). Strong evidence for negligible methyl chloroform (CH_3CCl_3) emissions from biomass burning. *Geophysical Research Letters* 34: L10805.

Singh, H.B. and Zimmerman, P.R. (1992). Atmospheric distribution and sources of nonmethane hydrocarbons. In: *Gaseous Pollutants: Characterisation and Cycling* (ed. J.O. Nriagu), 177–235. New York, NY: John Wiley & Sons.

Singsaas, E.L., Lerdau, M., Winter, K., and Sharkey, T.D. (1997). Isoprene increases thermotolernce of isoprene-emitting species. *Plant Physiology* 115: 1413–1420.

Six, K.D., Kloster, S., Ilyina, T. et al. (2013). Global warming amplified by reduced Sulphur fluxes as a result of ocean acidification. *Nature Climate Change* 3: 975–978.

Solomon, S., Qin, D., Manning, M. et al. (eds.) (2013). *Climate Change 2013: The Physical Science Basis. Contribution of Working Group 1 to the Fourth Assessment Report of the Intergovernmental Panel on Climate Change*. Cambridge, UK: Cambridge University Press.

Steele, L.P., Dlugokencky, E.L., Lang, P.M. et al. (1992). Slowing down of the global accumulation of atmospheric methane during the 1980s. *Nature* 358: 313–316.

Wang, K.-Y. and Shallcross, D.E. (2000). Modelling terrestrial biogenic isoprene fluxes and their potential impact on global chemical species using a coupled LSM–CTM model. *Atmospheric Environment* 34: 2909–2925.

Wanninkhof, R.J. (1992). Relationship between wind-speed and gas-exchange over the ocean. *Journal of Geophysical Research* 97: 7373–7382.

Watts, S.F. (2000). The mass budgets of carbonyl sulfide, dimethyl sulfide, carbon disulfide and hydrogen sulfide. *Atmospheric Environment* 34: 761–779.

Watts, S.F. and Roberts, C.N. (1999). H_2S from car catalytic convertors. *Atmospheric Environment* 33: 169–170.

WMO (2014). *Global Ozone Research and Monitoring Project*. Report No. 55, Scientific Assessment of Ozone Depletion. Geneva: World Meteorological Organization.

Xie, H., Moore, R.M., Miller, W.L., and Scarratt, M.G. (1997). A study of the ocean source of CS_2. *Abstracts, American Geophysical Union* 214 (Part 1): 86-GEOC.

Yokouchi, Y., Saito, T., Ooki, A., and Mukai, H. (2011). Diurnal and seasonal variations of iodocarbons (CH_2ClI, CH_2I_2, CH_3I, and C_2H_5I) in the marine atmosphere. *Journal of Geophysical Research* 116: D06301.

Further Reading

Brasseur, G.P., Orlando, J.J., and Tyndall, G.S. (1999). *Atmospheric Chemistry and Global Change*. New York: Oxford University Press.

Harper, D.B. (1995). *Naturally-Produced Organohalogens*, 235–244. Netherlands: Kluwer Academic Publishers.

Hewitt, C.N. (1999). *Reactive Hydrocarbons in the Atmosphere*. San Diego, CA: Academic Press.

5

Tropospheric Chemistry and Air Pollution

Paul Monks and Joshua Vande Hey

Department of Chemistry, University of Leicester, Leicester, United Kingdom

The thin gaseous envelope that surrounds our planet is integral to the maintenance of life on Earth. The vertical structure and standard nomenclature of the atmosphere is shown in Figure 5.1. The troposphere is the lowest region of the atmosphere extending from the Earth's surface to the tropopause at 10–18 km. About 90% of the total atmospheric mass resides in the troposphere and the greater part of the trace-gas burden is found there. The troposphere is well mixed and its bulk composition is 78% N_2, 21% O_2, 1% Ar, and 0.036% CO_2 with varying amounts of water vapour depending on temperature and altitude. The majority of the trace species found in the atmosphere are emitted into the troposphere from the surface and are subject to a complex series of chemical and physical transformations. It is becoming apparent that human activity is beginning to change the composition of the troposphere over a range of scales, leading to increased acid deposition, local and regional ozone episodes, and climate change. In this chapter, we will look at the fundamental chemistry of the 'natural' troposphere, including its biogenic and anthropogenic interactions, and explore the perturbed states including what is often termed as air pollution.

5.1   Sources of Trace Gases in the Atmosphere

Trace species emitted directly into the atmosphere are termed to have *primary* sources, e.g. trace gases such as SO_2, NO, and CO. Those trace species formed as a product of chemical and/or physical transformation of primary pollutants in the atmosphere, e.g. ozone, are referred to as having *secondary* sources or being *secondary* species. Emissions into the atmosphere are often broken down into broad categories of anthropogenic or 'man-made sources' and biogenic or natural sources with some gases also having geogenic sources. Table 5.1 lists a selection of the trace gases and their major sources. For the individual emission of a primary pollutant there are a number of factors that need to be taken into account in order to estimate the emission strength. These include the range and type of sources and the spatial and temporal distributions of the sources. Often, these factors are compiled into so-called emission inventories that combine the rate of emission of various sources with the number and type of each source and the

Atmospheric Science for Environmental Scientists, Second Edition. Edited by C.N. Hewitt and Andrea V. Jackson.

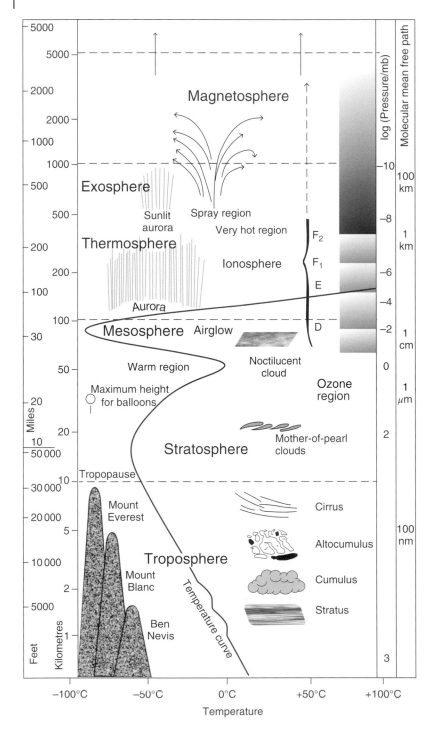

Figure 5.1 Vertical structure of the atmosphere. The vertical profile of temperature can be used to define the different atmospheric layers. Copyright, P. Biggs, University of Oxford.

Table 5.1 Natural and anthropogenic sources of a selection of trace gases.

	Compound	Natural sources	Anthropogenic sources
Carbon-containing compounds	Carbon dioxide (CO_2)	Respiration; oxidation of natural CO; destruction of forests	Combustion of oil, gas, coal and wood; limestone burning
	Methane (CH_4)	Enteric fermentation in wild animals; emissions from swamps, bogs, etc., natural wet land areas; oceans	Enteric fermentation in domesticated ruminants; emissions from paddy fields; natural gas leakage; sewerage gas; colliery gas; combustion sources
	Carbon monoxide (CO)	Forest fires; atmospheric oxidation of natural hydrocarbons and methane	Incomplete combustion of fossil fuels and wood, in particular motor vehicles, oxidation of hydrocarbons; industrial processes; blast furnaces
	Light paraffins, C_2–C_6	Aerobic biological source. Photochemical degradation of dissolved oceanic organic material	Natural gas leakage; motor vehicle evaporative emissions; refinery emissions
	Olefins, C_2–C_6	Insignificant	Motor vehicle exhaust; diesel engine exhaust
	Aromatic hydrocarbons		Motor vehicle exhaust; evaporative emissions; paints, gasoline, solvents
	Terpenes ($C_{10}H_{16}$)	Trees (broadleaf and coniferous); plants	
	CFCs and HFCs	None	Refrigerants; blowing agents; propellants
Nitrogen-containing trace gases	Nitric oxide (NO)	Forest fires; anaerobic processes in soil; electric storms	Combustion of oil, gas, and coal
	Nitrogen dioxide (NO_2)	Emissions from denitrifying bacteria in soil; oceans	Combustion of oil, gas, and coal; atmospheric transformation of NO
	Nitrous oxide (N_2O)	Aerobic biological source in soil	Combustion of oil and coal
	Ammonia (NH_3)	Breakdown of amino acids in organic waste material	Coal and fuel oil combustion; waste treatment
Sulphur-containing trace gases	Dimethyl sulphide (DMS)	Phytoplankton	Landfill gas
	Sulphur dioxide (SO_2)	Oxidation of H_2S; volcanic activity	Combustion of oil and coal; roasting sulphate ores
Other minor trace gases	Hydrogen	Oceans, soils; methane oxidation, isoprene and terpenes via HCHO	Motor vehicle exhaust; oxidation of methane via formaldehyde (HCHO)
	Ozone	In the stratosphere; natural NO–NO_2 conversion	Man-made NO–NO_2 conversion; supersonic aircraft
	Water (H_2O)	Evaporation from oceans	Insignificant

CFC, chlorofluorocarbons; HFC, hydroflurocarbons.

time over which the emissions occur. Figure 5.2 shows the UK emission inventory for a range of primary pollutants ascribed to different source categories (see figure legend). A number of key processes are evident in Figure 5.2; for instance, SO_2 has strong sources from public power generation, whereas ammonia has strong sources from agriculture. Figure 5.3 (Plate 5.3) shows the (2015) $1 \, km \times 1 \, km$ emission inventories for SO_2 and NO_2 for the United Kingdom. In essence, the data presented in Figure 5.2 have been apportioned spatially according to the magnitude of each source category (e.g. road transport, combustion

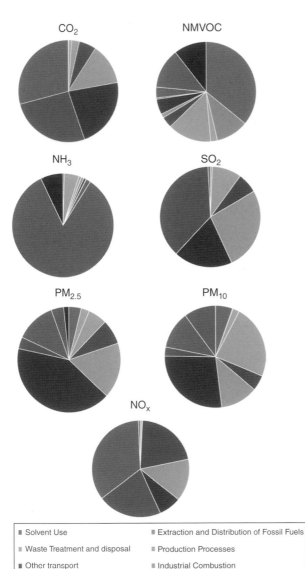

Figure 5.2 UK emission statistics by United Nations Economic Commission of Europe (UNECE) source category (1, combustion in energy production and transformation; 2, combustion in commercial, institutional, residential and agriculture; 3, combustion in industry; 4, production processes; 5, extraction and distribution of fossil fuels; 6, solvent use; 7, road transport; 8, other transport and mobile machinery; 9, waste treatment and disposal; 10, agriculture, forestry and land-use change; 11, nature). See colour plate section for the colour representation of this figure.

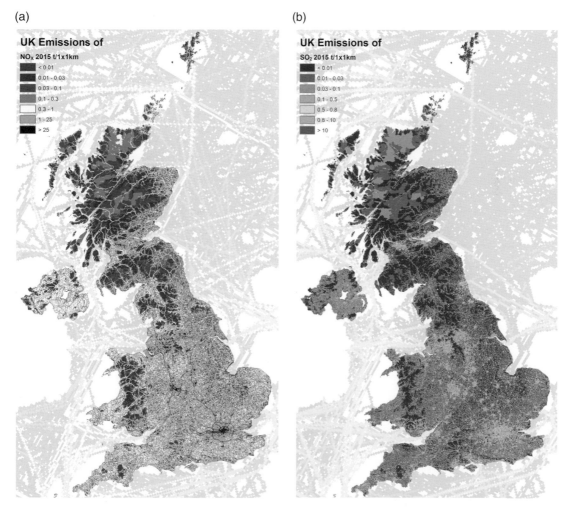

Figure 5.3 Emission maps (2002) for the UK on a 1 km × 1 km grid for (a) NO_2 and (b) SO_2. *Source:* data from UK National Atmospheric Emissions Inventory, www.naei.org.uk. See colour plate section for the colour representation of this figure.

in energy production and transformation, solvent use). For example, in Figure 5.3(a) (Plate 3a), the major road routes are clearly visible, showing NO_2 has a major automotive source (cf. Figure 5.2). It is possible to scale the budgets of many trace gases to a global scale.

It is worth noting that there are a number of sources that do not occur within the boundary layer (the decoupled lowest layer of the troposphere, see Figure 5.1), such as lightning production of nitrogen oxides and a range of pollutants emitted from the combustion taking place in aircraft engines. The non-surface sources often have a different chemical impact owing to their direct injection into the *free* troposphere (the part of the troposphere that overlays the boundary layer).

In summary, a range of trace species is present in the atmosphere, with a myriad of sources both natural and man-made that vary both spatially and temporally (Jackson 2007). It is the chemistry of the atmosphere that acts to transform the primary pollutants into simpler chemical species.

5.2   Key Processes in Tropospheric Chemistry

In general, tropospheric chemistry is analogous to a low-temperature combustion system, the overall reaction given by

$$CH_4 + 2O_2 \rightarrow CO_2 + 2H_2O \tag{5.1}$$

Unlike combustion, this is not a thermally initiated process but is process initiated and propagated by photochemistry. The chemistry that takes place in the troposphere, and in particular the photochemistry, is intrinsically linked to the chemistry of ozone (Monks et al. 2015). Tropospheric ozone acts as initiator, reactant, and product in much of the oxidation chemistry that takes place in the troposphere, and stratospheric ozone determines the amount of short wavelength radiation available to initiate photochemistry. Figure 5.4 shows a typical ozone profile through the atmosphere illustrating a number of interesting points. First, 90% of atmospheric ozone can be found in the stratosphere; on average, about 10% can be found in the troposphere. Second, the troposphere, in the simplest sense, consists of two regions. The lowest kilometre or so contains the planetary boundary layer (Stull 1988) and inversion layers that can act as pre-concentrators for atmospheric emissions from the surface and hinder exchange to the so-called *free* troposphere, the larger part by volume that sits above the boundary layer.

For a long time, transport from the stratosphere to the troposphere was thought to be the dominant source of ozone in the troposphere (Fabian and Pruchniewz 1977). Early in the 1970s it was first suggested (Chameides and Walker 1973; Crutzen 1973) that tropospheric ozone originated mainly from production within the troposphere by photochemical oxidation of CO and hydrocarbons catalysed by HO_x and NO_x. These sources are balanced by in-situ photochemical destruction of ozone and by dry

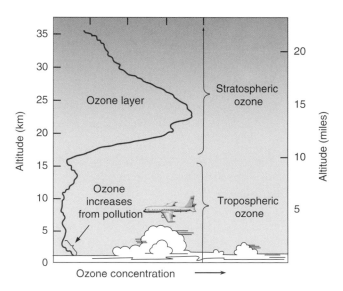

Figure 5.4   A typical atmospheric ozone profile through the atmosphere (WMO 2003). The concentration is expressed as a volume mixing ratio.

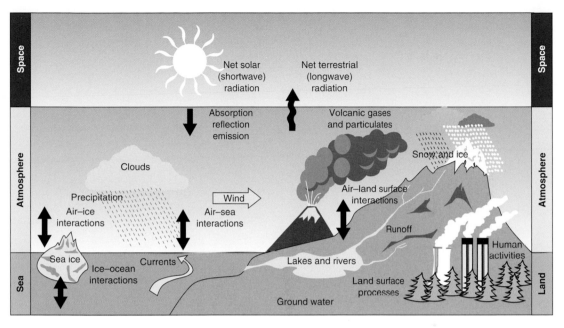

Figure 5.5 A schematic representation of the atmosphere's role in the Earth system.

deposition at the Earth's surface. Many studies, both experimental- and model-based, have set about determining the contribution of both chemistry and transport to the tropospheric ozone budget on many different spatial and temporal scales.

There is growing evidence that the composition of the troposphere is changing (Volz and Kley 1988). For example, analysis of historical ozone records has indicated that tropospheric ozone levels in both hemispheres have increased by a factor of three to four over the past 100 years. Methane concentrations have effectively doubled over the past 150 years and N_2O levels have risen by 15% since pre-industrial times (IPCC 2013). Measurements of halocarbons have shown that this group of chemically and radiatively important gases was increasing in concentration until relatively recently (IPCC 2013).

One of the difficulties in discussing tropospheric chemistry in general terms is that by the very nature of the troposphere being the lowest layer of the atmosphere, it has complex multiphase interactions with the Earth's surface, which can vary considerably between expanses of ocean to deserts (see Figure 5.5).

5.3   Initiation of Photochemistry by Light

Photodissociation, the breaking apart of molecules by solar radiation, plays a fundamental role in the chemistry of the atmosphere. The photodissociation of trace species such as ozone and formaldehyde contributes to their removal from the atmosphere, but probably the most important role played by these

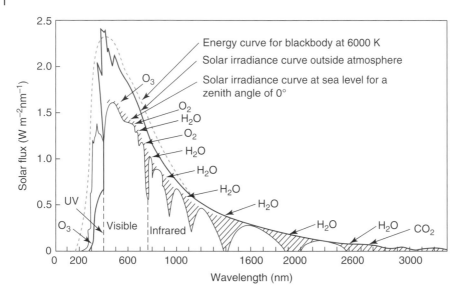

Figure 5.6 Solar flux outside the atmosphere and at sea level, respectively. The emission of a blackbody at 6000 K is included for comparison. The species responsible for light absorption in the various regions (O_2, H_2O, etc.) are also shown.

photoprocesses is the generation of highly reactive atoms and radicals. Photodissociation of trace species and the subsequent reaction of the photoproducts with other molecules is the prime initiator and driver for the bulk of atmospheric chemistry.

The light source for photochemistry in the atmosphere is the Sun. At the top of the atmosphere there is ca. $1370\,\mathrm{W\,m^{-2}}$ of energy over a wide spectral range, from X-rays through the visible to longer wavelengths. By the time the incident light reaches the troposphere much of the more energetic, shorter wavelength light has been absorbed by molecules such as oxygen, ozone and water vapour, or scattered higher in the atmosphere (see Chapter 2). Typically, in the surface layers, only light of wavelengths longer than 290 nm is available (see Figure 5.6). In the troposphere, the wavelength at which the intensity of light drops to zero is termed the *atmospheric cutoff*. For the troposphere, this wavelength is determined by the overhead stratospheric ozone column (absorbs ca. $\lambda \leq 310\,\mathrm{nm}$) and the aerosol loading. In the mid- to upper stratosphere, the amount of O_3 absorption in the 'window' region at 200 nm between the O_3 and O_2 absorptions controls the availability of short wavelength radiation that can photodissociate molecules that are stable in the troposphere. In the stratosphere (at 50 km), there is typically no radiation of wavelength shorter than 183 nm.

5.4 Tropospheric Oxidation Chemistry

Although atmospheric composition is dominated by both oxygen and nitrogen, it is not the amount of oxygen that defines the capacity of the troposphere to oxidize a trace gas. The *oxidizing capacity* of the troposphere is a somewhat nebulous term probably best described by Thompson (1992):

The total atmospheric burden of O_3, OH and H_2O_2 determines the '*oxidizing capacity*' of the atmosphere. As a result of the multiple interactions among the three oxidants and the multiphase activity of H_2O_2, there is no single expression that defines the earth's oxidizing capacity. Some researchers take the term to mean the total global OH, although even this parameter is not defined unambiguously.

Figure 5.7 gives a schematic representation of tropospheric chemistry, representing the links between emissions, chemical transformation and sinks for a range of trace gases.

Atmospheric photochemistry produces a variety of radicals that exert a substantial influence on the ultimate composition of the atmosphere (Monks 2005). Probably the most important of these in terms of its reactivity is the hydroxyl radical, OH. The formation of OH is the initiator of radical-chain oxidation. Photolysis of ozone by ultraviolet light in the presence of water vapour is the main source of hydroxyl radicals in the troposphere, namely

$$O_3 + h\nu \; (\lambda < 340\,\text{nm}) \rightarrow O\left(^1D\right) + O_2\left(^1\Delta_g\right) \tag{5.2}$$

$$O\left(^1D\right) + H_2O \rightarrow OH + OH \tag{5.3}$$

The fate of the bulk of the $O(^1D)$ atoms produced via reaction (5.2) is collisional quenching back to ground-state oxygen atoms, namely

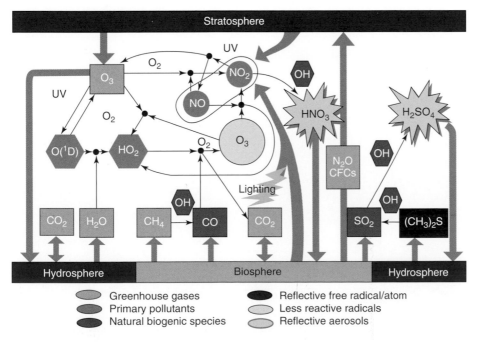

Figure 5.7 A simplified scheme of tropospheric chemistry. The figure illustrates the interconnections in the chemistry, as well as the role of sources, chemical transformations, and sinks.

$$O\left(^1D\right)+N_2 \rightarrow O\left(^3P\right)+N_2 \tag{5.4}$$

$$O\left(^1D\right)+O_2 \rightarrow O\left(^3P\right)+O_2 \tag{5.5}$$

The fraction of $O(^1D)$ atoms that form OH is dependent on pressure and the concentration of H_2O; typically in the marine boundary layer about 10% of the $O(^1D)$ generate OH. Reactions (5.2 and 5.3) are the primary source of OH in the troposphere, but there are a number of other reactions and photolysis routes capable of forming OH directly or indirectly. As these compounds are often products of OH-radical-initiated oxidation, they are often termed secondary sources of OH and include the photolysis of HONO, HCHO, H_2O_2, and acetone and the reaction of $O(^1D)$ with methane (see Figure 5.8). Table 5.2 illustrates the average contribution of various formation routes with altitude in a standard atmosphere.

Two important features of OH chemistry make it critical to the chemistry of the troposphere. The first is its inherent reactivity; the second is its relatively high concentration given its high reactivity. The hydroxyl radical is ubiquitous throughout the troposphere owing to the widespread nature of ozone and water. In relatively unpolluted regimes (low NO_x), the main fate for the hydroxyl radical is reaction with either carbon monoxide or methane to produce peroxy radicals such as HO_2 and CH_3O_2, namely

$$OH+CO \rightarrow H+CO_2 \tag{5.6}$$

$$H+O_2+M \rightarrow HO_2+M \tag{5.7}$$

and

$$OH+CH_4 \rightarrow CH_3+H_2O \tag{5.8}$$

$$CH_3+O_2+M \rightarrow CH_3O_2+M \tag{5.9}$$

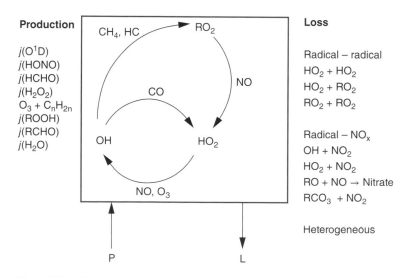

Figure 5.8 The sources, interconversions, and sinks for HO_x (and RO_x) in the troposphere.

Table 5.2 Calculated fractional contribution of various photolysis rates to radical production with altitude.

Altitude	$j(O(^1D))$ + H$_2$O	$j(O(^1D))$ + CH$_4$	j(Acetone)	j(H$_2$O$_2$)	j(HCHO)
Ground	0.68	0.0	Negligible	0.15	0.17
Mid-troposphere	0.52	Neg.	0.03	0.20	0.25
Upper-troposphere	0.35	0.02	0.1	0.25	0.28
Lower stratosphere	0.40	0.1	0.25	0.1	0.15

In low-NO$_x$ conditions, HO$_2$ can react with ozone leading to further destruction of ozone in a chain sequence involving production of hydroxyl radicals:

$$HO_2 + O_3 \rightarrow OH + 2O_2 \tag{5.10}$$

$$OH + O_3 \rightarrow HO_2 + O_2 \tag{5.11}$$

Alternatively, it can recombine to form hydrogen peroxide (H$_2$O$_2$):

$$HO_2 + HO_2 \rightarrow H_2O_2 + O_2 \tag{5.12}$$

or react with organic peroxy radicals such as CH$_3$O$_2$ to form organic hydroperoxides:

$$CH_3O_2 + HO_2 \rightarrow CH_3O_2H + O_2 \tag{5.13}$$

The formation of peroxides is effectively a chain termination reaction, as under most conditions these peroxides can act as effective sinks for HO$_x$. In more polluted conditions (high NO$_x$), peroxy radicals catalyse the oxidation of NO to NO$_2$

$$HO_2 + NO \rightarrow OH + NO_2 \tag{5.14}$$

leading to the production of ozone from the subsequent photolysis of nitrogen dioxide and reaction of the photoproducts, i.e.

$$NO_2 + h\nu \ \left(\lambda < 420\,nm\right) \rightarrow NO + O\left(^3P\right) \tag{5.15}$$

$$O + O_2 + M \rightarrow O_3 + M \tag{5.16}$$

Hydroxyl radicals produced in reaction (5.10) can go on to form more peroxy radicals (e.g. via reaction 5.11). Similarly to HO$_2$, CH$_3$O$_2$ can also oxidize NO to NO$_2$:

$$CH_3O_2 + NO \rightarrow CH_3O + NO_2 \tag{5.17}$$

The resulting methoxy radical reacts rapidly with O$_2$ to form formaldehyde and HO$_2$:

$$CH_3O + O_2 \rightarrow HCHO + HO_2 \tag{5.18}$$

The oxidation of methane is summarized schematically in Figure 5.9. The OH radical may have another fate – dependent on the concentration of NO$_2$, it can react with NO$_2$ to form nitric acid:

$$OH + NO_2 + M \rightarrow HNO_3 + M \tag{5.19}$$

The formation of HNO$_3$ represents an effective loss mechanism for both HO$_x$ and NO$_x$.

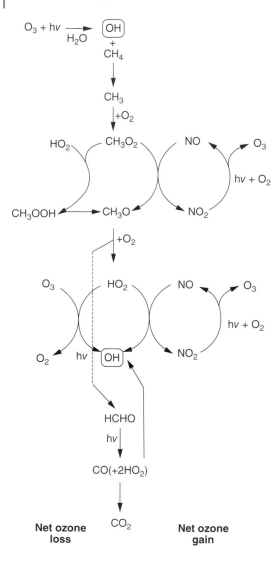

Figure 5.9 Simplified mechanism for the photochemical oxidation of CH_4 in the troposphere (Lightfoot et al. 1992).

Worked Example 5.1

Lifetime calculation

For the reaction $CH_4 + OH \rightarrow CH_3 + H_2O$, which is the key loss process for CH_4 (a greenhouse gas), the rate coefficient for the reaction at atmospheric temperatures is given by $k = 8.4 \times 10^{-15}$ cm^3 molecule^{-1} s^{-1}. Given the mean atmospheric concentration of $[OH] = 5 \times 10^5$ molecule cm^{-3}, what is the atmospheric lifetime of CH_4?

Answer

For a reaction of the type $A + B \rightarrow P$ (i.e. second-order), the atmospheric lifetime is given by

$$\tau_{OH}^{CH_4} = \frac{1}{k[OH]} = \frac{1}{(6.3\times10^{-15})(5\times10^5)}$$

$$= 3.1\times10^8 \ \text{seconds} = 9.9 \ \text{years}$$

These types of calculations are useful as they give an indication of the likely chemical lifetime (i.e. the amount of time it will take before a molecule is reacted away) of a molecule in the atmosphere. Clearly, this form of calculation does not take into account any other chemical loss routes other than reaction with OH or any other physical process that may remove a molecule.

Worked Example 5.2

Temperature dependence of a reaction
The lifetime of a compound in the atmosphere will depend on how fast it reacts with main atmospheric oxidants. Many reaction rates vary with temperature; therefore, in the atmosphere the lifetime will vary with altitude. For the reaction between OH and CH4, the temperature dependence of the reaction is given by $k = 1.85 \times 10^{-12} \exp(-1690/T)$. How does the reaction rate vary between 0.1 and 10 km and therefore affect the lifetime? For the lifetime calculation, see worked example 5.1.

Answer

Region	Altitude (km)	T (° C)
Boundary layer	0.1	25
Lower troposphere	1	9
Middle troposphere	6	−24
Lower stratosphere	10	−56

Region	k (cm^3 molecule^{-1} s^{-1})	τ_{OH} (years)
Boundary layer	6.3×10^{-15}	9.9
Lower troposphere	4.6×10^{-15}	13.8
Middle troposphere	2.0×10^{-15}	30.3
Lower stratosphere	7.6×10^{-16}	82.7

5.4.1 Nitrogen Oxides and the Photostationary State

From the preceding discussion it can be seen that the chemistry of nitrogen oxides are an integral part of tropospheric oxidation and photochemical processes. Nitrogen oxides are released into the troposphere from a variety of biogenic and anthropogenic sources, including fossil-fuel combustion, biomass burning, microbial activity in soils, and lightning discharges (see Figure 5.2). About 30% of the global budget of NO$_x$, i.e. (NO + NO$_2$), comes from fossil-fuel combustion, with almost 86% of the NO$_x$ emitted in one form or the other into the planetary boundary layer from surface processes. Typical NO/NO$_2$ ratios in surface air are 0.2–0.5 during the day, tending to zero at night. Over the time scales of hours to days, NO$_x$ is converted to nitric acid (reaction 5.19) and nitrates, which are subsequently removed by rain and dry deposition.

The photolysis of NO_2 to NO and the subsequent regeneration of NO_2 via reaction of NO with ozone is sufficiently fast, in the moderately polluted environment, for these species to be in dynamic equilibrium, namely

$$NO_2 + h\nu \rightarrow NO + O\left(^3P\right) \tag{5.15}$$

$$O\left(^3P\right) + O_2 + M \rightarrow O_3 + M \tag{5.20}$$

$$O_3 + NO \rightarrow NO_2 + O_2 \tag{5.21}$$

Therefore, at suitable concentrations, ambient NO, NO_2, and O_3 can be said to be in a photochemical steady-state or photostationary state (PSS; Leighton 1961), provided that they are isolated from local sources of NO_x and that sunlight intensity is relatively constant, therefore

$$\left[O_3\right] = \frac{j_{15}\left[NO_2\right]}{k_{21}\left[NO\right]} \tag{5.22}$$

Reactions (5.15, 5.20 and 5.21) constitute a cycle with no net chemistry. The PSS expression is sometimes expressed as a ratio, namely

$$\phi = \frac{j_{15}\left[NO_2\right]}{k_{21}\left[NO\right]\left[O_3\right]} \tag{5.23}$$

If ozone is the sole oxidant for NO to NO_2, then $\phi = 1$. This situation often pertains in urban areas where NO_x levels are high and other potential oxidants of NO to NO_2, such as peroxy radicals, are suppressed. In the presence of peroxy radicals, Eq. (5.23) has to be modified as the NO/NO_2 partitioning is shifted to favour NO_2 namely

$$\frac{\left[NO_2\right]}{\left[NO\right]} = \left(k_{21}\left[O_3\right] + k_{14}\left[HO_2\right] + k_{17}\left[RO_2\right]\right) / j_{15} \tag{5.24}$$

Although the radical concentrations are typically ca. 1000 times smaller than the $[O_3]$, the rate of the radical oxidation of NO to NO_2 is ca. 500 times larger than the corresponding oxidation by reaction with O_3. We shall return to the significance of the peroxy radical catalysed oxidation of NO to NO_2 when considering photochemical ozone production and destruction. From the preceding discussion, it can be seen that the behaviour of NO and NO_2 are strongly coupled through both photolytic and chemical equilibria. Because of their rapid interconversion, they are often referred to as NO_x; NO_x i.e. (NO + NO_2), is also sometimes referred to as 'active nitrogen'.

Worked Example 5.3

Photostationary state

At what NO_2/NO ratio will the photostationary state ratio be equal to 1 for midday conditions $(j_{15} = 1 \times 10^{-3}\ s^{-1})$ given that $k_{24} = 1.7 \times 10^{-14}\ cm^3$ molecule^{-1} s^{-1} and that $O_3 = 30\ ppbv$?

Answer

Using Eq. (5.23)

$$\phi = \frac{j_{15}[NO_2]}{k_{21}[NO][O_3]}$$

Converting O_3 from ppbv to molecule cm^{-3}, as $1\,ppbv = 2.46 \times 10^{10}\,molecule\,cm^{-3}$ (at $25°$ C and 1 atm) $30\,ppbv = 7.38 \times 10^{11}\,molecule\,cm^{-3}$. Given that $\phi = 1$, then

$$\frac{[NO_2]}{[NO]} = \frac{k_{21}[O_3]}{j_{15}}$$

$$[NO_2]/[NO] = 12.55$$

The extent of the influence of NO_x in any given atmospheric situation depends on its sources, reservoir species, and sinks. Therefore, an important atmospheric quantity is the lifetime of NO_x. If nitric acid formation is considered to be the main loss process for NO_x (i.e. NO_2), then the lifetime of NO_x (t_{NOx}) can be expressed as the time constant for reaction (5.19), the NO_2 to HNO_3 conversion

$$\tau_{NO_x} = \frac{1}{k_{23}[OH]}\left(1 + \frac{[NO]}{[NO_2]}\right) \tag{5.25}$$

Therefore, using this simplification, the lifetime of NO_x is dependent on the [OH] and $[NO]/[NO_2]$ ratio. Calculating t_{NOx} under typical upper tropospheric conditions gives lifetimes in the order of four to seven days and lifetimes in the order of days in the lower free troposphere. In the boundary layer, the situation is more complex, as there are other NO_x loss and transformation processes apart from those considered in Eq. (5.25), which can make t_{NOx} as short as one hour. Integrally linked to the lifetime of NO_x and therefore the role of nitrogen oxides in the troposphere is its relation to odd nitrogen reservoir species. The sum of total reactive nitrogen or total odd nitrogen is often referred to as NO_y and can be defined as $NO_y = NO_x + NO_3 + 2N_2O_5 + HNO_3 + HNO_4 + HONO + PAN + nitrate\ aerosol + alkyl\ nitrate$, where PAN is peroxyacetlynitrate (see urban chemistry); NO_y can also be thought of as NO_x plus all the compounds that are products of the atmospheric oxidation of NO_x. Owing to the potential for some of its constituents (e.g. HNO_3) to be efficiently removed by deposition processes, NO_y is not a conserved quantity in the atmosphere. Mixing of air masses may also lead to dilution of NO_y. The concept of NO_y is useful in considering the budget of odd nitrogen and evaluating the partitioning of NO_x and its reservoirs in the troposphere (Poisson et al. 2000).

In summary, the concentration of NO_x in the troposphere determines the following:

- Catalytic efficiency of ozone production;
- Partitioning of OH and HO_2;
- Amount of HNO_3 and nitrates produced;
- Magnitude and sign of net photochemical production or destruction of ozone.

5.4.2 Production and Destruction of Ozone

From the preceding discussion of atmospheric photochemistry and NO_x chemistry it can be seen that the fate of the peroxy radicals can have a marked effect on the ability of the atmosphere either to produce or to destroy ozone. Photolysis of NO_2 and the subsequent reaction of the photoproducts with O_2 (reactions 5.15 and 5.20) are the only known way of producing ozone in the troposphere. In the presence of NO_x the following cycle for the production of ozone can take place:

$$NO_2 + h\nu \rightarrow O\left(^3P\right) + NO \tag{5.15}$$

$$O\left(^3P\right) + O_2 + M \rightarrow O_3 + M \tag{5.20}$$

$$OH + CO \rightarrow H + CO_2 \tag{5.6}$$

$$H + O_2 + M \rightarrow HO_2 + M \tag{5.7}$$

$$HO_2 + NO \rightarrow OH + NO_2 \tag{5.14}$$

$$Net : CO + 2O_2 + h\nu \rightarrow CO_2 + O_3 \tag{5.26}$$

Similar chain reactions can be written for reactions involving RO_2. In contrast, when relatively little NO_x is present, as in the remote atmosphere, the following cycle can dominate over ozone production, leading to the catalytic destruction of ozone, namely

$$HO_2 + O_3 \rightarrow OH + 2O_2 \tag{5.10}$$

$$OH + CO \rightarrow H + CO_2 \tag{5.6}$$

$$H + O_2 + M \rightarrow HO_2 + M \tag{5.7}$$

$$Net : CO + O_3 \rightarrow CO_2 + O_2 \tag{5.27}$$

Clearly, there is a balance between photochemical ozone production and ozone loss dependent on the concentrations of HO_x and NO_x. Figure 5.10 shows the dependence of the production of ozone on NO_x taken from a numerical model. There are distinct regions in terms of $N[O_3]$ versus $[NO_x]$ on Figure 5.10. For example, in region A the loss of ozone ($L[O_3]$) is greater than the production of ozone ($P[O_3]$), hence the net product of this process i.e.

$$N\left(O_3\right) = P\left(O_3\right) - L\left(O_3\right) \tag{5.28}$$

leads to a net ozone loss. The photochemical loss of ozone can be represented as

$$L\left(O_3\right) = \left(f.j_7\left(O^1D\right).\left[O_3\right]\right) + k_{10}\left[HO_2\right] + k_{11}\left[OH\right] \tag{5.29}$$

where f is the fraction of $O(^1D)$ atoms that react with water vapour (reaction 5.3) rather than are deactivated to $O(^3P)$ (reactions 5.4 and 5.5). Evaluation of Eq. (5.29) is effectively a lower limit for the ozone loss rate as it neglects any other potential chemical loss processes for ozone such as cloud chemistry (Lelieveld and Crutzen 1990), NO_3 chemistry (see section on night-time chemistry) or halogen chemistry (see section on halogen chemistry). The balance point, i.e. where $N(O_3) = 0$, is often referred to, somewhat misleadingly, as the compensation point and occurs at a critical concentration of NO_x. Above the compensation point $P(O_3) > L(O_3)$ and therefore $N(O_3)$ is positive and the system is forming ozone. The in-situ formation rate for ozone is approximately given by the rate at which the peroxy radicals (HO_2 and

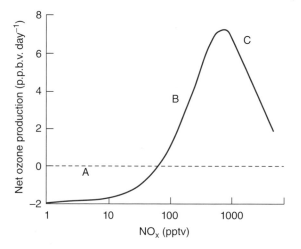

Figure 5.10 Schematic representation of the dependence of the net ozone ($N(O_3)$) production (or destruction) on the concentration of NO_x.

RO_2) oxidize NO to NO_2. This is followed by the rapid photolysis of NO_2 (reaction 5.15) to yield the oxygen atom required to produce O_3

$$P(O_3)=[NO].\left(k_{14}[HO_2]+\sum k_i[RO_2]_i\right) \tag{5.30}$$

It is worth noting that $P(O_3)$ can also be expressed in terms of the concentrations of NO_x, $j_2(NO_2)$, O_3, and temperature by substitution of Eq. (5.22) into Eq. (5.30) to give

$$P(O_3)= j_{15}[NO_2]-k_{21}[NO][O_3] \tag{5.31}$$

At some concentration of NO_x the system reaches a maximum production rate for ozone at $dP(O_3)/d(NO_x) = 0$ and even though $P(O_3)$ is still significantly larger than $L(O_3)$ the net production rate begins to fall off with increasing NO_x. Until this maximum is reached the system is said to be NO_x limited with respect to the production of ozone. The turnover, i.e. $dP(O_3)/d(NO_x) = 0$, is caused by the increased competition for NO_x by the reaction

$$OH + NO_2 + M \rightarrow HNO_3 + M \tag{5.19}$$

Worked Example 5.4

Photochemical production of ozone

Calculate the photochemical production rate of ozone at midday, given that $j_{15} = 1\times10^{-3}\,s^{-1}$, $O_3 = 30\,ppbv$ and $k_{24} = 1.7\times10^{-14}\,cm^3\,molecule^{-1}\,s^{-1}$ for a 2:1 ratio of NO_2: NO.

Answer

Using equation

$$P(O_3)= j_{15}[NO_2]-k_{21}[NO][O_3]$$

As the rate constant is in the units of $cm^3\,molecule^{-1}\,s^{-1}$, all the concentrations should be converted into molecule cm^{-3} (as per worked example 5.3). The answer should be converted into ppb $hour^{-1}$ by dividing the answer in $cm^3\,molecule^{-1}\,s^{-1}$ into ppbv and turning the seconds into hours. For $NO_2 = 2\,ppbv$, $NO = 1\,ppbv$, $P(O_3) = 7.1\,ppbv\,hour^{-1}$. The equation scales as per the absolute values of NO_2:NO used.

Figure 5.10 represents a slice through an *n*-dimensional surface where there should be a third axis to represent the concentration of volatile organic compounds (VOCs). The peak initial concentrations of ozone generated from various initial concentrations of NO_x and VOCs are usually represented as an 'O$_3$ isopleth diagram', an example of which is shown in Figure 5.11 (Sillman 1999). In an isopleth diagram, initial mixture compositions giving rise to the same peak O_3 concentration are connected by the appropriate isopleth. An isopleth plot shows that ozone production is a highly non-linear process in relation to NO_x and VOC, but picks out many of the features already highlighted in Figure 5.10, i.e. when NO_x is 'low' the rate of ozone formation increases with increasing NO_x in a near-linear fashion. On the isopleth, the local maximum in the ozone formation rate with respect to NO_x is the same feature as the turnover in $N(O_3)$ in Figure 5.10. The ridgeline along the local maximum separates two different regimes, the so-called NO_x-*sensitive* regime, i.e. $N(O_3) \propto (NO_x)$ and the VOC-*sensitive* (or NO_x *saturated* regime), i.e. $N(O_3) \propto (VOC)$ and increases with increasing NO_x. The relationship between NO_x, VOCs, and ozone embodied in the isopleth diagram indicates one of the problems in the development of air quality policy with respect to ozone. Reductions in VOC are effective only in reducing ozone under VOC-*sensitive* chemistry (high NO_x) and reductions in NO_x will be effective only if NO_x-*sensitive* chemistry predominates and may actually increase ozone in VOC-*sensitive* regions. In general, as an airmass moves away from emission sources, e.g. in an urban region, the chemistry tends to move from VOC-*sensitive* to NO_x-*sensitive* chemistry.

5.4.3 Role of Hydrocarbons

The discussion up to this point has focused in the main on the role of CO and CH_4 as the fuels for atmospheric oxidation. It is clear that there are many more carbon compounds in the atmosphere than just these two (Goldstein and Galbally 2007). One of the roles of atmospheric photochemistry is to 'cleanse' the troposphere of a wide-range of these compounds. Table 5.3 illustrates the global turnover of a range of trace gases including hydrocarbons and illustrates, for a number of trace gases, the primary role played by OH in their removal. Non-methane hydrocarbons (NMHC) have a range of both biogenic and

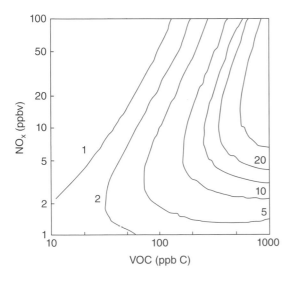

Figure 5.11 Isopleths giving net rate of ozone production (ppb hour^{-1}) as a function of volatile organic compounds (VOC; ppb C) and NO_x (ppbv) for mean summer daytime meteorology and clear skies (Sillman 1999).

Table 5.3 Global turnover of tropospheric gases and fraction removed by reaction with OH.

Trace gas	Global emission rate (Tg year^{-1})	Removal by OH (%)[a]
CO	2800	85
CH_4	530	90
C_2H_6	20	90
Isoprene	570	90
Terpenes	140	50
NO_2	150	50
SO_2	300	30
$(CH_3)_2S$	30	90
$CFCl_3$	0.3	0

Source: after Ehhalt (1999).
[a] Assuming mean global [OH] $= 1 \times 10^6$ molecule cm^{-3}.

anthropogenic sources. Carbon monoxide chemistry is not independent from NMHC chemistry as 40–60% of surface CO levels over the continents, slightly less over the oceans, and 30–60% of CO levels in the free troposphere, are estimated to come from NMHC oxidation (Poisson et al. 2000).

A major component of the reactive hydrocarbon loading are the biogenic hydrocarbons (Goldstein and Galbally 2007). As previously indicated, the hydrocarbon oxidation chemistry is integral to the production of ozone. Globally, the contribution of NMHC to net photochemical production of ozone is estimated to be about 40% (Houweling et al. 1998).

There are a number of inorganic molecules such as NO_2 and SO_2 (see Table 5.3) that are also lost *via* reaction with OH. A number of halocarbons also exist that possess insubstantial tropospheric sinks and have importance in the chemistry of the stratosphere (see Chapter 8).

The ultimate products of the oxidation of any hydrocarbon are carbon dioxide and water vapour, but there are many relatively stable partially oxidized organic species such as aldehydes, ketones, and carbon monoxide that are produced as intermediate products during this process, with ozone produced as a byproduct of the oxidation process. Figure 5.9 shows a schematic representation of the free radical catalysed oxidation of methane, which is analogous to that of a hydrocarbon. As previously discussed, the oxidation is initiated by reaction of the hydrocarbon with OH and follows a mechanism in which the alkoxy and peroxy radicals are chain propagators and OH is effectively catalytic, namely

$$OH + RH \rightarrow R + H_2O \tag{5.32}$$

$$R + O_2 + M \rightarrow RO_2 + M \tag{5.33}$$

$$RO_2 + NO \rightarrow RO + NO_2 \tag{5.34}$$

$$RO \rightarrow carbonyl\ products + HO_2 \tag{5.35}$$

$$HO_2 + NO \rightarrow OH + NO_2 \tag{5.14}$$

As both reactions (5.34) and (5.14) lead to the oxidation of NO to NO_2 the subsequent photolysis leads to the formation of ozone (see reactions 5.15 and 5.20). The individual reaction mechanism depends on the identity of the organic compounds and the level of complexity of the mechanism. Although OH is the main tropospheric oxidation initiator, reaction with NO_3, O_3, $O(^3P)$, or photolysis may be an important loss route for some NMHCs or the partially oxygenated products produced as intermediates in the oxidation (see reaction 5.35).

In summary, the rate of oxidation of VOCs and therefore by inference the production of ozone is governed by the concentration of the catalytic HO_x radicals. There are a large variety of VOCs with a range of reactivities; therefore, this remains a complex area.

5.5 Night-Time Oxidation Chemistry

Although photochemistry does not take place at night, it is important to note, within the context of tropospheric oxidation chemistry, the potential for oxidation chemistry to continue at night (Brown and Stutz 2012). This chemistry does not lead to the production of ozone (in fact the opposite), but has importance owing to the potential for the production of secondary pollutants. In the troposphere, the main night-time oxidant is thought to be the nitrate radical formed by the relatively slow oxidation of NO_2 by O_3, namely

$$NO_2 + O_3 \rightarrow NO_3 + O_2 \tag{5.36}$$

The time constant for reaction (5.36) is of the order of 15 hours at an ozone concentration of 30 ppbv and $T = 290\,K$. Other sources include

$$N_2O_5 + M \rightarrow NO_3 + NO_2 + M \tag{5.37}$$

but as N_2O_5 is formed from

$$NO_3 + NO_2 + M \rightarrow N_2O_5 + M \tag{5.38}$$

the two species act in a coupled manner. Dinitrogen pentoxide, N_2O_5, is potentially an important product, as it can react heterogeneously with water to yield HNO_3 and is a major atmospheric loss pathway for nitrogen oxides. During the daytime the NO_3 radical is rapidly photolysed as it strongly absorbs in the visible spectrum, namely

$$NO_3 + h\nu \rightarrow NO + O_2 \tag{5.39}$$

$$NO_3 + h\nu \rightarrow NO_2 + O\left(^3P\right) \tag{5.40}$$

having a lifetime in the region of five seconds for overhead sun and clear-sky conditions. Further, NO_3 will react rapidly with NO

$$NO_3 + NO \rightarrow NO_2 + NO_2 \tag{5.41}$$

which can have significant daytime concentrations in contrast to the night-time, where away from strong source regions, the NO concentrations should be near zero.

The nitrate radical has a range of reactivity towards volatile organic compounds. The nitrate radical is highly reactive towards certain unsaturated hydrocarbons such as isoprene, a variety of butenes and monoterpenes, as well as reduced sulphur compounds such as dimethylsulphide (DMS). In the case of DMS, if the NO_2 concentration is 60% that of DMS, then NO_3 is a more important oxidant than OH for DMS in the marine boundary layer. Figure 5.12 provides a simplified summary of the relevant night-time chemistry involving the nitrate radical.

One important difference between NO_3 chemistry and daytime OH chemistry is that NO_3 can initiate, but not catalyse, the removal of organic compounds. Therefore, its concentration can be suppressed by the presence of fast-reacting, with respect to NO_3, organic compounds and NO (reaction 5.41). Figure 5.13 shows some airborne measurements of both N_2O_5 and NO_3.

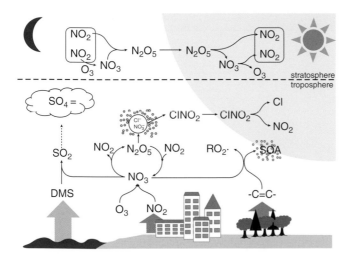

Figure 5.12 A simplified reaction scheme for night-time chemistry involving the nitrate radical (Brown and Stutz, 2012).

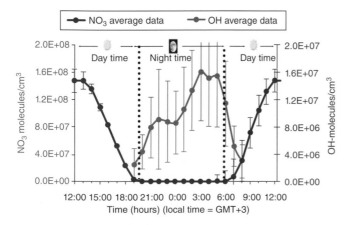

Figure 5.13 Concentrations of N_2O_5 and NO_3 from airborne measurements off the UK (Kennedy et al. 2011). See colour plate section for the colour representation of this figure.

Worked Example 5.5

Reactive lifetime with respect to different oxidants

Different oxidants react at different rates in the atmosphere. Further, the average concentrations of the main oxidants can vary. For the following table of data, calculate the relative lifetime with respect to reaction with OH and NO_3, given that the average $[OH] \approx 10^6$ molecule cm^{-3} and the $[NO_3] \approx 10^9$ molecule cm^{-3}.

Answer

Compound	$k(OH)$ (cm^3 molecule^{-1} s^{-1})	t_{OH} (days)
CH_4	8.5×10^{-15}	1361
C_2H_6	2.7×10^{-13}	43
C_2H_4	8.5×10^{-12}	1.3
$(CH_3)_2C = C(CH_3)_2$	1.1×10^{-10}	0.1
	$k(NO_3)$ (cm^3 molecule^{-1} s^{-1})	t_{NO3} (days)
CH_4	$<1 \times 10^{-19}$	115 740
C_2H_6	8×10^{-18}	1446
C_2H_4	2×10^{-16}	58
$(CH_3)_2C = C(CH_3)_2$	4.5×10^{-11}	2×10^{-4}

The data well illustrate that lifetime is a product of rate and the concentration of the radical species.

5.5.1 Ozone–Alkene Chemistry

Another *dark* source of HO_x in the atmosphere, more particularly in the boundary layer, is from the reactions between ozone and alkenes. The ozonolysis of alkanes proceeds in two distinct steps; the first being the 1,3 cycloaddition of ozone across a double bond to form a primary ozonide, followed by decomposition to a carbonyl and carbonyl oxide, otherwise known as a Criegee intermediate (named after Rudolph Criegee who first proposed the scheme in 1949) (see Figure 5.14). The excited Criegee intermediate can the undergo unimolecular decomposition or can form a stabilized Criegee intermediate. The decomposition of the Criegee intermediate can lead to the direct production of the OH and HO_2 radicals at varying

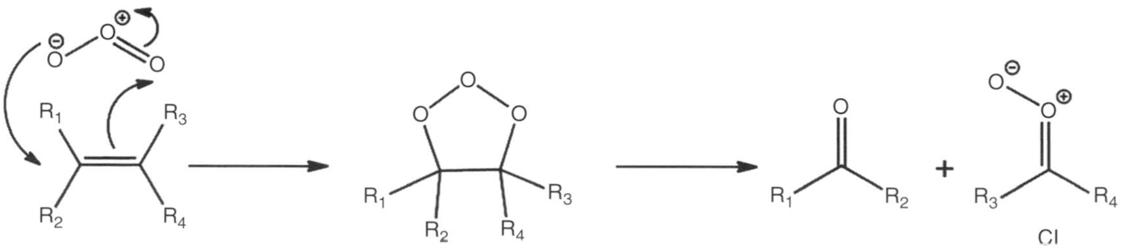

Figure 5.14 Formation of a primary ozonide followed by decomposition to a carbonyl and Criegee intermediate (CI) (after Khan et al. 2018).

yields (7–100%), depending on the structure of the alkene (see Table 5.4), normally accompanied by the co-production of an (organic) peroxy radical. Studies (e.g. Harrison et al. 2006) have shown that these reactions can be a significant source of OH to the tropospheric oxidation cycle, especially during night-time and the winter months (see Figure 5.15 and compare to the average state shown in Table 5.2).

As compared with both the reactions of OH and NO_3 with alkenes, the initial rate of the reaction of ozone with an alkene is relatively slow. This can be offset under regimes where there are high concentrations of alkenes and/or ozone. For example, under typical rural conditions the atmospheric lifetimes for the reaction of ethene with OH, O_3 and NO_3 are 20 hours, 9.7 days and 5.2 months respectively in contrast, for the same reactants with 2-methyl-2-butene the atmospheric lifetimes are 2.0 hours, 0.9 hours and

Table 5.4　Range of OH and HO_2 yields from the reaction of ozone with alkenes.

Alkene	OH yield	HO_2 yield
Ethene	0.17 ± 0.09	0.10 ± 0.03
Propene	0.36 ± 0.10	0.09 ± 0.02
2-Methylpropene	0.67 ± 0.18	0.34 ± 0.09
Isoprene	1.00	0.16 ± 0.04

Source: all data taken from Alam et al. (2015).

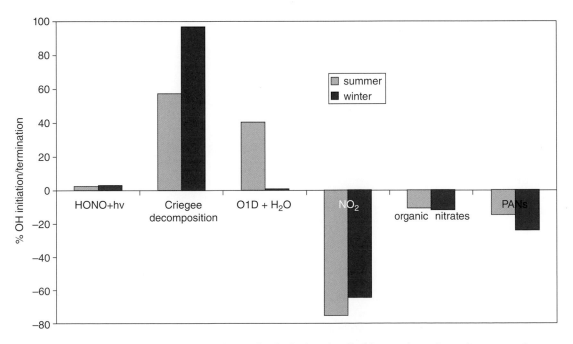

Figure 5.15　Main production and loss pathways for the hydroxyl radical in an urban winter-time atmosphere. *Source:* adapted from Harrison et al. (2006).

0.09 hours. More recently, there has been new evidence around that fate of the stabilized Criegee intermediate and its impact on tropospheric oxidation chemistry (see, e.g. Khan et al. 2018). It has been shown that reactions with water to form carbonyl compounds and with SO_2 and a range of oxygenated compounds, such as carboxylic acids and carbonyls, can lead to the formation of nucleation species of potential importance in cloud and aerosol formation processes.

5.6 Halogen Chemistry

In comparison to the atmospheric chemistry taking place in the stratosphere, where halogen chemistry is well known and characterized (see Chapter 8), there has been much debate as to the role of halogen species in the oxidative chemistry of the troposphere. Halogen chemistry is part of tropospheric photochemistry (Monks 2005; Saiz-Lopez and von Glasow 2012; Simpson et al. 2015).

Much of the proposed halogen chemistry is propagated through the reactions of a series of halogen atoms and radicals.

Bromine oxide species can be formed in the polar boundary layer (Barrie et al. 1988; Abbatt et al. 2012; Saiz-Lopez et al. 2004) and areas with high salt levels such as the Dead Sea (Heibestreit et al. 1990). The major source of gas-phase bromine in the lower troposphere is thought to be the release of species such as IBr, ICl, Br_2, and BrCl from sea-salt aerosol, following the uptake from the gas-phase and subsequent reactions of hypohalous acids (HOX, where X = Br, Cl, I).

$$HOBr + \left(Br^-\right)_{aq} + H^+ \rightarrow Br_2 + H_2O \tag{5.42}$$

The halogen release mechanism is autocatalytic and has become known as the 'bromine explosion'. The Br_2 produced in reaction (5.42) is rapidly photolysed, producing bromine atoms that can be oxidized to BrO by O_3 (5.44):

$$BrX + h\upsilon \rightarrow Br + X \tag{5.43}$$

$$Br + O_3 \rightarrow BrO + O_2 \tag{5.44}$$

the resultant BrO reacting with HO_2 (reaction 5.45) to reform HOBr.

$$BrO + HO_2 \rightarrow HOBr + O_2 \tag{5.45}$$

Thus, the complete cycle has the form

$$BrO + O_3 + \left(Br^-\right)_{aq} + \left(H^+\right)_{aq} \xrightarrow{\text{surface,HOx}} 2BrO + \text{products} \tag{5.46}$$

where effectively one BrO molecule is converted to two by oxidation of bromide from a suitable surface (Abbatt et al. 2012). Figure 5.16 shows a schematic representation of the bromine-explosion mechanism (Saiz-Lopez and von Glasow 2012). It is worth noting that the bromine explosion mechanism occurs only from sea salt with a pH < 6.5 therefore requiring acidification of the aerosol, potentially caused by the uptake of strong acids likely to be of anthropogenic origins or naturally occurring acids.

The bromine explosion

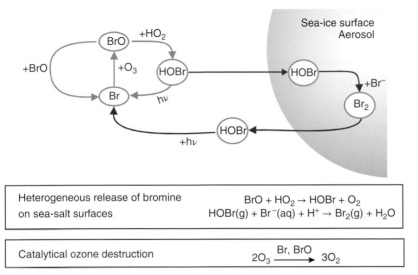

Heterogeneous release of bromine on sea-salt surfaces	$BrO + HO_2 \rightarrow HOBr + O_2$ $HOBr(g) + Br^-(aq) + H^+ \rightarrow Br_2(g) + H_2O$

Catalytical ozone destruction	$2O_3 \xrightarrow{Br,\ BrO} 3O_2$

Figure 5.16 A schematic representation of the so-called Bromine explosion mechanism where effectively one BrO molecule is converted to two by oxidation of bromide from a suitable aerosol surface (Frieß 2001).

Case study 5.1

Bromine Oxide in the Springtime Arctic

In the springtime in both the Arctic and Antarctic, large clouds of BrO-enriched air masses are observable from space (Wagner and Platt 1998; Simpson et al. 2007; Abbatt et al. 2012). These clouds cover several thousand square kilometres over the polar sea-ice, with BrO levels up to 30 pptv. The BrO is always coincident with low levels of ozone in the marine boundary layer (Barrie et al. 1988). In order to observe these events, there is a requirement for meteorological conditions that stop mixing between the boundary layer and the free troposphere and sunlight to drive the required photolysis of gaseous bromine (see reaction 5.43) released heterogeneously through chemical processes on the ice (Abbatt et al. 2012). Because of the prerequisites for strong surface inversions to confine the air and sunlight, episodes of bromine explosion events and boundary layer ozone depletion tend to be confined to spring.

With respect to iodine chemistry, the major sources of iodine release I_2, HOI, and organoiodine (including polyhalogenated iodine) compounds (Saiz-Lopez et al. 2012) (see Figure 5.17). Photolysis of the iodine and the organoiodine compounds releases the iodine.

$$RI_x + h\upsilon \rightarrow R + I \tag{5.47}$$

$$I_2 + h\upsilon \rightarrow 2I \tag{5.48}$$

$$I + O_3 \rightarrow IO + O_2 \tag{5.49}$$

Latterly, is has been suggested that the reaction between iodide and ozone in the surface ocean (Carpenter et al. 2013) may lead to the widespread release of I_2 and HOI viz

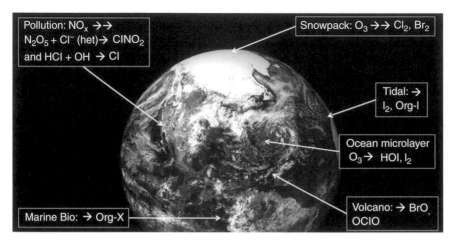

Figure 5.17 Primary sources of reactive halogen species or their precursor reservoir species (Simpson et al. 2015). See colour plate section for the colour representation of this figure.

$$H^+ + I^- + O_3 \rightarrow HOI + O_2 \qquad (5.50)$$

$$H^+ + HOI + I^- \Leftrightarrow I_2 + H_2O \qquad (5.51)$$

During daylight hours iodine monoxide, IO, exists in a fast photochemical equilibrium with I, namely

$$IO + h\upsilon \rightarrow I + O \qquad (5.52)$$

The aerosol 'explosion' mechanism, previously described for bromine, acts effectively to recycle the iodine back to the gas phase.

Case study 5.2

Iodine Monoxide Chemistry in the Coastal Margins

Figure 5.18 shows the measured concentration of iodine monoxide (IO) at Mace Head in Ireland (Carpenter et al. 2001). The data in Figure 5.18 show not only a clear requirement for radiation for photochemical production of IO but also a strong correlation with low-tide conditions. The correlation with low tidal conditions is indicative of the likely sources of IO, in that the photolysis of organoiodine compounds (reaction 5.47) such as CH_2I_2 and CH_2IBr emitted from macroalgae in the intertidal zone are possibly the candidate iodine sources. The photolysis lifetime of a molecule such as diodomethane is only a few minutes at midday. Recent experimental observations have suggested the I_2 can be emitted directly from the macroalgae at low tide, making this a substantial source of iodine. The chemistry can also lead to an enrichment of iodine in the aerosol in the form of iodate, providing a route to move biogenic iodine from ocean to land (Carpenter et al. 2001).

Tropospheric Halogens and Catalytic Destruction of Ozone

Potentially, the most important effect of reactive halogen species might be that their chemistry may lead to the catalytic destruction of ozone via two distinct cycles.

Cycle I:

$$XO + YO \rightarrow X + Y + O_2 \left(\text{or } XY + O_2 \right) \tag{5.53}$$

$$X + O_3 \rightarrow XO + O_2 \tag{5.54}$$

$$\text{Net}: O_3 + O_3 \rightarrow 3O_2 \tag{5.55}$$

In cycle I, the rate-limiting step involves reaction (5.53), the self- or cross-reaction of the halogen monoxide radicals. Cycle I has been identified to be the prime cause for polar boundary-layer ozone destruction (Barrie et al. 1988). The second ozone destruction cycle, which is more prevalent at low halogen levels, has the form

Cycle II:

$$XO + HO_2 \rightarrow HOX + O_2 \tag{5.56}$$

$$HOX + h\upsilon \rightarrow X + OH \tag{5.57}$$

$$X + O_3 \rightarrow XO + O_2 \tag{5.58}$$

$$OH + CO \rightarrow H + CO_2 \left(+M \right) \rightarrow HO_2 \tag{5.6, 5.7}$$

$$\text{Net}: O_3 + CO + hv \rightarrow CO_2 + O_2 \tag{5.59}$$

The rate-determining step in this reaction sequence is reaction (5.56), making ozone destruction linearly dependent on [XO]. The fraction of HOX that photolyses to give back OH depends critically on the accommodation coefficient of HOX on aerosols. Currently, there is a large uncertainty in this parameter. An important side effect of cycle 2 is the potential for the reduction of the $[HO_2]/[OH]$ ratio by consumption of HO_2.

Case study 5.3

The Potential Effect of Tropospheric Halogen Chemistry on Ozone

An example of the potential effect of the ozone depletion cycles (cycles I and II) has been assessed using a combination of measurements of IO and BrO and models in the marine boundary layer (Read et al. 2008). The measurements were taken in the remote tropical Atlantic Ocean. As shown in Figure 5.19, observationally driven modelling results show the need to add halogen chemistry to reconcile the observed ozone seasonal changes. The importance of these results is that they demonstrated the need to put halogen sources and their chemistry into atmospheric models or there may be significant errors in global ozone budgets and methane oxidation rates.

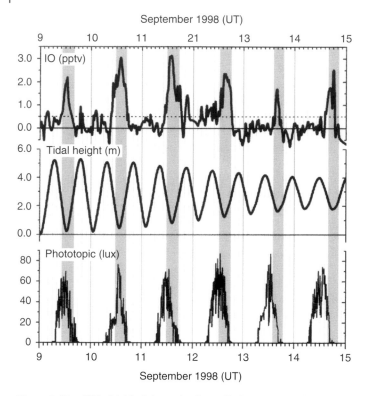

Figure 5.18 [IO], tidal height and solar radiation measured at Mace Head in Ireland. The grey areas represent the low tide areas during the day. The dotted line in the upper panel is the IO detection limit (Carpenter et al. 2001; Carpenter 2003).

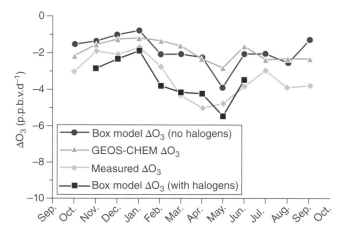

Figure 5.19 Monthly averages of the change of ozone to the mean (ΔO_3) in the remote marine boundary later compared to model with and without halogen chemistry. The data show the needs to add halogen chemistry to reconcile the observed ozone seasonal changes (Read et al. 2008).

5.7   Air Pollution and Urban Chemistry

Air pollution is defined by Weber (1982) as 'the presence of substances in the ambient atmosphere, resulting from the activity of man or from natural processes, causing adverse effects to man and the environment'. In this section we will examine the emissions and chemistry in typical urban environments with particular focus on processes that result directly or indirectly in adverse impacts on human health.

In some respects, the story of atmospheric chemistry and particularly ozone photochemistry begins with urban chemistry and photochemical smog. The term *smog* arises from a combination of the words smoke and fog. In the 1940s it became apparent that cities such as Los Angeles (LA) were severely afflicted with a noxious haze (Haagen-Smit 1952). Although at the time it was thought to be a relatively local phenomenon, with the understanding of its chemistry came the development of a photochemical theory for the whole of the troposphere. The LA smog is often termed *photochemical smog* and is quite different in origin to the London smogs of the nineteenth and twentieth centuries, which had their origins in abnormally high concentrations of smoke particles and sulphur dioxide. The London smogs were alleviated with the effective application of legislation that has reduced the burning of coal in the London area. The major features of photochemical smog are high levels of oxidant concentration, in particular ozone and peroxidic compounds, produced by photochemical reactions. The principal effects of smog are eye and bronchial irritation as well as plant and material damage (see Chapter 10). The basic reaction scheme for the formation of photochemical smog is

$$VOC + OH, h\upsilon \rightarrow VOC + R \tag{5.60}$$

$$R + NO \rightarrow NO_2 \tag{5.61}$$

$$NO_2 + h\nu\left(\lambda < 420\,nm\right) \rightarrow NO + O\left(^3P\right) \tag{5.15}$$

$$NO + O_3 \rightarrow NO_2 + O_2 \tag{5.21}$$

$$R + R \rightarrow R \tag{5.62}$$

$$R + NO_2 \rightarrow NO_y \tag{5.63}$$

There is a large range of available VOCs in the urban atmosphere, driven by the range of anthropogenic and biogenic sources (Jackson 2007). Table 5.5 illustrates the different loadings of the major classes of NMHC in urban and rural locations. These NMHC loadings must be coupled with measures of reactivity and the degradation mechanisms of the NMHC to give a representative picture of urban photochemistry. The oxidation of the VOCs drives, *via* the formation of peroxy radicals, the oxidation of NO to NO_2, where under the sunlit conditions the NO_2 can be dissociated to form ozone (reaction 5.15). Pre-existing ozone can also drive the NO to NO_2 conversion (reaction 5.21).

The basic chemistry responsible for urban photochemistry is essentially the same as that which takes place in the unpolluted atmosphere (see Section 5.4). It is the range and concentrations of NMHC fuels and the concentrations of NO_x coupled to the addition of some photochemical accelerants that can lead to the excesses of urban chemistry. For example, in the Los Angeles basin it is estimated that 3333 tons per day of organic compounds are emitted, as well as 890 tons per day of NO_x. In addition to the reactions forming OH in the background troposphere, i.e. *via* the reaction of $O(^1D)$ with H_2O, namely

Table 5.5 Percentage of non-methane hydrocarbon (NMHC) classes measured in the morning at various locations.

NMHC	Urban Los Angeles	Urban Boston	Rural Alabama
Alkanes	42	36	9
Alkenes	7	10	43^a
Aromatics	19	30	2
Other	33	24	46^b

Source: from Calvert et al. (2000).
[a] Large contribution from biogenic alkenes.
[b] Mainly oxygen containing.

$$O_3 + h\nu\left(\lambda < 340\,nm\right) \rightarrow O_2 + O\left(^1D\right) \tag{5.2}$$

$$O\left(^1D\right) + H_2O \rightarrow 2OH \tag{5.3}$$

under urban conditions, OH may be formed from secondary sources such as

$$HONO + h\nu\ \left(\lambda < 400\,nm\right) \rightarrow OH + NO \tag{5.64}$$

where the HONO can be emitted in small quantities from automobiles or formed from a number of heterogeneous pathways, as well as gas-phase routes (Kleffmann 2007). Production of OH from HONO has been shown to be the dominant OH source in the morning under some urban conditions, where the HONO has built up to significant levels overnight. Another key urban source of OH can come from the photolysis of the aldehydes and ketones produced in the NMHC oxidation chemistry, in particular formaldehyde, namely

$$HCHO + h\nu\left(\lambda < 334\,nm\right) \rightarrow H + HCO \tag{5.65}$$

$$H + O_2 + M \rightarrow HO_2 + M \tag{5.7}$$

$$HCO + O_2 \rightarrow HO_2 + CO \tag{5.66}$$

$$Net : HCHO + 2O_2 + h\nu \rightarrow 2HO_2 + CO \tag{5.67}$$

Smog chamber experiments have shown that the addition of aldehyde significantly increases the formation rates of ozone and the conversion rates of NO and NO_2 under simulated urban conditions (Wayne 2000).

A marked byproduct of oxidation in the urban atmosphere, often associated, but not exclusive to, urban air pollution is peroxyacetyl nitrate (PAN). Peroxyacetyl nitrate is formed by

$$OH + CH_3CHO \rightarrow CH_3CO + H_2O \tag{5.68}$$

$$CH_3CO + O_2 \rightarrow CH_3CO.O_2 \tag{5.69}$$

addition of NO_2 to the peroxyacetyl radical ($RCO \cdot O_2$) leading to the formation of peroxyacetyl nitrate

$$CH_3CO.O_2 + NO_2 + M \leftrightarrow CH_3CO.O_2.NO_2 + M \tag{5.70}$$

Peroxyacetyl nitrate is often used as an unambiguous marker for tropospheric chemistry. The lifetime of PAN in the troposphere is very much dependant on the temperature dependence of the equilibrium in reaction (5.70), the lifetime varying from 30 minutes at $T = 298\,K$ to 8 hours at $T = 273\,K$. At mid-troposphere temperature and pressures, PAN has thermal decomposition lifetimes in the order of 46 days. It is also worth noting that peroxyacyl-radical-like peroxy radicals can oxidize NO to NO_2

$$CH_3CO.O_2 + NO \rightarrow CH_3CO.O + NO_2 \tag{5.71}$$

Peroxyacetyl nitrate can be an important component of NO_y in the troposphere (see Section 5.4.1) and has the potential to act as a temporary reservoir for NO_x. In particular, PAN has the potential to transport NO_x from polluted regions into the background/remote atmosphere.

5.7.1 Sulphur Chemistry

Sulphur chemistry is an integral part of life, owing to its role in plant and human metabolism. Sulphur compounds have both natural and anthropogenic sources. In modern times, the atmospheric sulphur budget has become dominated by anthropogenic emissions, particularly from fossil-fuel burning. It is estimated that 75% of the total sulphur emission budget is dominated by anthropogenic sources, with 90% of it occurring in the Northern Hemisphere. The natural sources include volcanoes, plants, soil, and biogenic activity in the oceans (Jackson 2007). In terms of photochemistry, the major sulphur oxide, sulphur dioxide (SO_2) does not photodissociate in the troposphere (cf. NO_2), i.e.

$$\begin{aligned} SO_2\left(X^1A_1\right) + h\nu\left(240 < \lambda < 330\,nm\right) \\ \rightarrow SO_2\left(^1A_2,\ ^1B_1\right) \end{aligned} \tag{5.72}$$

$$SO_2\left(X^1A_1\right) + h\nu\left(340 < \lambda < 400\,nm\right) \rightarrow SO_2\left(^3B_1\right) \tag{5.73}$$

The oxidation of sulphur compounds in the atmosphere has implications in a number of different atmospheric problems such as acidification, climate balance and the formation of a sulphate layer in the stratosphere, the so-called Junge layer. By far the largest sulphur component emitted into the atmosphere is SO_2. Figure 5.20 shows the spatial distribution of SO_2 emissions in 2005 and 2015 (Krotkov et al. 2016). In Europe, the source regions for SO_2 are quite apparent in the East centred on mining, industrial areas, and the burning of high S-lignite (brown) coal for power generation. The absolute maximum in emissions is in southern Italy around Sicily, where the largest single source of both natural, the volcano Mount Etna, and anthropogenic SO_2 is found. Sulphur dioxide can be detected from space-borne sensors (Krotkov et al. 2016) as a product of volcanic activity and fossil-fuel burning. Figure 5.20 also illustrates another interesting point in that over much of Europe between 1990 and 2011, there has been a decrease of about 80% in SO_2 emissions owing to legislative limits. Interestingly, with decreasing land emissions, the importance of ship emissions has increased.

The atmospheric oxidation of SO_2 can take place by a number of different mechanisms, both homogeneously and heterogeneously in the liquid and gas phases (see Figure 5.16). The gas-phase oxidation of SO_2, namely

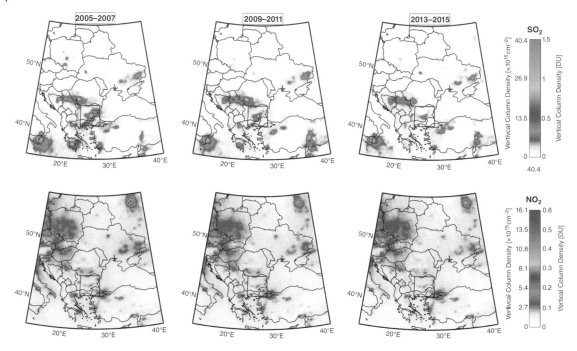

Figure 5.20 Three year average satellite instrument measured SO_2 over Eastern Europe. The largest source is Mt. Etna, Sicily, Italy. The blue box is around the coal mining and burning power stations in Bulgaria, showing a 50% reduction owing to the installation of flue gas desulphurization. (Krotkov et al. 2016). See colour plate section for the colour representation of this figure.

$$SO_2 + OH + M \rightarrow HSO_3 + M \tag{5.74}$$

$$HSO_3 + O_2 \rightarrow HO_2 + SO_3 \tag{5.75}$$

$$SO_2 + H_2O + M \rightarrow H_2SO_4 \tag{5.76}$$

can lead to the formation of sulphuric acid, which owing to its relatively low vapour pressure can rapidly attach to the condensed phase such as aerosol particles. The bulk of the H_2SO_4 is lost *via* wet deposition mechanisms in cloud droplets and precipitation. There is another potential gas-phase loss route for SO_2 that can lead to the formation of sulphuric acid in the presence of H_2O: that is, the reaction of SO_2 with Criegee intermediates (see Section 5.5.1). The aqueous phase oxidation of SO_2 is more complex, depending on a number of factors such as the nature of the aqueous phase (e.g. clouds and fogs), the availability of oxidants (e.g. O_3 and H_2O_2) and the availability of light. An overview of the mechanism is given in Figure 5.21. The key steps include the transport of the gas to the surface of a droplet, transfer across the gas–liquid interface, the formation of aqueous-phase equilibria, the transport from the surface into the bulk aqueous phase, and subsequent reaction. In brief, the SO_2 gas is dissolved in the liquid phase, establishing a set of equilibria for a series of S(IV) species, i.e. $SO_2.H_2O$, HSO_3^-, and SO_3^{2-}:

$$SO_2(g) + H_2O \rightleftharpoons SO_2.H_2O(aq) \tag{5.77}$$

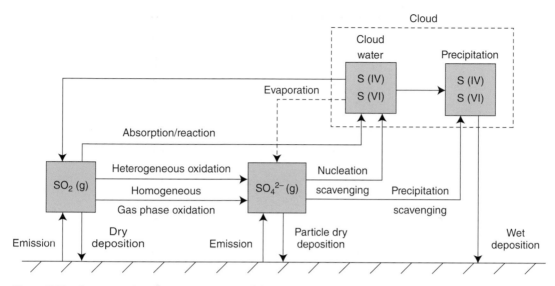

Figure 5.21 Summary of emission, oxidation, and deposition of S(IV) and S(VI). *Source:* after Lamb et al. (1987).

$$SO_2.H_2O(aq) \rightleftharpoons HSO_3^- + H^+ \tag{5.78}$$

$$HSO_3^- \rightleftharpoons SO_3^{2-} + H^+ \tag{5.79}$$

The solubility of SO_2 is related to the pH of the aqueous phase, decreasing at lower values of pH. The oxidation of sulphur (IV) to sulphur (VI) is a complex process dependent on many physical and chemical factors. The main oxidants seem to be O_2 (catalysed/uncatalysed), O_3, H_2O_2, the oxides of nitrogen, and free radical reactions in clouds and fogs. For example, H_2O_2 is highly soluble in solution, so even at relatively low gas-phase concentrations (typically ca. 1 ppbv) there is a significant concentration of H_2O_2 present in solution. The oxidation proceeds as

$$HSO_3^- + H_2O_2 \rightleftharpoons \begin{array}{c} O^- \\ \diagdown \\ O \diagup \end{array} S\text{–OOH} + H_2O \tag{5.80}$$

$$\begin{array}{c} O^- \\ \diagdown \\ O \diagup \end{array} S\text{–OOH} + HA \rightleftharpoons H_2SO_4 + A^- \tag{5.81}$$

where HA is an acid. The ubiquitous occurrence of H_2O_2, its solubility, its high reactivity, and pH independence (under atmospheric conditions) of the rate constant for the reaction with SO_2 makes H_2O_2 one of the most important oxidants for SO_2 in the troposphere. A more detailed description of aqueous-phase oxidation of SO_2 is given by Finlayson-Pitts and Pitts Jr. 2000.

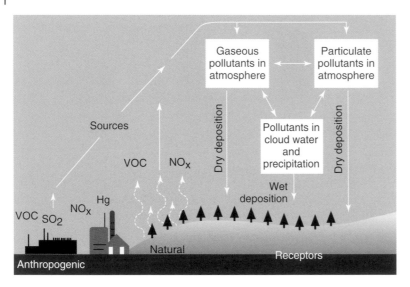

Figure 5.22 Schematic representation of dominant acidification proceses; VOC, volatile organic compounds. (Redrawn from image provided by the US Environmental Protection Agency.)

5.7.2 Acidification Processes – Acid Rain

The key atmospheric loss processes of wet and dry deposition also influence land surface chemistry and the hydrosphere. Some of these processes influence the pH of precipitation and the surface and lead to a number of effects, including corrosion and damage to aquatic ecosystems, forests, and crops. The key acidification pathways are shown in Figure 5.22.

The pH scale of acidity is a measure of the concentration of the hydrogen ion (H^+) according to the relationship $pH = -\log[H^+]$. Pure distilled water has a neutral pH of 7.0. Acidic solutions have higher H^+ concentrations, and therefore a pH below 7. The pH scale is shown in Figure 5.22, with some notable levels indicated. A key feature of rain is the natural acidity of even 'pure' rainwater owing to its interaction with atmospheric carbon dioxide through the following mechanisms

$$H_2O + CO_2 \rightarrow H_2CO_3 \tag{5.82}$$

$$H_2CO_3 \rightarrow HCO_3^- + H^+ \tag{5.83}$$

At 15 °C and an atmospheric concentration of 340 ppmv of CO_2, the acidity of rainwater reaches equilibrium at a pH of 5.6. Rainwater of pH 5.6 is considered 'pure', whereas acid rain is defined as having a pH below 5.6. The primary atmospheric contributors to acid rain are sulphur dioxide and nitrogen dioxide *via* their conversions to sulphuric acid (H_2SO_4) and nitric acid (HNO_3), respectively. Rainfall with pH as low as 1.5 has been recorded in Wheeling, West Virginia, in 1979 (Pawlick 1984), with Los Angeles smogs regularly in the 2.2 to 4.0 region during the 1980s. Recent reductions in SO_2 emissions in the United States and Europe through reductions in the use of coal and the removal of sulphur from transport fuel have significantly reduced the prevalence of acid rain episodes in these regions. Emissions from coal-fired power stations are still a significant and, in some cases, expanding factor in certain developing nations.

5.7.3 Other Key Air Pollution Species

In addition to species with fundamental roles in tropospheric chemistry, there are also a number of key compounds worthy of particular note owing to their proven or suspected impact on human health. These include benzene, 1,3-butadiene, polycyclic aromatic hydrocarbons (PAHs), dioxins, polychlorinated biphenyls (PCBs), and heavy metals such as lead, mercury, and chromium.

Benzene is a known human carcinogen and is primarily emitted from the combustion and distribution of petrol, with minor contributions from a number of industrial processes. Elevated concentrations of benzene have been connected with increased cases of leukaemia. The compound 1,3-butadiene is a suspected human carcinogen emitted predominantly from motor vehicle exhausts, where it is formed from the cracking of higher alkenes; it is also used in the production of synthetic rubber for tyres. Owing to reductions in the benzene content of petrol and the uptake of catalytic converters, emissions of benzene and 1,3-butadiene have both fallen significantly in the period from 1990 to 2005.

The PAHs, hydrocarbons with two or more benzene rings, are predominantly produced through the incomplete combustion of organic materials. Primary anthropogenic sources include motor vehicles, coal and oil-fired power plants, and deliberate biomass burning. Natural biomass burning and volcanic emissions also provide a relatively minor contribution. The importance of PAHs lies in their impact on human health, which varies widely across the more than 100 known atmospheric species. Some arc carcinogenic, the most potent including benzo(a)pyrene, benzoflouranthenes, benz(a)anthracene, dibenzo(ah)anthracene, and indeo(1,2,3-cd)pyrene (Jackson 2007). Studies of occupational exposure to PAHs have shown an increased incidence of tumours of the lung, skin, and possibly bladder. Lung cancer is most obviously linked to exposure to PAHs through inhaled air.

The most significant source of dioxins is the incineration of municipal solid waste and clinical waste, with minor contributions from domestic coal burning and power stations. They possess a number of toxicological properties but the main concern is regarding their possible role in immunologic and reproductive effects. The PCBs are classified as probably carcinogenic to humans and have been linked with subtle chronic effects such as reduced male fertility and long-term behavioural and learning impairment (Jackson 2007). Prior to the mid-1970s they were used in the manufacture of electrical components, with the majority of emissions arising from leakage from poor maintenance of in-service large capacitors and transformers. Total estimated emissions of dioxins in the UK reduced from 1142 to 346 g year[1] during the period 1990–1999, whilst emissions of PCBs decreased from 6976 to 2071 kg year[1] over the same period (Jackson 2007).

5.7.4 Particulate Matter

In addition to trace gases, solid particles and liquid droplets of various sizes are emitted into and formed within the troposphere and are collectively termed particular matter. This material varies significantly in size and composition. Large particles include sand, smoke, soot, and dust, which in some cases can remain suspended in air even at diameters >0.1 mm. Small-scale particulate matter includes organic aerosols with diameters <100 nm, only a few molecules across. Small-particle formation plays a direct role in tropospheric chemistry, whilst larger particles have a significant indirect effect, providing surfaces for heterogeneous reactions and attenuating actinic flux, as well as performing the role of cloud condensation nuclei.

The size distribution of a typical urban load of particular matter is shown schematically in Figure 5.23. The major sources of particulate matter are shown in Table 5.6. The dominance of soil dust and sea salt to total aerosol load can be clearly seen, contributing about 90% of total mass. Significant sources of anthropogenic emissions of primary particles include transportation, coal combustion, cement manufacturing, metallurgy, and waste incineration (Jackson 2007). Further detail on particulate matter in the atmosphere can be found in Chapter 7.

5.7.5 Air Pollution and Climate

Air quality and climate are often treated as separate science and policy areas. Air quality encompasses the here-and-now of pollutant emissions, atmospheric transformations, and their direct effect on human and ecosystem health. Climate change (see Chapter 11) deals with the drivers leading to a warmer world and the consequences of that. These two science and policy issues are inexorably linked by (i) emissions, (ii) atmospheric properties, processes, and chemistry, and (iii) mitigation options.

It is clear that many of the same sources emit both greenhouse gases and air pollutants. For example, emissions from vehicles include particulate matter, nitrogen oxides, carbon monoxide, and carbon dioxide (CO_2). Once in the atmosphere, the emitted species have a variety of atmospheric properties that determine whether or not they have a direct (e.g. ozone and black carbon) or indirect influence on radiative forcing (climate change), their lifetime in the atmosphere, the atmospheric chemistry processes they are involved in, and their influence on human health and ecosystems (see Figure 5.24). For example, particulate matter has a direct influence on radiative forcing by scattering or absorbing incoming radiation, depending on the composition, as well as an adverse effect on human health, in addition to indirect effect where particles can act as cloud condensation nuclei and thereby affect radiative forcing, as well as weather patterns. Finally, many mitigation options offer the possibility to both improve air quality and

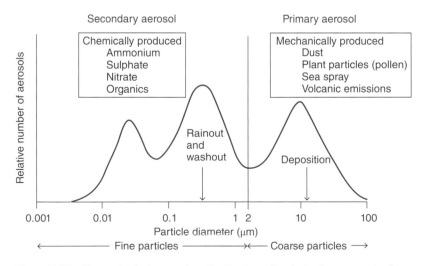

Figure 5.23 Size and relative number distribution of typical urban aerosols. *Source:* adapted from Whitby (1978). See colour plate section for the colour representation of this figure.

Table 5.6 Source strength of various types of aerosol particles.

Source		Global flux (Tg year^{-1})	Low	High
Primary	Carbonaceous aerosols:			
	organic matter (0–2 μm)			
	–biomass burning	54	45	80
	–fossil fuel	28	10	30
	biogenic (>1 μm)	56	0	90
	Black carbon (0 to 2 μm):			
	biomass burning	5.7	5	9
	fossil fuel	6.6	6	8
	aircraft	0.006		
	Industrial dust	100	40	130
	Soil dust (mineral aerosol)	2150	1000	3000
	Sea salt	3340	1000	6000
Secondary	Sulphate (as NH_4HSO_4):			
	anthropogenic	122	69	214
	biogenic	57	28	118
	volcanic	21	9	48
	Nitrate (as NO_3^-):			
	anthropogenic	14.2	9.6	19.2
	natural	3.9	1.9	7.6
	Anthropogenic organic compounds	0.6	0.3	1.8
	Biogenic VOC	16	8	40

VOC = volatile organic compounds.
Range reflects estimates reported in the literature (Jackson 2007). The actual range of uncertainty may encompass values larger and smaller than those reported here.

mitigate climate change, such as improvements in energy efficiency, or a switch to wind or solar power, all of which reduce emissions across the board.

It has become clear through recent research that there are opportunities for 'win-win' scenarios that would benefit both air quality and climate, whilst there are also measures that would benefit one or the other, but not both (Williams 2012). A better understanding of effective mitigation measures, emission sectors, and compounds to target to achieve these air quality and climate co-benefits has emerged.

5.8 Summary

The chemistry of the atmosphere is diverse, driven in the main by the interaction of light with a few molecules that drive a complex array of chemistry. The type and impact of atmospheric chemistry varies in concert with the physical and biological change throughout the atmosphere. An integral understanding

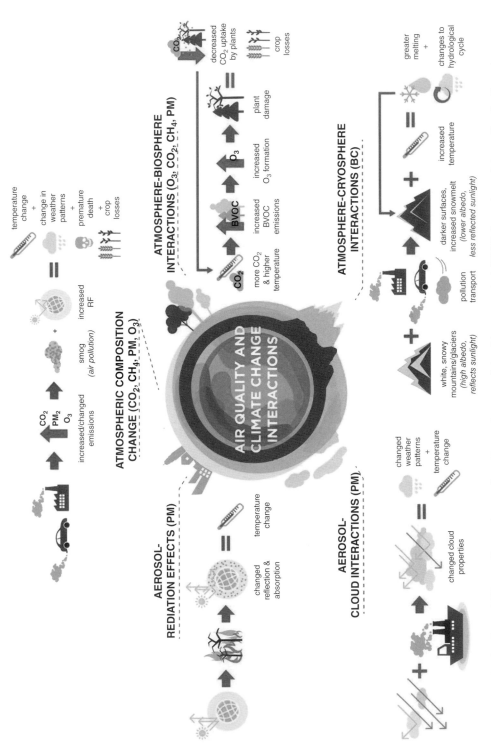

Figure 5.24 Air Quality and Climate Change Interactions. *Source:* from von Schneidemesser et al. (2015). See colour plate section for the colour representation of this figure.

of atmospheric chemistry within the Earth system context underpins many contemporary global environmental problems and is therefore vital to sustainable development.

This chapter has outlined the main features of the chemistry of the troposphere and introduced some key aspects of air pollution. The effects of photochemical smog/urban air pollution remain on the political agenda owing to their potential impact on human health and the economy. In summary, urban photochemistry is not substantially different from tropospheric photochemistry. It is the range and concentrations of the VOCs involved in oxidation coupled to the concentration of NO_x and other oxidants that lead to a larger photochemical turnover.

In some senses it must be remembered that our understanding of atmospheric chemistry is still evolving, but the science expressed gives entry into the rich and changing world of atmospheric science. In the coming years we need to understand the interaction of atmospheric chemistry with climate as well as better understand the multiphase impacts of the chemistry. An expanding global population in conjunction with an evolving regulatory and technology base produces a set of constantly changing air pollution input factors. However, the principles presented in this chapter provide the basis for the majority of currently understood processes. There is still much to discover, rationalize, and understand in atmospheric chemistry.

Questions

1 What are the main reactions that contribute to the formation of so-called photochemical smog?

2 What physical, chemical, and social effects can aggravate the formation of photochemical smog?

3 Peroxyacetyl nitrate (PAN) is of interest as a characteristic product of tropospheric photochemistry.
   A How is PAN formed in the atmosphere?
   B The unimolecular decomposition of PAN to peroxy acetyl radicals and NO_2 is strongly temperature dependent. The following expression gives the temperature dependence of the unimolecular rate constant $k = 1.0 \times 10^{17} \exp(-14000 / T)$
   C If the lifetime is given by $1/k$, calculate the atmospheric lifetime of PAN at $T = 310, 298, 290$ and 280 K.
   D In the light of your answer to part (b), comment on the potential of PAN as a reservoir for reactive nitrogen in remote regions.

4 Explain the following.
   A In the shadow of a thick cloud, the concentration of OH falls to nearly zero.
   B On a global scale, the concentration distribution of CO_2 is about the same, whereas the concentration of isoprene varies from one location to the next.
   C There is more than one chemical pathway to the initiation of night-time chemistry.
   D Both HONO and HCHO are detected in abundance in areas suffering from photochemical smog and may lead to the acceleration of the production of smog.
   E The effect that NO_x from aeroplanes and lightning can have on ozone can be quite different from NO_x emitted from Earth's surface.

5 The ozone continuity equation below is a reasonable representation of the processes that controls ozone in the boundary layer, namely

$$\frac{d[O_3]}{dt} = C + \frac{E_v\left([O_3]_{ft} - [O_3]\right)}{H} + \frac{v_d[O_3]}{H}$$

C is a term representative of the photochemistry (production or destruction), E_v is the entrainment velocity, $[O_3]_{ft}$ is the concentration of free tropospheric ozone, v_d is the dry deposition velocity, and H the height of the boundary layer. The ozone budget shown in Table Q5.5 has been calculated both in the summer and winter using the ozone continuity equation for a site in the marine boundary layer.

A Describe the chemistry involved in calculating the term C in the ozone continuity equation.

B Given the magnitude of the photochemical terms in Table Q5.5, comment on the amount of available NO_x.

C How does the ability to make or destroy ozone change with season?

D What factors affect the magnitude of a deposition velocity?

E What role does entrainment play in the boundary layer?

F What are the weaknesses of using this approach for the calculation of chemical budgets?

G Predict the likely diurnal cycle of ozone and hydrogen peroxide in the remote marine boundary layer.

6 Nitric acid has many potential loss processes in the troposphere. Given that the reaction

$$OH + HNO_3 \rightarrow H_2O + NO_3$$

has a rate constant of 2.0×10^{-13} cm^3 molecule^{-1} s^{-1}, the hydroxyl radical concentration is 4×10^6 molecule cm^{-3}, the deposition velocity for HNO_3 is 0.005 cm s^{-1} over a 1 km well-mixed boundary layer, and the photoprocess data are given in Table Q5.6 – what is the atmospheric lifetime of HNO_3?

7 The following reactions are important in the tropospheric chemistry of iodine.

$I + O_3 \rightarrow IO + O_2$	$k = 1.3 \times 10^{-12}$ cm^3 molecule^{-1} s^{-1}
$IO + h\nu \rightarrow I + O$	$j = 0.15$ s^{-1}
$O + O_2 + M \rightarrow O_3 + M$	fast

i By applying arguments based on photochemical steady-state, calculate the ratio of iodine atoms to IO radicals (i.e. the ratio $[I]/[IO]$) for a typical tropospheric ozone concentration of 35 ppbv (parts per billion by volume). Assume a temperature of 298 K and a pressure of 101 325 Nm^{-2}.

ii How do these three reactions affect the concentration of ozone in the troposphere?

Calculated average ozone removal and addition rates according to pathway, ppbv day^{-1} (upper part) and fractional contributions to overall production or destruction pathways (lower part)

	Pathway			
	O_3 removal		O_3 addition	
Season	Photochemistry	Deposition	Photochemistry	Entrainment
Summer	1.19	0.18	0.56	2.1
Winter	0.61	0.35	0.29	0.1
Summer	87%	13%	21%	79%
Winter	64%	36%	74%	26%

Photoprocess data for HNO_3.

Wavelength (nm)	$F(\lambda)$ (photon cm^{-2} s^{-1})	σ (cm^2 mol^{-1})	ϕ
295–305	2.66×10^{13}	0.409×10^{-20}	1
305–315	4.20×10^{14}	0.146×10^{-20}	1
315–325	1.04×10^{15}	0.032×10^{-20}	1
325–335	1.77×10^{15}	0.005×10^{-20}	1
335–345	1.89×10^{15}	0	1
345–355	2.09×10^{15}	0	1

References

Abbatt, J.P.D., Thomas, J.L., Abrahamsson, K. et al. (2012). Halogen activation via interactions with environmental ice and snow in the polar lower troposphere and other regions. *Atmospheric Chemistry and Physics* 12 https://doi.org/10.5194/acp-12-6237-2012.

Alam, M.S., Rickard, A.R., Camredon, M. et al. (2015). Radical product yields from the Ozonolysis of short chain alkenes under atmospheric boundary layer conditions. *The Journal of Physical Chemistry. A* 117: 12468–12483. https://doi.org/10.1021/jp408745h.

Barrie, L.A., Bottenheim, J.W., Schnell, R.C. et al. (1988). Ozone destruction and photochemical-reactions at polar sunrise in the lower Arctic atmosphere. *Nature* 334: 138.

Brown, S.S. and Stutz, J. (2012). Nighttime radical observations and chemistry. *Chemical Society Reviews* 41: 6405–6447. https://doi.org/10.1039/c2cs35181a.

Calvert, J.G., Atkinson, R., Kerr, J.A. et al. (2000). *The Mechanisms of Atmospheric Oxidation of the Alkenes*. Oxford, UK: Oxford University Press.

Carpenter, L.J. (2003). Iodine in the marine boundary layer. *Chemistry Review* 103: 4953.

Carpenter, L.J., Hebestreit, K., Platt, U., and Liss, P.S. (2001). Coastal zone production of IO precursors: a 2-dimensional study. *Atmospheric Chemistry and Physics* 1: 9–18.

Carpenter, L.J., MacDonald, S.M. et al. (2013). Atmospheric iodine levels influenced by sea surface emissions of inorganic iodine. *Nature Geoscience* 6: 108–111. https://doi.org/10.1038/ngeo1687.

Chameides, W.L. and Walker, J.C.G. (1973). Photochemical theory of tropospheric ozone. *Journal of Geophysical Research* 78: 8760.

Crutzen, P.J. (1973). A discussion on the chemistry of some minor constituents in the stratosphere and troposphere. *Pure and Applied Geophysics* 106–108: 1385–1399.

Ehhalt, D. (1999). Photooxidation of trace gases in the troposphere. *Physical Chemistry Chemical Physics* 24: 5401.

Fabian, P. and Pruchniewz, P.G. (1977). Meridional distribution of ozone in troposphere and its seasonal-variations. *Journal of Geophysical Research* 82: 2063.

Finlayson-Pitts, B.J. and Pitts, J.N. Jr. (2000). *Chemistry of the Upper and Lower Atmosphere*. San Diego, CA: Academic Press.

Frieß, U. (2001). Spectroscopic measurements of atmospheric trace gases at Neumayer station, Antarctica. PhD thesis. University of Heidelberg.

Goldstein, A.H. and Galbally, I.E. (2007). Known and unexplored organic constituents in the Earth's atmosphere. *Environmental Science & Technology* 41: 1515–1521.

Haagen-Smit, A.J. (1952). Ozone formation in photochemical oxidation of organic substances. *Industrial and Engineering Chemistry* 44: 1342.

Harrison, R.M., Yin, J., Tilling, R.M. et al. (2006). Measurement and modelling of air pollution and atmospheric chemistry in the U.K. west midlands conurbation: overview of the PUMA consortium project. *Science of the Total Environment* 360: 5–25. https://doi.org/10.1016/j.scitotenv.2005.08.053.

Heibestreit, K., Stutz, J., Rosen, D. et al. (1990). DOAS measurements of tropospheric bromine oxide in mid-latitudes. *Science* 283: 55.

Houweling, S., Dentener, F., and Lelieveld, J. (1998). The impact of nonmethane hydrocarbon compounds on tropospheric photochemistry. *Journal of Geophysical Research (Atmosphere)* 103: 106730.

IPCC (2013). *Climate Change 2013: The Scientific Basis*. Cambridge, UK: Cambridge University Press.

Jackson, A.V. (2007). Sources of air pollution. In: *Handbook of Atmospheric Science* (eds. C.N. Hewitt and A.V. Jackson), 124–155. Oxford, UK: Blackwell Publishing.

Kennedy, O.J., Ouyang, B. et al. (2011). An aircraft based three channel broadband cavity enhanced absorption spectrometer for simultaneous measurements of NO_3, N_2O_5 and NO_2. *Atmospheric Measurement Techniques* 4: 1759–1776. https://doi.org/10.5194/amt-4-1759-2011.

Khan, M.A.H., Percival, C.J., Caravan, R.L. et al. (2018). Criegee intermediates and their impacts on the troposphere. *Environmental Science: Processes and Impacts* 20: 437–453. https://doi.org/10.1039/c7em00585g.

Kleffmann, J. (2007). Daytime sources of nitrous acid (HONO) in the atmospheric boundary layer. *Chemphyschem: A European Journal of Chemical Physics and Physical Chemistry* 8: 1137–1144. https://doi.org/10.1002/cphc.200700016.

Krotkov, N.A., McLinden, C.A., Li, C. et al. (2016). Aura OMI observations of regional SO_2 and NO_2 pollution changes from 2005 to 2015. *Atmospheric Chemistry and Physics* 16: 4605–4629. https://doi.org/10.5194/acp-16-4605-2016.

Lamb, D., Miller, D.F., Robinson, N.F., and Gertler, A.W. (1987). Importance of liquid water concentration in the atmospheric oxidation of SO_2. *Atmospheric Environment* 21: 2333–2344.

Leighton, P.A. (1961). *The Photochemistry of Air Pollution*. New York, NY: Academic Press.

Lelieveld, J. and Crutzen, P.J. (1990). Influences of cloud photochemical processes on tropospheric ozone. *Nature* 343: 227.

Lightfoot, P.D., Cox, R.A., Crowley, J.N. et al. (1992). Organic peroxy radicals: kinetics, spectroscopy and tropospheric chemistry. *Atmospheric Environment* 10: 1805–1964.

Monks, P.S. (2005). Gas-phase radical chemistry in the troposphere. *Chemistry Society Reviews* 34: 376.

Monks, P.S., Archibald, A.T., Colette, A. et al. (2015). Tropospheric ozone and its precursors from the urban to the global scale from air quality to short-lived climate forcer. *Atmospheric Chemistry and Physics* 15: 8889–8973. https://doi.org/10.5194/acp-15-8889-2015.

Pawlick, T. (1984). *A Killing Rain: The Global Threat of Acid Precipitation*. San Francisco, CA: Sierra Club.

Poisson, N., Kanakidou, M., and Crutzen, P.J. (2000). Impact of non-methane hydrocarbons on tropospheric chemistry and the oxidizing power of the global troposphere: 3-dimensional modelling results. *Journal of Atmospheric Chemistry* 36: 157.

Read, K.A., Mahajan, A.S., Carpenter, L.J. et al. (2008). Extensive halogen-mediated ozone destruction over the tropical Atlantic Ocean. *Nature* 453: 1232–1235.

Saiz-Lopez, A. and von Glasow, R. (2012). Reactive halogen chemistry in the troposphere. *Chemical Society Reviews* 41: 6448–6472. https://doi.org/10.1039/c2cs35208g.

Saiz-Lopez, A., Plane, J.M.C., and Shillito, J.A. (2004). Bromine oxide in the mid-latitude marine boundary layer. *Geophysical Research Letters* 31: L03111.

Saiz-Lopez, A., Plane, J.M.C., Baker, A.R. et al. (2012). Atmospheric chemistry of iodine. *Chemical Reviews* 112: 1773–1804. https://doi.org/10.1021/cr200029u.

von Schneidemesser, E., Monks, P.S., Allan, J.D. et al. (2015). Chemistry and the linkages between air quality and climate change. *Chemical Reviews* 115: 3856–3897. https://doi.org/10.1021/acs.chemrev.5b00089.

Sillman, S. (1999). The relation between ozone, NOx and hydrocarbons in urban and polluted rural environments. *Atmospheric Environment* 33: 1821.

Simpson, W.R., von Glasow, R. et al. (2012). Halogens and their role in polar boundary-layer ozone depletion. *Atmospheric Chemistry and Physics* 7: 4375–4418.

Simpson, W.R., Brown, S.S., Saiz-Lopez, A. et al. (2015). Tropospheric halogen chemistry: sources, cycling, and impacts. *Chemical Reviews* 115: 4035–4062. https://doi.org/10.1021/cr5006638.

Stull, R.R. (1988). *An Introduction to Boundary Layer Meteorology*. Amsterdam, the Netherlands: Kluwer Academic Publishers.

Thompson, A.M. (1992). The oxidizing capacity of the earth's atmosphere – probable past and future changes. *Science* 256: 1157.

Volz, A. and Kley, D. (1988). Evaluation of the Montsouris series of ozone measurements made in the 19th-century. *Nature* 332: 240.

Wagner, T. and Platt, U. (1998). Satellite mapping of enhanced BrO concentrations in the troposphere. *Nature* 395: 486–490.

Wayne, R.P. (2000). *Chemistry of Atmospheres*, 3rde. Oxford University Press.

Weber, E. (1982). *Air Pollution: Assessment Methodology and Modelling*, vol. 2. New York, NY: Plenum Press.

Whitby, K.T. (1978). The physical characteristics of sulfur aerosols. *Atmospheric Environment* 12: 135.

Williams, M. (2012). Tackling climate change: what is the impact on air pollution? *Carbon Management* 3: 511–519.

WMO (2003). Twenty questions and answers about the ozone layer. In: *WMO, Scientific Assessment of Ozone Depletion: 2002. Report no*, 47. Geneva, Switzerland: Global Ozone Research and Monitoring Project, World Meteorological Organization.

Further Reading

Brassuer, G., Orlando, J.J., and Tyndall, G.S. (eds.) (2000). *Atmospheric Chemistry and Global Change*. Cambridge University Press.

Brimblecombe, P. (1996). *Air Composition and Chemistry*. Cambridge University Press.

Fuller, G. (2018). *The Invisible Killer, the Rising Threat of Air Pollution – And how we Can Fight Back*. Melville House, UK.

Greadel, T. and Crutzen, P. (1993). *Atmospheric Change*. Freeman.

Holloway, A.M. and Wayne, R.P. (2010). *Atmospheric Chemistry*. RSC Publishing.

Ritchie, G. (2017). *Atmospheric Chemistry: From the Surface to the Stratosphere*. Singapore: World Scientific Publishing Company.

Seinfeld, J.H. and Pandis, S.N. (1998). *Atmospheric Chemistry and Physics*. Wiley.

Vallero, D. (2014). *Fundamentals of Air Pollution*. Elsevier.

6

Cloud Formation and Chemistry

Peter Brimblecombe

School of Energy and Environment, City University of Hong Kong, Hong Kong

Clouds are a key component of atmospheres and drive a wide range of physical and chemical processes. They are so much a part of human experience that their form is well represented in poetry and painting. Their vivid appearance has ensured that artists such as Constable (1776–1837) took an interest in the emerging subject of meteorology. The brilliant white clouds on a summer day cannot go unnoticed, but even today the role that clouds play in controlling the temperature of a changing Earth remains uncertain.

Yet clouds are insubstantial. Children imagine them to take on animal shapes or we can claim that our perceptions have become cloudy. They are tenuous in reality also; aircraft are unimpeded by the droplets that make up clouds, although strong turbulence can buffet flight. There is very little water in even the heaviest of thunderclouds – just a gram or two of liquid in every cubic metre. Despite such small quantities of water, clouds are an essential factor in the removal of materials from the atmosphere. Cloud droplets can remove directly dust particles, salts, and soluble gases from the atmosphere. In recent years droplet chemistry has been recognized as a further pathway for removing materials from the atmosphere. This chemistry has revealed itself to be every bit as complex as that found in the gas phase.

6.1 Clouds

Clouds are visible aggregates of minute droplets or particles in suspension, which are typically thought of as water or ice. As they form, enough latent heat is released to make the air of clouds warmer than that which surrounds it. Additionally, they have a strong influence on the radiative temperature balance and can affect the intensity and spectra of incoming light.

Clouds are a very general property of planetary atmospheres and are found throughout the solar system. On the giant planets, these are very deep and dominate the planet; Jupiter's interior is liquid hydrogen and helium but the atmosphere contains water droplets and ice and ammonium hydrosulphide and ammonia crystals. On bodies with better-defined solid surfaces clouds are also highly variable. Venus has

Atmospheric Science for Environmental Scientists, Second Edition. Edited by C.N. Hewitt and Andrea V. Jackson.

clouds of sulphuric acid. On Mars they consist of particles of carbon dioxide or water, whilst on Saturn's moon Titan they seem to be derived from methane or ethane. There is evidence of haze, the product of reactions and perhaps even clouds on distant Pluto.

The clouds on other planets may seem rather unusual, yet even in the Earth's atmosphere the composition can be exotic. Here at heights 15–25 km of over the winter poles where temperatures are low (−80 to −90 °C) nacreous or polar stratospheric clouds (PSCs) form. These consist of nitric acid and water, with sulphuric and hydrochloric acid amongst a range of subsidiary components. There are two types of PSC: Type 1 PSCs of nitric acid and water as crystals or solution droplets or less common Type 2 PSCs which are water-ice crystals. PSCs influence the nitrogen and chlorine chemistry over the poles and play an important role in ozone depletion. At even higher altitudes at altitudes (around 85 km in the mesosphere) noctilucent clouds can be found between 50° and 60° latitudes in the summer months. Mars also has its own version of these faint clouds that are so thin that they can only be seen at night.

In our troposphere we find seemingly more familiar clouds. The highest (5–14 km) of these are cirrus, which are thread-like, hairy or curled. They consist of ice crystals. Cumulus are fluffy fair weather clouds typical of summer, forming as pockets of warms moist air rises from the Earth's surface. They often have flattened bases and the large clouds show a distinctive cauliflower shape like cotton wool. These clouds are found at a height of about 500 m and are composed of water droplets. The cumulonimbus is another cloud seen as typical of the summer in the temperate continents. It is a cumulus form that extends as to great heights and takes on an anvil shape. At the top of the cloud, water droplets freeze and appears rather fibrous. These clouds are frequently associated with intense thunderstorms.

The other main cloud form is the stratus cloud found in sheets or layers, although they seem like grey shapeless masses that can be a kilometre thick but as much as 1000 km wide. They can often cover the whole sky and are responsible for dull, wet days. These clouds form as a layer of warm, moist air rises slowly over a mass of colder air. They can extend right to ground level and lead to fog. Fog is typically defined in terms of visibility, often where visibility is less than a kilometre. It is sometimes called mist or haze where the visibility is greater. The term *mist* tends to be applied to liquid particles, whilst *haze* is often used for solid aerosols.

Ice is present in the atmosphere as snow, hail, graupel, and rime (on surfaces). The chemistry of ice is more difficult to describe, because the composition may be heterogeneous and the extent to which they are in equilibrium with the atmosphere is often uncertain. Sleet is melting snow.

Although water can be present in all three phases, as ice, liquid water, and water vapour, most water is in the vapour phase, with liquid water as cloud water on average being only some 4% of the total available. The amount of water vapour in the atmosphere is typically referred to by meteorologists as *absolute humidity* ($g m^{-3}$) or perhaps more conveniently, as *specific humidity* (g kg[air]$^{-1}$). More commonly, we find the water vapour content expressed as relative humidity, which is the ratio of the amount of water relative to its saturation value, at the temperature under consideration. It is usually expressed as a percentage.

6.2 Cloud Formation

There are two main sink or loss processes for compounds in the atmosphere, namely depositional (sometimes referred to as physical) loss processes and photochemical loss processes.

6.2.1 Cloud Droplets

Clouds form by condensation of water vapour from the atmosphere. As rising air cools, its relative humidity increases. Although condensation might be expected once the relative humidity exceeded 100%, this does not happen. It requires a high degree of supersaturation of water vapour to form small droplets. This arises because the surface curvature of small drops enhances their vapour pressure. The idea that the vapour pressure over a curved surface of radius r (p_r) is greater than that over a plane surface (p_∞) was represented in the Kelvin equation, first derived by W. Thomson:

$$\ln\left(p_r/p_\infty\right) = 2\sigma M/\rho_L RTr \tag{6.1}$$

where σ is the surface tension, M the molecular weight, ρ_L the density of the liquid, and R is the gas constant and T the temperature in Kelvin. This can be simplified.

Problem 6.1

The vapour pressure of water (p_∞) in millibars can be estimated from the equation:

$$p_\infty = 6.112 \exp\left(17.67\, Tc/\left(Tc + 243.5\right)\right)$$

Answer

Simple solution gives 13.6324 mb for the water pressure, but as a mixing fraction it would be 0.013457, or just over 1%. A curved droplet would have an increased vapour pressure due to the Kelvin effect.

How much water would be found in an atmosphere at 80% relative humidity and 15 °C and atmospheric pressure of 1013 millibars? How would vapour pressure of water over a small droplet differ from the pressure calculated here?

This water vapour pressure is reduced by the presence of particles in the atmosphere. In particular, hygroscopic particles (notably salts) can start to absorb water below 100% relative humidity, whilst non-hygroscopic particles (dusts) require some degree of supersaturation. The effectiveness of salts as condensation nuclei arises because water activity can be much less than unity in concentrated electrolyte solutions. Salts thus lower the water vapour pressure over small droplets and prevent them from evaporating. The effect of sodium chloride on the equilibrium relative humidity over water droplets is shown in Figure 6.1. The Kelvin effect is significant when droplet sizes are less than a micron, so it becomes relevant in processes such as the formation of cloud droplets. However, these initial droplets have low volume and will grow rapidly, so the effect is generally neglected after the droplets have developed into clouds.

Sodium chloride and sulphuric acid are typical components of cloud condensation nuclei. However, it is possible to postulate that a range of organic materials can act as cloud condensation nuclei. In the remote Arctic, nucleation has been attributed to the oxidation products of the amino acid, L-methionine (Leck and Bigg 1999). Dicarboxylic acids (such as oxalic acid) or humic acids (derived from soils or more likely the oxidation of carbonaceous materials in the atmosphere) may also affect the ability of aerosols to act as cloud condensation nuclei (Gelencser et al. 2000).

The presence of an electrolyte core in a cloud droplet controls its initial composition. Composition will obviously vary with size as the cloud droplets grow. Prupacher and Klett (1997) have offered a conceptual model developed from Ogren (Figure 6.2) to illustrate the change in concentration with radius. In the

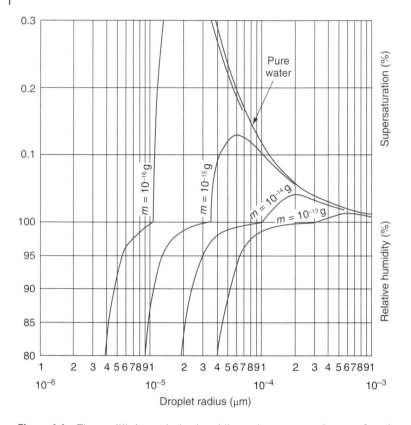

Figure 6.1 The equilibrium relative humidity and supersaturation as a function of droplet radii for sodium chloride droplets containing different masses of salt. *Source:* after Mason (1975).

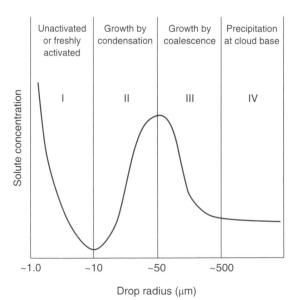

Figure 6.2 The conceptual model of Ogren, as developed by Prupacher and Klett (1997) showing the change in solute concentration as a function of droplet size.

first stage, as a droplet grows between 1 and 10 μm, the salt concentration, derived from the original cloud condensation nucleus, decreases. Then, up to 50 μm the concentration rises with increasing drop size. In this size range, larger droplets grow by vapour diffusion more rapidly than the smaller ones, and these larger drops have generally arisen from larger initial salt particles. Above 50 μm, droplet growth occurs via coalescence, which is a process where large droplets accumulate smaller ones and thus become diluted. Beyond a few hundred microns in size, cloud water is well mixed and concentration constant with size.

6.2.2 Dew Formation

Dew forms through condensation on surfaces as they cool by radiating to the night sky. Most measurements tend to be made on artificial surfaces, but these are not always good models of the dew chemistry, which is so often associated with vegetation. *Leaf wetness* is different from the water that condenses on nonvegetative surfaces because it is derived from two processes: condensation and guttation (Hughes and Brimblecombe 1994). Condensation can occur onto particles on the leaf surface or small irregularities. Guttation is exuded through the leaf stomata and can exert a strong control on the potassium-rich composition of *leaf wetness*.

6.2.3 Ice Particles

Ice can nucleate on hygroscopic particles, but solids such as the clays kaolinite and montmorillonite seem more effective (Prupacher and Klett 1997). Snow, graupel, and hail have ionic compositions that are similar to that of rainwater. However, where riming occurs (i.e. growth by collision with droplets that then freeze), the particles have a much higher scavenging efficiency (3–20 times) for solutes such as sulphate (Barrie 1991) and the nitrate ion, which may be about double the concentration in rainwater under similar conditions. Domine (1999) has shown that diffusive growth models of the incorporation of nitric and hydrochloric acid into snow crystals can describe their uptake in laboratory experiments, but Greenland snow appears to be somewhat undersaturated.

6.3 Particle Size and Water Content

6.3.1 Clouds

Cloud droplets are tens of microns across and a cubic metre of air could easily contain a hundred million droplets. The droplets are so small that they can remain in liquid form to temperatures around −30 °C. The amount of liquid water and other characteristics of these forms of precipitation are listed in Table 6.1. The quantity of water is important because it controls the amount available as a solvent. Nevertheless, the absolute amounts are rather small, and even in large cumulus clouds phase volume ratios of anything greater than one in a million are difficult to achieve (i.e. 1 g[water] m^{-3}).

In the case of rainfall, we can see from the table it is the droplet size that changes with rainfall intensity (rather than the number of drops in the rainfall). Large droplets are found only at the high relative humidity in or below clouds and are short-lived (hours) because of the need for updrafts to keep them suspended.

Table 6.1 Some typical size and water contents of liquid water in the atmosphere (precipitation: Barry and Chorley 1998, clouds: Mason 1975; Heitnzenberg 1998, dew: Brimblecombe and Todd 1977; Hughes and Brimblecombe 1994).

Water	Particle radius	Liquid water content g m^{-3}	Ionic strength molal
precipitation			10^{-4}
rain (0.1 cm hour^{-1})	0.1 cm		
rain (1.3 cm hour^{-1})	0.2 cm		
rain (10 cm hour^{-1})	0.3 cm		
cloud, mist, and aerosol droplets			
cumulonimbus	20 μm	2	
cumulus congestus	10 μm	1	10^{-3}
continental *cumulus*	6 μm	0.45	
stratus (Hawaii)	–	0.35	
mist, fog	–	0.05–0.5	5×10^{-3}
hygroscopic aerosols	<1 μm	10^{-5}–10^{-4}	>1
leaf wetness			
dew (over grass)	0.2 mm	100 g/m^{-2}	
guttation (over grass)	1.5 mm	100 g/m^{-2}	2×10^{-2}

6.3.2 Aerosol Water

Another form of liquid water in the atmosphere is associated with aerosol particles. This water is found even in air under cloud-free conditions (Pilinis et al. 1989), and much of it arises from the ability of salts to absorb water below equilibrium saturation vapour pressures. This hygroscopic water is associated with the most common deliquescent salts in the atmosphere, typically sodium chloride, ammonium bisulphate, and sulphuric acid. Sulphuric acid, being a liquid, does not crystallize at temperatures found in the lower atmosphere. The amount of water associated with aerosols is a function of humidity and the amount of saline material in the aqueous atmospheric aerosol. As noted in the row that describes hygroscopic aerosols in Table 6.1, this amounts to some 10^{-5}–10^{-4} g m^{-3}.

The relative humidity (in percentage) of water vapour in equilibrium can be described by the formula:

$$RH/100 = a_w = \exp\left(-0.018 \, \Sigma v_i m_i \phi_i\right) \tag{6.2}$$

where a_w is water activity and v_i, m_i, and ϕ_i are the stoichiometric number (i.e. number of ions from one molecule of electrolyte), molality (concentration in mole per kilogram of solvent), and osmotic coefficient of the salt i in the solution. The osmotic coefficient can be obtained from tables (e.g. *Electrolyte Solutions: Second Revised Edition* of R.A. Robinson and R.H. Stokes). In the case of a 1 mol kg^{-1} solution of sodium chloride, the osmotic coefficient is 0.9355, so the relative humidity is: 100 * exp. (−0.018 * 2 * 1 * 0.9355), i.e. 96.7%

Problem 6.2

Determine the humidity of air in equilibrium with an ammonium sulphate aerosol at various concentrations at 25 °C

$$RH/100 = a_w = \exp\left(-0.018 \, \Sigma v_i m_i \phi_i\right)$$

The stoichiometric number is v_i. The osmotic coefficient (φ_i) has the values 0.767, 0.677, 0.640, 0.623, and 0.672, at 0.1, 0.5, 1.0, 2.0, and 5.0 mol kg^{-1}, respectively, at 25 °C.

Answer

Use 3 as a stiochiometric number (i.e. $2NH_4$ and $1SO_4$):

molality	0.1	0.5	1	2	5
osmotic coefficient	0.767	0.677	0.64	0.623	0.672
RH	99.6	98.2	96.6	93.5	83.4

Physical chemists usually prefer to describe the lowered vapour pressure of water as water activity (a_w). When the concentration of a solution is high, the osmotic coefficient can be difficult to determines, but high concentrations are relevant to the atmosphere where small aerosol droplets can be supersaturated. Such systems depart from ideality, so their thermodynamic properties require quite complex descriptions. Over the last few decades the formalism of Pitzer has been successfully adapted to aerosols (e.g. Clegg et al. 1998a, b). Online tutorials on this method are available from the *On-Line Aerosol Inorganics Model* at the AIM Website (http://www.aim.env.uea.ac.uk/aim/aim.htm).

6.4 Dissolved Solids in Cloud Water and Rainfall

Liquid water, although a minor component of the troposphere, plays important roles in atmospheric chemistry. It is responsible for removing gases and solid trace components from air through dissolution. This process is usually followed by sedimentation or more characteristically wet removal in rainwater and chemical reactions in the droplet can convert volatile species to nonvolatile ones. Because the droplet phase in the atmosphere has a much lower volume than the gas phase, the dissolution process represents a great increase in concentration.

The most common inorganic anions in solution in the atmosphere are sulphates, chlorides, and nitrates. These are typically associated with the sodium, ammonium, and hydrogen ion and to a lesser extent the alkaline earths: calcium and magnesium. Hydrogen chloride and nitric acid are fairly volatile, but extremely soluble, so found dissolved in atmospheric water. Ammonium chloride and nitrate solids are fairly volatile and can also be lost from aerosols:

$$NH_4Cl_{(s)} \leftrightarrow NH_{3(g)} + HCl_{(g)}$$

$$NH_4NO_{3(s)} \leftrightarrow NH_{3(g)} + HNO_{3(g)}$$

These are exceptions, and in general most soluble salts and sulphuric acid are not particularly volatile, so they partition into aerosols very effectively. Such dissolved salts provide a background electrolyte in aqueous aerosols and rainwater. In aerosols they are highly concentrated and as noted before can even be supersaturated, but are diluted in clouds or rain, where their concentrations are typically below the millimolar range. Once in the droplets and transferred to rain, these solutes can be removed to the ground.

Numerous analyses of water-soluble particles have been collected from the atmosphere by pumping large volumes of air through filters. Simple analysis of the ions only gives clues to their original chemical forms, but these are often taken to be sodium chloride, ammonium sulphates, bisulphates, and sulphuric acid. Examination of dry particles tends to confirm the importance of these compounds. In remote oceanic air, the materials are generally identified as sea salt (sodium chloride), ammonium sulphates, and sulphuric acid. Sea salt is typically predominant in the larger-size particles (>2 μm), whilst of secondary sulphates from the oxidation of sulphur dioxide are found in smaller-size ranges (<2 μm). Nitrates are a product of the oxidation of nitrogen oxides. From the point of view of aquatic chemistry in the atmosphere, the original association between anions and cations in the initial salt is often not usually relevant.

Organic compounds in aerosols include a great many hydrocarbons derived from combustion processes and biological emissions. They frequently contribute to a substantial fraction of the material in remote aerosols, but this material may not be very soluble in water. However, atmospheric transformations, most particularly oxidation, convert compounds from volatile low solubility compounds into more soluble forms. Once again, there are issues that limit their presence in solution. Many oxidation products (e.g. the alcohols, aldehydes, ketones, and acids) are still rather volatile if they have few carbon atoms and can remain in the gas phase.

Although a significant fraction of the total organic matter present in aerosols dissolves in water, there are still many uncertainties concerning organic materials in aerosols (e.g. Huebert and Charlson 2000). In fog water, almost 80% of the water-soluble organic carbon can be identified as: (i) neutral/basic compounds; (ii) mono- and di-carboxylic acids; and (iii) polyacidic compounds (Descari et al. 2000). Some biological compounds, although not volatile, seem to find their way into the atmosphere. These are typified by sugars, amino acids, uric acid urea, and humic substances.

Typical candidate oxidation products that have a low volatility yet high solubility in water include bifunctional acids such as the dicarboxylic acids (oxalic, malonic, succinic, and hydroxy acids such as lactic acid) and some ring compounds represented by furancarboxylic acid (2-furoic acid). Saxena and Hildemann (1996) attempted a more rigorous exploration of likely water soluble organic compounds based on estimates of their estimated partition into aqueous particles. In addition to the acids noted above, they recognized the potential for glyoxal, alkyldiols and polyols, glycols, oxoalkanoic acids (e.g. pyruvic acid), and multifunctional acids (e.g. citric, lactic, tartaric acid) to be found in aerosol droplets. Some nitrogen-containing organic compounds such as nitrophenols, ethanolamine, and the amino acids have physical properties that would suggest they should be found in the aerosol phase. Kames and Schurath (1992) have also shown that alkyl nitrates and other bifunctional nitrates should be soluble in aqueous atmospheric systems.

Some long-chain hydrocarbons, although not soluble, may behave as surfactants on aerosol droplets. The most obvious surfactants would be the carboxylic acids and nitrates (Seidl 2000). Levoglucosan is a tracer of wood smoke, but it is not very soluble. However, soluble sugars such as α- and β-glucose, α- and β-fructose, sucrose, and mycose present in the atmosphere are probably derived from wind-blown soils. Terpenes can be volatile enough to evaporate from plants, and we know terpenes such as pinene and limonine from the subtle smells of pine forests in summer. These compounds are not especially soluble in clouds and aerosols, but their double bonds mean that they are readily oxidized to carbonyl compounds. Pinonaldehyde and pinic acid are more polar and soluble in aqueous systems and could affect the hygroscopic character of aerosols over forests (Figure 6.3).

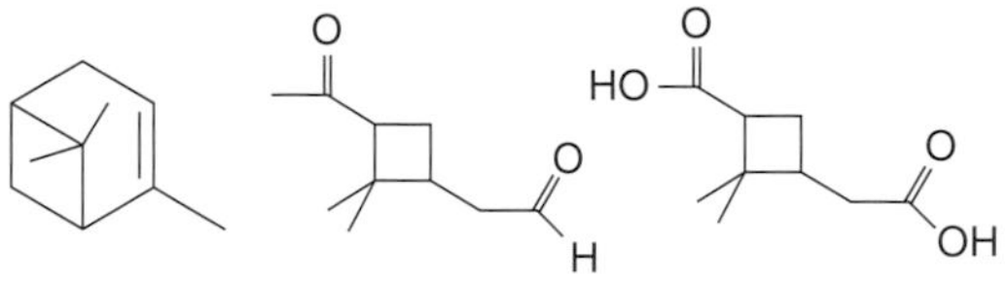

Figure 6.3 The structural formulae of α-pinene (with its double bonds), the oxidation products pinonaldehyde and pinic acid.

It is difficult to assign exact chemical formulae to much of the organic matter in the atmosphere. However, it is possible to recognize broad classes such as the polyols, sugars, terpenes, and polcyclic aromatic hydrocarbons. A significant fraction that has been difficult to resolve resembles humic acids in soils. This brown material is known as HULIS (humic-like substance) and is associated with organic-laden soot and its weathering products. HULIS is surface active and likely to be involved in cloud nucleation. It will also bind with metal ions and by analogy with humic acid in soils should be able to promote aerosol photochemistry as a photosensitizer.

Surface active compounds may also be associated with vegetation. Leaf wetness has an unusual composition because so much can be exuded as guttation through the stomata of plants. There are often high concentrations of dissolved carbon dioxide and plant derived ions such as potassium (Hughes and Brimblecombe 1994). Once on the surface of vegetation, 'dew' can represent an important sink for soluble trace gases such as sulphur dioxide, which can be converted to sulphates. In regions such as Japan, where high concentrations of yellow dust are deposited from the air in spring, dew may become neutral or alkaline even where there is a high probability of acidification from sulphates and nitrates (Chung et al. 1999). Concentrations of ammonium and nitrite also tend to be higher in dew than in rain. However, as the leaf wetness dries, nitrite and ammonium appear to be lost (Takenaka et al. 1999). The low volumes of dew (as with fog water) mean high concentrations of aldehydes and the dominance of glyoxal (ethanedial, $C_2H_2O_2$) in many dew-water samples. The degradation of biological materials such leaf waxes (Fruekilde et al. 1998) has the potential to yield water-soluble organic compounds such as 4-oxopentanal in leaf wetness.

6.5   Dissolution of Gases

Gases can also be found dissolved in cloud or rainwater. These gases must be very soluble gases to dissolve effectively in the liquid phase in the atmosphere, because the amount of liquid water is so low even in thick thunder clouds.

6.5.1   Henry's Law

The dissolution of a gas in water can be written as an equilibrium between the gas ($X_{(g)}$) and aqueous phases $X_{[aq]}$, that can be represented as an equilibrium constant (the Henry's law constant, K_H):

$$X_{(g)} \leftrightarrow X_{(aq)}, K_H = \gamma X \, mX/pX, \text{or } K_H = aX/pX \tag{6.3}$$

where γX is the activity coefficient, mX the concentration (in molar, mol l^{-1} or molal mole $kg[H_2O]^{-1}$ units), pX the pressure and aX the activity (where $aX = \gamma X \; mX$). Strictly speaking, pX should be as fugacity rather than pressure, but these quantities are almost identical for most gases under ambient conditions. In rainwater, the activity coefficient, γX is fairly close to unity, so Eq. (6.1) reduces to:

$$K_H = mX/pX \qquad (6.4)$$

but in aerosols there are very significant departures from ideality and, as mentioned earlier, these can be treated using the formalism of Kenneth Pitzer that has become widely adopted for atmospheric chemistry (e.g. Clegg et al. 1998a, b).

We should also note that Eq. (6.3) represents the solubility of a gas that does not undergo extensive hydrolysis. Typically, the constant K_H is given the units mol l^{-1} atm^{-1} (or mol kg^{-1} atm^{-1}), but this strays from strict SI conventions. There are arguments for presenting the constant in dimensionless terms. Nevertheless, the dimensioned form is rather convenient for atmospheric chemists and typical values are given in these units in Table 6.2. Henry's law constant is a sensitive function of temperature, typically becoming larger as the temperature falls, so there is the need to tabulate enthalpy related parameters. An excellent tabulation is maintained by Rolf Sander (available at https://www.atmos-chem-phys.net/15/4399/2015).

Henry's law constant for oxygen is small (1.2×10^{-3} mol l^{-1} atm^{-1}), so it partitions only slightly into water in the atmosphere. Nevertheless, its concentration in water in equilibrium with air is still comparatively large, (~ 0.25 mmol l^{-1} i.e. $mO_2 = K_{H,O2} \; pO_2 = 1.2 \times 10^{-3} \; 0.21$), relatively to other trace gases in droplets.

Table 6.2 Henry's law constants ($K_H[T_0]$ at 298 K) and temperature dependencies for gases that do not undergo extensive hydrolysis. These are from Sander (1999), but the very extensive tabulation lists a great range of values and substances. Values for other temperatures T, may be calculated from $K_H(T) = K_H(T_0)$ exp $\cdot$ $(-dlnK_H/d[1/T]\{1/T - 1/T_0\})$.

Gas	$K_H(T_0)$ mol l^{-1} atm^{-1}	$-dlnK_H/d(1/T)$ K
Acetone	32	5800
Chloroform	0.25	4500
Ethanol	0.019	6600
Benzene	0.16	4100
Hydrogen peroxide	8.3×10^4	7400
Methane	1.4×10^{-3}	1600
Methanol	0.022	5200
Methylchloride	0.094	3000
Nitrous oxide	0.024	2800
Nitric oxide	1.9×10^{-3}	1700
Nitrogen dioxide	1.2×10^{-2}	2500
Oxygen	1.2×10^{-3}	1700
Ozone	9.4×10^{-3}	2500

6.5.2 Partitioning

Table 6.2 shows hydrogen peroxide to be a remarkably soluble gas. In fact, it is so soluble that it partitions effectively into the liquid phase of clouds. This can be seen from a simple equation for equipartitioning of a trace gas between the air and suspended aqueous droplets. The number of moles in the gas phase is pX/RT, whilst the number of moles in the liquid phase will be the concentration times the volume of water, i.e. $K_H pX L$, where L is the volume of liquid water present in air. The situation where equal amounts are present in the gas and liquid phase can be written:

$$pX/RT = K_H pX L \tag{6.5}$$

This can be used to define a critical value of the Henry's law constant where this equipartitioning will occur (i.e. $K_{H,crit} = RTL$). If there is a gramme of liquid water (i.e. 0.001 l) in every cubic metre of air, the critical Henry's law constant amounts to 4×10^4 mol l^{-1} atm^{-1}. Thus, in Table 6.2 only hydrogen peroxide appears soluble enough to partition into cloud water, as its K_H exceeds the critical equipartitioning value ($K_{H,crit}$).

6.5.3 Hydrolysis

However, this does not mean that hydrogen peroxide is the only gas to partition strongly into cloud water. Gases that undergo subsequent hydrolysis reactions can also partition into the liquid phase. These processes, and their associated equilibria, can be typified by the dissolution of formaldehyde:

$$HCHO_{(g)} \leftrightarrow HCHO_{(aq)} \text{ i.e. } K_{Hy} = aH_2C(OH)_2 /aHCHO \tag{6.6}$$

$$H_2O + HCHO_{(aq)} \leftrightarrow H_2C(OH)_{2(aq)} \text{ i.e. } K_H = aHCHO/pHCHO \tag{6.7}$$

6.5.4 Weak Acids

We can write the dissolution of atmospheric formic acid (a weak acid) in a similar way:

$$HCOOH_{(g)} \leftrightarrow HCOOH_{(aq)}, K_H = aHCOOH/pHCOOH \tag{6.8}$$

$$HCOOH_{(aq)} \leftrightarrow H^+_{(aq)} + HCOO^-_{(aq)}, K' = aH^+ aHCOO^- /aHCOOH \tag{6.9}$$

Where gases hydrolyse or undergo other equilibrium reactions, partition between air and droplets requires additional dissolved species to be considered. These can be derived for individual cases, but as an example take the dissolution of a weak acid, such as formic acid. We can write the total formic acid concentration in solution, $T_{HCOOH(aq)}$:

$$T_{HCOOH(aq)} = HCOOH_{(aq)} + HCOO^-_{(aq)} \tag{6.10}$$

$$= K_H pHCOOH + K_H K' pHCOOH/H^+ \tag{6.11}$$

The pseudo-Henry's law constant ($K^{\ddagger}$) to describe this dissolution in terms of the total amount of dissolved formate can be written:

$$K^! = K_H(1 + K'/H^+) = T_{HCOOH(aq)}/pHCOOH_{(g)} \tag{6.12}$$

6.5.5 The pH of Rainwater

The reactions of dissolving formaldehyde and formic acid are different. The formic acid gives a proton as a product, so the reaction involves an increase in the acidity of water droplets. This explains the acidification of water in the atmosphere. Let us take the dissolution of carbon dioxide (along with the equilibria and equilibrium constants, the latter being taken from Table 6.3):

$$CO_{2(g)} \leftrightarrow H_2CO_{3(aq)}, K_H = aH_2CO_3/pCO_2 = 3.43 \times 10^{-2}\, \text{mol}\, l^{-1}\, \text{atm}^{-1} \tag{6.13}$$

$$H_2CO_{3(aq)} \leftrightarrow H^+_{(aq)} + HCO_3^-{}_{(aq)}, K' = aH^+\, aHCO_3^-/aH_2CO_3 = 4.3 \times 10^{-7}\, \text{mol}\, l^{-1} \tag{6.14}$$

$$HCO_3^-{}_{(aq)} \leftrightarrow H^+_{(aq)} + CO_3^{2-}{}_{(aq)}, K'' = aH^+\, aCO_3^{2-}/aHCO_3^- = 4.7 \times 10^{-11}\, \text{mol}\, l^{-1} \tag{6.15}$$

The principle of charge neutrality allows us to equate the concentrations of positive and negative ions such that:

$$H^+_{(aq)} = HCO_3^-{}_{(aq)} + 2 * CO_3^{2-}{}_{(aq)} + OH^-{}_{(aq)} \tag{6.16}$$

Table 6.3 Henry's law constants and hydrolysis constants for gases that undergo reaction in water (at 298 K). The values of K_{Hy} are hydrolysis constants and K′ and K″ are first and second dissociation constants. Henry's law constants can be found in Sander (1999), but hydrolysis constants and information on aldehydes (Findlayson-Pitts and Pitts 2000) and organic acids (Clegg et al. 1996; Khan et al. 1996) can be found in other sources.

Simple hydrolysis	$K_H(T_o)$ mol l^{-1} atm^{-1}	K_{Hy}	
Formaldehyde	1.3	2.3×10^{-3}	
Glyoxal	≥1.4	2.2×10^{-5}	
Methylglyoxal	1.4	2.7×10^{-3}	
Weak acid or base	$K_H(T_o)$ mol l^{-1} atm^{-1}	K' mol l^{-1}	K'' mol l^{-1}
Formic acid	5.53×10^3	1.77×10^{-4}	
Acetic acid	5.50×10^3	1.75×10^{-5}	
Pyruvic acid	3.11×10^5	3.4×10^{-3}	
Oxalic acid	7.2×10^8	5.29×10^{-2}	5.33×10^{-5}
Sulphur dioxide	1.23	1.3×10^{-2}	6.6×10^{-8}
Carbon dioxide	3.43×10^{-2}	4.3×10^{-7}	4.7×10^{-11}
Hydrogen sulphide	0.12	1.8×10^{-7}	
Ammonia	62	1.7×10^{-5}	
Strong acid	$K_H(T_o)$ mol kg^{-2} atm^{-1}		
Hydrochloric acid	2.04×10^6		
Methanesulfonic acid	6.5×10^{13}		
Nitric acid	2.45×10^6		

Note that the double charge on $CO_3^{2-}{}_{(aq)}$ requires it to be multiplied by 2. This equation can be expanded from the carbon dioxide equilibria and the dissociation of water (i.e. $K_w = aH^+aOH^-$) as:

$$H^+{}_{(aq)} = K_H K' pCO_2 / H^+{}_{(aq)} + 2 * K_H K' pCO_2 / \left(H^+{}_{(aq)}\right)^2 + K_w / H^+{}_{(aq)} \tag{6.17}$$

For the moment, assuming rainwater to be rendered slightly acidic by dissolving carbon dioxide, we can neglect both the terms for $CO_3^{2-}{}_{(aq)}$ and $OH^-{}_{(aq)}$, which are small at pH < 7, so the equation simplifies:

$$H^+{}_{(aq)} = \left(K_H K' pCO_2\right)^{1/2} \tag{6.18}$$

Problem 6.3

An industrial think tank once wrote a letter to me that read: '...one of your colleagues kindly gave me your name as he said you would be able to answer my query. I believe over 75% of CO_2 emissions are encapsulated by cloud formation and brought back to earth and encapsulated in the soil due to lime being a catalyst. Is this true?'

Ascertain if it is possible for 75% of the CO_2 emissions to come down in rain. The Henry's law constants of various gases are given in the table associated with the questions. The annual emissions from human activity are about 5×10^{15} g(C) per year. The partial pressure of CO_2 is about 380 ppm (i.e. 3.8×10^{-4} atm at ground level). Total rainwater amounts to 0.5×10^{18} kg and has a pH of about 5.5. What are likely problems affecting the reliability of this simple calculation?

What is the pH of a droplet in equilibrium with a pure aqueous droplet surrounded by air with the following gas concentrations: ammonia 10 ppb, sulphur dioxide 20 ppb, carbon dioxide 380 ppm?

Important equilibrium constants are found in Table 6.3. The dissociation constant of water may be taken as 1×10^{-14}.

You will need to write the charge balance equation:

$$H^+ + NH_4^+ = OH^- + HSO_3^- + 2 * SO_3^- + HCO_3^- + 2 * CO_3^-$$

in terms of partial pressures (see equations in text). Once you have done this, you will find that you need to solve a cubic equation in H^+, or make some simplifying assumptions. Students who know MATLAB or SCILAB will find extracting the roots relatively simple using the function ROOTS, although this gives three solutions, but the correct one is immediately obvious.

Answer

$$CO_{2(g)} = H_2CO_{3(aq)}, K_H = aH_2CO_3/pCO_2 = 3.43 \times 10^{-2} \, mol\,l^{-1}\,atm^{-1}$$

If dissociation is neglected and this is small at pH 5.5, then the concentration of dissolved CO_2 will be given by $K_H^* pCO_2$. We then need to multiply this by the annual rainfall 0.5×10^{18} kg to find that there are 6.52E+12 mol in a year's rain. This is $= 2.87 \times 10^{14}$ g CO_2, which is much less than 75%. Additionally, much of that removed would be the natural CO_2 and a more thoughtful calculation would need to consider pH effects and the equilibration processes.

Problem 6.4

What is the pH of a droplet in equilibrium with a pure aqueous droplet surrounded by air with the following gas concentrations: ammonia 10 ppb, sulphur dioxide 20 ppb, carbon dioxide 380 ppm?

Important equilibrium constants are found in Table 6.3. The dissociation constant of water may be taken as 1×10^{-14}.

You will need to write the charge balance equation:

$$H^+ + NH_4^+ = OH^- + HSO_3^- + 2*SO_3^- + HCO_3^- + 2*CO_3^-$$

in terms of partial pressures (see equations in text). Once you have done this, you will find that you need to solve a cubic equation in H^+, or make some simplifying assumptions. Students who know MATLAB or SCILAB will find extracting the roots relatively simple using the function ROOTS, although this gives three solutions, but the correct one is immediately obvious.

Answer

My MATLAB code was as follows:

Setting the constants
KHa = 90;K1a = 1.85e-5%(ammonia equilibrium)
KHc = 0.04;K1c = 3.0e-6;K2c = 4.0e-10%(carbon dioxide equilibrium)
KHs = 2;K1 s = 0.02;K2 s = 1.0e-7%(sulphur dioxide equilibrium)
pc = 0.00038;ps = 20.e-9;pa = 10.e-9%(partial pressures)
Kw = 1.e-14

Setting the coefficients of the third-order polynomial:
c3 = (1+K1a*KHa*pa/Kw)
c2 = 0
c1 = −(Kw+ps*KHs*K1 s+pc*KHc*K1c)
c0 = −2*(ps*KHs*K1 s*K2 s+pc*KHc*K1c*K2c)
pcoefs = [c3 c2 c1 c0];

Using the MATLAB routine to get the roots of the polynomial
roots(pcoefs)

Only one is positive so take the log to get pH:
−log10(0.7921e-6) = 6.1012

Solving this with the constants from Table 6.3 and for pCO_2 at 3.8×10^{-4} atm (i.e. 380 ppm) gives $H^+_{(aq)}$ of 2.37×10^{-6} mol l^{-1}, which is equivalent to a pH value of 5.63 (i.e. $-\log(H^+_{(aq)})$. We should note that activity and concentration have been treated as identical in this calculation, which is a reasonable approximation for the dilute solutions typical of rainwater. This is typical of the acidity of remote rain or snowfall such as that found in the Canary Islands (see Figure 6.4). If we were to repeat this calculation for sulphur dioxide rather than carbon dioxide, note that both K_H and K' are considerably greater, so even at low partial pressures of sulphur dioxide the pH is lower. This partly explains the acidification of rainfall so well known as an environmental issue. In some locations, such as Norway in the 1980s, the rain and snow became distinctly acidic. By contrast, over the Atlas Mountains of Morocco, Sahelian

Figure 6.4 The pH of a set of melted snow samples is collected from Norway, Tenerife, and Morocco, showing that the acidity of remote Canary Islands is close to the pH expected for equilibrium with carbon dioxide acid. The precipitation in Norway is more acid, whilst alkaline dust contributed to the higher pH values found in Morocco. *Source:* from Brimblecombe, P. *Weather* 35 79–84 (1980).

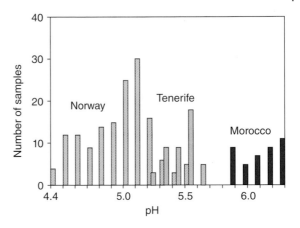

dusts contribute to alkalinity of the rainfall. This is also true in China, where despite intense energy generation from coal, the presence of dust blown from its central plateau helps neutralize the deposition.

6.5.6   Strong Acids

Strong acids can be treated as dissociating completely, so the dissolution of hydrochloric acid can be written:

$$HCl_{(g)} \leftrightarrow H^+_{(aq)} + Cl^-_{(aq)} \text{ i.e.} K_H = aH^+ \, aCl^- / pHCl \tag{6.19}$$

This is often done because it is difficult to assign equilibrium constants to the dissociation of strong electrolytes. This equilibrium step can be ignored without loss of generality, because the undissociated species in a strong electrolyte are low in relative concentration. However, we should note that Henry's law constant needed here takes on different units (in the case of the acids in Table 6.3, mol kg^{-2} atm^{-1}). These strong acids usually have effective Henry's law constants that are so large that they partition completely into the liquid phase.

6.5.7   Other Reactions and Surface Activity

Aldehydes are especially notable in the atmosphere, as they readily hydrolyse in water to form glycols (e.g. Montonya and Mellado 1995), which significantly increases their solubility. In the case of formaldehyde, this is by a factor of almost 2000. This hydrolysis is particularly significant for glyoxal (CHOCHO), which has an effective Henry's law constant of about 3×10^5 mol l^{-1} atm^{-1}. This means that it largely partitions from the gas phase into rainwater.

Aldehydes undergo further reactions in solution particularly with dissolved S[IV] anions:

$$HCHO + HSO_3^- \leftrightarrow CH_2(OH)SO_3^-$$

The product here is the hydroxymethansulfonate ion. The formation constant is strongest with formaldehyde, but glyoxal and hydroxyacetaldehyde are also likely to have a substantial sulfonate formation in aqueous solutions in the atmosphere, which can be responsible for enhancing the dissolution of sulphur dioxide. The reaction of S(IV) with aldehydes tends to be relatively slow compared with oxidation to sulphuric acid by oxidants found in atmospheric water (most typically aqueous hydrogen peroxide). Thus, it is only under relatively alkaline conditions greater than pH 5, and where oxidation rates by H_2O_2 are low, that the formation of these adducts with aldehydes becomes important. Such conditions can lead to as much as an order of magnitude increase in S(IV) concentrations in droplets, over the value expected simply on the basis of hydration of dissolving SO_2.

Dissolution within droplets can also be affected by other types of equilibrium reactions, most notably the chelation with metal ions. This requires ligands that have large stability constants with metal ions, because concentrations will typically be low in atmospheric water. Oxo-ligands are the most likely, and the polycarboxylic acids in particular seem to be strong acids that ensure they highly charged. The few studies of metal complexes in atmospheric water have focussed on oxalate complexes with iron and their potential for photochemistry (Zuo and Holgne 1992). Humic substances in rainwater probably form exceedingly strong metal complexes (Spokes et al. 1996) because they contain many oxy and hydroxy groups, along with nitrogen atoms are also potential electron donors.

Surface active compounds change the surface tension of water droplets in the atmosphere (Seidl 2000). There is also experimental evidence that Henry's law constants for organic compounds increased in the presence of surfactants via micelle formation (Vane and Giroux 2000). This is hardly surprising, and it is a potential route for a solubility increase amongst compounds that are not highly polar. Commercial surfactants used in these kinds of experiments are often derived from strong organic acids, so differ from the weak carboxylic acids most often discussed in atmospheric systems. There seems to be about 100 pmol m^{-3} of surfactant in continental air. There is the potential for surface activity on atmospheric aerosols to enhance the solubility of low-polarity organic compounds and surfactants can also affect the rate of gas dissolution in aerosols and perhaps even global albedo (Facchini et al. 1999).

6.5.8 Kinetics of Dissolution

The considerations in the sections above have assumed that the gas and liquid phases reach equilibrium relatively quickly. This seems reasonable as most of the hydrolysis reactions we have discussed above proceed fairly rapidly. However, such an approach neglects the transfer of a trace gas to a droplet in the atmosphere. This process can be broken down into a number of steps: (i) diffusion of the gas to the surface of the liquid, (ii) transfer across the interface, which also involves moving that gas through a film, (iii) hydration or ionization, (iv) diffusion through the liquid phase and ultimately, and (v) any other chemical reaction. The speed of each of these processes if often described in terms of characteristic or relaxation times. For many processes, the chemical reaction is slower than the transfer processes and hydration:

i) The time for gas phase diffusion to achieve equilibrium with the surface of a liquid droplet depends on drop radius, but even with large droplets 100 µm radius it is less than a millisecond.

ii) The time to attain interfacial equilibrium is relatively rapid and is a function of the accommodation coefficient (i.e. the fraction of molecules that strike the surface which dissolve) and Henry's law constant of the gas. Soluble gases have much longer characteristic times than less soluble ones. In the case of hydrogen peroxide this may be as long as a second.

iii) The hydration of aqueous SO_2 proceeds rapidly with rate, a constant of 3.4×10^6 s^{-1}. On the other hand, the hydrolysis of some of the aldehydes such as glyoxal are slow and may take many hours.

iv) The characteristic time for diffusion in a spherical droplet is given as $r^2/D\pi^2$ where r is the droplet radius and D the diffusion constant of the species in solution. As D is typically 1.8×10^{-9} m^2 s^{-1}, characteristic times for large 100 μm radius droplets will be close to a second. This does not consider whether droplets are mixed by internal circulation as they fall through the air. This process can increase the rate of achieving a homogeneous distribution of solutes in droplets larger than 100 μm in radius.

In general, the slower chemical reactions require droplets with longer lifetimes to be significant. Long lifetimes require the droplet to be small so as to have a low fall velocity, which in turn means that the distances over which diffusion takes place are also small, so characteristic times can be relatively short.

6.5.9 Wet Removal

The overall process through which rain removes materials in the atmosphere is described as wet deposition. This distinguishes it from dry deposition, which is the direct transfer of gases or particles to the surface of the Earth. The removal process by rain can be further subdivided into in-cloud scavenging (rain-out) and below-cloud scavenging (wash-out).

In-cloud scavenging is particularly effective for hygroscopic particles, which are incorporated into cloud water as a part of nucleation. In addition to this, particles can be incorporated into droplets through impaction scavenging. If the droplet is small, then Brownian diffusion or thermophoresis can be important, but at larger drop sizes, the falling drop can effectively sweep the particle from the atmosphere.

Typically, only highly soluble gases (gases that undergo extensive hydrolysis or reaction in solution) that partition into water droplets are effectively removed by wet deposition. The gases that are most typically transferred, in substantial fractions, into atmospheric water are carbon dioxide, sulphur dioxide, and ammonia, along with the stronger mineral acids.

The overall rate at which materials are removed from the atmosphere by precipitation is often described in terms of a wet removal or washout coefficient (λ), which is effectively a first-order removal constant, such that:

$$c_t = c_o \exp(-\lambda t)$$

(6.20)

where c_o and c_t is the initial concentration in the air and the concentration after time t. The value of lambda depends on a range of properties of the material being scavenged and the rainfall characteristics (droplet size, rate, etc.). It takes on the value of reciprocal seconds.

6.6 Reactions and Photochemistry

6.6.1 Sequestration

Rainwater removes materials from the atmosphere, but can also have quite subtle effects on atmospheric chemistry. The partition of gases into the droplet phase can alter the rate of reactions in the gas phase via a process that might be thought of as sequestration. Here species can be separated with soluble ones going into solution and less soluble ones remaining in the gas phase.

An example of this is in the chemistry of nitrophenol. The ortho- or 2-nitrophenol has a Henry's law constant of about 150 mol l^{-1} atm^{-1} whilst that of the para- or 4-nitrophenol is some two orders of greater magnitude at 10^4 mol l^{-1} atm^{-1}. This would mean that a significant fraction of 4-nitrophenol would be found in droplets, while the 2 nitrophenol equilibrate to the gas phase. This means that 2-nitrophenol would be sensitive to gas phase oxidation, whilst the other isomer would be less vulnerable in become proportionately the more dominant form in the atmosphere.

6.6.2 Reactions in Droplets

Droplets in the atmosphere also serves as a type of reactor, because as gases dissolve in aqueous aerosols species can reach much higher concentrations. They can potentially achieve faster reaction rates than typical of the gas phase. The liquid phase can promote reactions that may not occur in the gas phase, thus allowing additional reaction pathways to become significant. Ravishankara (1997) has argued that two rather unreactive gas phase species can potentially react effectively in the aqueous aerosol. The argument goes that reactions between filled shell molecules are slow in the gas phase because of the high energy barriers, but can be faster in the liquid phase reactions because of ionic reaction pathways.

6.6.3 Hydrolysis

Hydrolysis was seen as an important control on dissolution of gases in the sections above. The reactions discussed there were essentially reversible, but some gases will hydrolyse rapidly and are virtually irreversibly in water. This happens in the oxidative chemistry of HCFC-124 (CF_3CFClH), a fire extinguishant and refrigerant that replaced chlorofluorocarbons that deplete stratospheric ozone. The gas phase oxidation of HCFC-124 leads to trifluoroacetylfluoride in the atmosphere, which readily dissolves in water and rapidly hydrolyses:

$$CF_3CFO_{(g)} + H_2O_{(l)} \rightarrow CF_3COOH_{(aq)} + HF_{(aq)}$$

The trifluoroactic acid produced by this process is a very stable compound and may well accumulate in surface waters.

Another halogen-containing gas phosgene ($COCl_2$) is present in the atmosphere at a few tens of parts per trillion. It can be produced in the atmosphere during the photolysis or oxidation of chlorinated hydrocarbons, such as trichloroethene and tetrachloroethene, chloral (trichloroethanal, a precursor to DDT), and dichlorvos (2,2-dichlorovinyl dimethyl phosphate, a widely used insecticide). In the gas phase, it has a lifetime of many years and is not very soluble in water. However, once dissolved, it hydrolyses rapidly:

$$COCl_2 + H_2O \rightarrow HCl + CO_2$$

It is also possible for gases to be produced in droplet phases. This can be seen in the production carbonyl sulphide in rainwater. It requires that the product is a low-solubility gas such that it is lost from the droplet phase.

6.6.4 Reactions of Saline Droplets

Sea-salt particles are frequently found to have chloride concentrations much lower than expected from maritime ionic ratios. This can be easily explained in terms of a displacement by dissolved acids at high concentrations in aqueous marine aerosols:

$$HNO_{3(g)} + Cl^- \rightarrow HCl_{(g)} + NO_3^-$$

$$H_2SO_4 + Cl^- \rightarrow HCl_{(g)} + HSO_4^-$$

One would also expect fluoride to be even more strongly depleted in the marine aerosol, as it has a lower Henry's law constant than HCl. Reactions can also occur directly with solid sea-salt particles. Various nitrogen oxides can react with sodium chloride to give ClNO:

$$2NO_{2(g)} + NaCl_{(s)} \rightarrow NaNO_{3(s)} + ClNO_{(g)}$$

with the potential for an analogous bromine chemistry.

6.6.5 Sulphur Oxidation

The oxidation of sulphur dioxide (S(IV)) to sulphuric acid in the atmosphere has long been of interest, first because of the importance of this process in urban air. In the 1980s, it became relevant to the production of acid rain. The importance of droplet chemistry was obvious in the oxidation of this soluble gas, which is typically more important than its gas phase chemistry. Chemical engineers had been interested in aqueous phase oxidation, and their work suggested radical chain reactions contributed to the autoxidation of sulphites. Neglecting the details for a moment allows us to treat the reaction in a simplified manner. At typical atmospheric pH values 2–6, most of the S(IV) is present as bisulphite anions and the oxidation can be represented:

$$0.5O_2 + HSO_3^- \rightarrow H^+ + SO_4^{2-}$$

The oxidation of sulphur(IV) by molecular oxygen is very slow, although in polluted environments there is the potential for the reaction to become catalysed by a range of transition metals, such as iron and manganese (Warneck 1999b). The mechanisms are seen as occurring through an electron transfer such that the metal is reduced in solution, so we might represent it:

$$M(III)(OH)_n + HSO_3^- \rightarrow M(II)(OH)_{n-1} + SO_3^- + H_2O$$

Essentially initiating a radical chain:

$$SO_3^- + O_2 \rightarrow SO_5^-$$

$$SO_5^- + SO_3^{2-} \rightarrow SO_4^- + SO_4^{2-}$$

$$SO_4^- + SO_3^{2-} \rightarrow SO_3^- + SO_4^{2-}$$

This was a mechanism originally proposed by Bäckström in the 1930s. Metal catalysed pathways are likely to be more important at relatively modest acidity because the reactions slow down as pH decreases and the reaction generates acidity overall with the conversion of sulphurous to sulphuric acid. Organic materials such as terpenes may decrease the chain length of the radical reaction sequence and thus reduce the overall oxidation rate (Ziajka and Pasuik-Bronikowska 1999).

Penkett et al. (1979) showed that other oxidants, most particularly ozone and hydrogen peroxide, are capable of dissolving and oxidizing bisulphite to sulphate. The hydrogen peroxide route is a particularly significant one, as this reaction pathway is faster in acid solution. This means that it would not slow down as the system became more acidic through the production of sulphuric acid. The oxidation by ozone can readily be represented:

$$HSO_3^- + O_3 \rightarrow H^+ + SO_4^{2-} + O_2$$

The oxidation enclosing hydrogen peroxide can be simplified to:

$$HSO_3^- + H_2O_2 \rightarrow H^+ + SO_4^{2-} + H_2O$$

The peroxide oxidation can involve both hydrogen peroxide and organo-peroxides (written as R below) in the following way:

$$ROOH + HSO_3^- = ROOSO_2^- + H_2O$$

$$ROOSO_2^- \rightarrow ROSO_3^-$$

$$ROSO_3^- + H_2O \rightarrow ROH + SO_4^{2-}$$

Alternatively, one can imagine OH in atmospheric droplets initiating an opening step and subsequent reaction chain:

$$OH + HSO_3^- \rightarrow SO_3^- + H_2O$$

Thus, there are a range of routes to the oxidation of sulphur dioxide in droplets in the atmosphere. Warneck (1999b) investigated the efficiency of various reactions contributing to the oxidation of sulphur dioxide and nitrogen dioxide in cloud water, which gave an idea of the fraction of sulphate produced through various mechanisms. Ozone and hydrogen peroxide are the most important oxidants, in the aqueous phase, but the reaction of peroxynitric acid with the bisulphite anion:

$$HOONO_2 + HSO_3^- \rightarrow 2H^+ + NO_3^- + SO_4^{2-}$$

can also make a significant contribution (Warneck 1999a).

In more polluted situations there is insufficient hydrogen peroxide to oxidize the large amounts of sulphur dioxide in the air. In such situations, oxidation by hydrogen peroxide and ozone will be typically be less important than oxidation by OH, Br_2^-, and Cl_2^- (Herrmann et al. 2000).

6.6.6 Nitrogen Compounds

The chemistry of the nitrogen oxides in the atmosphere has been studied in great detail since it was realized in the 1950s that they played a key role in the formation of photochemical smog. As the use of coal in cities has declined, sulphur dioxide chemistry has been less important, whilst the chemistry of the nitrogen oxides, from automobiles and industry, has become more evident. The acidification of rainwater has increasingly been controlled by nitric rather than sulphuric acid in Europe and North America. The simplest route to the production of nitric acid is the addition of a hydroxyl radical:

$$OH_{(g)} + NO_{2(g)} \rightarrow HNO_{3(g)}$$

The chemistry of nitrogen compounds in droplets has been less extensively studied than sulphur chemistry, possibly because NO and NO_2 are less soluble than SO_2. However, the role of peroxynitric acid in the formation of sulphuric acid was mentioned in the section above. This is a very soluble product of atmospheric oxidation (Macleod et al. 1988).

Peroxynitric acid (HNO_4 i.e. $HOONO_2$) is produced by a three body reaction:

$$HO_{2(g)} + NO_{2(g)} \rightarrow HOONO_{2(g)}$$

It is a moderately strong acid ($pKa \approx 5$), and if the pH is low enough, most of the acid is in the molecular form and it is relatively stable and so available to react with S(IV). As seen in the equation above, this also represents a source of nitric acid in droplets. In addition to oxidizing S(IV), peroxynitric acid contributes to nitrite in cloud water (Warneck 1999b):

$$HOONO_2 = H^+ + NO_4^-$$

$$NO_4^- \rightarrow NO_2^- + O_2$$

It should also be mentioned reactions that nitrite oxidation seems to proceed very quickly in freezing particles (Takenaka et al. 1998).

6.6.7 Organic Solutes

Cloud and fog water processes are potentially important contributors to secondary organic aerosol formation (Blando and Turpin 2000). Organic vapours dissolve within suspended droplets and participate in aqueous-phase reactions. Typically, aldehydes, ketones, alcohols, monocarboxylic acids, and organic peroxides can be oxidized to carboxylic acids (especially poly-functional ones), such as polyols, glyoxal, and esters. The organic compounds that dissolve in atmospheric water most likely react via photochemically induced oxidation reactions, which are covered in the section below.

Dimerization might well occur with dissolved compounds such as methacrylic acid (Khan et al. 1992), which forms from the oxidation of isoprene. However, the low concentrations of organic compounds expected in most atmospheric droplets limit the potential for polymerization reactions.

In systems with a rich biology, there is an opportunity for biochemical reactions. Thus, urea might be expected to degrade biologically:

$$NH_2CONH_2 + H_2O \rightarrow 2NH_3 + CO_2$$

In guttation, fluids on leaves enzymes such as the peroxidases could catalyse the oxidative cross-linking and polymerization of organic compounds, utilizing hydrogen peroxide and other organic peroxides that are present in dew (Kerstetter et al. 1998). Legrand (1998) showed that at coastal Antarctic sites, bacterial decomposition of uric acid is a source of ammonium, oxalate, and cations (such as potassium and calcium) in aerosols, in addition to a subsequent large ammonia loss from ornithogenic soils to the atmosphere.

6.7   Radical and Photochemical Reactions

In the 1950s, the work of Haagen-Smit and later that of Leighton's group established the importance of radical chemistry in the production of Los Angeles smog. These studies helped unravel the intricate chemistry of the gas phase. In the last few decades, the photochemistry and radical chemistry of atmospheric water droplets has been seen as increasingly important (Herrmann et al. 1999). A rigorous treatment of photon fluxes within cloud droplets using Mie theory shows that photolysis frequencies in the aqueous phase will be twice as rapid as might be expected on the basis of the actinic flux in the interstitial air (Ruggaber et al. 1997). In addition, we can expect different types of photochemistry in aqueous systems compared with that which has become familiar in the gas phase (Faust 1994). Spectral shifts can stabilize some molecules, so, for example, HCHO is photochemically degraded in the gas phase, but when hydrolysed to $CH_2(OH)_2$ in droplets, it is not sensitive to photolysis. There is also the potential for some reactions to be photosensitized, perhaps through the presence of the ferric ion. Despite the fact that the importance of these processes in the liquid phase was pointed out more than 30 years ago, the development of detailed mechanisms was at first rather slow, and study of these reactions in-situ is not easy. The understanding of the balance of reactions in the liquid phase remains an important area of study.

6.7.1   Peroxides and Hydroxyl Radicals

The production and loss of hydrogen peroxide, and the related processes with the HO_2 and OH radical, lie at the heart of this droplet phase chemistry. Hydrogen peroxide partitions very effectively into the liquid phase because of a large Henry's law constant, so is an important oxidant. Although hydrogen peroxide can be photo dissociated in solution to give OH, this is less effective than the direct transfer of OH into aqueous solution.

However, we can see that just as trace gases dissolve in droplets, radical species can also be absorbed into solution. Some of the most notable for droplet chemistry are: OH, HO_2, NO_3, and CH_3O_2. This dissolution process represents the most important source of aqueous HO_2. The dissolved hydroperoxide ion is a moderately strong acid (pKa 4.88) and gives the O_2^-:

$$HO_2 \leftrightarrow H^+ + O_2^-$$

The hydroperoxide radical can also be produced in solution. This can be through the reaction of hydrogen peroxide with the OH radical:

$$H_2O_2 + OH \rightarrow HO_2 + H_2O$$

In addition, to transfer into solution from the gas phase, hydrogen peroxide can be produced in solution via a $Fe(II)_{(aq)}$ mediated photo-production. This has been observed in simulated cloud-water experiments. Potential electron donors for these types of processes are as oxalate, formate, or acetate commonly found in cloud water (Siefert et al. 1994).

When H_2O_2 is abundant in solution, a significant fraction of aquous OH can come from an iron(II)-HOOH photo-Fenton reaction mechanism:

$$Fe(II) + HOOH \rightarrow Fe(III) + OH + OH^-$$

Iron(III) can also produce the OH radical via a photochemical reaction that involves the hydroxide, which is common at cloud-water pH (3–5):

$$Fe(OH)^{2+} + h\nu \rightarrow Fe^{2+} + OH$$

The O_2^- derived from aqueous HO_2 can react rapidly with dissolved ozone to give O_3^-:

$$O_2^- + O_3 \rightarrow O_2 + O_3^-$$

$$O_3^- \rightarrow O_2 + O^-$$

$$O^- + H^+ \rightarrow OH$$

There are a range of sinks for the OH radical in solution. Where formaldehyde is present, its hydrated form methyleneglycol can act as an important sink of OH:

$$CH_2(OH)_2 + OH \rightarrow CH(OH)_2 + H_2O$$

$$CH(OH)_2 + O_2 \rightarrow HCOOH + HO_2$$

The formic acid produced can react further with OH to oxidize to carbon dioxide.

The alkylperoxy radical CH_3O_2 can also dissolve effectively, but it is probably converted to the peroxide:

$$CH_3O_2 + HO_2 \rightarrow O_2 + CH_3OOH$$

and as CH_3OOH is not very soluble, it can be lost from the droplet phase.

6.7.2 Nitrite Radical Chemistry

The nitrate radical NO_3 is an important species in night-time oxidation in the gas phase. It is able to dissolve in droplets and can react with chloride or the bisulfite anion, which means that the presence of Cl tends to moderate the chemistry of NO_3 through the reversible reactions (Buxton et al. 1999b):

$$NO_3 + Cl^- \rightarrow NO_3^- + Cl$$

The concentration may be much modified by the presence of aldehydes, which have a high rate of reaction with the nitrate radical, converting them via hydrogen abstraction to the acids. The nitrate radical can also react by addition with dissolved organic material. The nitrite radical can also have an aqueous phase chemistry and there has been interest in dinitrogen tetroxide the weakly bound dimer of NO_2, which has both a symmetric and asymmetric form ($ONONO_2$). On aerosols and in strong acid solutions, ionization can lead to strongly nitrating cations such as NO^+ and NO_2^+.

6.7.3 Halogen Chemistry

Saline droplets in the atmosphere, especially where they are concentrated, represent an important source of atomic bromine and chlorine that have a range of gas phase reactions. In addition to the processes with nitrogen and sulphur radicals noted in the section above, there is the potential for reactions via OH (e.g. Knipping et al. 2000):

$$OH + Cl^- \rightarrow HOCl^-$$

$$HOCl^- + H^+ \rightarrow Cl + H_2O$$

A further important sequence is:

$$O_3^- + H^+ + Br^- \rightarrow HOBr + O_2^-$$

$$HOBr + Cl^- \rightarrow BrCl + H_2O$$

which represents a source of BrCl to the gas phase (Vogt et al. 1996). Alternate drivers for the same types of processes arise from the dissolution of the hypochlorous and hypobromous acids:

$$HOCl_{(g)} \rightarrow HOCl_{(aq)}$$

$$HOCl_{(aq)} + Cl^- + H^+ \rightarrow Cl_2 + H_2O$$

$$Cl_2 + HO_2 \rightarrow Cl_2^- + H^+ + O_2$$

$$Cl_2^- = Cl^- + Cl$$

The equilibrium constant is of the order of $1/10^5$, which means that in typical cloud water, Cl dominates over Cl_2^-. Reactions with SO_4^- represent a further source (Buxton et al. 1999a):

$$SO_4^- + Cl^- \rightarrow Cl + SO_4^{2-}$$

Chlorine atom can be lost with the production of OH (Buxton et al. 2000):

$$Cl + H_2O \rightarrow Cl^- + H^+ + OH$$

These radicals undergo interconversion, being equilibrated by chloride, sulphate, and hydrogen ions such that no one radical acts as a sink. Thus, at high Cl^- concentrations, for example, the radical chemistry of Cl_{2-} and Cl will be important, whilst at high sulphate that of $SO_4 \bullet^-$ will predominate (Buxton et al. 1999a).

6.7.4 Organic Chemistry

There is also an active organic chemistry in sunlit droplets. The aqueous-phase photolysis of biacetyl is an important source of organic acids and peroxides to aqueous aerosols, and fog and cloud drops. The half-life of aqueous-phase biacetyl with respect to photolysis is an hour or two with a solar zenith angle of 36°. Major products of aqueous biacetyl photolysis are acetic acid, peroxyacetic acid, and hydrogen peroxide, with pyruvic acid and methylhydroperoxide as minor photoproducts. Typical reducing agents in atmospheric waters are likely to be formate, formaldehyde, glyoxal, phenolic compounds, and carbohydrates (Faust et al. 1997).

In recent years there has been much interest in nitration of organic compounds in aerosols, as these are potentially carcinogenic. Phenol can be nitrated by the dissolved NO_3 radical or N_2O_4. In acidic aerosols, it could be nitrated by NO_2^+, which results in the 2 or 4 nitrophenol:

$$C_6H_5OH + NO_2^+ \rightarrow C_6H_4OHNO_2 + H^+$$

6.7.5 Other Systems

Continental aerosol chemistry is likely to be a little different from that of marine clouds as several processes are likely to consume more OH. There are three most notable differences:

1) Larger formic acid concentrations in the gas phase can dissolve and consume OH.
2) Transition metals can scavenge HO_2/O_2^- from the system.
3) Higher concentrations of SO_2 can reduce H_2O_2.

The chemistry of dew is likely to be different from that of bulk water in the atmosphere. Solutes can be at higher concentration, and the guttation fluids typically have a higher pH than rainwater and may contain a novel range of organic and nitrogen compounds, along with biologically active compounds. Peroxidase in dew on plant leaves can enzymatically promote the degradation of hydrogen peroxide. This would mean that dissolved sulphur dioxide might have a longer lifetime than in dew that collects on inert materials where hydrogen peroxide is an effective oxidizing agent.

6.8 Summary

Liquid water in the atmosphere represents an important part of the process of removing trace substances from the atmosphere. The residence time of water in the atmosphere is a matter of days (4–10) and the lifetime of rain drops and dew drops considerably shorter. However, the effectiveness of aqueous systems in removing gases from the atmosphere requires these gases to be very soluble in water or undergo rapid reactions.

Chemistry within the droplets may serve to reprocess materials. This is particularly evident where reactions within the droplet produce species of a low solubility and become lost from the liquid phase.

The influence of droplet phases on atmospheric chemistry leads to sequestration and alters the chemistry of the gas phase – for example, through the high solubility of peroxy radicals (Monod and Carlier 1999), which has an impact of ozone concentrations. Lelieveld and Crutzen (1990) suggested that the removal of HO_2 from the gas phase into cloud water will prevent it from oxidizing NO to NO_2. This limits the gas phase production of O_3 from photolysis of NO_2 and thus effectively leads to a loss of ozone from the troposphere:

$$OH + O_3 \rightarrow HO_2 + O_2$$
$$HO_2 = O_2^- + H^+$$
$$O_2^- + O_3 \rightarrow O_2 + O_3^-$$

$$O_3^- \rightarrow O_2 + O^-$$

$$O^- + H^+ \rightarrow OH$$

This sums to give:

$$2O_3 \rightarrow 3O_2$$

The chemistry of the liquid phase has been typically more difficult to resolve than that in the gas phase. Gas phase chemistry is affected by the liquid phase, so it has been possible to treat the liquid phase simply as a sink as a first approximation. However, as our understanding of aqueous phase aerosol chemistry has begun to involve multiphase considerations and radicals, the sink approximation becomes less tenable. We currently witness considerable advances in the radical and photochemistry of droplets in the atmosphere.

References

Barrie, L.A. (1991). Snow formation and processes in the atmosphere that influence its chemical composition. In: *Seasonal Snowpacks* (eds. T.D. Davies, M. Tranter and H.G. Jones), 1–20. Berlin: Springer-Verlag.

Barry, R.G. and Chorley, R.J. (1998). *Atmosphere, Weather and Climate*. London: Routledge.

Blando, J.D. and Turpin, B.J. (2000). Secondary organic aerosol formation in cloud and fog droplets: a literature evaluation of plausibility. *Atmospheric Environment* 34: 1623–1632.

Brimblecombe, P. and Todd, I.J. (1977). Sodium and potassium in dew. *Atmospheric Environment* 11: 649–650.

Buxton, G.V., Bydder, M., and Salmon, G.A. (1999a). The reactivity of chlorine atoms in aqueous solution. Part II. The equilibrium $SO_4^- + Cl^- Cl^{Nsbd} + SO_4^{2-}$. *Physical Chemistry Chemical Physics* 1: 269–273.

Buxton, G.V., Salmon, G.A., and Wang, J.Q. (1999b). The equilibrium $NO_3^{\bullet} + Cl^- = NO_3^- + Cl^{\bullet}$: a laser flash photolysis and pulse radiolysis study of the reactivity of NO_3 center dot with chloride ion in aqueous solution. *Physical Chemistry Chemical Physics* 1: 3589–3593.

Buxton, G.V., Bydder, M., Salmon, G.A., and Williams, J.E. (2000). The reactivity of chlorine atoms in aqueous solution. Part III. The reactions of cl with solutes. *Physical Chemistry Chemical Physics* 2: 237–245.

Chung, Y.S., Kim, H.S., and Yoon, M.B. (1999). Observations of visibility and chemical compositions related to fog, mist and haze in South Korea. *Water Air & Soil Pollution* 111: 139–157.

Clegg, S.L., Brimblecombe, P., and Khan, I. (1996). The Henry's law constant of oxalic acid and its partitioning into the atmospheric aerosol. *Idojárás* 100: 51–68.

Clegg, S.L., Brimblecombe, P., and Wexler, A.S. (1998a). A thermodynamic model of the system H-NH4-Na-SO4-NO3-Cl-H2O at 298.15 K. *The Journal of Physical Chemistry* 102A: 2155–2171.

Clegg, S.L., Brimblecombe, P., and Wexler, A.S. (1998b). A thermodynamic model of the system H-NH4-SO4-NO3-H2O at tropospheric temperatures. *The Journal of Physical Chemistry* 102A: 2137–2154.

Descari, S., Facchini, M.C., Fuzzi, S., and Tagliavini, E. (2000). Characterization of water-soluble organic compounds in atmospheric aerosol: a new approach. *Journal of Geophysical Research-Atmospheres* 105: 1481–1489.

Domine, F. (1999). Incorporation of trace gases into ice particles. In: *Transport and Chemical Transformation in the Troposphere*, vol. 1 (eds. P.M. Borrell and B.P. Borrell), 435–443. Southampton: WIT Press.

Facchini, M.C., Mircea, M., Fuzzi, S., and Charlson, R.J. (1999). Cloud albedo enhancement by surface-active organic solutes in growing droplets. *Nature* 401: 257–259.

Faust, B.C. (1994). Photochemistry of clouds, fogs, and aerosols. *Environmental Science & Technology* 28: A217–A222.

Faust, B.C., Powell, K., Rao, C.J., and Anastasio, C. (1997). Aqueous-phase photolysis of biacetyl (an alpha-dicarbonyl compound): a sink for biacetyl, and a source of acetic acid, peroxyacetic acid, hydrogen peroxide, and the highly oxidizing acetylperoxyl radical in aqueous aerosols, fogs, and clouds. *Atmospheric Environment* 31: 497–510.

Findlayson-Pitts, B.J. and Pitts, J.N. (2000). *Chemistry of the Upper and Lower Atmosphere*. San Diego, CA: Academic Press.

Fruekilde, P., Hjorth, J., Jensen, N.R. et al. (1998). Ozonolysis at vegetation surfaces: a source of acetone, 4-oxopentanal, 6-methyl-5-hepten-2-one, and geranyl acetone in the troposphere. *Atmospheric Environment* 32: 1893–1902.

Gelencser, A., Sallai, M., Krivacsy, Z. et al. (2000). Voltammetric evidence for the presence of humic-like substances in fog water. *Atmospherioc Research* 54: 157–165.

Heitnzenberg, J. (1998). Condensed water aerosols. In: *Atmospheric Particles* (eds. R.M. Harisson and R. van Grieken), 509–542. Chichester, UK: Wiley.

Herrmann, H., Ervens, B., Nowacki, P. et al. (1999). A chemical aqueous phase radical mechanism for tropospheric chemistry. *Chemosphere* 38: 1223–1232.

Herrmann, H., Ervens, B., Jacobi, H.W. et al. (2000). CAPRAM2.3: a chemical aqueous phase radical mechanism for tropospheric chemistry. *Journal of Atmospheric Chemistry* 36: 231–284.

Huebert, B.J. and Charlson, R.J. (2000). Uncertainties in data on organic aerosols. *Tellus* 52B: 1249–1255.

Hughes, R.N. and Brimblecombe, P. (1994). Dew and guttation: formation and environmental significance. *Agricultural and Forest Meteorology* 67: 173–190.

Kames, J. and Schurath, U. (1992). Alkyl nitrates and bifunctional nitrates of atmospheric interest: Henry's law constants and their temperature dependencies. *Journal of Atmospheric Chemistry* 15: 79–95.

Kerstetter, R., Zepp, R.G., and Carreira, L. (1998). Peroxidases in grass dew derived from guttation: possible role in polymerization of soil organic matter. *Biogeochemistry* 42: 311–323.

Khan, I., Brimblecombe, P., and Clegg, S.L. (1992). The Henry's law constants of pyruvic and methacrylic acids. *Environmental Technology* 13: 587–593.

Khan, I., Brimblecombe, P., and Clegg, S.L. (1996). Solubilities of pyruvic acid and the lower (C1-C6) carboxylic acids. Experimental determination of equilibrium vapour pressures above pure aqueous and salt solutions. *Journal of Atmospheric Chemistry* 22: 285–302.

Knipping, E.M., Lakin, M.J., Foster, K.l. et al. (2000). Experiments and simulations of ion-enhanced interfacial chemistry on aqueous NaCl aerosols. *Science* 288: 301–306.

Leck, C. and Bigg, E.K. (1999). Aerosol production over remote marine areas – a new route. *Geophysical Research Letters* 26: 3577–3580.

Lelieveld, J. and Crutzen, P.J. (1990). Influences of cloud photochemical processes on tropospheric ozone. *Nature* 343: 227–233.

Macleod, H., Smith, G.P., and Golden, D.M. (1988). Photodissociation of pernitric acid (HO2NO2) at 248 nm. *Journal of Geophysical Research* 93: 3813–3823.

Mason, B. (1975). *Clouds and Rainmaking*. Cambridge, UK: Cambridge University Press.

Monod, A. and Carlier, P. (1999). Impact of clouds on the tropospheric ozone budget: direct effect of multiphase photochemistry of soluble organic compounds. *Atmospheric Environment* 33: 4431–4446.

Montonya, M.R. and Mellado, J.M.R. (1995). Hydration constants of carbonyl and dicarbonylcompounds. Comparison between electrochemical and no electrochemical technique. *Portugaliae Electrochimica Acta* 13: 299–303.

Penkett, S.A., Jones, B.M.R., Brice, K.A., and Eggleton, A.E.J. (1979). The importance of atmospheric ozone and hydrogen peroxide in oxidizing sulphur dioxide in cloud and rain water. *Atmospheric Environment* 13: 323–337.

Pilinis, C., Seinfeld, J.H., and Grosjean, D. (1989). Water content of atmospheric aerosols. *Atmospheric Environment* 23: 1601–1606.

Prupacher, H.R. and Klett, J.D. (1997). *Microphysics of Clouds and Precipitation*. Dordrecht, the Netherlands: Kluwer.

Ravishankara, A.R. (1997). Heterogeneous and multiphase chemistry in the troposphere. *Science* 276: 1058–1065.

Ruggaber, A., Dlugi, R., Bott, A. et al. (1997). Modelling of radiation quantities and photolysis frequencies in the aqueous phase in the troposphere. *Atmospheric Environment* 31: 3135–3148.

Sander, R. (1999). Compilation of Henry's law constants for inorganic and organic species of potential importance in environmental chemistry. http://www.mpch-mainz.mpg.de/~sander/res/henry.html.

Saxena, P. and Hildemann, L.M. (1996). Water-soluble organics in atmospheric particles: a critical review of the literature and application of thermodynamics to identify candidate compounds. *Journal of Atmospheric Chemistry* 24: 57–109.

Seidl, W. (2000). Model for a surface film of fatty acids on rainwater and aerosol particles. *Atmospheric Environment* 34: 4917–4932.

Siefert, R.L., Pehkonen, S.O., Erel, Y., and Hoffmann, M.R. (1994). Iron photochemistry of aqueous suspensions of ambient aerosol with added organic-acids. *Geochimica et Cosmochimica Acta* 58 (15): 3271–3279.

Spokes, L.M., Lucia, M., Campos, A.M., and Jickells, T.D. (1996). The role of organic matter in controlling copper speciation in precipitation. *Atmospheric Environment* 30: 3959–3966.

Takenaka, N., Daimon, T., Ueda, A. et al. (1998). Fast oxidation reaction of nitrite by dissolved oxygen in the freezing process in the tropospheric aqueous phase. *Journal of Atmospheric Chemistry* 29: 135–150.

Takenaka, N., Suzue, T., Ohira, K. et al. (1999). Natural denitrification in drying process of dew. *Environmental Science and Technology* 33: 1444–1447.

Vane, L.M. and Giroux, E.L. (2000). Henry's law constants and micellar partitioning of volatile organic compounds in surfactant solutions. *Journal of Chemical & Engineering Data* 45: 38–47.

Vogt, R., Crutzen, P.J., and Sander, R. (1996). A mechanism for halogen release from sea-salt aerosol in the remote marine boundary layer. *Nature* 383: 327–330.

Warneck, P. (1999a). *Chemistry of the Natural Atmosphere*, 2e, vol. 71. San Diego, CA: Academic Press.

Warneck, P. (1999b). The relative importance of various pathways for the oxidation of sulfur dioxide and nitrogen dioxide in sunlit continental fair weather clouds. *Physical Chemistry Chemical Physics* 1: 5471–5483.

Ziajka, J. and Pasuik-Bronikowska, W. (1999). Effect of alpha-pinene and cis-verbenol on the rate of S(IV) oxidationcatalysed by Fe. In: *Transport and Chemical Transformation in the Troposphere*, vol. 1 (eds. P.M. Borrell and B.P. Borrell), 756–761. Southampton, UK: WIT Press.

Zuo, Y.G. and Holgne, J. (1992). Formation of hydrogen-peroxide and depletion of oxalic-acid in atmospheric water by photolysis of iron(III) oxalato complexes. *Environmental Science & Technology* 26: 1014–1022.

Further Reading

Notably chapter 8 ofFindlayson-Pitts, B.J. and Pitts, J.N. (2000). *Chemistry of the Upper and Lower Atmosphere*. San Diego, CA: Academic Press.

Sukhapan, J. and Brimblecombe, P. (2002). Ionic surface active compounds in atmospheric aerosols. *The Scientific World Journal* 2: 1138–1146.

Websites

1 http://www.aim.env.uea.ac.uk/aim/aim.php

2 Henry's law constants – tabulation by Rolf Sander (https://www.atmos-chem-phys.net/15/4399/2015).

7

Particulate Matter in the Atmosphere

Paul I. Williams

School of Earth and Environmental Sciences & National Centre for Atmospheric Science, The University of Manchester, Manchester, United Kingdom

The atmosphere around us is filled with microscopic particles known as aerosols. Whilst reading this sentence you are likely to breathe in anything from 10 000 to more than 10 000 000 of these tiny particles. Aerosol particles are any solid or liquid material (or a combination of both) suspended in the atmosphere. They are found in a multitude of forms and sizes, and like the aerosol filling the room around you, generally go unnoticed in everyday life. Small they may be, but their effects can be dramatic.

In the past several decades, atmospheric aerosols have increasingly been recognized as constituting one of the major uncertainties in the current understanding of climate change (IPCC 2013, http://www.ipcc.ch). The uncertainty is mainly due to the large variability in their chemical and physical properties, as well as their temporal and spatial distributions. Atmospheric aerosols may originate from either naturally occurring sources or anthropogenic processes. Major natural aerosol sources include sea-spray emissions from oceans, volcanic emissions, and mineral dust from arid regions, whilst major anthropogenic sources include emissions from industry and combustion processes. These sources are further classified into primary and secondary categories. Aerosols directly emitted into the atmosphere constitute primary sources, whilst secondary sources arise from a chemical transformation. These include gas-to-particle conversion of compounds such as nitric acid and nitrogen oxide (collectively known as NO_x), sulphur dioxide (SO_2), and hydrocarbons. Clouds are a source of secondary aerosol. Cloud droplets form on preexisting aerosol particles, which can modify the chemical composition of the aerosol. If the cloud droplet is not rained out, when the droplet evaporates it leaves behind a modified aerosol.

Characterization of the life cycle of atmospheric aerosols is a complex and much-faceted issue. The schematic illustration of Figure 7.1 summarizes: (i) aerosol sources; (ii) transformation mechanisms; and (iii) aerosol sink processes. Aerosol size is one of the most important parameters in describing aerosol properties and their interactions with the atmosphere and each other, and its determination and use is of fundamental importance. With reference to Figure 7.1, the size fraction with diameter $d > 1–2\,\mu$m is usually referred to as the coarse mode, and the fraction $d < 1–2\,\mu$m is the fine mode. The latter mode can be further divided into the accumulation mode ($d \approx 0.1–1.0\,\mu$m), Aitken ($d \approx 0.01–0.1\,\mu$m), and nucleation ($d < 0.01\,\mu$m) modes. The volume (and

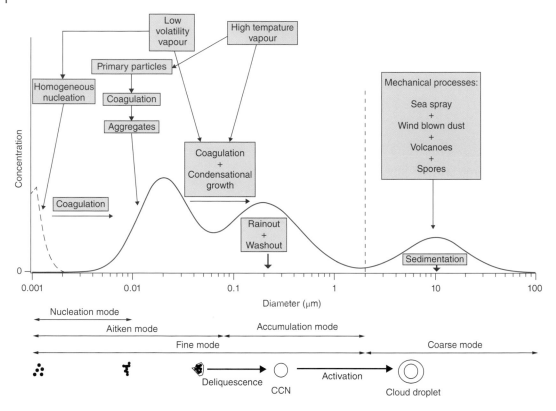

Figure 7.1 Schematic of the aerosol formation, processes and removal within the atmosphere.

mass) of a particle is proportional to d^3; hence, the coarse mode is typified by a maximum volume concentration. The surface area of the aerosol is proportional to d^2 and has a maximum in the accumulation mode, whereas the Aitken and nucleation modes are typified by maximum number of concentrations.

Primary particle formation occurs when a particle is emitted directly into the atmosphere, and is often driven by the influence of wind (e.g. desert dust). Secondary aerosol formation occurs when an aerosol is formed or processed in the atmosphere, such as cloud processing described above. Aerosol arising from homogeneous or heterogeneous nucleation are further examples of secondary processes. The former refers to the formation of new nuclei, and the latter to the condensational growth on existing nuclei. Condensation onto a host surface occurs as a critical supersaturation, which is substantially lower than for homogeneous nucleation. For example, water vapour condenses onto aerosol particles at supersaturations of 1–2% (heterogeneous) to form cloud droplets, whereas if water was to self-nucleate (homogeneous) it would require supersaturations of 300%. Examples of gas-to-particle conversion are combustion processes and the ambient formation of nuclei from gaseous organic emissions. Examples of the latter are the observations of new particle formation involving iodine in the Artic (Allan et al. 2015) and from coastal sites where seaweed exposed at low tide emits iodine-containing compounds that nucleate to form particles (McFiggans et al. 2006). In these processes, high numbers of nucleation mode particles are emitted, which rapidly reduce through coagulation, resulting in aerosol lifetimes of the order of minutes for these smallest particles.

Particles in the Aitken and accumulation mode typically arise from either (i) the condensation of low-volatility vapours; or (ii) coagulation. Atmospheric aerosol may be composed of a range of chemical species. When particles are chemically distinct from one another, they are termed as *externally mixed*. Alternatively, when particles have a similar composition, they are known as *internally mixed*. The degree of internal mixing of ambient particles is complex, and can range from a complete chemical mixing (e.g. all components dissolved in a cloud droplet) to less well mixed (e.g. sulphate coating on soot aerosol). It is worth noting that in the two examples given, were a cloud drop to form on the sulphate-coated soot aerosol, the sulphate would go into solution (dissolve), but the soot would not. Particles in the accumulation mode have a longer atmospheric lifetime than other modes, as there is a minimum efficiency in the sink (removal) processes. Of these processes, wet deposition (e.g. rain) is the major sink process.

Particles in the coarse mode are usually produced by weathering and wind-erosion processes and are generally primary particles. Dry deposition (primary sedimentation) is the dominant removal process. Chemically, their composition reflects their sources, as demonstrated by mineral dust from deserts and sea salt from oceans. Organic compounds such as biological particles (spores, pollens, and bacteria) as well as biogenic particles resulting from direct emission of hydrocarbons into the atmosphere may also be constituents of the course mode. As the sources and sinks of coarse and fine modes are different, there is only a weak association of particles in both modes.

The above brief summary of atmospheric aerosol matter serves to illustrate the complex processes involved in modelling their behaviour and accessing their influence on climate. It is mainly for the latter reason that the interest in aerosols has grown and hence this chapter focuses on the climate effects of aerosols. Aerosol types and composition are considered first, followed by their interactions with radiation through the so-called direct and indirect aerosol effects. Further reading on the aerosol fundamentals may be found in the following list of reprinted and new books: Baron and Willeke (2011), Charlson and Heintzenberg (1999), Singh (1995), Hinds (1999), and Seinfeld and Pandis (2006).

7.1   Aerosol Properties

Knowledge of the different modes of an aerosol size distribution and its composition are of primary importance, as most other parameters of interest may be deduced from this information. The atmospheric aerosol may be classified into several categories, according to source and geographical location. However, before these are explored, it is worth considering some of the basic properties of aerosol particles and what these physically mean. The most fundamental of these is size.

The following sections serve as a brief introduction to some of the properties of aerosol particles. A thorough coverage of all aspects would be a book in itself. The rigorous treatments and derivations and all the approximations for the equations are beyond the scope of this book, but can be found in the references given. The reader is encouraged not to be daunted by these equations, and to simply use them, where necessary, as tools in the first steps to understanding aerosol science.

7.1.1   Particle Size

What is the size of an aerosol particle? Throughout this chapter, the size of a particle is classified by its diameter, d. Diameter immediately implies that a particle is spherical. However, dry aerosol particles are rarely spherical. Figure 7.2 shows images from an electron microscope of four different aerosol types. It can clearly

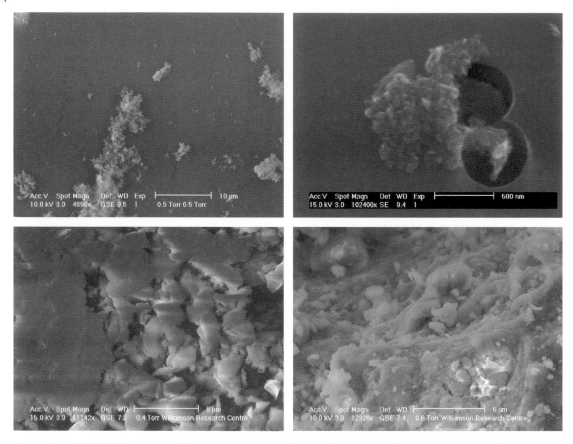

Figure 7.2 SEM/ESEM images of different aerosol types from the University of Manchester supplied by Rachel Burgess: top left, diesel fuel; top right, product from *Laminaria digitata* (seaweed) + ozone; bottom left, Arizonan dust; bottom right, Saharan dust.

be seen that the particles are not spherical, so how should they be categorized? This is where the notion of *equivalent diameters* is used. An equivalent diameter is the diameter a sphere of defined physical properties that behaves the same as the particle being studied. One such property is the settling velocity, which is sometimes called the *terminal settling velocity*, the speed a particle reaches when gravitational forces are balanced by drag forces. It can be thought of as dropping a tennis ball from a very high tower. The tennis ball does not accelerate to an infinite velocity; rather, it reaches a maximum when the drag forces equal the gravitational forces, its terminal settling velocity. Aerosol size covers several decades in diameter and as a result, a variety of instruments are required for its determination. These instruments will report a variety of equivalent diameters, depending on the technique used. Some examples of different equivalent diameters are:

- *Aerodynamic equivalent diameter.* The diameter of a spherical particle that has the same terminal velocity as the one of interest and assuming it has a standard density of $1\,\mathrm{g\,cc^{-1}}$.
- *Stokes diameter.* The diameter of a spherical particle that has the same settling velocity and density of the particle of interest.

- *Optical equivalent diameter.* The diameter of a spherical particle that has the same optical properties as the one of interest.

This list is by no means extensive, and a further discussion can be found in Hinds (1999) and Baron and Willeke (2011). The questions arises, why is there a range of different equivalent diameters? The answer is because no instrument, other than a microscope, measures particle size directly. They measure another property of the particle (e.g. terminal settling velocity) and calculate a diameter.

When comparing measurements from different instruments, it is important to know which diameter is being reported and, where necessary, how to convert between diameters. For large particles ($d > 1\,\mu m$), conversion between the aerodynamic equivalent diameter and the Stokes diameter can be approximated by

$$\frac{g\rho_0 d_a^{\,2}}{18\eta} = \frac{g\rho_b d_s^{\,2}}{18\eta} = V_{TS} \tag{7.1}$$

where ρ_0 is the standard density ($1\,\text{g cc}^{-1}$ or $1000\,\text{kg m}^{-3}$), d_a is the aerodynamic diameter, ρ_b is the density of the bulk material, d_s the Stokes diameter, g is the acceleration due to gravity ($9.81\,\text{m s}^{-2}$), η the viscosity of air and V_{TS} is the terminal settling velocity.

Worked Example 7.1

An irregularly shaped, solid particle has a settling velocity of $0.0025\,\text{m s}^{-1}$ and a density of $1.72\,\text{g cc}^{-1}$. What are the aerodynamic equivalent (d_a) and Stokes (d_s) diameters? (assume $\eta = 1.83\text{e}^{-5}\,\text{kg m}^{-1}\,\text{s}^{-1}$).

Answer

Using Eq. (7.1), we can rearrange to obtain the diameter, d_s:

$$d_s = \sqrt{\frac{18 \times 1.83 e^{-5} \times 0.0025}{9.81 \times 1720}}$$

which produces d_s of $6.99\,\mu m$. Similar rearrangement gives $d_a = 9.16\,\mu m$.

Equation (7.1) is only applicable to particles with a diameter larger than $1\,\mu m$. However, atmospheric particles extend to much smaller sizes. The expression for the settling velocity can be extended to include smaller particles by the addition of the Cunningham slip correction factor C_c. Equation (7.1) now becomes:

$$V_{TS} = \frac{g\rho d^2 C_c}{18\eta} \tag{7.2}$$

The Cunningham slip correction factor is applied because the interaction of gas molecules with large particles is different than for smaller particles. Hinds (1999) gives the expression for C_c as:

$$C_c = 1 + \frac{\lambda}{d}\left[2.34 + 1.05\exp\left(-0.39\frac{d}{\lambda}\right)\right] \tag{7.3}$$

where λ is the mean free path of the gas the aerosol is in. For air at standard pressure and $293\,\text{K}$, this is $6.64\text{e}^{-8}\,\text{m}$.

7.1.2 Particle Motion

How particles move in the atmosphere compared with how they behave in a sampling pipe of an instrument is of fundamental importance. Having determined the settling velocity, it is then possible to estimate how quickly a particle can sediment out in still air or indeed in a flow in a pipe. Gravitational settling affects large particles whereas for smaller particles, Brownian motion and diffusion have the largest influence. Figure 7.3 summarizes the effects of particle motion. For large particles ($d > 1\,\mu$m), sedimentation will occur quicker than diffusion. For smaller particles, diffusion is dominant and will cause the greatest movement of the particles.

Brownian motion was first observed in 1827 when pollen grains in water appeared to move in a random fashion. This movement caused by bombardment of water molecules is what is known as Brownian motion. For aerosol particles, Brownian motion is caused by the constant bombardment of air molecules on the aerosol surface. Diffusion is the net transport of particles from a high concentration to a low concentration and the movement is driven by the Brownian motion. Both processes are related to what is called the diffusion coefficient, D. The higher the value of D, the larger the Brownian motion and the quicker a particle will diffuse. The diffusion coefficient is given by:

$$D = \frac{kTC_c}{3\pi\eta d} \qquad (7.4)$$

where K is the Boltzmann's constant ($1.38 \times 10^{-23}\,\mathrm{Nm\,K^{-1}}$) and T is the temperature in Kelvin. The diffusion coefficient, D, has units of $\mathrm{m^2\,s^{-1}}$. For small particles where C_c is important, D is approximately proportional to d^{-2}, whereas for large particles D is inversely proportional to d.

Worked Example 7.2
Calculate the diffusion coefficient for a $0.1\,\mu$m and a $1.0\,\mu$m sphere at $293\,$K.

Answer
First, we need to calculate the C_c for each sphere from Eq. (7.3):

$$C_c = 1 + \frac{6.64\mathrm{e}^{-8}}{0.1\mathrm{e}^{-6}}\left[2.34 + 1.05\exp\left(-0.39\frac{0.1\mathrm{e}^{-6}}{6.64\mathrm{e}^{-8}}\right)\right]$$
$$= 2.94$$

$$C_c = 1 + \frac{6.64\mathrm{e}^{-8}}{0.1\mathrm{e}^{-6}}\left[2.34 + 1.05\exp\left(-0.39\frac{1\mathrm{e}^{-6}}{6.64\mathrm{e}^{-8}}\right)\right]$$
$$= 1.16$$

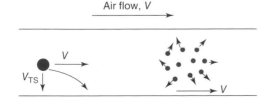

Figure 7.3 Summary of the dominant effects on particles of different sizes with no external forces except gravity.

From this, use Eq. (7.4):

$$D_{0.1\mu m} = \frac{1.38e^{-23} \times 293 \times 2.94}{3\pi \times 1.83e^{-5} \times 0.1e^{-6}} = 6.89e^{-10} \, m^2 \, s^{-1}$$

and a value of $D = 2.72e^{-11} \, m^2 \, s^{-1}$ is obtained for the $1.0\,\mu m$ particle.

The worked example above shows that the diffusional effects of a $0.1\,\mu m$ particle are 10 times greater than a $1.0\,\mu m$ particle, but how does this relate to the motion of the particles? The diffusion coefficient can be used to approximate the average distance travelled by a particle in a time, t, by:

$$x = \sqrt{2Dt} \tag{7.5}$$

so in any given time interval, a $0.1\,\mu m$ particle will move five times further, on average, than the $1.0\,\mu m$ particle by diffusion.

Another use of the diffusion coefficient is to calculate the penetration of particles through a pipe. That is, for a given number of particles entering a pipe, what percentage is transmitted through the tube? For a cylindrical pipe, the transmission, T_p can be approximated by:

$$T_p = 1 - 5.50\mu^{2/3} + 3.77\mu \text{ for } \mu < 0.009 \tag{7.6a}$$

and

$$T_p = 0.819\exp\left(-11.5\mu\right) + 0.0975\exp\left(-70.1\mu\right)$$
$$\text{for } \mu \geq 0.009 \tag{7.6b}$$

where, for cylindrical pipes

$$\mu = \frac{DL}{Q} \tag{7.7}$$

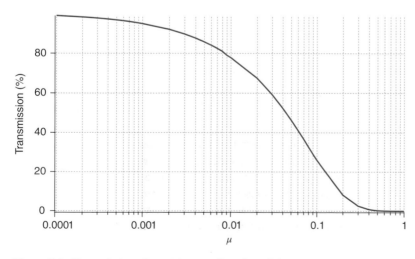

Figure 7.4 Transmission of particles as a function of size.

with D the diffusion coefficient, L the length of the pipe and Q the volumetric flow rate. The values for T_p are taken from Hinds (1999). Other approximations are also given in Baron and Willeke (2011). Figure 7.4 shows T_p for different values of μ.

Therefore, the smaller the particle diameter, the larger the diffusion coefficient, the larger μ, and the more particles will be lost.

Worked Example 7.3

How many 10 nm particles will be transmitted down a pipe 500 cm long with a diameter of 2 mm in an air stream travelling at a velocity of $10\,m\,s^{-1}$ at 293 K?

Answer

The first thing we need to calculate is the volumetric flow rate. If we assume this is the average velocity, we can say that:

$$\text{Velocity} = \frac{Q}{A}$$

where A is the area of the tube. Therefore, Q equals $10\,m\,s^{-1} \times \pi \times (2e^{-3}/2)^2$, which gives $Q = 3.14e^{-5}\,m^3\,s^{-1}$ $(1.881\,min^{-1})$. Using Eqs. (7.3) and (7.4) we obtain:

$$C_c = 1 + \frac{6.64e^{-8}}{10\,nm}\left[2.34 + 1.05\exp\left(-3.09\frac{10\,nm}{6.64e^{-8}}\right)\right]$$
$$= 23.1$$

$$D = \frac{1.38e^{-23} \times 293 \times 23.1}{3\pi \times 1.83e^{-5} \times 10\,nm} = 5.42e^{-8}\,m^2\,s^{-1}$$

which yields a value of $\mu = 0.0086$. Reading off the value from Figure 7.4, this gives a transmission of about 80%. A better approximation can be derived from Eq. (7.6b):

$$T_p = 0.819\exp(-11.5 \times 0.0086) + 0.0975\exp(-70.1 \times 0.0086) = 79.5\%$$

7.1.3 Hygroscopic Properties

As the relative humidity (RH) of the atmosphere changes, then the diameter of particle can also change, depending on its chemical composition. This is due to the ability of a particle to take up water, its *hygroscopicity*, which is a very important property as a change in size of the particle can change its light-scattering properties. Of course, not all particles will take up water. For example, soot particles directly emitted from a car exhaust are very poor at taking up water. Particles that can easily take up water are classed as *hygroscopic*; those that cannot are classed as *hydrophobic*. If a particle can take up water, it then becomes important to distinguish what diameter is being reported. The *dry* diameter is the diameter of a solid particle that has not absorbed any water. The *wet* diameter is the diameter of a particle that has grown by absorbing water. The ratio of the wet diameter to the dry diameter is called the *growth factor* (GF). It is important when reporting the wet diameter to know what the RH was at the time of measurement.

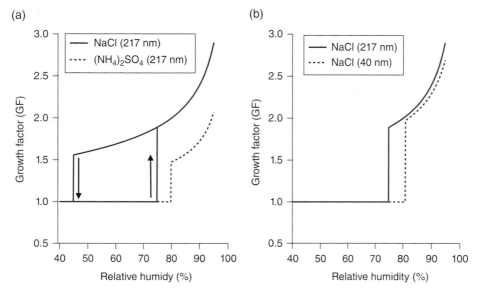

Figure 7.5 Growth curves for two different salts (a) and two different sizes for the same salt (b). Results from the ADDEM model by Topping et al. (2005).

Figure 7.5 illustrates the growth curves for (a) different salts at the same size and (b) different sizes of the same salt. With reference to (a), a dry, solid particle exposed to increasing RH does not begin to grow immediately, there is a minimum amount of free water in the atmosphere that is required to dissolve the particle and cause it to grow. When it reaches this point, the GF increases from 1 (no growth) and the particle begins to grow. This point is called the *deliquescence point*. It can be seen that for NaCl this is at about 75% and for $(NH_4)_2SO_4$ it is about 80% for 217 nm particles. Once past the deliquescence point, the particles grow with increasing RH. If the RH then begins to drop, the particle does not return to the solid phase at the deliquescence point. Instead, the RH has to drop further, to the *efflorescence point* before the particle becomes solid (dry). Only the efflorescence curve for NaCl is shown. Figure 7.5b highlights the effect of particle diameter on the growth curves, namely that the smaller the particle, the higher its deliquescence point. The values shown here for the different sizes are approximate and are for illustration purposes only. The exact point at which different size particles and indeed chemically different particles deliquesce and effloresce is a complex function of composition and other properties.

7.1.4 Aerosol Size Distribution

Figure 7.6 illustrates typical size distributions for number, surface, and volume concentration for various aerosol types. The number size distribution of the atmospheric aerosol may be approximated by an empirical power law equation for radii $r > 0.1\,\mu m$. However, many aerosol particle number size distributions tend to follow *log-normal distributions*:

$$dN = \frac{N}{\sqrt{2\pi}\,\ln\sigma_g}\exp\left\{\frac{-\left(\ln(d)-\ln d_g\right)^2}{2\ln\sigma_g^{\,2}}\right\}d\ln(d) \tag{7.8}$$

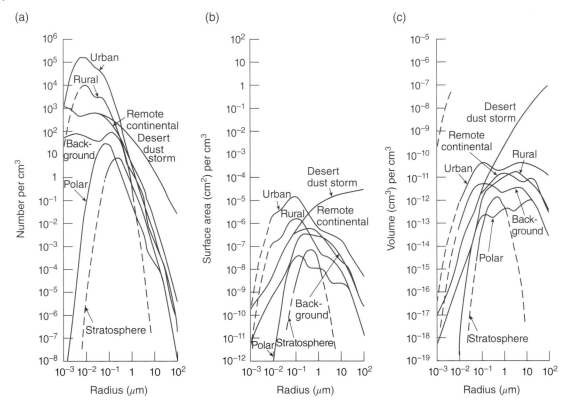

Figure 7.6 (a) Number, (b) surface, and (c) volume distributions for various atmospheric aerosols. Aerosol types are self-explanatory apart from background aerosol, which refers to the tropospheric aerosol 5 km above the continents and 3 km above the oceans. *Source:* from Jaenicke (1988).

where dN is the concentration of particles in the bin $d\ln(d)$, N is the total aerosol number concentration, d_g is the count mean diameter and σ_g is the standard deviation (or width) of the distribution. Typical values of d_g and σ_g for the Aitken, accumulation, and coarse mode aerosol for different aerosol types can be found in Baron and Willeke (2011).

The large range in magnitude of N is demonstrated by the values $<20\,\mathrm{cm^{-3}}$ for the polar regions and $>10^5\,\mathrm{cm^{-3}}$ for an urban environment. Not only is there a large geographical variation in aerosol concentrations, but the vertical extent also varies substantially. For instance, a background aerosol at 3 km over the oceans or 5 km over the continents is typified by $N \approx 150\,\mathrm{cm^{-3}}$, and the stratospheric aerosol at 20 km by $N \approx 10\,\mathrm{cm^{-3}}$. The chemical composition of the urban, remote continental, and remote marine can be described to a first approximation by the components listed in Table 7.1. Sulphate and organics are the major aerosol components of urban and remote continental regions and sodium chloride of remote marine regions. The aerosol composition and geographical type are thus recognized as being fairly specific to an aerosol source and are discussed in greater detail later.

7.1.5 Aerosol Optical/Radiative Properties

As discussed in Chapter 3, the interaction of aerosol particles with radiation is one of the key parameters when considering the effects of aerosol particles on climate change. Considering how much an aerosol can either scatter or/and absorb radiation determines whether it has a positive forcing (warming) or

Table 7.1 Typical mass composition ($\mu g\,m^{-3}$) of various chemical species in urban, remote continental and remote marine aerosol types.

Element or compound	Urban aerosol – photochemical smog	Remote continental aerosol	Remote marine aerosol
SO_4^{2-}	16.5	0.5–5.0	2.6
NO_3^-	10.0	0.4–1.4	0.05
Cl^-	0.7	0.08–0.14	4.6
Br^-	0.5	–	0.02
NH_4^+	6.9	0.4–2.0	0.16
Na^+	3.1	0.02–0.08	2.9
K^+	0.9	0.03–0.01	0.1
Ca^{2+}	1.9	0.04–0.30	0.2
Mg^{2+}	1.4	–	0.4
Al_2O_3	6.4	0.08–0.40	–
SiO_2	21.1	0.2–1.3	
Fe_2O_3	3.8	0.04–0.40	0.07
CaO	–	0.06–0.18	–
Organics	30.4	1.1	0.9
Elemental carbon	9.3	0.04	0.04
Total	112.9	2.99–12.44	12.04

Selected mass fractions and molar ratios

Element or compound	Urban aerosol – photochemical smog	Remote continental aerosol	Remote marine aerosol
SO_4^{2-} (%)	15.9	30.2–45.7	22.6
NO_3^- (%)	9.6	13.3–22.7	0.44
NH_4^+/SO_4^{2-}	2.2	2.1–3.4	0.47

Source: adapted from Pueschel (1995).

negative forcing (cooling) effect on the atmosphere. Scattered or absorbed radiation from an incident beam (e.g. the Sun) is defined by the *total scattering* (k_s) and *absorption* (k_a) *coefficients*, which are a measure of the fractional change in beam intensity per metre. The sum of k_s and k_a is the *extinction coefficient*, k_e and when integrated over the beam length gives the *aerosol optical depth* (AOD: Ω).

$$k_e = k_s + k_a \tag{7.9}$$

$$\Omega = \int k_e dl \tag{7.10}$$

In other words, the AOD represents the sum of all the scattering and absorption within an aerosol population or layer. The AOD relates to the visibility of the atmosphere. On a clean day, the Sun will appear bright in the sky and the AOD will be low. On a polluted day, the Sun will seem duller and the AOD will be higher. It is important to note that the coefficients given in Eqs. (7.9) and (7.10) are wavelength dependent.

An important parameter in global aerosol models is the ratio of the scattering to extinction, otherwise known as the *single-scattering albedo*, ω_0, and is a measure of the total light extinction due to scattering:

$$\omega_0 = \frac{k_s}{k_e} \tag{7.11}$$

When looking at objects, our eyes are generally detecting the scattered light. For example, if looking down from space at the Earth, clouds appear bright, whereas the oceans appear dark. This is because clouds are very good at scattering light and will have a high single-scattering albedo. The oceans absorb incoming solar radiation more and therefore have a lower single-scattering albedo and appear darker.

Worked Example 7.4
A satellite orbiting the Earth is measuring the single-scattering albedo. An aerosol layer comes into its field of view. The layer has a mix of absorbing and scattering aerosol. Will this increase or decrease the single-scattering albedo recorded by the satellite?

Answer
The answer is that it depends on what surface it was looking at in the first place. Take the example of the ocean and cloud tops that is represented in Figure 7.7. The dark square represents the ocean surface, the white square the cloud tops. As the aerosol layer is transported into the field of view of the satellite, it absorbs and scatters the incoming radiation. As the ocean absorbs most of the radiation, the aerosol layer will increase the overall scattering and the surface will become brighter and ω_0 will increase. Conversely, as the aerosol passes over the cloud, this will increase the amount of absorption and decrease ω_0.

The absorption and scattering coefficients are also used to determine the *refractive index* of a particle, which is a *complex function* with the *real* contribution from the scattering coefficient and the *imaginary* from the absorbing coefficient. The coefficients are also dependent on the RH. As discussed above, as the RH increases and a particle begins to take up water, the diameter, d, and hence k_s, increases. The fractional change in the k_s is denoted f(RH), similar to the fractional change in the particles diameter being defined as the *GF*. When the RH increases beyond 100%, cloud-aerosol interactions become important. Table 7.2

summarizes the range of optical properties representative of polluted continental, clean continental and clean marine aerosol types. Polluted continental aerosol concentrations are nearly a factor 10 larger than for remote regions, whilst a lower value of ω_0 indicates a higher proportion of absorbing species. Large efforts are presently under way to obtain detailed aerosol databases for various representative locations around the globe. The Global Atmosphere Watch (GAW) programme of the World Meteorological Organization has, for instance, been established to collect data over decadal time scales to monitor long-term changes in atmospheric composition.

7.2 Aerosol Sources

Aerosols sources are generally classified as anthropogenic (man-made) or natural. An estimate of the annual atmospheric contributions to major aerosol sources is given in Table 7.3. Estimates suggest that the anthropogenic emissions of sulphate are greater than from natural sources. Owing to the short atmospheric lifetime of aerosols, it should be remembered that globally averaged figures in Table 7.3 are not necessarily indicative of local concentrations. For instance, natural sources will dominate on a global scale due to their large-area emission sources (e.g. deserts and oceans). In contrast, anthropogenic emissions from industrialized regions in Europe, the USA, and East Asia, which are relatively smaller, are likely to exceed the contributions from natural sources.

7.2.1 Natural Sources – Primary Emissions

7.2.1.1 Sea-Salt Aerosol

Sea-salt aerosol results from the bursting of bubbles, formed by wave and wind action at the ocean surface (e.g. O'Dowd et al. 1997). As a result, sea-spray droplets are ejected and either return to the water surface or evaporate to form inorganic/organic aerosols, which

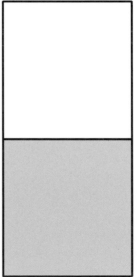

Figure 7.7 Effect of a mixed aerosol layer on an absorbing (low ω_0) and a reflective (high ω_0) surface, on the top and bottom, respectively.

may then be entrained into the marine boundary layer (MBL) by wind turbulence. These aerosols can then act as an effective cloud condensation nucleus (CCN), the site upon which marine clouds form. The wind-driven sea-salt size distribution is generally composed of three lognormal modes: a film drop and jet mode caused by bubble bursting, and the spume mode formed by particles being

Table 7.2 Representative values of observed aerosol optical properties in the lower troposphere for $\lambda = 0.5-0.55\,\mu m$ and RH < 60%.

Parameter	Polluted continental	Clean continental	Clean marine
Optical depth (Ω)	0.2–0.8*	0.02–0.1*	0.05–0.1*
Single-scattering albedo (ω_0)	0.92†	0.97†	0.97–0.99†
Total scattering coefficient (σ_{sp}, m^{-1})	$50-300\times10^{-6}$*	$5-30\times10^{-6}$*	$5-20\times10^{-6}$*
Absorption coefficient (σ_{ap}, m^{-1})	$5-50\times10^{-6}$*	$1-10\times10^{-6}$*	$<0.05\times10^{-6}$*
Fine mass concentration ($\mu g\,m^{-3}$)	5–50*	1–10*	1–5*
CN number concentration (cm^{-3})	10^3-10^5*	10^2-10^3*	$<10^2$*
CCN number concentration (cm^{-3}, 0.7–1%)	1000–5000*	100–1000*	10–200*

CN, condensation nuclei; CCN, cloud condensation nuclei.
Sources: adapted from * IPCC (1999) and † IPCC (2001).

Table 7.3 Global emission source strengths for atmospheric aerosols.

Aerosol component		Tg year^{-1}	Aerosol size mode
Natural	Primary:		
	Sea salt	1400–6800[a]	Mainly coarse
	Mineral dust	1000–4000[a]	Mainly coarse
	Primary Biological Aerosol Particles (PBAP)	50–1000[a]	Coarse
	Volcanic ash	33[b]	Coarse
	Secondary:		
	Biogenic sulphate	57[c]	Fine
	Volcanic sulphate	21[c]	Fine
	Nitrate	3.9[c]	Fine/coarse
	Organics from all biogenic VOCs	20–380[a, d]	Fine
Anthropogenic	Primary:		
	Industrial dust	363[e]	Fine/coarse
	Organics	6.3 – 15.3[a]	Fine
	Black carbon	3.6–6.0[a]	Mainly fine
	Biomass burning	29.0–85.3[a]	Fine
	Secondary:		
	Sulphate	122[c]	Fine
	Nitrate	14.2[c]	Fine/coarse
	Organics from VOC	24.6–100[f]	Fine

VOC, volatile organic compounds
[a] Adapted from IPCC (2013) (min–max)
[b] Best estimate
[c] Adapted from IPCC (2001)
[d] Expressed as TgC year^{-1}
[e] Ginoux et al. (2012)
[f] Kelly et al. (2018), Spracklen et al. (2011)

Table 7.4 Summary of sea-salt aerosol size distribution above 100 nm and constants for calculating the number as a function of wind speed.

Model	Dry diameter (μm)	a	b	Formation
Film drop	0.2–0.4	0.095	0.283	Bubble bursting
Jet	2–4	0.0422	−0.288	Bubble bursting
Spume	12	0.069	−5.81	Wind action

Source: taken from O'Dowd et al. (1997).

ripped from the ocean surface at high wind speeds. The number concentration, N, in each mode can be approximated by the following equation:

$$\text{Log} N = aU_{10} + b \tag{7.12}$$

where U_{10} is the wind speed at 10 m above the surface in $m\,s^{-1}$. Table 7.4 summarizes the above and gives an example of the values for the mean dry modal diameters and the constants a and b for one particular experiment.

The fairly global uniformity of seawater composition is reflected in the composition of sea-salt aerosol. The bulk composition of sea-salt aerosol is given in Table 7.1 and is mainly salts NaCl, KCl, $CaSO_4$, and Na_2SO_4. As a result of the general uniform trace aerosol chemical composition over the oceans, source regions cannot be easily identified. However, water-soluble and insoluble organic compounds may also be an important component, the fraction depending on a number of parameters, such as location and time of year. Cavalli et al. (2004) attributed an increase in organic mass in marine aerosol to biological activity and O'Dowd et al. (2004) suggested under these conditions that over 60% of the fine fraction mass could be organic in nature, although some of these organics may be secondary in nature. The presence of organics may affect the cloud-forming potential of these aerosol, which in turn can affect the single-scattering albedo.

Worked Example 7.5

A scientist measures sea-salt particles around the globe to try and estimate a mean production rate (PR) of particles ($N\,cm^{-2}\,s^{-1}$) and determines that the mean size of the sea-salt mass mode is 10 μm. If he assumes that all particles are this size, what average global PR would he calculate?

Answer

First, we know that the emission source strength of sea-salt particles is $1400 - 6800e^{12}\,kg\,year^{-1}$ from Table 7.3. If we take the mean value as the average flux ($4100e^{12}\,kg\,year^{-1}$), we can convert this to an average mass production rate by dividing it by the area covered by the Earth's oceans, which is ≈ 70% of the Earth's surface. This assumes the average production is uniformly spread over all the oceans. The radius of the Earth is ≈ 6400 km (most basic physics handbooks contains this information) and the surface area of a sphere is given by $4\pi r^2$. Therefore, the mass produced per unit area per year is:

$$\frac{4100e^{12}}{0.7 \times 4\pi \times \left(6400e^3\right)^2} = 11.40\,kg\,m^2\,year^{-1}$$

$$= 3.61e^{-11}\,kg\,cm^{-2}\,s^{-1}$$

We assume, in this example, that all particles are spherical and $10\,\mu m$ in diameter. The density of pure NaCL is $2.16\,\mathrm{g\,cm^{-3}}$, and so we can work out the average mass per particle, which is:

$$M_p = \frac{4}{3}\pi\left(\frac{d}{2}\right)^3\rho$$
$$= 1.13e^{-12}kg\ per\ particle$$

To calculate the number of particles per unit area per second (PR), we divide the mass production rate by the mass per particle ($3.61e^{-11}/1.13e^{-12}$), which gives ≈ 32 particles $\mathrm{cm^{-2}\,s^{-1}}$.

In reality, this number is likely to be too small for the total number produced and too large for $10\,\mu m$ particle flux. A $100\,\mu m$ particle will have 1000 times more mass than a $10\,\mu m$ particle, so the contribution from even a few of these larger particles will be significant and reduce the number of $10\,\mu m$ required. Similarly, for smaller particles, especially those below $1\,\mu m$, the number generated is a function of wind speed and is significantly more than $10\,\mu m$ particles (Eq. 7.12), but their contribution to the overall mass is negligible.

7.2.1.2 Mineral Dust

Wind-blown mineral dust from desert and semi-arid regions is an important source of tropospheric aerosols. Mineral dust arises from the physical and chemical weathering of rock and soils. The wind speed is the main controlling factor in entraining particles into the atmosphere, followed by other factors, such as soil moisture and surface composition and land use. Atmospheric size distributions in the vicinity of soil sources are generally bimodal, in which the range $d \approx 10$–$200\,\mu m$ consists mainly of quartz grains, and the range $d < 10\,\mu m$ of clay particles. Quartz grains will preferentially sediment close to their source, resulting in a fractionation process from quartz/clay to a clay aerosol with increasing distance. The size distribution will also change with downwind distance as a result of sedimentation. Measurements have indicated that the modal diameter changes from $d \approx 60$ to $100\,\mu m$ to $d \approx 2\,\mu m$ at a distance of $5000\,km$, where it appears to stabilize.

The principle elemental constituents of mineral dust (oxides and carbonates of Si, Al, Ca, Fe) bear a close resemblance to the average crustal composition and may be used to identify source regions. Due to the inert nature of mineral dust, chemical transformation processes in the atmosphere are thus considered minor, although surface chemical reactions with species such as sulphate and nitrate may be important.

In a project based in Africa in 2006, samples were taken of Saharan dust out-flowing from Africa to the Atlantic Ocean. It was proposed that as the dust deposited to the ocean, the iron content of the dust caused a nitrification of the ocean, which could lead to an increase in the level of certain algae. These algae in turn would remove CO_2 from the atmosphere. Following on from the previous section, this increase in biological activity could lead to an increase in the organic fraction of marine aerosol. Although this is a hypothesis and may be a minor effect, it highlights the complex interactions that can exist between different aerosol within the atmosphere.

Mineral dust is estimated to contribute between 1000 and $4000\,\mathrm{Tg\,year^{-1}}$ to the global atmospheric emissions (Table 7.3), which originated from areas totalling about 10% of the Earth's surface. This compares to a similar sea-salt emission from the oceans covering an area about 70% of the Earth's surface. Principle source regions of mineral dust cover about one-third of the land surface and include the Saudi Arabian peninsula, the USA Southwest and the Saharan and Gobi deserts. Images of two different mineral dusts are shown in Figure 7.2. Although mineral dust is considered a natural emission, it has been suggested that 20–50% of the current atmospheric burden may arise from disturbed soils through human activity (Tegen et al. 1996; Moulin and Chiapello 2006).

7.2.1.3 Primary Organic Aerosols/Biological Debris

Continental sources mainly arise from vegetation (plant waxes and fragments, pollen, spores, fungi, and decaying material), whereas marine sources consist of organic surfactants formed via bubble bursting. Typical size distributions are dominated by the coarse mode, although appreciable amounts of organic material can be found in submicron marine aerosol. Continental and marine source burdens are similar, with a total global emission of 50–1000 Tg year^{-1}, although the scarcity of data makes such an estimate very uncertain.

7.2.1.4 Volcanic Emissions

Although volcanic activity occurs on a sporadic basis and is mainly located in the Northern Hemisphere, outbreaks such as Mount St Helens (USA, 1980), El Chichón (Mexico, 1982), Mount Pinatubo (Philippines, 1991), and the Soufrière Hills volcanoes (Montserrat, 1997) have highlighted the importance of volcanic emissions to the atmosphere. These emissions are composed of ash (principally SiO_2, Al_2O_3, and Fe_2O_3), gases (SO_2, H_2S, CO_2, HCL, and HF) and water vapour. Mount Pinatubo was estimated to have emitted 9 Tg sulphur (S) compared with 3.5 Tg (S) for El Chichón, of which a large fraction in the form SO_2 was buoyantly injected into the stratosphere. Table 7.3 gives an estimated emission of 33 Tg year^{-1} for ash particles in the coarse mode, which mainly limits their impact to the regional scale. The minimum flux reported due to background eruptions is 4 Tg year^{-1}, whereas the maximum due to a large explosive eruption can be as high as 10 000 Tg year^{-1}. Of greater long-term importance is the emission of SO_2 into the stratosphere and the subsequent formation of H_2SO_4 aerosol droplets, which generally exhibit a bimodal structure in the Aitken accumulation mode size. Whereas aerosols in the troposphere have an average lifetime of five to seven days, due mainly to wet scavenging by clouds, stratospheric aerosols have lifetimes of up to several years, due to thermal stratification of the stratosphere and the absence of wet scavenging.

Stratospheric aerosol removal mechanisms are primarily sedimentation, subsidence and exchange with the upper troposphere. Although Table 7.3 indicates that explosive volcanic emissions may only contribute up to 10–20% of the total natural sulphur emission to the atmosphere, the radiative impact of stratospheric aerosols may be quite significant. Figure 7.8 illustrates the effect that El Chichón and Mount Pinatubo had on stratospheric optical properties. Peak values of the integrated aerosol backscatter, as measured with a ground-based LiDAR (Osborn et al. 1995), occurred about six months after each eruption. During this time, H_2SO_4 droplets of a sufficient size to interact with radiation were formed. A non-volcanic background aerosol is also evident in Figure 7.8 and is attributed to the formation of H_2SO_4 from the upward flux of COS (carbonyl sulphide) emitted from the oceans. A good review of the formation and properties of stratospheric aerosols is given by Pueschel (1996).

7.2.2 Natural Sources – Secondary Emissions

Secondary natural aerosols may be formed from a number of natural precursor gas sources, containing sulphur, nitrogen, and hydrocarbons. The main natural sources are the release of sulphur-containing species such as dimethyl sulphide, or DMS (CH_3SCH_3), from the oceans and biogenic VOCs. Dimethyl

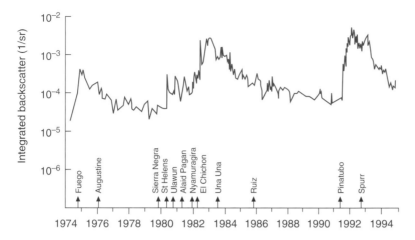

Figure 7.8 The integrated aerosol backscatter (a measure of the total stratospheric column of aerosol) using a ground-based LiDAR at l = 0.694 mm. Injections of aerosols associated with El Chichón and Mount Pinatubo are seen to be superimposed on a natural background. *Source:* from Osborn et al. (1995).

sulphide is formed from the biological activity of phytoplankton and eventually forms sulphate aerosol via the photooxidation to methanesulphonic acid and SO_2. The contribution to sulphate from DMS and other sources, except seawater, is known as NSS (no sea salt) sulphate to differentiate it from seawater as a source. The seasonal variation of DMS follows the ocean productivity cycle and may be a magnitude higher in the summer than in the winter season. Concentrations of NSS sulphate in the remote MBL generally range from 20 to 800 ng m^{-3} for the Southern Hemisphere oceans to 400–3000 ng m^{-3} for the northern Atlantic and illustrate the enhanced contribution from anthropogenic sources in the Northern Hemisphere (Heintzenberg et al. 2000). Typical global concentrations above the MBL rapidly decline with altitude to several ng m^{-3} in the troposphere. In comparison, typical values over the continents are <100 ng m^{-3}. These concentrations may be put into context by considering that <50% of sulphate in the MBL is of marine origin, whilst the rest may be attributed to soil dust and anthropogenic sulphate. The formation of nitrate aerosols from nitrogen precursor gases has two main natural sources: (i) oxides of nitrogen (NO_x) from lightning and soils; and (ii) nitrous oxide (N_2O) from bacterial activity in soils and the oceans. An emission rate of 3.9 Tg year^{-1} is estimated in Table 7.3. Secondary organic aerosols or SOAs are formed by oxidation products of volatile organic compounds (VOCs). These reactions are mainly initiated by reactions with ozone and OH and NO_3 radicals. Some products of these oxidation reactions have a low enough vapour pressure for partitioning between the gas phase and the aerosol phase to become significant. The equilibrium constant, K, describing the partitioning of a compound between the gas and the particle phase, is mainly a function of its vapour pressure and of its activity in the particulate phase.

$$K = \frac{R_G T}{\gamma \rho M_{OM}} \tag{7.13}$$

where R_G is the gas constant, T is the temperature, M_{OM} is the mean molecular weight of the organic particulate phase into which the compound partitions, γ is the activity coefficient of the compound in the particulate phase, and p is its vapour pressure. Thus, the formation of SOA is described by K for all compounds that are found in the organic aerosol phase. This, however, is a challenging task, because generally only a small fraction of the organic aerosol mass can be analysed on a molecular level.

Monoterpenes and isoprene emitted from plants are considered to be the largest class of natural VOCs able to form SOAs. Arneth et al. (2008) estimated the global emission of monoterpenes and isoprene as between 30 and 600 TgC year^{-1} (expressed as Tg of carbon), with isoprene dominating the flux, and total BVOC of approximately 1000 Tg (Guenther et al. 2012). From modelling work constrained by measurements, the global formation rate of SOA from BVOCs has been estimated at 20–380 Tg year^{-1} (Kanakidou et al. 2005; Spracklen et al. 2011). Information on the molecular composition of biogenic SOA is still largely unavailable. In general, the small amounts of sample available for analysis and the highly oxidized compounds make chemical analysis difficult. Studies concerning the composition of laboratory-generated aerosols at the molecular level have begun to resolve an appreciable amount of the organic particle mass. In ambient aerosol samples, only 10–20% of the organic aerosol mass can usually be analysed on a molecular level. In ambient samples it is difficult to distinguish between primary and secondary organic mass, as no simple experimental methods exist to separate these two fractions. It is assumed that SOA mass grows mainly on existing aerosol particles (such as salt aerosols or sulphuric acid nuclei). However, some events have been observed in field measurements where nucleation of new particles was attributed to low-volatility organic compounds (Marti et al. 1997). In addition, anthropogenic emissions may enhance the formation of SOA from BVOCs (Heald et al. 2011), which increases the uncertainty in the quantification of SOA formation.

7.2.3 Anthropogenic Sources – Primary Emissions

7.2.3.1 Biomass Burning
Natural wildfires and anthropogenic fires (e.g. for agricultural clearing) are commonly termed *biomass burning* (Levine 1991). The latter source has grown so quickly in the past two decades that it is estimated to account for 95% of biomass burning. Table 7.3 gives a range of anthropogenic components of 29–85.3 Tg year^{-1}, which consists of soot, sulphate, nitrate and incomplete combustion products containing carbonaceous compounds. Release of biomass products to the atmosphere occurs mainly in the tropics during the dry seasons, i.e. December to March in the Northern Hemisphere and June to September in the Southern Hemisphere. Although Table 7.3 considers only particle emissions, gaseous emissions of CO_2, CO, CH_4, and VOCs are also important, and the latter may result in SOA formation.

7.2.3.2 Industrial Aerosols
Aerosol emissions from industrial processes, estimated at 363 Tg year^{-1}, have a diverse number of sources. Major sources in industrialized countries include: coal and mineral dust from mining, aerosols formed from incombustible inorganic compounds in oil and coal fuels, stone crushing, cement manufacture, metal foundries and grain elevators. However, decreasing emissions in industrialized nations are occurring against rapidly rising emissions in emerging nations, where the implementation of modern technologies is not keeping pace with rapid economic and industrial development.

7.2.3.3 Black Carbon Aerosol
Soot is a ubiquitous component of the atmospheric aerosol. As a result of it being chemically relatively inert and having poor hygroscopic properties, it may be used as an anthropogenic tracer. For instance,

a recent study of soot in an ice-core from the Alps exhibited enhanced soot concentrations since the beginning of industrialization in the mid-nineteenth century and has been attributed to anthropogenic activity in Europe (Lavanchy et al. 1999). Soot aerosols are generated in incomplete combustion processes and consist mainly of an organic fraction and an inorganic graphite-like carbon fraction. These two fractions are often not clearly distinguished in the literature, which is in large part due to technical difficulties in distinguishing them. The graphite-like carbon fraction is also often known as black carbon (BC) or elemental carbon (EC), the distinction between the two normally reflecting the technique used to detect it. Depending on the source and burning conditions, the amount of soot is highly variable. For instance, biomass soot has an organic carbon content, in contrast to diesel soot with a high EC content. Emissions of anthropogenic black carbon aerosols from fossil-fuel combustion are estimated at $3.6–6.0\,Tg\,year^{-1}$.

7.2.4 Anthropogenic Sources – Secondary Emissions

7.2.4.1 Sulphate and Nitrate Aerosol

The main atmospheric source of secondary particles is the oxidation of SO_2 and NO_x. It is estimated that about 50% of SO_2 and NO_x is oxidized before being deposited (Langner and Rodhe 1991), although this value is highly uncertain and globally variable. Oxidation may occur in either the gas or condensed phases. Gas-phase oxidation to both H_2SO_4 and HNO_3 is dominated by the OH radical. For the condensed phase, about 50% of NO_x and $\geq$80% of SO_2 is oxidized to HNO_3 and H_2SO_4, respectively, by heterogeneous reactions. Oxidation of SO_2 always results in the formation of aerosol mass, due to the low H_2SO_4 vapour pressure, and is in contrast to HNO_3, which is distributed between the gas and aerosol phases. The chemical transformation of gases into particles depends on many factors, including chemical reaction kinetics and physical factors such as plume mixing and dispersion, oxidant concentration, sunlight and pre-existing aerosol surfaces. Despite this, the conversion rates of SO_2 are generally around 1–2% per hour and somewhat higher for nitrate. Table 7.3 indicates that current sulphate and nitrate emissions from anthropogenic sources exceed natural sources.

The largest source of secondary anthropogenic aerosol comes from fossil-fuel emissions of SO_2 and subsequent conversion to H_2SO_4. Over continental surfaces in the PBL or above, where gaseous ammonia is present, H_2SO_4 forms ammonium hydrogen sulphate (NH_4HSO_4) and ammonium sulphate (($NH_4)_2SO_4$). These components may exist simultaneously and are illustrated by varying molar ratios of NH_4^+/SO_4^{2-} according to the atmospheric aerosol type, as in Table 7.1. In regions of the stratosphere and upper troposphere, H_2SO_4 is found to be the major aerosol component. Atmospheric emissions from fossil-fuel combustion have been increasing since the beginning of industrialization in about 1850. Approximately 90% of these emissions arise in industrialized regions of the Northern Hemisphere. Little mixing occurs into the Southern Hemisphere due to a long interhemispheric mixing time of $\approx$1 year, compared with aerosol lifetimes of $\approx$1 week.

Current estimates of total NO_x emissions from natural sources ($\approx$11.3 Tg N year^{-1}; IPCC 2013) are lower than the total anthropogenic sources ($\approx$37.5 Tg N year^{-1}; IPCC 2013). The formation of HNO_3 from NO_x is a major removal mechanism for tropospheric NO_x as most HNO_3 is subsequently lost through wet and dry deposition.

Sulphate aerosol concentrations are more stable to fluctuations in H_2SO_4 concentration, temperature, and humidity conditions, in contrast to ammonium nitrate and chloride aerosols. For these aerosols, the reversible reactions shown below form the parent gaseous components under conditions of low atmospheric ammonia concentration, high temperature, and low humidity:

$$NH_4NO_3(s) \leftrightarrow NH_3(g) + HNO_3(g)$$
$$NH_4Cl(s) \leftrightarrow NH_3(g) + HCl(g)$$

where (g) and (s) denote the gaseous and solid phases, respectively. The main source of atmospheric HCl is from refuse incineration and coal combustion. Ammonia plays an important role in the neutralization of acid species, as it is the most common atmospheric alkaline gas. Conversion to ammonium salts is a function of not only altitude, but also of temperature and humidity. Major natural and anthropogenic sources respectively include: (i) soils and organic decomposition, (ii) fertilizers and animal farming, and (iii) catalysed vehicle emissions. The aqueous-phase production of aerosol material on cloud droplets is an important mechanism in nonprecipitating clouds. Cloud droplets, which form on CCN, may undergo on average 10 evaporation–condensation cycles before precipitable droplets are formed. During this process, gaseous species are scavenged and undergo chemical transformation, whilst aerosols and other droplets are scavenged by coagulation and phoresis mechanisms. As a result, the aerosol mass and hygroscopicity increases, in turn increasing the CCN activity. The conversion of SO_2 and NH_3, which are dissolved in droplets, appears to be an efficient process for the production of NH_4HSO_4 and $(NH_4)_2SO_4$. Such processes have been postulated as responsible for the rather uniform composition of the background tropospheric aerosol.

7.2.4.2 Secondary Organic Aerosol

Fossil-fuel combustion and biomass burning, caused by human activities, are the main sources of anthropogenic VOCs that can lead to SOAs. Understanding SOA, from their formation to evolution, provides one of the greatest challenges in modern aerosol science, especially in the urban environment where up to 70% of the organic matter can be secondary in smog events. Computer models, which are not constrained by measurement, consistently underpredict the concentrations in the atmosphere, and as a result, the ratio of secondary to primary organic matter may be much larger than previously reported.

The same principles apply to anthropogenic SOA formation as for SOA formed from biogenic precursor compounds. Aromatic compounds are the main class of anthropogenic VOCs that lead to significant aerosol formation. Pandis et al. (1992) estimated that aromatics of anthropogenic emissions can be responsible for about two-thirds of the total SOA formation in an urban atmosphere. Aerosol yields of different single aromatic compounds, when compared with the SOA yield of whole gasoline vapour, show that aromatics are mainly responsible for SOA from fossil-fuel VOC emissions. As in the case of biogenic aerosol particles, little is known about the molecular composition of anthropogenic SOA.

Within the urban environment where concentrations of SOA are significant, the effects of these pollutants have been linked to human health through epidemiological studies. Although the precise mechanisms of how these particles affect human health is still relatively unclear, several studies have shown a clear link between increasing levels of SOA and an increase in morbidity.

7.3 The Role of Atmospheric Particles

The most noticeable effect aerosols have on the atmosphere is in the production of clouds. When a particle reaches a certain size, it can become a CCN (see Figure 7.1), the core upon which all cloud droplets form. The properties of the CCN directly affect the properties of the cloud, which in turn affects the radiative properties of the cloud. This section summarizes how aerosol particles within the atmosphere affect heterogeneous chemistry and then looks in some detail at aerosol and climate forcing.

7.3.1 Heterogeneous Chemistry

Heterogeneous reactions are those involving more than one phase, such as reactions of gaseous molecules with compounds on a solid or liquid surface or in bulk liquids. Some reactions that are unfavourable in the gas phase are able to occur on the surfaces or in the bulk of atmospheric aerosol particles.

There are two different implications when considering heterogeneous chemistry, they modify the aerosol composition and they influence gas-phase chemistry. One of the major difficulties in investigating heterogeneous reactions in laboratory experiments is to accurately simulate relevant aerosols, as ambient particles are often complex mixtures of organic and inorganic compounds, and water.

The most basic example of a heterogeneous reaction, the adsorption of a gaseous molecule onto a solid particle surface, can be described with the following equation:

$$A(g) + \{B\} \rightarrow \{AB\}$$

where species {AB} denote surface-bound compounds. The rate with which gaseous molecules adsorb to aerosol particles is described by:

$$J = K_t(d) N_{mol} \qquad (7.14)$$

where J is the flux, $K_t(d)$ is the condensational loss rate and N_{mol} is the number concentration of the molecules of interest. Variable K_t can be written as:

$$K_t = \sum_{i=0}^{i=\infty} N_i(d) B(d) \qquad (7.15)$$

where $N_i(d)$ is the particle number concentration at diameter d and $B(d)$ is the attachment coefficient. Within the literature there exists several expressions for K_t and to consider them fully here is beyond the scope of this book; rather the reader is directed to Schwartz (1986). The expression given in Eq. (7.15) is a function of a dimensionless parameter, γ, known as the *uptake coefficient* or the *sticking coefficient*. A value of $\gamma = 1$ means that all the gas molecules will be uptaken by the aerosol. A value of $\gamma = 0.03$ means that approximately 3% of gas molecules interacting with the aerosol surface area will stick.

For values of $\gamma \leq 0.01$ and $d \leq 10\,\mu m$, $B(d)$ can be given by:

$$B(d) = \frac{1}{4}\pi \overline{c} d^2 \gamma \tag{7.16}$$

The adsorbed molecule may also undergo a chemical reaction on/in the aerosol, resulting in either an altered aerosol composition or a release of a reaction product into the gas phase, or both. Depending on the aerosol chemical and physical (i.e. solid or liquid) composition, heterogeneous reactions may be quite diverse, and are governed by many factors. A schematic of one cycle that is thought to be present in the atmosphere is given in Figure 7.9, a model of the processing of reactive halogen species in the MBL that illustrates the complexity of heterogeneous chemistry. In this proposed scheme, algae within the ocean release gaseous species containing bromine (Br), chloride (Cl), and iodine (I). These react with sunlight ($h\nu$) and gases such as ozone (O_3) to produce an intermediary that is taken up by the aerosol. These halogen-containing compounds undergo further chemical reactions within the aerosol, and are re-emitted into the atmosphere, but with increased number. It has been hypothesized that this scheme may lead to the depletion of O_3 in the MBL.

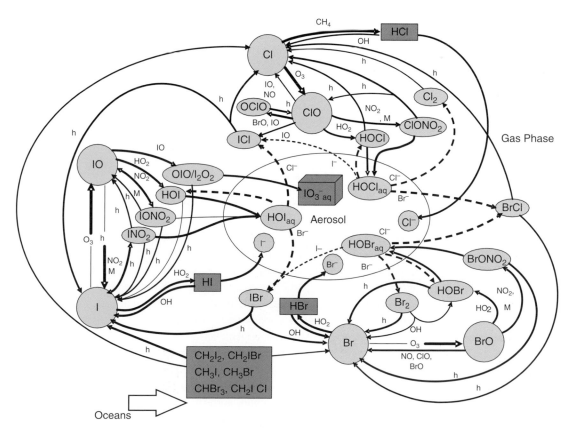

Figure 7.9 Schematic of proposed heterogeneous reactions of marine halogens and other gases and aerosol particles.

Equations (7.14)–(7.16) can be applied equally to dry and liquid aerosol. An important example of heterogeneous reactions in cloud droplets is the aqueous oxidation of SO_2. The main oxidation paths are reactions with aqueous H_2O_2 and ozone:

$$SO_2(g) \rightarrow SO_2(aq) \qquad \left(\text{uptake of gaseous } SO_2 \text{ to the droplet}\right)$$

$$SO_2(aq) + H_2O_2(aq) \rightarrow H_2SO_4 \qquad \left(\text{reaction within the droplet}\right)$$

$$SO_2(aq) + O_3(aq) + H_2O \rightarrow H_2SO_4 + O_2 \quad \left(\text{reaction within the droplet}\right)$$

Another important example of a heterogeneous reaction scheme in liquid aerosols is the night-time formation of nitric acid:

$$NO_2 + O_3 \rightarrow O_3 + NO_3 \qquad \left(\text{in the gas phase}\right)$$

$$NO_3 + NO_2 + M \rightarrow N_2O_5 + M \quad \left(\text{in the gas phase}\right)$$

$$N_2O_5(aq) + H_2O \rightarrow 2HNO_3 \qquad \left(\text{reaction within the droplet}\right)$$

In conclusion, heterogeneous reactions are important in atmospheric chemistry and much work – from atmospheric measurements to chamber studies to computer simulations – is continuing to further understand this complex subject.

7.3.2 Climate Forcing

Aerosols are recognized as being a major source of uncertainty in studies of global climate change (IPCC). Emissions of anthropogenic aerosols to the atmosphere may explain the lower observed temperature increase than is otherwise predicted for greenhouse gas emissions. Aerosols are considered to be responsible for a negative forcing or cooling of the Earth–atmosphere system, in contrast to a positive forcing, i.e. 'warming,' from greenhouse gases. In this definition, a forcing refers to a natural or anthropogenic perturbation in the radiative energy budget of the Earth's climate system.

As introduced in Chapter 3, aerosols may influence the atmosphere in two important ways, by direct and indirect effects. The climate effects of aerosols are still poorly quantified and it is at present still unclear what magnitudes are involved. Figure 7.10 illustrates recent estimates of the mean annual radiative forcing for various climate-change mechanisms, averaged globally since 1750–2011 (IPCC 2013). When viewed on a global scale, shortwave radiative forcing from anthropogenic aerosols is considered to offset part of the longwave radiative forcing due to greenhouse gases. However, the same argument cannot be applied on a regional scale for a number of reasons. First, greenhouse gases have lifetimes measured in decades to centuries and are globally well mixed, in contrast to aerosols, which have lifetimes of less than one week on average and are geographically variable in extent. Second, aerosol forcing responds more rapidly to changes in aerosol emissions than greenhouse forcing, which is still influenced by accumulated past emissions. Third, aerosol radiative forcing is more restricted to source and downwind regions, which may then experience an overall negative forcing. Last, greenhouse-gas forcing is not as diurnally and seasonally

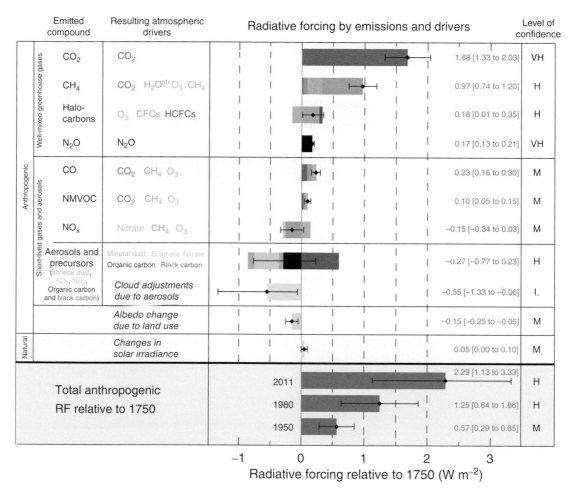

Figure 7.10 Figure SPM.5 from IPCC (2013). Radiative forcing estimates in 2011 relative to 1750 and aggregated uncertainties for the main drivers of climate change. Values are global average radiative forcing (RF). The best estimates of the net radiative forcing are shown as black diamonds with corresponding uncertainty intervals. Volcanic forcing is not included as its episodic nature makes is difficult to compare to other forcing mechanisms. Total anthropogenic radiative forcing is provided for three different years relative to 1750.

variable as aerosol forcing, which has a greater influence during: (i) daylight, (ii) cloudless conditions, and (iii) summer. Such difficulties greatly hinder the modelling of aerosol direct and indirect effects.

7.3.3 Direct Aerosol Effects: Estimates of Direct Forcing

Direct effects refer to the scattering and absorption of shortwave radiation and the subsequent influence on the climate system and planetary albedo. The application of different models in assessing the influence of atmospheric aerosols is made difficult by the large spatial and temporal variation in their properties. Most models predict a regional offset of greenhouse forcing by sulphate aerosols in the industrialized regions of the eastern USA, central Europe, and eastern China. However, the regional forcing is not expected to be indicative of regional climate response, as atmospheric circulation may result in a nonlocal response to local forcing.

The IPCC 2013 estimate the direct radiative forcings to be: -0.4 ($-0.6 - -0.2$) Wm^{-2} for sulphate, $+0.4$ ($+0.05 - +0.8$) Wm^{-2} for black carbon, -0.12 ($-0.4 - +0.1$) Wm^{-2} for primary and SOA, 0.0 ($-0.2 - +0.2$) Wm^{-2} biomass burning, -0.11 ($-0.3 - -0.03$) Wm^{-2} for nitrate aerosol and -0.1 ($-0.3 - +0.1$) Wm^{-2} for mineral dust. In general, the variation in the above estimates is mainly due to the differing sophistication of various models, whilst the uncertainty and confidence levels (Figure 7.10: VH – very high, H – high, M – medium, L – low, VL – very low) depends on assumed optical properties and the global modelled distribution of aerosol species. Sulphate and biomass aerosols have been most widely modelled to date, but additional aerosol types, such as volcanic emission, mineral dust, and carbonaceous aerosols, are being increasingly included.

Volcanic emissions to the stratosphere influence the climate by warming the lower stratosphere through aerosol absorption and by reducing the net radiation transmitted to the troposphere and surface. However, as stratospheric aerosol residence times are of the order of one to two years, the radiative influence is also restricted to similar time scales. The El Chichón (1982) and Mount Pinatubo (1991) eruptions allowed the climate effect of a large transient forcing to be studied. Model temperature predictions over short time scales were found to be in reasonable agreement with observations. Mount Pinatubo is estimated to have contributed a maximum forcing of $-4Wm^{-2}$ and about $-1Wm^{-2}$ up to two years later, which illustrates the large transient cooling effect when compared with other radiative forcing mechanisms. It is estimated that Mt. Pinatubo lead to a temporary decrease in the mean global temperature of around 0.1 K.

7.3.4   Indirect Aerosol Effects: Aerosol Effects on Clouds

Indirect effects refer to a complex positive feedback system, whereby an increase in CCN arises from an increase in aerosol concentration. As a consequence, cloud droplets become smaller for a given cloud liquid water content (LWC), thereby increasing cloud albedo and resulting in a negative forcing. The principal aerosol–cloud interactions are summarized schematically in Figure 7.11, which illustrates the life cycle in marine and continental air, respectively. Many of the physical and chemical aspects of the life cycle have already been considered in previous sections.

The formation of NSS sulphate via the emission of DMS has been proposed as a cloud–climate feedback mechanism. The present greenhouse warming of the Earth's surface–atmosphere is considered to warm the ocean surface waters, which in turn leads to an increase in phytoplankton activity and hence DMS emissions. As a result of increased CCN concentration due to aerosol formation from DMS, the albedo of marine stratiform clouds may increase (Twomey 1977), thereby offsetting a global temperature rise. Although the mechanisms and different pathways are complex, the large aerial extent of stratiform clouds, covering 25% of the oceans, renders such a feedback mechanism of potential importance. Cloud lifetimes and precipitation frequencies are also thought to be affected. However, the aerosol–climate–DMS feedback theory has not been conclusively proven to date.

In order to assess the indirect effects of aerosols, cloud properties are considered further. The critical supersaturation at which CCNs become activated depends on aerosol composition, size and age, and may be described by the *Köhler theory*. As RH increases beyond 100% the aerosol droplets continue to grow until a critical supersaturation S_C is reached, corresponding to a critical diameter d_{CRIT} at which the aerosol

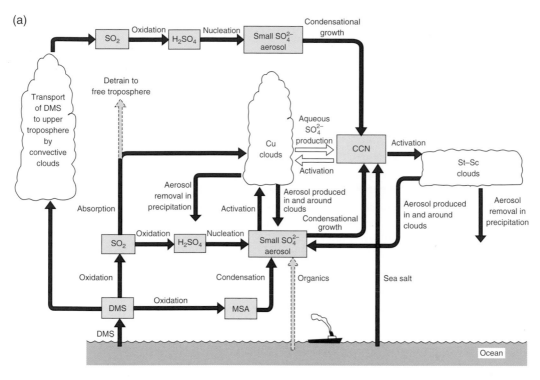

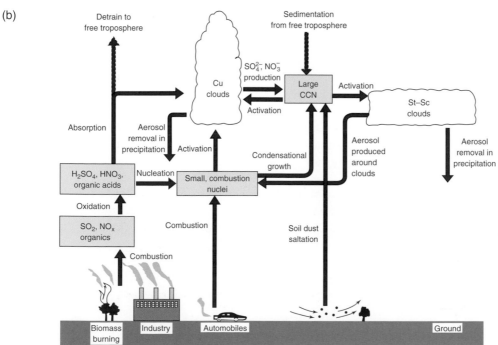

Figure 7.11 Schematic of aerosol–cloud interactions for (a) marine air and (b) continental air. Cloud types: Cu, cumulonimbus; St, stratus; Sc, stratocumulus. CCN, cloud condensation nuclei; DMS, dimethyl sulphide. *Source:* from Hobbs (1993).

becomes *activated*. At this point droplets are in an unstable equilibrium with their environment and may grow uncontrollably or return to a stable equilibrium, where they will exist as unactivated droplets or haze.

The value of S_C depends on the solubility of chemical components and the dry diameter of the aerosol. The larger both parameters are, the lower the critical supersaturation required to activate the aerosol. As a result the larger, hygroscopic aerosols tend to form CCN first. Typical supersaturations in marine stratus clouds, estimated at 0.1%, imply that dry aerosol diameters with $d > 0.1\,\mu m$ will be activated to produce cloud droplets, whereas for $d < 0.1\,\mu m$ the aerosol will remain interstitial to cloud droplets.

Effective CCN sources are, in general, secondary aerosols of either natural or anthropogenic origin. Non-sea-salt sulphate is considered to be the major natural source, whilst biomass/organic and sulphate aerosols are major anthropogenic sources. The ability of pure inorganic salts and acids such as nitrates, sulphates or sulphuric acid to act as CCN is relatively well investigated in experimental and modelling work. The role of organic aerosols or the mixture of organic and inorganic particles, however, is less well characterized and largely dependent on the hygroscopic behaviour of the organic compounds. Studies have shown that smoke particles from biomass burning exhibit a CCN activation of up to 100%, in contrast to an activation of less than several percent for fresh soot from petroleum fuel. Other studies suggest that organics mixed with inorganic salt aerosols alter their hygroscopic behaviour. Organics were found to reduce the critical supersaturation needed to activate inorganic aerosols (Topping et al. 2006). Others found that even large amounts of hydrophobic organics did not affect the hygroscopic growth of inorganic particles below 100% humidity. Aerosol from biomass burning containing large parts of organics have been shown to act as CCN. Field measurements, for example in the Amazon region, found significant influence of biomass burning events and CCN concentrations (Kaufman et al. 1998). The role of organics seems to be an important but as yet unresolved aspect in the activation of ambient aerosols to cloud droplets.

Cloud physical properties over continental regions differ from those over oceans. Over land, larger CCN concentrations result in increased cloud droplet concentrations (N_{CLOUD}), and since the LWC of both cloud types are similar, continental clouds have a smaller average droplet size than marine clouds. Observations indicate that marine cumulus clouds have a median value $N_{CLOUD} \approx 45\,cm^{-3}$ and a broad droplet size spectrum with median at $d \approx 30\,\mu m$, whereas continental cumuli have a median value $N_{CLOUD} \approx 230\,cm^{-3}$ and a narrower size spectrum with median $\approx 10\,\mu m$. Parameterizations of clouds use the *cloud optical thickness* τ_c, which is defined as the attenuation of light (radiation) passing through a cloud due to absorption and scattering by the cloud droplets and may be given as:

$$\tau_c = \frac{L}{r_e} \tag{7.17}$$

where L is the cloud LWC and r_e is the effective radius of cloud droplets. The value of L may be approximated from Eq. (7.18) and, when solving for r_e and differentiating, Eq. (7.19) results in the following:

$$L \approx \frac{4}{3}\pi r_e^{3} N_{CLOUD} \tag{7.18}$$

$$\frac{\Delta \tau_c}{\tau_c} = \frac{1}{3}\frac{\Delta N_{CLOUD}}{N_{CLOUD}} \tag{7.19}$$

Equation (7.18) calculates the volume of the cloud drop with radius r_e, multiples it by the number of droplets, and assumes a density of $1\,g\,m^{-3}$ to approximate the LWC.

Worked Example 7.6

A layer of aerosol is advected off the coast of southern England into the English Channel where it mixes with the naturally occurring sea-salt particles and is then lofted up and clouds begin to form. What effect do these anthropogenic aerosols have on the cloud optical thickness? Is it more or less likely to be a precipitating cloud with the addition of these aerosol to the naturally occurring sea-salt particles? (Assume that all the particles are effective CCN.)

Answer

The first thing to consider is the cloud LWC. Assuming that the amount of free water available to condense onto the particles has not changed, then that implies the LWC of the cloud will be the same. However, the number of aerosol particles has increased, so the number of droplets will also increase. For Eq. (7.18) to remain constant, this means the effective radius must decrease. Therefore, the cloud optical thickness will increase (Eq. 7.17). As for the probability of the cloud precipitating, to a first approximation it is less likely to precipitate. A raindrop is effectively a cloud drop that has grown too large to be suspended in the atmosphere. If the effective radius, and hence mass, has decreased, then it is likely to remain suspended in the atmosphere longer.

A further cloud parameter of importance is the albedo, A_0, which is the ratio of the scattered to incident solar radiation, and can be approximated by:

$$A_0 \approx \frac{\tau_c}{\tau_c + 6.7} \tag{7.20}$$

Hence, a greater change in albedo occurs when τ_c is low. Using the above equations, and assuming that the cloud LWC and depth remain constant, then a parameter known as the *susceptibility* may be derived:

$$\frac{\Delta A_0}{\Delta N_{CLOUD}} = \frac{A_0(1 - A_0)}{3N_{CLOUD}} \tag{7.21}$$

The susceptibility $\Delta A_0/\Delta N_{CLOUD}$, which defines the change in albedo with a change in the number of cloud droplets, is most sensitive to change when N_{CLOUD} is small and when A_0 lies between ≈ 0.25 and 0.75, conditions that are both typical of marine clouds. Hence, small increases in anthropogenic CCN concentrations are likely to have a greater influence in marine than continental regions, i.e. in the Southern Hemisphere. As clouds are already optically thick to longwave radiation, only shortwave radiation is influenced by cloud properties. As a consequence, enhancement of the shortwave albedo is considered to result in increased reflection of solar radiation back to space and a cooling of the Earth's surface.

The susceptibility of marine stratiform clouds to increased CCN concentrations has been observed in satellite images where ship-stack exhausts have resulted in an increase in cloud albedo. The increase in droplet concentration and reduction in size have been confirmed simultaneously by in-situ and remote-sensing measurements. Observational evidence suggests that for a factor 10 increase in the aerosol size range $d \approx 0.1$–$0.3\,\mu m$, a two- to fivefold increase in droplet concentration results. An additional important effect, as a result of the above processes, is a reduction in precipitation efficiency and increased cloud lifetime (Albrecht 1989). Such observations are at present difficult to quantify and the complex processes involved are, as yet, still very poorly understood.

7.3.5 Indirect Aerosol Effects: Estimates of Indirect Forcing

As a consequence of the large uncertainties in the interaction of aerosols with clouds, an estimate of the sign and magnitude of indirect forcing is fraught with assumptions. The 2007 IPCC report estimates a negative global mean indirect forcing, with the range of model results varying widely, from -0.22 to $-1.85\,\mathrm{W\,m^{-2}}$. There are considerable differences in the treatment of aerosol, cloud processes, and aerosol–cloud interaction processes in these models.

Amongst the numerous new uncertainties and findings that have been highlighted in recent studies, several investigations have provided evidence that the incorporation of soot aerosols from biomass burning may be influencing cloud albedo. For example, a satellite study of cumulus and stratocumulus clouds over the Amazon basin during the burning season showed that the average droplet diameter decreased from 14 to $9\,\mu\mathrm{m}$, with a corresponding increase in cloud reflectance from 0.35 to 0.45 when smoke aerosols were present (Kaufman and Fraser 1997).

7.4 Aerosol Measurements

Having discussed some of the key properties and sources of aerosols, it is important to understand the techniques used to sample and measure these particles. It is evident when considering all the instruments and techniques that exist to quantify the different properties of aerosol that no one instrument will tell you everything about the ambient particles; it is therefore vital to know what properties you wish to study in what size range to decide which path to take when observing particles. Presented in this section is a selection of techniques used to determine the size, number, chemical composition, and optical properties of ambient aerosol particles. As well as the references already cited, a good review of the state of the art in aerosol mass spectrometry is given by Coe and Allan (2006).

7.4.1 Aerosol Size and Number

The most basic properties of ambient particles are size and number (number of particles per unit volume). As the size range covered by ambient particles spans several orders of magnitude, several different types of measurement instrument are required.

7.4.1.1 Optical Particle Counters

For many years, the most common method of counting and sizing particles has been achieved using optical particle counters (OPCs). An OPC consists of a light source, focusing optics, and photodetectors. A stream of aerosol particles is passed through a light source and scatters the light onto a photodetector. The photodetector converts this light signal into a voltage. By recording the number of events, it is possible to count the number of particles passing through the light beam. By monitoring the amplitude of the voltage, it is possible to determine the size of the particles. The sizing of OPCs is based on Mie theory (Mie 1908) (see Chapter 3), which states that the intensity of scattered light at a given angle is proportional to the diameter and applies to particles in the size range of $\approx 100\,\mathrm{nm}$ to $100\,\mu\mathrm{m}$. OPCs are single particle, fast-response instruments. They can provide a complete number–size distribution every second. However, the intensity of scattered light is not only dependent on the size of the particle, it is also dependent on the shape and refractive index. In converting the light signal to a diameter, the refractive index of the particles must be known or assumed. For ambient particles, where there can be a range of refractive indices, this may cause errors in the sizing if only one value is used.

7.4.1.2 Condensation Particle Counters

Figure 7.12 shows a schematic of a TSI Inc. model 3790 Condensation Particle Counter, which is a modified OPC. The section housing the laser diode, lenses, and photodetector in Figure 7.12 is essentially an OPC. The laser diode (light source) is focused into the aerosol stream, and the scattered light is then directed onto the photodetector. The photodetector records the number of scattering events and hence counts the number of particles. The instrument shown in Figure 7.12 does not report the size of the particles, merely the number concentration. Unlike an OPC, the CPC can count the number of particles with sizes down to 10 nm. This is achieved by first growing the particles to a detectable size. The particles are drawn in and exposed to a supersaturated environment of reagent-grade n-butyl alcohol (denoted butanol). The butanol is absorbed into the wick and warmed by the heated saturator and enters the aerosol stream as a vapour. The butanol vapour condenses onto the particles, causing them to grow into the micron size range, which are then easily counted in the detector region. By first growing the particles, the initial size information is lost. Other CPCs similar to the TSI 3790 can count particles as small as 2.5 nm.

Figure 7.12 Schematic of a TSI Inc. Model 3790 condensation particle counter.

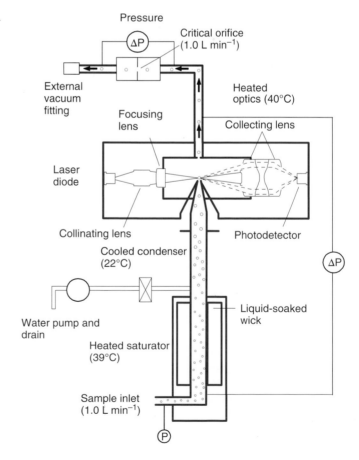

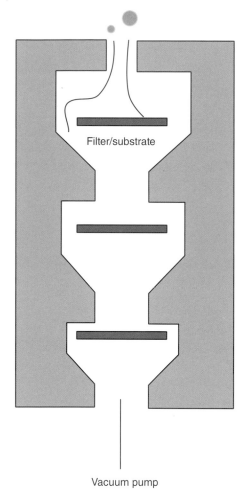

Filter/substrate

Vacuum pump

Figure 7.13 Schematic of an inertial cascade impactor.

7.4.2 Aerosol Composition

The instrumentation required to determine the chemical composition is equally varied, ranging from laser-based devices to samples collected on a filter. When considering the advances in aerosol technology in recent years, the development of instruments and procedures to determine the chemical composition of a particle has been the greatest. With the advancement of this technology, the ability to link other properties such as single-scattering albedo and hygroscopic GFs to chemical composition is now possible. In general, there are two approaches for determining the chemical components that make up an aerosol particle, and within each approach, several methods of extracting the chemical information.

The first of these approaches is to use what is known as *offline analysis*. This method samples aerosol particles onto a variety of filter papers or substrates collected over several hours, which are later analysed in a laboratory. The simplest method for collection is to attach a substrate or filter mounted in a holder to a pump to suck air through, effectively a large vacuum cleaner with a filter on the inlet. The filter will have a certain pore size, and all particles larger than that size will be retained on the filter. A more common method is to size segregate the samples by either having a series of filters with different pore sizes or have the substrates arranged in an *impactor*, as shown in Figure 7.13, which is a type of inertial classifier known as a *cascade impactor*. Air is drawn through the impactor and round the substrates, the air accelerating through each successive stage. At the first stages, the largest particles have too much momentum to follow the air streamline, so slip through and impact onto the substrate. As the particle laden air travels to the lower stages, the size at which particles will impact, the cut-off size, decreases. The substrates can then be analysed using a number of techniques such as gravimetric analysis or ion chromatography, depending on what information is required. Inertial classifiers separate particles based on their aerodynamic diameter.

The processes described above are amongst the easiest and cheapest techniques for sampling aerosol particles and have been established for a long time. More recently, work has been advancing in the field of online aerosol characterization, in particular in the field of online aerosol mass spectrometry. Online aerosol mass spectrometry allows analysis of ambient or laboratory particles on time scales of several minutes rather than several hours. The basic principles of operation are to sample aerosol into a low-pressure chamber, vaporize them to produce a kinetic gas, ionize the gas, and analyse with a mass spectrometer. The method of vaporization and ionization is quite different, depending on the instrument used. One method is to impact the aerosol on a heated surface to vaporize the particles and pass them over a tungsten filament emitting electrons at 70 eV. This method is favoured by the Aerodyne Research Inc Aerosol Mass Spectrometer (AMS) (http://www.aerodyne.com). Other methods combine the vaporization and ionization processes by ablating the particles with a high-powered laser. Although both techniques are based on mass spectrometry, they yield different information about the aerosol. The AMS provides quantitative, size resolved, chemical composition of the non-refractory aerosol. Non-refractory refers to anything that can vaporize on the heater, which is nominally set to 600 °C, e.g. nitrates and organics, but not crystal materials, metals, and soot. The laser ablation techniques cannot provide quantitative information, but it does yield a complete mass spectrum on a single particle basis, which includes the material the AMS cannot detect, and can determine if particles are internally or externally mixed.

7.4.3 Aerosol Optical/Radiative Properties

Nephelometry is a technique used to study the direct radiative properties of particles and integrating nephelometers are used to measure the scattering coefficient of ambient particles. The instruments in many ways are similar to OPCs, in so much as they have a light source and photodetectors, but instead of determining the size of the particles, a nephelometer measures the forward and back scatter. Particles are drawn into a sample volume, where they are illuminated by the light source. By knowing the intensity of the light source with no particles present, the scattering can be determined.

Another useful atmospheric monitoring device is a sun photometer. A sun photometer measures the direct solar irradiance reaching the detector, normally located at the Earth's surface. The intensity of sunlight at the top of the atmosphere is constant, so under cloud-free conditions, measuring the intensity of light at the surface and knowing the depth of the atmosphere reveals information about the amount of particles present and yields the total extinction. This, in turn, can be used to calculate the AOD. These techniques are used in the global AERONET network (https://aeronet.gsfc.nasa.gov), which provides global (current and historic) AOD measurements and other metrics. The operation of both the sun photometers and nephelometers are wavelength dependent, and this feature can be used to reveal information about the size of the particles. The relative scattering at visible red versus blue wavelengths, for example, is size dependent. So if the instruments can resolve these different wavelengths, then information of the size is gained.

7.5 Summary

Atmospheric aerosol particles (which are any solid or liquid or combination of both suspended in the atmosphere) exist in all regions of the globe and from sea level into the stratosphere. They have a multitude of sizes, shapes, compositions, and sources, and all of these factors contribute to how these particles interact with each other, the environment, and indeed, ourselves. Their size can range from a few nanometres, where the particles are nothing more than a cluster of molecules, to several hundred microns. The composition of aerosols varies significantly, with pure sulphuric acid particles being formed in the stratosphere to sodium chloride particles being generated over oceans to organic-rich particles emanating from our cities. The relative importance of the different source regions often depends on the scale of interest. A city such as London may be producing over $100\,000\,cm^{-3}$ and when compared with a similar sized area in the mid-Atlantic, which may be producing only several $100\,cm^{-3}$, then regionally the emissions from the city dominate. However, as oceans cover approximately 70% of the Earth's surface, globally the burden of marine aerosol is significant.

Understanding the physical and chemical properties of aerosol particles is vital if you are to understand their impact on human health and climate change, and this chapter presented a brief introduction to aerosol measurements with an overview of some of the techniques available to probe aerosol properties. Not included was a discussion on the limitations and problems one encounters when measuring ambient particles. For instance, sampling aerosol down a pipe into an instrument leads to diffusional loses to the walls of small particles, and gravitational settling of large particles. Similarly, in the impactor, an artefact that is often observed is particle bounce. Certain particles can bounce between the stages. That is why is it important to have a good understanding of the aerosol properties as described at the beginning of this chapter. This allows you to look at the sampling system or instrument in an objective manner and ascertain whether you are measuring real particles or artefacts. It is also important to know what physical or chemical properties you want to observe, and why, so you can choose the correct instrument for the task.

Finally, it is worth remembering that aerosol particles still present the largest uncertainty in understanding climate change, both in our level of scientific understanding and in the size of the forcing. Only by linking the chemical properties, such as composition and interaction with gases, to the physical properties will we be able to better quantity their effects on our climate.

Acknowledgement

This chapter is based on Chapter 7 in *Atmospheric Science for Environmental Scientists,* first edition, 2009, by P. I Williams and U. Baltensperger.

Questions

1   Using Eq. (7.3), plot a graph of C_c versus diameter and use this to determine above which size the correction factor is negligible (i.e. $C_c \approx 1$).

2   Compare the diffusional coefficient and the average distance travelled by a 10 nm particle with the 0.1 and 1.0 μm particles given in worked example 7.2.

3   Estimate the number of particles at the peaks in the three-mode sea-salt number distribution at $15\,m\,s^{-1}$ (refer to Eq. (7.12) and Table 7.4).

4   Assume that all mineral dust can be categorized into either Saharan-like or Gobi-like dust and that the Saharan-like comprises 65% of the Earth's sources and the Gobi-like 35% (this is a simplification that in reality does not hold). From the literature, determine a mean diameter for both sources and estimate an average global flux of mineral dust. A good starting point is the 'Web of Science' where you can search scientific journals. How does this compare with the reported mineral dust flux from different regions?

References

Albrecht, B.A. (1989). Aerosols, cloud microphysics, and fractional cloudiness. *Science* 245: 1227–1230.

Allan, J.D., Williams, P.I., Najera, J. et al. (2015). Iodine observed in new particle formation events in the Arctic atmosphere during ACCACIA. *Atmospheric Chemistry and Physics* 15: 5599–5609.

Arneth, A., Monson, R.K., Schurgers, G. et al. (2008). Why are estimates of global terrestrial isoprene emissions so similar (and why is this not so for monoterpenes)? *Atmospheric Chemistry and Physics* 8: 4605–4620.

Baron, P.A. and Willeke, K. (eds.) (2011). *Aerosol Measurement*, 3e. New York, NY: John Wiley & Sons.

Cavalli, F., Facchini, M.C., Decesari, S. et al. (2004). Advances in characterization of size-resolved organic matter in marine aerosol over the North Atlantic. *Journal of Geophysical Research* 109: D24215.

Charlson, R.J. and Heintzenberg, J. (eds.) (1999). *Aerosol Forcing of Climate*. Chichester, UK: John Wiley & Sons.

Coe, H. and Allan, J.D. (2006). Mass spectrometric methods for aerosol composition measurements. In: *Analytical Techniques for Atmospheric Measurement* (ed. D.E. Heard), 265–310. Oxford, UK: Blackwell Publishing.

Ginoux, P., Prospero, J.M., Gill, T.E. et al. (2012). Global-scale attribution of anthropogenic and natural dust sources and their emission rates based on MODIS deep blue aerosol products. *Reviews of Geophysics* 3 (50): RG3005.

Guenther, A.B., Jiang, X., Heald, C.L. et al. (2012). The model of emissions of gases and aerosols from nature version 2.1 (MEGAN2.1): an extended and updated framework for modeling biogenic emissions. *Geoscientific Model Development* 5: 1471–1492.

Heald, C.L., Coe, H., Jimenez, J. et al. (2011). Exploring the vertical profile of atmospheric organic aerosol: comparing 17 aircraft field campaigns with a global model. *Atmospheric Chemistry and Physics* 11: 12673–12696.

Heintzenberg, J., Covert, D.C., and Van Dingenen, R. (2000). Size distribution and chemical composition of marine aerosols: a compilation and review. *Tellus* 52B: 1104–1122.

Hinds, W.C. (1999). *Aerosol Technology: Properties, Behavior, and Measurements of Airborne Particles*. New York, NY: John Wiley & Sons.

Hobbs, P.V. (ed.) (1993). *Aerosol–Cloud–Climate Interactions*. San Diego, CA: Academic Press.

IPCC (1999). *Aviation and the Global Atmosphere*. Cambridge, UK: Cambridge University Press.

IPCC (2001). *Climate Change 2001: The Scientific Basis*. Cambridge, UK: Cambridge University Press.

IPCC (2013). Summary for policymakers. In: *Climate Change 2013: The Physical Science Basis. Working Group I Contribution to the Fifth Assessment Report of the Intergovernmental Panel on Climate Change* (eds. T.F. Stocker, D. Qin, G.-K. Plattner, et al.), 373. Cambridge, UK, and New York, NY: Cambridge University Press.

Jaenicke, R. (1988). Atmospheric physics and chemistry. In: *Meteorology: Physical and Chemical Properties of Air* (ed. G. Fischer), 391–451. Berlin, Germany: Springer-Verlag.

Kanakidou, M., Seinfeld, J.H., and Pandis, S.N. (2005). Organic aerosol and global climate modelling: a review. *Atmospheric Chemistry and Physics* 5: 1053–1123.

Kaufman, Y.J. and Fraser, R.S. (1997). The effect of smoke particles on clouds and climate forcing. *Science* 277: 1636–1639.

Kaufman, Y.J., Hobbs, P.V., Kirchhoff, V.W.J.H. et al. (1998). Smoke, clouds, and radiation – Brazil (SCARB) experiment. *Journal of Geophysical Research* 103: 31783–31808.

Kelly, J.M., Doherty, M.R., C'Connor, F.M., and Mann, G.W. (2018). The impact of biogenic, anthropogenic, and biomass burning volatile organic compound emissions on regional and seasonal variations in secondary organic aerosol. *Atmospheric Chemistry and Physics* 18: 7393–7422.

Langner, J. and Rodhe, H. (1991). A global threedimensional model of the tropospheric sulfur cycle. *Journal of Atmospheric Chemistry* 13: 255–263.

Lavanchy, V.M.H., Gäggeler, H.W., Schotterer, U. et al. (1999). Historical record of carbonaceous particle concentrations from a European high-alpine glacier (Colle Gnifetti, Switzerland). *Journal of Geophysical Research* 104: 21227–21236.

Levine, J.S. (ed.) (1991). *Global Biomass Burning: Atmospheric, Climatic and Biospheric Implications.* Cambridge, MA: MIT Press.

Marti, J.J., Weber, R.J., McMurry, P.H. et al. (1997). New particle formation at a remote continental site: assessing the contributions of SO2 and organic precursors. *Journal of Geophysical Research* 102: 6331–6339.

McFiggans, G.P., Artaxo, P., Baltensperger, U. et al. (2006). The effect of physical and chemical aerosol properties on warm cloud droplet activation. *Atmospheric Chemistry and Physics* 6: 2593–2649.

Mie, G. (1908). Beigrade zur optic truber medien, speziell kolloidaler metallosumgen. *Annalen der Physik* 4 (25): 377–446.

Moulin, C. and Chiapello, I. (2006). Impact of human-induced desertification on the intensification of Sahel dust emission and export over the last decades. *Geophysical Research Letters* 33 (18) Art. No. L18808 SEP 23.

O'Dowd, C., Smith, M.H., Consterdine, I.E., and Lowe, J.A. (1997). Marine aerosol, sea-salt, and the marine sulfur cycle: a short review. *Atmospheric Environment* 31: 73–80.

O'Dowd, C.D., Facchini, M.C., Cavalli, F. et al. (2004). Biogenically driven organic contribution to marine aerosol. *Nature* 431 (7009): 676–680.

Osborn, M.T., DeCoursey, R.J., Trepte, C.R. et al. (1995). Evolution of the Pinatubo volcanic cloud over Hampton, Virginia. *Geophysical Research Letters* 22: 1101–1104.

Pandis, S.N., Harley, R.A., Cass, G.R., and Seinfeld, J.H. (1992). Secondary organic aerosol formation and transport. *Atmospheric Environment* 26: 2269–2282.

Pueschel, R.F. (1995). Atmospheric aerosols. In: *Composition, Chemistry and Climate of the Atmosphere* (ed. H.B. Singh), 120–175. New York, NY: Van Nostrand Reinhold.

Pueschel, R.F. (1996). Stratospheric aerosols: formation, properties, effects. *Journal of Aerosol Science* 27: 383–402.

Schwartz, S.E. (1986). Mass-transport considerations pertinent to aqueous phase reactions of gases in liquid-water clouds. In: *Chemistry of Multiphase Atmospheric Systems*, NATO ASI Series G6 (ed. W. Jaeschke), 415–471. Berlin, Germany: Springer-Verlag.

Seinfeld, J.H. and Pandis, S.N. (2006). *Atmospheric Chemistry and Physics: From Air Pollution to Climate Change.* New York, NY: John Wiley & Sons. ISBN-10: 0471720186.

Singh, H.B. (ed.) (1995). *Composition, Chemistry and Climate of the Atmosphere.* New York, NY: Van Nostrand Reinhold.

Spracklen, D.V., Jimenez, J.L., Carslaw, K.S. et al. (2011). Aerosol mass spectrometer constraint on the global secondary organic aerosol budget. *Atmospheric Chemistry and Physics* 11: 12109–12136.

Tegen, I., Lacis, A.A., and Fung, I. (1996). The influence on climate forcing of mineral aerosols from disturbed soils. *Nature* 380: 419–422.

Topping, D.O., McFiggans, G.B., and Coe, H. (2005). A curved multi-component aerosol hygroscopicity model framework: part 1 – inorganic compounds. *Atmospheric Chemistry and Physics* 5: 1205–1222.

Topping, D.O., McFiggans, G.B., Kiss, G. et al. (2006). Surface tensions of multi-component mixed inorganic/ organic aqueous systems of atmospheric significance: measurements, model predictions and importance for cloud activation predictions. *Atmospheric Chemistry and Physics Discussions* 6: 12057–12120.

Twomey, S.A. (1977). The influence of pollution on the short-wave albedo of clouds. *Journal of Atmospheric Science* 34: 1149–1152.

8

Stratospheric Chemistry and Ozone Depletion

Martyn P. Chipperfield[1] and A. Rob MacKenzie[2]

[1] *School of Earth and Environment, University of Leeds, Leeds, United Kingdom*
[2] *School of Geography, Earth and Environmental Sciences, University of Birmingham, Birmingham, United Kingdom*

Ozone is formed via the photolysis of oxygen by ultraviolet radiation in the solar spectrum. Since the source of sunlight is at the top of the atmosphere, but atmospheric density increases exponentially from the top of the atmosphere to the Earth's surface, the majority of ozone is found in a layer in the middle atmosphere, between about 10 and 30 km. Once formed, ozone itself absorbs ultraviolet radiation from the Sun. The energy trapped by this absorption is converted into heat. Hence, the almost inexorable, and readily experienced, decrease of temperature with height that characterizes the lower atmosphere, the troposphere, is halted, and is replaced by the increasing temperatures and high static stability that characterize the lower stratosphere (see Chapter 3). In absorbing solar ultraviolet radiation, ozone not only gives the stratosphere its principal structure, but it also protects living organisms from the chemical-bond-breaking danger of ultraviolet light. These roles of ozone – as controlling influence in stratospheric stability and as UV shade for the biosphere – explain why changes in its abundance are monitored closely and have been the cause of international concern and legislation.

Although chemical measurements of ozone had been available since the middle of the nineteenth century, it was the discovery of the spectroscopy of ozone that allowed it to be detected remotely. By the end of the nineteenth century, spectroscopic measurements had determined that most of the ozone in a column of atmosphere was in the middle of that column. Measurements made through the first three decades of the twentieth century identified the exact altitude of the ozone maximum, showed that a great deal of the variability in ozone column from day-to-day can be attributed to changes in tropopause height, and demonstrated the existence of a slow overturning circulation in the stratosphere and mesosphere (see Section 8.2), which is as important to the middle atmosphere as the better-known Hadley circulation is to tropospheric weather and climate (Chapter 2).

Atmospheric Science for Environmental Scientists, Second Edition. Edited by C.N. Hewitt and Andrea V. Jackson.
© 2020 John Wiley & Sons Ltd. Published 2020 by John Wiley & Sons Ltd.

8.1 Ozone Column Amounts

It is usual in the troposphere to measure the local concentrations, or volume mixing ratios, of trace chemical constituents (see Chapters 4 and 5). Such measurements of stratospheric *trace gases* – i.e. gases present as a few molecules per million, billion, or even trillion molecules of air – are also frequently done in the stratosphere, as discussed in subsequent sections. However, the remoteness of the stratosphere makes it easier to measure the total number of ozone, or other chemical, molecules in a vertical column reaching from the ground to space. These *column amounts* can be measured from the ground, looking up, or from space, looking down, and they carry a good deal of information about the chemistry and wind-driven transport in the stratosphere. Column ozone amounts are usually expressed in *Dobson units* (DU), which are dimensionally milli-atmosphere centimetres, or the depth as pure gas, in thousandths of a centimetre, that the ozone column amount would have if it was all brought to the standard temperature and pressure (STP) of 273 K and 1 atm.

Mean column ozone abundances as a function of latitude and season are shown in Figure 8.1. Values range from below 240 DU to above 440 DU; that is, taking all the ozone in a vertical column that reached from the ground to space, and bringing it to STP would result in a layer of ozone about 2–4 mm thick. Looking now at the variation across the globe, there is a broad minimum at equatorial latitudes at all seasons, with maxima at middle-to-high latitudes in the late winter and spring of each hemisphere. At high southern latitudes, embedded in the Southern Hemisphere maximum, is a severe local minimum centred on September. This is the Antarctic 'ozone hole', described below.

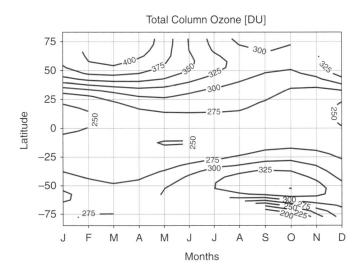

Figure 8.1   The mean column ozone as a function of latitude and season based on satellite observations. Note the relatively constant values in the tropics, and the much greater seasonal variation nearer the poles. Arctic ozone columns are maximum in late winter and early spring. Antarctic ozone columns are largest in mid-winter, and show a decrease in the spring as a result of ozone hole chemistry. There is a marked collar region of larger ozone columns in the southern middle latitudes in winter and spring. *Source:* plot courtesy of Sandip Dhomse, University of Leeds.

This distribution of global ozone is somewhat counterintuitive. The tropics comprise a region where solar input per unit area is greatest, so one would expect ozone production to be efficient there, and indeed it is. Conversely, the high latitudes are regions where ozone production would be expected to be inefficient, and this again is found to be the case. In fact, net ozone destruction commonly occurs at higher latitudes, due to the chemistry described in subsequent sections. The distribution shown in Figure 8.1 cannot be explained, therefore, by photochemistry alone, but must also be the result of transport of ozone by winds in the stratosphere. Wind flow in the stratosphere is discussed in detail in Section 8.3. Here we note simply that an 'overall' circulation – technically, the 'residual' circulation or the *Brewer–Dobson circulation* – can be deduced from consideration of the distribution of trace gases such as ozone and water vapour (Section 8.2). The circulation is not thermally direct – i.e. it is not due to hot air rising, even although the upward branch of the circulation is in the tropics – but rather due to the action of waves internal to the atmosphere (see, e.g. McIntyre 1992; Shepherd 2003). The general sense of the residual circulation is upward in the tropics and downward at the poles.

One important consequence of the Brewer-Dobson circulation is the distribution of trace gases in the stratosphere. Gases such as chlorofluorocarbons (CFCs), N_2O, and water vapour, which are chemically inert in the troposphere, enter the stratosphere in the tropics and are transported upwards (Table 8.1). Exposure to short-wave, high-energy solar radiation, above the altitudes where these wavelengths are filtered out, can convert these compounds into ozone-depleting *free radicals*, and the *reservoir compounds* for these radicals (see below) as the air moves slowly poleward. In the downward branch of the circulation at high latitudes, then, the downward transport of the radical-reservoir compounds means that the potential for severe ozone depletion exists. This is why ozone depletion is most severe at the poles, especially the southern pole, even though the source of the ozone-depleting chemicals is mostly in the northern middle latitudes, and even though the most active ozone photochemistry is in the tropical stratosphere.

Early concerns about ozone depletion focused on low latitudes, and on the middle and upper stratosphere. It is here that CFCs and N_2O are broken down into ozone-depleting radicals. In the middle of the 1980s, however, attention switched abruptly to the polar lower stratosphere (Farman et al. 1985). Ground-based measurements – made by Joe Farman, Brian Gardiner, Jonathan Shanklin, and colleagues at the British Antarctic Survey – showed a marked decline in the October monthly-mean ozone column

Table 8.1 Example source gases which are decomposed in the stratosphere to release radical species.

Compound	Formula	Approximate stratospheric lifetime (years)	Source
Nitrous oxide	N_2O	320	Biological sources in soil and water
Water vapour	H_2O	640	Evaporation and transpiration
CFC-11	$CFCl_3$	50	Industry
CFC-12	CF_2Cl_2	102	Industry
Halon-1211	$CBrClF_2$	20	Industry
Halon-1301	$CBrF_3$	65	Industry

Source: from Brasseur and Solomon (2005).
Photochemical lifetimes are estimated for typical conditions at 20 km.

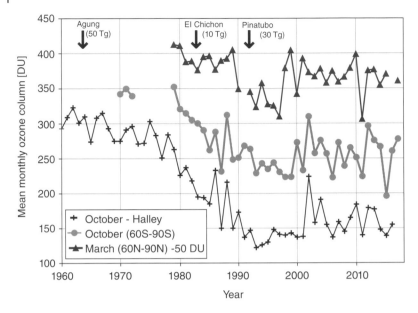

Figure 8.2 October mean column ozone over Halley Bay (76° S) and average over the Antarctic (60–90° S). Also shown are March mean ozone columns over the Arctic (60–90° N). Note that 50 DU has been subtracted from the Arctic means to aid graph plotting. The dates of major volcanic eruptions are also marked. *Source:* Halley ozone data courtesy of J. Shanklin, British Antarctic Survey, Cambridge, UK.

over the Faraday research station, Antarctica (Figure 8.2). Satellite measurements later confirmed that this was a continent-wide phenomenon.

Chemical ozone depletion, similar in scale to that seen in Antarctica in the early 1980s, has occurred in the Arctic, for example in the winters of 1996–1997, 1999–2000, 2004–2005, and 2010–2011 in particular. Figure 8.2 also shows satellite observations of March-mean Arctic total ozone columns (note that 50 DU have been subtracted from the Arctic columns, to make them convenient to plot on the same graph as the Antarctic columns). The greater variability in the Arctic compared with the September-mean Antarctic ozone columns is clear: this is due to the greater variability in transport by the residual circulation in the Northern Hemisphere. Some recent Northern Hemisphere winters have shown only moderate ozone depletion, but others have shown column depletions of up to 30%.

Worked Example 8.1

Ozone is the only atmospheric trace gas which is measured in Dobson units. The column abundance of most gases is measured in molecules cm^{-2}. Convert an ozone column of 300 DU to units of molecules cm^{-2}. As noted above, 1 DU is equivalent to 1 m.atm.cm of pure ozone at STP (273 K, 1 atm).

Answer

We can use the ideal gas equation to estimate the concentration of a gas under any pressure and temperature:

$$p V = N k_b T$$

where p is pressure, V is volume, N is the number of molecules, k_b is Boltzmann's constant, and T is temperature. (Nb the ideal gas equation can also be written in terms of moles of a gas (n) – pV = nRT, where R is the gas constant, but here we need the form written in terms of number of molecules.)
The concentration of a gas is therefore:

$$N/V = p/(k_b T)$$

Using SI units for T = 273 K and p = 101 325 Pa (1 atm) gives $N/V = 2.69 \times 10^{25}$ molecules m^{-3}. Converting this to molecules cm^{-3}, a column of 300 DU can therefore be calculated as:

$$\text{Column} = 300\,\text{m.atm.cm} = 300 \times 10^{-3} \times 2.69 \times 10^{19} \times 1 = 8.1 \times 10^{18}\,\text{molecules}\,cm^{-2}$$

8.2 Physical Structure of the Stratosphere

The stratosphere, as its name suggests, is a stably stratified layer of the Earth's atmosphere, extending very roughly from 10 to 50 km altitude (see Chapter 3). Temperature generally increases throughout the stratosphere, due to the absorption of solar radiation by ozone. Consequently, potential temperature increases rapidly. Because air must gain or lose energy to cross surfaces of constant potential temperature – *isentropes* – and because this is a process that occurs slowly relative to other transport processes occurring in the stratosphere, it is often convenient to use potential temperature as a vertical coordinate instead of height or pressure. It is also often convenient to follow air motion along isentropes – i.e. *adiabatic* air motion. The residual circulation, which is discussed next, is a *diabatic circulation*, involving the motion of air across isentropes.

8.2.1 The Residual Circulation

Figure 8.3 bears closer examination. The thick line curving upwards from pole to Equator represents the tropopause, the boundary between troposphere and stratosphere. At low latitudes the average position of this boundary is best represented by the mean position of the 380 K isentrope. Once in the stratosphere, the residual circulation carries anthropogenic and natural long-lived source gases (such as CFCs, N_2O, and H_2O) into a regime of intense ultraviolet irradiation from the sun. These compounds, which are chemically stable in the lower atmosphere, then begin to break down, e.g.:

$$N_2O + O(^1D) \rightarrow NO + NO \tag{8.1}$$

$$H_2O + O(^1D) \rightarrow OH + OH \tag{8.2}$$

$$CF_2Cl_2 + h\nu\ (\lambda < 227\,\text{nm}) \rightarrow CF_2Cl + Cl \tag{8.3}$$

where λ is the wavelength of the photon responsible for the photolysis and $O(^1D)$ is an electronically excited oxygen atom, produced from the photolysis of ozone:

$$O_3 + h\nu\ (\lambda < 310\ \text{nm}) \rightarrow O_2 + O(^1D), \tag{8.4}$$

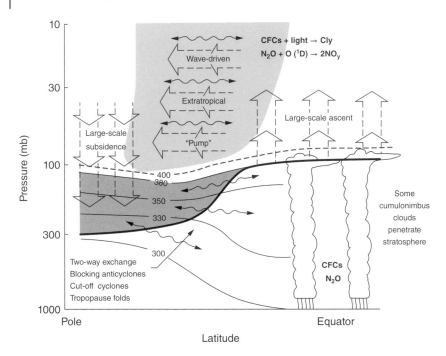

Figure 8.3 The residual circulation in the stratosphere and associated transport processes. The tropopause is shown by the thick line. Thin lines are isentropes, labelled in Kelvin. Wiggly double-headed arrows denote meridional transport by eddy motions. This eddy transport is not necessarily symmetric in and out of the stratosphere. The broad arrows show transport by the global-scale circulation (Holton et al. 1995). CFCs, chlorofluorocarbons.

which itself requires ultraviolet sunlight. (See Chapter 5 for more details of atmospheric chemistry notation.) Note that the wavelength limit given here is approximate, see, e.g. Holloway and Wayne (2010) for details. In particular, small but significant production of $O(^1D)$ from ozone photolysis can occur at wavelengths longer than 310 nm. The release of reactive compounds from their sources can cause ozone depletion, through the cycles outlined below but, in the tropical upper stratosphere, these cycles are soon terminated by production of radical-reservoir compounds.

The mean upward motion in the tropics is balanced by mean downward motion in the extratropics. Descent in the polar regions in winter is especially dramatic (Figure 8.4), bringing upper stratospheric and mesospheric air down to lower stratospheric altitudes (roughly 25 km and below). The air that descends in the wintertime stratosphere is depleted in CFCs, N_2O, CH_4, and other long-lived gases, and is correspondingly enriched in chlorine, nitrogen, and hydrogen radical-reservoir compounds. With little or no warming available from the Sun, the wintertime polar regions cool, to temperatures at which aerosols take up water and other condensable vapours (see Section 8.5 below). The combination of high concentrations of radical-reservoir compounds, high surface areas of aerosol, and low meridional mixing (i.e. north-to-south or south-to-north flow, see below) provides the ideal chemical and physical conditions for rapid ozone depletion.

Connecting the upward and downward portions of the residual circulation is a mean-meridional transport, towards the pole. The residual circulation as a whole is superimposed on much more rapid zonal

CLAES: Global stratospheric CF₂Cl₂
Combined 20–27 March 1992

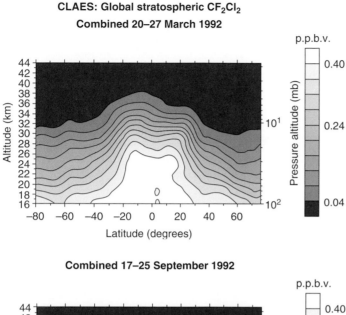

Combined 17–25 September 1992

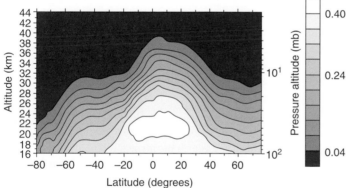

Figure 8.4 A latitudinal cross-section of CF_2Cl_2 from the satellite instrument CLAES. Note the steeply sloping isopleths in the subtropics and the wintertime polar circle. The relatively shallow slopes of the isopleths in the middle latitudes shows the effects of stirring and mixing in the Rossby-wave surf zone. *Source:* after Dessler et al. (1998); this diagram courtesy of A.E. Roche, Lockheed Palo Alto Research Laboratory.

(i.e. west-to-east or east-to-west) transport. This zonal flow is eastward in the extratropics in winter; westward in the extratropics in summer; and oscillating between eastward and westward with a quasi-biennial period in the deep tropics.

Alongside the transport processes that move air wholesale from place to place, diffusive processes mix air parcels. Mixing blurs the chemical and dynamical characteristics imparted to air parcels by their transport through the stratosphere. Molecular diffusion is, of course, ubiquitous, and increases with altitude. Magnitudes of molecular diffusion vary by a factor of two to three for different chemical constituents, but indicative values are those for water vapour: $0.2\,cm^2\,s^{-1}$ at $0\,km$; $1.2\,cm^2\,s^{-1}$ at $16\,km$; and $75\,cm^2\,s^{-1}$ at $42\,km$. Molecular diffusion rates are important in the calculation of the flux of trace gases to aerosol surfaces, but other mixing processes are much more important in the bulk mixing of stratospheric air parcels. Turbulent mixing – i.e. mixing due to unresolved eddies, hence often called 'eddy diffusivity' – although much reduced compared with the troposphere, is still important in determining the fine structure of chemical distributions.

Because the winds in the stratosphere produce horizontal strain fields and strong vertical shear, mixing processes are most likely to act on vertical, cross-isentrope, contrasts in trace gas concentrations rather than on horizontal, along-isentrope, contrasts. The exact value of vertical diffusivity, as a function of space and time, therefore becomes important. Vertical diffusivity has been estimated to reach $2000\,cm^2\,s^{-1}$ in patches of three-dimensional turbulence, although other analyses suggest an upper limit of $200\,cm^2\,s^{-1}$. In any event, turbulent mixing may always be expected to be much larger than molecular diffusion. This explains, incidentally, why heavy trace gases do not 'pile up' at our feet; eddy diffusion makes the molecular weight of trace gases almost irrelevant for mixing up to the homopause (c. 100 km).

8.2.2 Barriers to Stirring and Mixing

Even with an appreciation of the basic Brewer–Dobson circulation, early models of stratospheric transport failed to reproduce the rather sharp gradients in trace gas and aerosol concentrations seen in satellite observations such as Figure 8.4. Concentrations of source gases such as CF_2Cl_2 decrease with height due to photolysis (see above), as expected, but there are also striking horizontal gradients in concentration, particularly in the springtime hemisphere. At any given altitude, the sharpest changes in horizontal mixing-ratio gradient occur in the subtropics and near the polar circle. These sharp changes are indicative of mixing barriers, which inhibit the transport of air *along* isentropic surfaces.

High-resolution models of the stratosphere show these transport barriers clearly. In Figure 8.5, the polar region is bounded by a region of steep meridional gradients (i.e. rapid changes in the north–south direction). This is the wintertime *polar vortex*, within which severe ozone depletion can occur. Another region of steep gradients occurs in the subtropics throughout the year. This is the subtropical mixing barrier. The region bounded by the subtropical mixing barrier is sometimes known as the *tropical pipe*. Between the steep gradients at the polar and subtropical regions is a region composed of many interleaved filaments known, for reasons outlined below, as the *Rossby-Wave Surf Zone* or, simply, the 'Stratospheric Surf Zone'.

There are many implications arising from this view of the stratosphere, which emphasizes internal waves and their interaction with the mean atmospheric flow. McIntyre (1992) and Shepherd (2003) provide exemplary introductions to stratospheric flow; only a few of the practical results will be dealt with here. Vortices, mixing barriers, and the stirring of the surf zone can all be understood as stemming from the behaviour of *potential vorticity waves* on isentropic surfaces. This makes understanding and interpreting stratospheric flows much easier, so long as we have a clear idea of what potential vorticity is.

To start with a definition: potential vorticity – sometimes known as Rossby–Ertel potential vorticity – is calculated as:

$$Q = \frac{1}{\rho}\left(f+\zeta\right)\frac{\partial\theta}{\partial z}$$

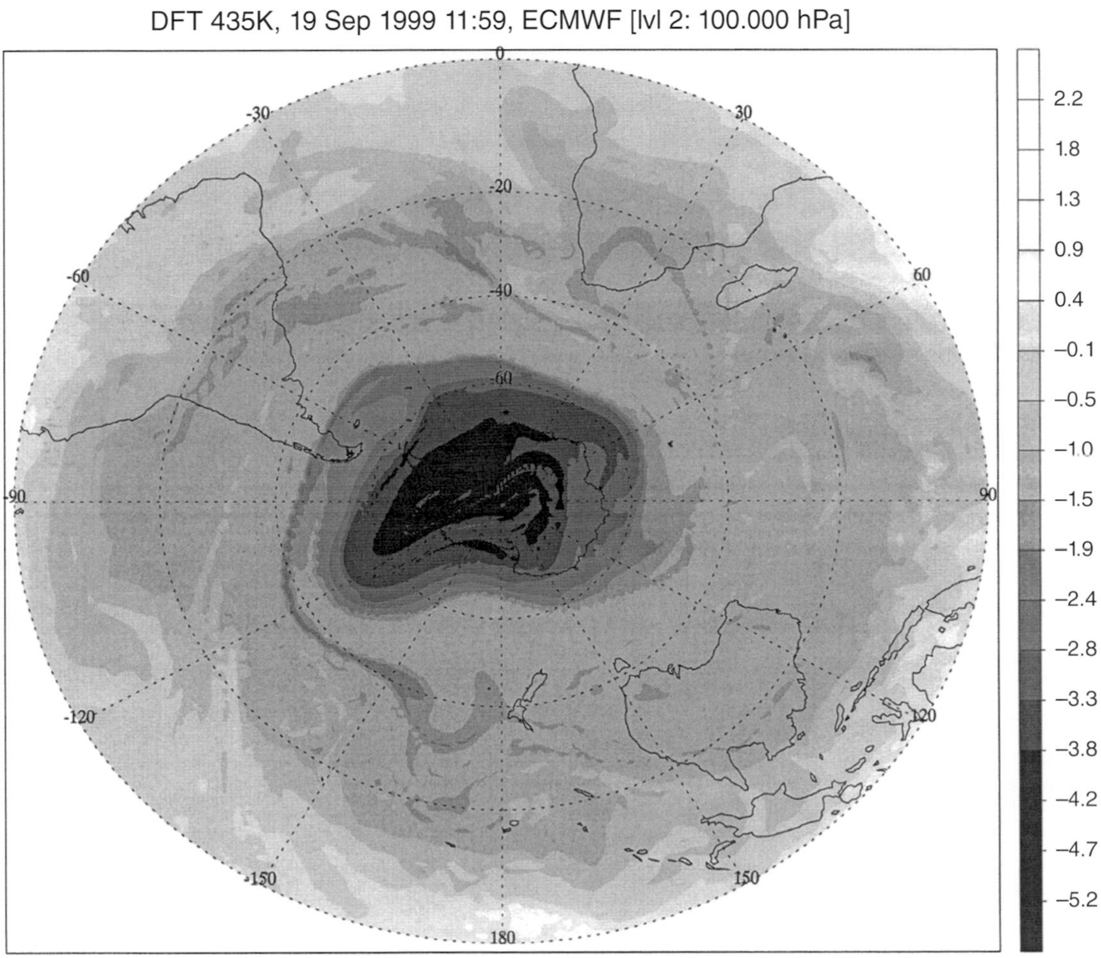

DFT 435K, 19 Sep 1999 11:59, ECMWF [lvl 2: 100.000 hPa]

Figure 8.5 A high-resolution simulation of the lower stratosphere showing mixing barriers. (Calculation and graphic by Gianluca Redaelli, University of l'Aquila. See, for further work, Dragani et al. 2002, and references therein.) A regular grid of several thousand points is constructed for the domain shown. A backward air mass trajectory from each point is then calculated. The potential vorticity, for the position of the trajectory's origin, is then interpolated onto each trajectory from a meteorological analysis. *Source:* from the European Centre for Medium-Range Weather Forecasts, in this case. Finally, the potential vorticity for each trajectory is mapped onto the original regular grid, i.e. transported forward in time under the assumption that the potential vorticity is conserved.

where ρ is the atmospheric density, f is the Coriolis parameter ($2\Omega \sin \phi$) also known as the *planetary vorticity*, Ω is the planetary angular velocity, ϕ is the latitude, ζ is the *relative vorticity*, θ is the potential temperature, and z is height. The gradient of potential temperature with height is really three-dimensional, but here it has been assumed to be directed exclusively in the vertical. Note that Q has units $K\,m^2\,kg^{-1}\,s^{-1}$ (and not s^{-1}, which is the unit of vorticity, e.g. for ζ). These units are clumsy so the PV unit ($1\,PVU = 10^{-6}\,K\,m^2\,kg^{-1}\,s^{-1}$) is widely used.

The planetary vorticity measures the rotation about the local vertical due to the rotation of the Earth on its axis. The relative vorticity is a measure of the 'swirliness' of the fluid around a local centre of rotation. So, the planetary vorticity is large in magnitude when the latitude is large; the relative vorticity is large when the fluid contains intense cyclonic flow. Potential vorticity, Q, as a whole can be thought of as the *cyclonicity* of an air parcel. One of the amazing properties of rotating fluids is that, in the absence of frictional and other diabatic effects, Q is conserved. In the lower stratosphere, adiabatic motion is a reasonable approximation for 5–10 days, and so the conservation of Q is a reasonable assumption over similar time scales. Following the distribution of Q on isentropic surfaces gives a much clearer picture of the stirring and mixing in the stratosphere than does following the distribution of winds or pressures at a particular height, and so visualizations of Q have become important for interpreting stratospheric flow. In winter, horizontal gradients of Q are concentrated at the subtropical mixing barrier and the edge of the polar vortex (Figure 8.5). Undulations in these mixing barriers are known as Rossby waves. Figure 8.5 shows the superposition of wave-three (three cycles around a latitude circle) and wave-two Rossby waves, leading to a pronounced equatorward distortion of the vortex around 120 °W. When this distortion is very large, air is transported outwards in filaments across the mixing barriers into the middle latitudes. One such filament is obvious in Figure 8.5. The formation of filaments of vortex air, or subtropical air, in the middle latitudes is called *Rossby-wave breaking*, by analogy to the breaking of water waves on a beach. Transport inwards – i.e. towards the pole – is much less common.

8.2.3 The Tropopause

Exchange of mass between the troposphere and the stratosphere is of central importance to several problems in atmospheric pollution. The rate of source gas (e.g. CFC) transport into the stratosphere affects the likely duration and extent of ozone depletion into the future. The rate of downward transport of ozone to the troposphere is also important because of the contribution of stratospheric ozone to the oxidizing capacity of the troposphere (see Chapter 5). As would be expected from the mean circulation shown in Figure 8.3, stratosphere-to-troposphere transport (STT) occurs mainly in the middle latitudes, whereas troposphere-to-stratosphere transport (TST) occurs mainly in the tropics.

Sudden bursts of ozone-rich, very dry air have been measured at ground air pollution stations, particularly in spring in middle latitudes. It can be shown straightforwardly that these 'ozone episodes' are not photochemical in origin, but are instead due to rapid downward transport of air from the stratosphere.

These rapid events are called *tropopause folds*, and are an extreme example of tropopause undulation. The pressure-altitude of the tropopause moves according to the weather systems in the troposphere. In particular, low-pressure systems lower the tropopause. Jet streaks are often associated with steep N–S changes in tropopause height. These streaks of high horizontal wind speeds initiate vertical circulation patterns at their beginning and end. Under particular conditions, closely related to the formation of a surface low pressure system with its attendant fronts, the vertical circulations at the entrance to a jet bring a thin layer of stratospheric air to very low altitudes, as sketched in Figure 8.6. The stratospheric air is then mixed into the lower troposphere by turbulence. Between about 1950 and 1970, rapid STT in tropopause folds constituted a potential health risk, due to the downward transport of radioactivity that had been released into the atmosphere in above-ground nuclear weapons tests (e.g. Holton et al. 1995).

The stratosphere is extremely dry. Water vapour mixing ratios in the stratosphere generally range from 1 to 6 ppmv; in the troposphere mixing ratios of ppth (parts per thousand by volume) are normal. Much

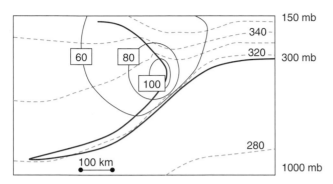

Figure 8.6 A schematic horizontal–distance/pressure cross-section of stratosphere–troposphere exchange in a tropopause fold. The thick line shows the tropopause. Grey, solid lines are isotachs – i.e. lines of equal wind speed – showing a jet that is causing the deformation of the tropopause. Broken lines show isentropes: note the rapid change in the vertical gradient of potential temperature (i.e. isentrope spacing) across the tropopause. Although the tropopause fold is reversible to some extent, turbulence in the middle and lower troposphere will cause irreversible mixing of stratospheric air into the troposphere.

of the stratospheric water vapour comes from the oxidation of methane, a fact that is easily demonstrated by the near-constancy of total hydrogen, ΣH_2, in the stratosphere:

$$\sum H_2 = 2CH_4 + H_2O + H_2.$$

(H_2 itself is always a minor component of ΣH_2 in the troposphere and stratosphere). At the tropopause $\Sigma H_2 \approx 7.5\,\mathrm{ppmv}$, $CH_4 \approx 1.7\,\mathrm{ppmv}$, and $H_2 \approx 0.5\,\mathrm{ppmv}$, from which we deduce an 'entry level' $H_2O \approx 3.5\,\mathrm{ppmv}$. Because there are no chemical sinks for water capable of removing hundreds of ppmv (Chapter 5), there must be a physical sink that causes intense drying of air as it passes upward across the tropopause. The only conceivable method for this dehydration to take place is by cloud processing.

When a cloud forms, the total mass of water substance in the air parcel is partitioned between condensed and vapour phases, according to the vapour pressure equation for water. If the ice particles formed in the cloud are large enough to precipitate out of the airmass then, on evaporation of the cloud, the resulting total water vapour mixing ratio will be less than the original total water. This process is also known as *freeze-drying*. Only the tropical tropopause is cold enough to dry air, in this way, to stratospheric mixing ratios (Holton et al. 1995; Ren et al. 2007).

8.2.4 The Mesoscale and the Stratosphere

In addition to the planetary-scale residual circulation, and the synoptic-scale influence of tropospheric weather systems, it has recently become apparent that smaller-scale processes – with typical horizontal scales of tens of kilometres – are also important to the distribution of trace gases in the stratosphere. The most important of these small-scale processes are *gravity waves*. These waves are internal to the atmosphere, generated by forced upward movement of air parcels, and have gravity as a restoring mechanism. The forced upward movement can be due to a solid obstruction to the flow, such as a mountain, or can be due to up-welling of buoyant air in convection. As gravity waves propagate upwards into the stratosphere

Table 8.2 Comparison of gravity and Rossby waves in the stratosphere.

Characteristic	Gravity waves	Rossby waves
Season in extratropical stratosphere	All seasons	Wintertime
Time scale	Minutes to hours	Days
Restoring mechanism	Buoyancy	Q-induced horizontal motion
Forced by	Vertical undulations of θ surfaces	Quasi-horizontal undulations of Q contours on θ surfaces
Requires for presence	Vertical gradient in θ	Isentropic gradient in Q
Character	High vertical velocity	Layer-wise two-dimensional Small vertical velocity
Consequences of wave 'breaking'	Three-dimensional turbulence Vertical mixing of chemicals	Irreversibly deformed θ surfaces Two-dimensional turbulence Irreversibly deformed Q contours Isentropic mixing of chemicals

and mesosphere, the amplitude of the wave must increase to compensate for the decrease in the density of the atmosphere, and eventually the waves become unstable and dissipate, or break, producing a torque (i.e. a turning-force acting on a rotating body) applied to the mean flow. This torque has a significant effect on the flow in the stratosphere and mesosphere, contributing to the Brewer–Dobson residual circulation. This wave-forcing of the residual circulation explains why it cannot be regarded as thermally direct (i.e. hot air rising). The importance of the wave forcing is also a practical problem, because much of the gravity-wave spectrum responsible for the torque on the mean flow is too small-scale to be resolved by current global models of stratospheric climate and chemistry.

Lower in the stratosphere, gravity waves have an impact on chemistry by subjecting air parcels to rapid, adiabatic, temperature oscillations. When the mean temperature is already low (i.e. below 200 K), the induced adiabatic oscillation can lead to uptake of water on the stratospheric aerosol, and even to cloud formation (see Section 8.4). Ozone depletion by gravity-wave-induced stratospheric clouds is a significant contribution to ozone depletion in the Arctic. The gravity waves in this case are forced by the underlying topography: the Scandinavian Alps, the Greenland ridge, and the Urals (Waibel et al. 1999). Gravity waves also produce cloud formation, and thence dehydration, at the tropical tropopause (e.g. Ren et al. 2007).

Table 8.2 summarizes the properties of gravity and Rossby waves in the stratosphere.

8.3   Gas-Phase Chemistry of the Stratosphere

The chemical conversion of trace gases along the course of the Brewer–Dobson circulation has been mentioned in passing above; below is a somewhat more detailed account. A complete review of chemical reactions, their rate coefficients, and their mechanisms, is beyond the scope of this chapter. The reader is directed to the continuing series of assessments by NASA Jet Propulsion Laboratory (JPL; https://jpldataeval.jpl.nasa.gov) and the International Union of Pure and Applied Chemistry (IUPAC; http://iupac.pole-ether.fr).

8.3.1 Oxygen-Only Chemistry

Ozone is an *allotrope* of oxygen, that is, an alternative form of the element. Given that oxygen makes up one-fifth of the atmosphere, by volume, and that the upper atmosphere is bathed in high-energy radiation from the Sun, it is not surprising that ozone is produced in the atmosphere:

$$O_2 + h\nu \; (\lambda < 243\,\text{nm}) \rightarrow O + O, \text{ followed by} \tag{8.5}$$

$$O + O_2 + M \rightarrow O_3 + M, \tag{8.6}$$

where M is any atmospheric molecule, acting as a heat sink for the exothermic combination of O and O_2. That is, the reaction produces heat, which is converted into kinetic energy by collision of the O_3 with M. Without the heat sink, M, the newly formed O_3 molecule would contain too much internal energy and so would immediately fall apart.

Clearly, unless ozone is to accumulate in the atmosphere, there must also be loss processes. The Oxford scientist, Sidney Chapman, first closed the loop of the ozone life cycle by suggesting

$$O_3 + h\nu \; (\lambda < 1180\,\text{nm}) \rightarrow O + O_2 \tag{8.4a}$$

$$O + O_3 \rightarrow 2O_2 \tag{8.7}$$

(Note that a third possible oxygen-only loss, from the recombination of O atoms is too slow to be significant in the stratosphere. Note also that reaction (8.4a) differs from reaction (8.4) only in the wavelength threshold: light at the longer wavelengths can produce ground-state oxygen atoms, $O(^3P)$, from O_3). Reactions (8.4a) and (8.6) swap oxygen rapidly between *odd oxygen* (1-atom and 3-atom) allotropes, so that it is convenient to discuss the 'odd-oxygen family' and to treat O and O_3 as in chemical equilibrium with each other. This grouping of reactive intermediates into families is an important concept in reducing the complexity of stratospheric chemistry, as exemplified by the use of other families, below.

The oxygen-only *Chapman mechanism* for formation and destruction of ozone has the correct properties to generate an ozone layer in the stratosphere, as is observed. That is, the source of the light to photolyse oxygen originates above the atmosphere but the density of oxygen molecules per unit volume increases with decreasing altitude (see, e.g. Holloway and Wayne 2010). However, careful measurements reveal that the rate of reaction (8.7) is too slow to account for the ozone concentrations observed; i.e. the Chapman mechanism predicts higher peak ozone concentrations than are observed.

8.3.2 Classic Radical Cycles: Ozone-Depleting and Null Cycles, Termination Reactions

Several additional loss processes for ozone are now known. In the upper stratosphere, *homogeneous catalysis* of reaction (8.7) is important. Homogeneous catalysis is the promotion of a chemical reaction by a compound that is in the same physical state as the reactants (e.g. gas phase), and that is not consumed by the reaction:

$$X + O_3 \rightarrow XO + O_2 \tag{8.8}$$

$$XO + O \rightarrow X + O_2 \tag{8.9}$$

$$\text{Net}: O + O_3 \rightarrow 2O_2$$

where X is H, OH, NO, Cl, and, to a minor extent, Br. The source of these catalysts is discussed above. The rapid interconversion of radical intermediates – for example, the X/XO radicals above, make it convenient to consider families of radical intermediates. Hence, the HO_x family represents H, OH, and HO_2; the NO_x family represents NO, NO_2, and NO_3; and so on. The concept of chemical families makes modelling of the chemistry considerably simpler.

For all the radicals listed above, oxidation by ozone, and reduction of the oxidized radical by O, are extremely rapid reactions, so that, if the catalysts are present in sufficient concentration, the rate of the net reaction is increased. A second channel is available to OH radicals, since they can be reduced by O atoms, as well as oxidized by O_3:

$$OH + O \rightarrow H + O_2 \tag{8.10}$$

$$H + O_2 + M \rightarrow HO_2 + M \tag{8.11}$$

$$HO_2 + O \rightarrow OH + O_2 \tag{8.12}$$

$$\text{Net}: O + O \rightarrow O_2$$

The catalytic reaction cycles above are in competition with null cycles, which simply tie-up a proportion of the catalyst in reactions that do not destroy ozone, or with holding cycles, which remove the catalysts temporarily by forming short-lived reservoir compounds. An important null cycle is that involving NO_2:

$$NO + O_3 \rightarrow NO_2 + O_2 \tag{8.13}$$

$$NO_2 + h\nu \left(\lambda < 420\,\text{nm}\right) \rightarrow NO + O \tag{8.14}$$

$$O + O_2 + M \rightarrow O_3 + M \tag{8.6}$$

$$\text{Net}: \text{null}.$$

Reaction (8.14) can close the null cycle channels of other radicals, since NO can be oxidized by many of the other XO species, e.g.:

$$OH + O_3 \rightarrow HO_2 + O_2 \tag{8.15}$$

$$HO_2 + NO \rightarrow NO_2 + OH \tag{8.16}$$

$$NO_2 + h\nu \left(\lambda < 420\,\text{nm}\right) \rightarrow NO + O \tag{8.14}$$

$$O + O_2 + M \rightarrow O_3 + M \tag{8.6}$$

$$\text{Net}: \text{null}.$$

An important holding cycle involves the production of N_2O_5, which is stable during the night, thereby locking up two NO_x radicals for the hours of darkness:

$$NO_2 + O_3 \rightarrow NO_3 + O_2 \tag{8.17}$$

$$NO_2 + NO_3 + M \rightarrow N_2O_5 + M \tag{8.18}$$

$$N_2O_5 + h\nu \rightarrow NO_2 + NO_3 \tag{8.19}$$

$$NO_3 + h\nu \left(\lambda < 640\,nm\right) \rightarrow NO_2 + O \tag{8.20a}$$

$$O + O_2 + M \rightarrow O_3 + M \tag{8.6}$$

Net : null, but accumulation of N_2O_5 during darkness.

When reactions remove the radicals to form products that are more stable, then the 'reservoir' product can be transported long distances through the stratosphere before being broken down to re-form the radicals, and so the radical reaction chain can be considered to be terminated. Examples are:

$$OH + NO_2 + M \rightarrow HNO_3 + M \tag{8.21}$$

$$Cl + CH_4 \rightarrow HCl + CH_3 \tag{8.22}$$

$$ClO + NO_2 + M \rightarrow ClONO_2 + M \tag{8.23}$$

$$HO_2 + NO_2 + M \rightarrow HO_2NO_2 + M \tag{8.24}$$

(The conversion of NO_x to reservoirs is sometimes called *denoxification*.)

The efficiency of a catalytic cycle is measured by the chain length, i.e. the average number of times a particular cycle is executed before termination. However, the impact of a catalytic cycle on ozone concentrations depends not only on the efficiency of the cycle but also on the concentration of the radicals. For example, the efficiency of the Br/BrO cycle is much larger than that of the Cl/ClO cycle, but the abundance of bromine is smaller than chlorine.

Reactivation of the radical chemistry is by photolysis, e.g.

$$HNO_3 + h\nu \rightarrow OH + NO_2, \tag{8.25}$$

reaction with OH, e.g.

$$HNO_3 + OH \rightarrow H_2O + NO_3, \tag{8.26}$$

or by heterogeneous processes (see Section 8.5).

The existence of important cross-family null cycles, holding cycles and termination reactions tells us that the chemistry of the stratosphere is very *nonlinear* with respect to changes in the concentrations of radical families; we should not expect changes in ozone to be proportional to changes in the concentration of one of the radical families. Predicting the response of ozone to changes in any one family is therefore virtually impossible without numerical modelling of the full system.

8.3.3 Ozone Depletion Cycles Without O Atoms

The classic ozone-destroying cycles, described in the previous section, require O atoms for the reduction of the XO species. However, other catalytic cycles exist that do not depend on the presence of the O atom. Such cycles are particularly important in the lower stratosphere, where the ratio of O atoms to ozone molecules is much less than in the upper stratosphere. There are three general types of ozone-specific catalytic cycles. The first involves formation of an XO species that itself reacts with ozone: e.g.

$$OH + O_3 \rightarrow HO_2 + O_2 \tag{8.15}$$

$$HO_2 + O_3 \rightarrow OH + 2O_2 \tag{8.27}$$

$$Net : 2O_3 \rightarrow 3O_2$$

or

$$NO + O_3 \rightarrow NO_2 + O_2 \tag{8.13}$$

$$NO_2 + O_3 \rightarrow NO_3 + O_2 \tag{8.17}$$

$$NO_3 + h\nu \left(584 < \lambda < 640 \, nm\right) \rightarrow NO + O_2 \tag{8.20b}$$

$$Net : 2O_3 + h\nu \rightarrow 3O_2$$

Note the second channel for photolysis of NO_3 (reaction (8.20b), cf. reaction (8.20a)).

The second type of ozone-specific cycle involves formation of a compound from two XO species, with elimination of O_2, e.g.:

$$Br + O_3 \rightarrow BrO + O_2 \tag{8.28}$$

$$OH + O_3 \rightarrow HO_2 + O_2 \tag{8.15}$$

$$HO_2 + BrO \rightarrow HOBr + O_2 \tag{8.29}$$

$$HOBr + h\nu \rightarrow OH + Br \tag{8.30}$$

$$Net : 2O_3 + h\nu \rightarrow 3O_2$$

The third type of ozone-specific cycle involves formation of a compound from two XO species, and subsequent elimination of O_2 by photolysis of a different bond to the one initially formed, e.g.:

$$Cl + O_3 \rightarrow ClO + O_2 \tag{8.31}$$

$$NO + O_3 \rightarrow NO_2 + O_2 \tag{8.13}$$

$$ClO + NO_2 + M \rightarrow ClONO_2 + M \tag{8.32}$$

$$ClONO_2 + h\nu \rightarrow Cl + NO_3 \tag{8.33}$$

$$NO_3 + h\nu \left(584 < \lambda < 640 \, nm\right) \rightarrow NO + O_2 \tag{8.20b}$$

$$Net : 2O_3 + 2h\nu \rightarrow 3O_2$$

An important example of this third type of ozone-specific cycle is the *dimer cycle*:

$$2\left(Cl + O_3 \rightarrow ClO + O_2\right) \tag{8.31}$$

$$2ClO + M \rightarrow Cl_2O_2 + M \tag{8.34}$$

$$Cl_2O_2 + h\nu \rightarrow Cl + ClOO \tag{8.35}$$

$$ClOO + M \rightarrow Cl + O_2 + M \tag{8.36}$$

$$Net : 2O_3 + h\nu \rightarrow 3O_2$$

which is responsible for over half of the ozone destruction in the springtime polar vortices, leading to the formation of ozone 'holes'. Note that the second and third type of ozone-specific cycles, above, require sunlight. These cycles do not take place during polar night, therefore, even when the polar vortices are 'primed' for ozone destruction by heterogeneous chemistry (see below). In the Northern Hemisphere, air in the Arctic vortex is sloshed about by planetary waves, so that vortex air is exposed to periods of sunlight throughout the winter. In the Southern Hemisphere, the vortex is less perturbed and ozone loss can take place only at the edge of the polar vortex until sunlight returns to the Antarctic, in spring.

Figure 8.7 gives an example of a model calculation for the Arctic lower stratosphere, showing chemical conversion of chlorine from reservoirs to active radicals, and the ozone destruction that results (Becker et al. 2000). For this idealized, but still reasonably realistic, calculation, temperature, and sunlight fluxes vary as the air parcel oscillates around 70 °N, simulating the Rossby-wave dynamics of the Arctic vortex. The trajectory slowly descends in altitude (and potential temperature), simulating the diabatic cooling that occurs in the vortex (remember Figures 8.3 and 8.4). Near the start of the three-month model simulation, chlorine is released rapidly from the reservoir compounds, HCl and $ClONO_2$, leading to high mixing ratios of chlorine radicals (Cl_x: this process is usually called *chlorine activation*). Ozone loss occurs steadily until increased solar irradiation begins to deactivate the chlorine, converting it back into $ClONO_2$ and, more slowly, HCl. The HOCl mixing ratios increase in the middle of the model run: i.e. HOCl acts like a classic reactive intermediate. By the end of the model run, ozone is reduced to about 30% of its initial value. Ozone losses of this magnitude have been deduced from observations (Rex et al. 2004; Manney et al. 2011).

8.4 Aerosols and Clouds in the Stratosphere

The thermal stability and dryness of the stratosphere (Section 8.2) make cloud formation unlikely, but it does occur. Nacreous clouds have been recorded over the Arctic for more than a century-and-a-quarter; they are a dramatic and colourful sight in otherwise dark skies (google 'nacreous cloud' to see). Satellite observations in the last 20 years have greatly improved our understanding of the frequency and distribution of these clouds, now known as *polar stratospheric clouds* (PSCs). Detailed in-situ measurements have shown a continuum of particle sizes, from the sub-micrometre background aerosol to ice particles with radii of many micrometres. Keep in mind, however, that there is only 4–5 ppmv of water in the stratosphere, so these clouds are much more tenuous than tropospheric clouds (even high cirrus clouds in the middle latitudes typically contain ≈100 ppmv condensed water).

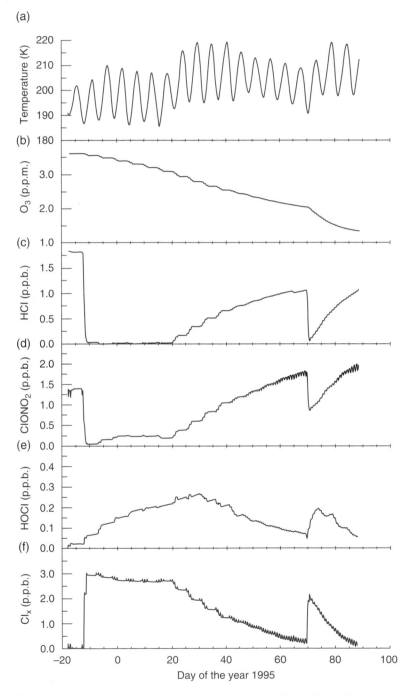

Figure 8.7 A model of the chemical evolution in the winter-time polar lower stratosphere, following an idealized vortex air trajectory *Source:* after Becker et al. 2000, redrawn by the authors of that work. Plots against day of the year: (a) temperature, (b) ozone, (c) HCl, (d) ClONO$_2$, (e) HOCl, (f) Cl$_x$.

8.4.1   The Volcanic Aerosol and the Junge Layer

A layer of aerosol, often called the *Junge layer*, blankets the globe. The centre of mass of the Junge layer is about 5 km above the tropopause, although the aerosol is present in a wide altitude band throughout the lower stratosphere (Thomason and Peter 2006). The stratospheric aerosol is composed primarily of concentrated sulphuric acid droplets but can also contain mineral elements derived from meteorites and, in the very lowermost parts of the stratosphere, soot and other material injected from the troposphere.

The primary source of the stratospheric aerosol is oxidation of the SO_2 injected by volcanic eruptions. An additional, very small but constant, source may be the percolation to the stratosphere of carbonyl sulphide (OCS), emitted by microbes, with subsequent oxidation to sulphuric acid. The most effective injection of volcanic aerosol in the past decades occurred in June 1991 with the eruption of Mount Pinatubo in the Philippines (McCormick et al. 1995). The global enhancement of the stratospheric aerosol loading due to the eruption has been estimated at up to 30 Tg (i.e. 30 million tonnes). Atmospheric opacity, or optical depth, peaked at values greater than 0.2 in the tropics shortly after the eruption. The largest injection of volcanic gas and ash into the stratosphere, for which we have evidence, is the eruption of Toba, in Sumatra, 73 500 years before present. It is estimated that the total stratospheric aerosol loading following Toba reached 1000 Tg (Rampino and Self 1992).

8.4.2   Polar Stratospheric Clouds

The composition of the stratospheric aerosol changes dramatically as an air mass is cooled to temperatures in the range of 180–200 K. Initially, the sulphuric acid aerosol takes up water vapour, to become more dilute. On further cooling, below about 192 K, the sulphuric acid aerosol takes up nitric acid and more water vapour, to form a PSC. Further cooling below the frost point ($\approx$188 K) results in freezing of the particles (Lowe and MacKenzie 2008).

The formation and evolution of PSCs in the atmosphere is rather uncertain (Tolbert and Toon 2001; Lowe and MacKenzie 2008). In particular, although there is ample evidence for solid PSC particles at temperatures above the frost point, the mechanism by which they form is not known. Liquid–solid phase change is not a well-quantified process in any system, including the concentrated aqueous solutions that are present in the lower stratosphere. Laboratory studies show widely varying freezing temperatures for PSC-like liquid compositions and the theory of freezing contains many uncertain parameters (see, e.g. MacKenzie et al. 1998).

As discussed in Section 8.5, PSCs have important chemical effects in the stratosphere, especially the polar wintertime stratosphere. They also have an important physical effect, called *denitrification*. As the aerosol particles grow into PSCs by uptake of nitric acid and water, their terminal settling velocity becomes significant. By settling out of cold air masses, PSC particles remove NO_y, in the form of nitric acid. The settling out of very cold PSC particles, containing mostly water, brings about *dehydration* of the polar vortex. This occurs in most Antarctic winters, but only rarely in Arctic winters. Note the specialized use of the term *denitrification* in this paragraph; there is another common use of this term in biogeochemical cycling of nitrogen, referring to the conversion of soil nitrate back to molecular nitrogen (Chapter 4).

8.5 Heterogeneous Chemistry of the Stratosphere

In the atmospheric science community, the term *heterogeneous chemistry* is generally taken to include any reactions occurring on or *in* particles in the atmosphere. Reactions occurring in liquid particles are, of course, homogeneous liquid-phase reactions, *sensu stricto,* but atmospheric chemistry models usually fold the gas–particle partitioning into the rate expression, so there is some justification for the terminology.

Table 8.3 lists some of the most important reactions occurring on sulphuric acid aerosol and PSC particles. The overall effect of heterogeneous reactions is to convert the chlorine and hydrogen in reservoir compounds into more active forms, whilst converting the NO_x in temporary reservoir compounds into nitric acid (i.e. denoxification and, if the nitric acid remains in the particles and sediments, denitrification). Many of the heterogeneous reactions in the sulphuric acid aerosol are strongly temperature dependent, due to the variation of aerosol composition, and hence reactant solubility, with temperature. For aerosol loadings typical of a volcanically quiescent period, heterogeneous reactions begin to activate chlorine at temperatures below about 200 K. This activation is relatively slow, however. When aerosol loadings are much increased following a volcanic eruption, the temperature at which chlorine activation begins is increased by up to 5 K and aerosol reactions can effectively compete with reactions on PSCs. One heterogeneous reaction that is not temperature dependent is the hydrolysis of N_2O_5.

8.5.1 Rates of Heterogeneous Reactions in Liquids

Diffusion is important for reactions involving particles, and reactants partition between the gas and condensed phases according to their solubilities or adsorptivities. In liquid particles, if reaction is slow, the reactants will have time to continuously adjust to their equilibrium partitioning, and the rate of reaction has the form

$$\frac{-d\left[R_1\right]}{dt} = k\left[R_1\right]\left[R_2\right]$$

where k is the rate of reaction in solution ($M^{-1}\,s^{-1}$) for reactants R_1 and R_2. By Henry's law and the ideal gas law

Table 8.3 Some heterogeneous reactions of importance in the lower stratosphere.

Reaction	Effect
$N_2O_5 + H_2O \rightarrow 2HNO_3$	Decreases NO_x concentration throughout lower stratosphere (denoxification)
$ClONO_2 + H_2O \rightarrow HOCl + HNO_3$	Chlorine activation at low temperatures, but HOCl is less photolabile than Cl_2 (denoxification)
$ClONO_2 + HCl \rightarrow Cl_2 + HNO_3$	Chlorine activation at low temperatures (denoxification)
$HOCl + HCl \rightarrow Cl_2 + H_2O$	Chlorine activation at low temperatures
$BrONO_2 + H_2O \rightarrow HOBr + HNO_3$	Indirectly affects ClO_x and HO_x concentrations
$HOBr + HCl \rightarrow BrCl + H_2O$	Indirectly affects ClO_x and HO_x concentrations

Heterogeneous reactions are those that occur on sulphuric acid aerosol and polar stratospheric clouds (PSCs).

$$\left[R_n\right] = H * p_n; \text{ Henry s law}$$

$$p_n V = n_n RT \Rightarrow p_n = \left(\frac{n_n N_A}{V}\right) k_B T; \text{ ideal gas law}$$

$$\left(\text{and Dalton s law of partial pressures}\right)$$

$$\Rightarrow \left[R_n\right] = 10 H * k_B T c_n$$

where $[R_n]$ is either the molar concentration (M) of R_1 or R_2, H^* is the effective Henry's law coefficient (M atm^{-1}) defining the reactant's solubility, k_B is Boltzmann's constant (J K^{-1} molec^{-1}), T is the temperature (K), V is gas volume (m^3), R is the universal gas constant (J K^{-1} mol^{-1}), p_n is the partial pressure of reactant (N m^{-2}), c_n is the gas phase number density of the reactant (molec cm^{-3}), and the factor 10 in the final expression ensures the units are correct (10^{-5} atm per N m^{-2}, times 10^6 cm^3 m^{-3}). So the reaction rate can be given in terms of gas-phase concentrations, which is important for connecting the heterogeneous chemistry to the gas-phase chemistry that is occurring simultaneously. The reaction of HCl and HOCl in solution can be treated this way. If the rate coefficient for reaction is faster than about 10^5 M^{-1} s^{-1}, then the rate of reaction becomes limited by transport of reactants into the particle and the effective volume available for reaction is reduced. In the limit of instantaneous reaction, the rate is dependent on the aerosol surface area, rather than the volume, and the heterogeneous chemistry on the particle is analogous to that on solid particles (see below). The reaction of ClONO$_2$ with HCl is an example of a reaction requiring a more complete treatment of multiphase transport and chemistry.

8.5.2 Rates of Heterogeneous Reactions on Solids

For solid particles, reaction rates depend linearly on the particle surface area (but then, of course, surface area is not as easily measured or calculated as for spherical liquid drops). Assuming that one can meaningfully define the surface of solid PSCs, the rate of reaction is given by

$$-\frac{d\left[X\right]}{dt} = \gamma \bar{v} \frac{A}{4} \left[X\right]$$

where γ is the dimensionless *reaction uptake coefficient*, $\bar{v}$ is the mean molecular velocity in the gas phase (cm s^{-1}), and A is the surface area density of the PSC cloud (cm^2 cm^{-3}).

It has yet to be established whether the surface of solid PSC particles is indeed solid, or if it consists of a 'quasi-liquid layer' a few molecular diameters deep. Without a robust theoretical model of solid PSC particle surfaces, calculated chlorine activation rates, and hence ozone depletion rates, remain uncertain, at least when the chlorine activation is occurring on solid PSCs.

8.6 Future Perturbations to the Stratosphere

8.6.1 Current Inputs, Future Scenarios

The emissions of ozone-depleting gases have been reduced significantly since the implementation of the Montreal Protocol on Substances that Deplete the Ozone Layer (which came into force in 1989) and its six subsequent amendments. The total organic chlorine in the troposphere peaked at about 3.7 ppbv in the early 1990s. In the stratosphere the time at which total chlorine peaks is a function of height, since the transport of air to the uppermost stratosphere, via the Brewer–Dobson circulation, is slow. The peak in chlorine therefore occurred later in the stratosphere than in the troposphere. At 22 km, for example, the peak was measured in 1998. One might be tempted to regard ozone depletion as a problem solved, but some important uncertainties remain, and these may yet cause the atmosphere to surprise us.

Bromine is estimated to be about 50 times more efficient than chlorine in destroying stratospheric ozone on an atom-for-atom basis (see Section 8.3 above). Bromine is carried into the stratosphere in various forms such as halons and substituted hydrocarbons, of which methyl bromide is the predominant form. Three major anthropogenic sources of methyl bromide have been identified: soil fumigation; biomass burning; and the exhaust of automobiles using leaded petrol. Recent measurements have shown that there is more methyl bromide in the Northern Hemisphere than in the Southern Hemisphere. Halon 1211 ($CBrClF_2$) and 1301 ($CBrF_3$) have been widely used in fire protection systems. Their production has now ceased, but emissions from existing fire protection systems are expected to continue for decades. Global background levels are about 3.5×10^{-3} ppbv (H-1211) and 3.4×10^{-3} ppbv (H-1301). Concentrations of H-1211 are now declining, whilst those of H1301 have stabilized (WMO/UNEP 2018).

Concern has grown in the past decade that so-called very-short-lived substances (VSLS) may be reaching the stratosphere in appreciable quantities (Law and Sturges 2006). VSLS are compounds that can be broken down in the troposphere at a rate comparable to, or shorter than, the average rate of mixing through the troposphere. This means that the distribution of VSLS in the troposphere is highly non-uniform. Table 8.4 lists some of the more abundant VSLS. The contribution of VSLS to the total chlorine source to the stratosphere is believed to be a few percent given the current input from long-lived chlorine source gases (e.g. CFCs) but, as long-lived source gases decline, the relative importance of VSLS is expected to increase. The contribution of VSLS to the total bromine source to the stratosphere is much more significant: around 5 pptv (5×10^{-3} ppbv) in a total bromine source of around 20 pptv. There has been debate about a source of iodine to the stratosphere from iodine-containing VSLS, but the current consensus is that the source is very small.

Table 8.4 Examples of very-short-lived substances (VSLS) containing bromine and chlorine.

Compound	Formula	Approximate tropospheric lifetime (days)	Source
Bromochloromethane	CH_2BrCl	150	Natural
Dibromomethane	CH_2Br_2	120	Natural
Bromodichloromethane	$CHBrCl_2$	78	Natural (minor anthropogenic source)
Tribromomethane (bromoform)	$CHBr_3$	26	Natural (minor anthropogenic source)
Trichloromethane (chloroform)	$CHCl_3$	150	Anthropogenic and natural
Dichloromethane (methylene chloride)	CH_2Cl_2	140	Anthropogenic (minor natural source)
Tetrachlorocthene (perchloroethylene, PCE)	C_2Cl_4	99	Anthropogenic
1,2-Dichloroethane	CH_2ClCH_2Cl	70	Anthropogenic
Chloroethane (ethyl chloride)	C_2H_5Cl	30	Anthropogenic and natural

Tropospheric lifetimes are estimated for typical conditions at 5 km.

This is important, because iodine is an even more efficient homogeneous catalyst of ozone depletion than bromine. Overall, the largest impact of VSLS is a greater role for bromine cycles in the radical chemistry of the stratosphere, particularly those cycles also involving HO_x (e.g. reactions (8.28)–(8.30)) and ClO_x.

In the 1970s, it was postulated that NO_x, emitted in the exhausts of supersonic aircraft (SST), would result in large ozone depletions. Since then there have been intermittent concerns over the possible impact of aircraft flights, both supersonic and subsonic, on ozone levels in the lower stratosphere. Production of a new generation of SSTs has not been ruled out by the aircraft industry. Total ozone changes, for cruise altitudes of 16 and 20 km, are calculated by models to be a few percent for reasonable estimates of the size of the supersonic fleet. Emissions higher in the stratosphere lead to larger local ozone losses because the NO_x emitted remains in the stratosphere for longer. Although changes in NO_x have the largest impact on ozone, the effects of H_2O emissions contribute about 20% to the calculated ozone change.

Many subsonic flights, cruising at 10–12 km (c. 220–290 hPa), pass through the lowermost parts of the stratosphere, especially at high latitudes (recall Figure 8.3). Subsonic aircraft flying in the North Atlantic flight corridor emit 44% of their exhaust emissions into the stratosphere. Models predict a resultant ozone decrease in the lower stratosphere of less than 1%, but modelling the lower stratosphere is particularly difficult as a number of chemical processes are of comparable importance to each other, and to the transport processes. Industry projections out to 2050 are for a 1–3% per year increase in air traffic, so the impact of aircraft on the atmosphere is likely to grow significantly.

8.6.2 World Avoided by the Montreal Protocol

The Montreal Protocol is rightly regarded as one of the most successful environmental treaties ever. Given this success at reducing the atmospheric levels of ozone-depleting chlorine and bromine compounds, it is nowadays easy to forget how serious a concern ozone depletion was in the final decades of the twentieth century, and how catastrophic the situation could have been without policy action. Models can be used to predict the state of the atmosphere under the assumption of no Montreal Protocol – i.e. the so-called 'world avoided'. Chipperfield et al. (2015), used a detailed 3-D chemical transport model to show that even by 2010–2011, uncontrolled increased in CFCs and similar source gases could have led to the regular occurrence of an Arctic ozone hole in cold stratospheric winters. Newman et al. (2009), used a coupled chemistry-climate model (CCM) to simulate the impact of very large increases in chlorine and bromine should CFC emissions continue unchecked to 2065. Their model, in agreement with other similar studies, predicted a collapse of the ozone layer. Such a collapse would have led to extremely large increases in biologically damaging UV radiation at the surface.

8.6.3 Chemistry–Climate Feedbacks

The overwhelming majority of expert opinion is now that human perturbation of the climate is taking place, and will increase (Chapter 1). Greenhouse gases cause a warming of the troposphere but a cooling of the stratosphere, because the greenhouse-gas 'thermal blanket' reduces the upwelling of infrared radiation to the stratosphere. Changes in stratospheric temperature lead to changes in stratospheric chemistry and stratospheric winds. Ozone columns, and so ultraviolet radiation penetration of the stratosphere, change, leading, ultimately, to further changes in climate. Quantifying the effects of this climate–ozone feedback requires integration and interpretation of coupled global CCMs, which operate at the limit of current computer power. Nevertheless, many research groups worldwide have created such models, and their results are combined to give the most robust predictions (Dhomse et al. 2018). Figure 8.8 shows column ozone from such CCM simulations for different latitude bands from 1960 to 2100.

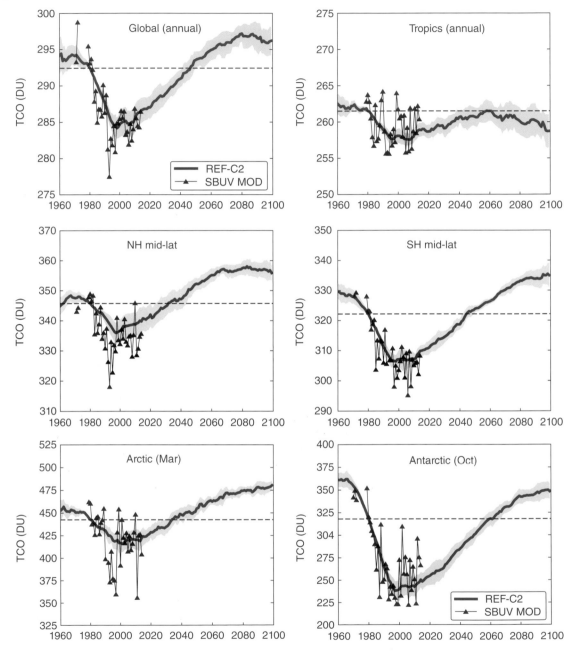

Figure 8.8 Past and future column ozone. Multi-model mean total column ozone (TCO) time series (DU) from coupled chemistry-climate model (CCM) simulations (REF-C2, grey line) and 1σ uncertainty (shading) for five latitudinal bands and the near-global (60° S–60° N) mean (see main text). The dashed black lines show the 1980 reference value for each latitude band. Also shown are satellite (merged solar backscatter ultraviolet [SBUV]) observations (triangles). *Source:* adapted from Dhomse et al. (2018).

The general picture is one of ozone depletion through 2000 followed by recovery. This pattern follows inversely the atmospheric loading of chlorine and bromine, which together peaked in the late 1990s. Note that the recovery is not predicted to be uniform in all regions. Notably, models predict that in the future column ozone in the tropics will decrease due to the effects of climate change increasing the speed of the upwelling branch of the Brewer-Dobson circulation. In other regions, column ozone recovers to values greater than in 1980 due to stratospheric cooling, which slows down the catalytic removal of ozone (Section 8.3). The interplay between CFC-induced ozone depletion and greenhouse-gas-induced climate change is of central concern to policy makers, as it demonstrates the connectedness of the Montreal and Kyoto protocols (and Paris Agreement), and should be of great interest to students of environmental science, because it demonstrates the links between environmental problems.

8.7 Summary

1) The ozone layer is the inevitable consequence of an external source of O_2-bond-breaking photons and an exponential decrease of atmospheric pressure with height.
2) Ozone is often quantified as a vertical column, measured in DU.
3) Stratospheric winds move ozone from its source region in the tropics towards the poles.
4) The Brewer–Dobson circulation describes the slow circulation of air through the stratosphere, and explains why the natural (unperturbed) ozone layer is thickest at high latitudes, with a large annual cycle.
5) Severe ozone depletion in the Antarctic occurs every year; severe ozone depletion in the Arctic is more sporadic due to meteorological variability and occurs during winters with cold stratospheric temperatures.
6) Substances that deplete stratospheric ozone do so via catalytic reaction cycles; this means that substances present at concentrations much smaller than that of ozone can have a significant effect on ozone concentrations.
7) The stratospheric aerosol and PSCs provide media for heterogeneous chemistry; this heterogeneous chemistry generally increases ozone depletion by chlorine and bromine.
8) Concentrations of many man-made ozone-depleting substances are declining, but significant ozone depletion is expected to continue for many years to come due to the long atmospheric lifetimes of CFCs and other source gases.
9) Recovery of the ozone layer is expected during this century. Due to climate change, the rate and extent of this recovery will vary with latitudinal region. Models predict that a faster stratospheric circulation will lead to decrease in column ozone in the tropics, despite recovery elsewhere.

Questions

1   What (i) dynamical information and (ii) chemical information could be used to show that high ozone mixing ratios in the lower troposphere are due to a tropopause fold?

2   Gravity waves move air up and down. The air is cooled by adiabatic expansion at the wave crests, and warmed by adiabatic compression at the wave troughs (Chapter 10). What amplitude of gravity wave is required to cool air at the tropical tropopause from 190 to 185 K?

3 Using your knowledge of the periodic table, suggest other possible catalysts for reaction (8.7).

4 Write an ozone-only catalytic cycle involving OH, and a cycle including OH, O, and O_3.

5 Write a reaction sequence for ozone depletion due to N_2O, showing (i) an initiation reaction, (ii) at least one catalytic cycle destroying ozone, (iii) one null cycle, and (iv) a termination reaction.

6 Test your dimensional analysis and units: show that $[R_n] = 10H^* \, kb \, Tc_n$ balances dimensionally.

7 Compare the chemical structures in Table 8.1 and Table 8.4. What chemical bond makes the difference between long-lived and short-lived organic source gases? What reactions will destroy VSLS compounds in the troposphere? *Hint:* Refer to the tropospheric chemistry outlined in Chapter 5.

References

Becker, G., Muller, R., McKenna, D.S. et al. (2000). Ozone loss rates in the Arctic stratosphere in the winter 1994/1995: model simulations underestimate results of the match analysis. *Journal of Geophysical Research – Atmospheres* 105 (D12): 15175–15184.

Brasseur, G.P. and Solomon, S. (2005). *Aeronomy of the Middle Atmosphere.* Berlin, Germany: Springer-Verlag.

Chipperfield, M.P., Dhomse, S.S., Feng, W. et al. (2015). Quantifying the ozone and UV benefits already achieved by the Montreal Protocol. *Nature Communications* 6: 7233. https://doi.org/10.1038/ncomms8233.

Dessler, A.E., Burrage, M.D., Groos, J.-U. et al. (1998). Selected science highlights from the first 5 years of the Upper Atmosphere Research Satellite (UARS) program. *Reviews of Geophysics* 36 (2): 183–210.

Dhomse, S.S., Kinnison, D., Chipperfield, M.P. et al. (2018). Estimates of ozone return dates from chemistry-climate model initiative simulations. *Atmospheric Chemistry and Physics* 18: 8409–8438. https://doi.org/10.5194/acp-18-8409-2018.

Dragani, R., Redaelli, G., Visconti, G. et al. (2002). High-resolution stratospheric tracer fields reconstructed with Lagrangian techniques: a comparative analysis of predictive skill. *Journal of the Atmospheric Sciences* 59 (12): 1943–1958.

Farman, J.C., Gardiner, B.G., and Shanklin, J.D. (1985). Large losses of total ozone in Antarctica reveal seasonal ClO_x/NO_x interaction. *Nature* 315 (6016): 207–210.

Holloway, A.M. and Wayne, R.P. (2010). *Atmospheric Chemistry.* Oxford, UK: Oxford University Press. ISBN: 978-1-84755-807-7.

Holton, J.R., Haynes, P.H., McIntyre, M.E. et al. (1995). Stratosphere–troposphere exchange. *Reviews of Geophysics* 33 (4): 403–439.

Law, K.S. and Sturges, W.T. (2006). Halogenated very short-lived substances. In: *Scientific Assessment of Ozone Depletion: 2006* (ed. C.A. Ennis), 2.1–2.46. Geneva, Switzerland: UNEP/WMO.

Lowe, D. and MacKenzie, A.R. (2008). Polar stratospheric cloud microphysics and chemistry. *Journal of Atmospheric and Solar – Terrestrial Physics* https://doi.org/10.1016/j.jastp.2007.09.011.

MacKenzie, A.R., Laaksonen, A., Batris, E., and Kulmala, M. (1998). The Turnbull correlation and the freezing of stratospheric aerosol droplets. *Journal of Geophysical Research – Atmospheres* 103 (D9): 10875–10884.

Manney, G.L., Santee, M.L., Rex, M. et al. (2011). Unprecedented Arctic ozone loss in 2011. *Nature* 478: 469–475.

McCormick, M.P., Thomason, L.W., and Trepte, C.R. (1995). Atmospheric effects of the Mt-Pinatubo eruption. *Nature* 373 (6513): 399–404.

McIntyre, M. (1992). Atmospheric dynamics: some fundamentals, with observational implications. In: *The Use of EOS for Studies of Atmospheric Physics* (eds. J. Gille and G. Visconti), 313–386. North Holland, Amsterdam: Proceedings of the International School of Physics.

Newman, P.A., Oman, L.D., Douglass, A.R. et al. (2009). What would have happened to the ozone layer if chlorofluorocarbons (CFCs) had not been regulated? *Atmospheric Chemistry and Physics* 9: 2113–2128.

Rampino, M.R. and Self, S. (1992). Volcanic winter and accelerated glaciation following the Toba super-eruption. *Nature* 359 (6390): 50–52.

Ren, C., MacKenzie, A.R., Schiller, C. et al. (2007). Diagnosis of processes controlling water vapour in the tropical tropopause layer by a Lagrangian cirrus model. *Atmospheric Chemistry and Physics* 7: 5401–5413. https://doi.org/10.5194/acp-7-5401-2007.

Rex, M., Salawitch, R.J., von der Gathen, P. et al. (2004). Arctic ozone loss and climate change. *Geophysical Research Letters* 31 (4).

Shepherd, T. (2003). Large-scale atmospheric dynamics for atmospheric chemists. *Chemistry Review* 103: 4509–4532.

Thomason, L. and Peter, T. (Eds) (2006). Assessment of Stratospheric Aerosol Properties (ASAP). SPARC Report, stratospheric processes and their role in climate.

Tolbert, M.A. and Toon, O.B. (2001). Solving the PSC mystery. *Science* 292: 61–63.

Waibel, A.E., Peter, T., Carslaw, K.S. et al. (1999). Arctic ozone loss due to denitrification. *Science* 283 (5410): 2064–2069.

WMO/UNEP (2018). *Scientific Assessment of Ozone Depletion: 2018*. Geneva: World Meteorological Organization.

9

Boundary Layer Meteorology and Atmospheric Dispersion

Janet Barlow and Natalie Theeuwes

Department of Meteorology, University of Reading, Reading, United Kingdom

The boundary layer is the lowest kilometre or so of the atmosphere where the biosphere, hydrosphere, and lithosphere interact with the air above. It is also the part of the atmosphere where we live, and where changes we make to the surface have their greatest impacts. In particular, this chapter focuses on the physical nature of surfaces covered in vegetation and urban canopies, and discusses their influence on the exchange of gases with the atmosphere.

Vegetation plays an essential role in modifying the atmosphere by uptake of carbon dioxide, thus making the air breathable for respiring creatures such as human beings. It also directly affects the atmosphere in a physical sense, changing the flow, temperature, and humidity of the local environment. This chapter takes a closer look at these processes from the scale of a leaf up to an entire forest.

Humans choose to live increasingly urban lives, and although cities constitute a tiny fraction of Earth's surface, the microclimate that cities create can influence millions of people. By learning about the physical impact of cities on the atmosphere, we can devise sustainable building and planning practices to make city microclimates more bearable in the face of impending climate change.

The urban sources of pollution have already been discussed in a previous chapter, but the atmospheric processes influencing the dispersion of pollutant gases once released into the atmosphere are outlined in this chapter.

9.1 The Atmospheric Boundary Layer

Garratt (1992) provides the following 'working definition' of the atmospheric boundary layer:

> ... the boundary layer [is] the layer of air directly above the Earth's surface in which the effects of the surface (heating and cooling, friction) are felt directly on time scales less than a day, and in which significant fluxes of momentum, heat or matter are carried by turbulent motions on a scale of the order of the depth of the boundary layer or less.

The boundary layer is approximately 1–2 km deep, its top being marked by a sharp increase in temperature (*inversion*), above which is the so-called *free atmosphere*.

9.1.1 Surface Energy Balance

Air motions within the boundary layer are strongly influenced by energy inputs from the surface. As the lower atmosphere is largely transparent to shortwave radiation, most of the energy in the Sun's rays reaches the surface. The amount of shortwave radiation actually absorbed by the surface depends on its reflectivity, or *albedo*, which varies between 0 and 1. For example, dense forests have an albedo of 0.1, whereas fresh snow has an albedo of approximately 0.8. Longwave radiation from the sky (clouds, water vapour) is also absorbed at the surface. The surface emits longwave radiation at a rate according to its temperature and *emissivity*, ε, which is the ratio of the energy radiated by an object to the amount radiated by a black body (i.e. a perfect radiator) at the same temperature. The *net radiation*, Q^*, is defined as the difference between incoming and outgoing short- and longwave radiation and is measured in units of $W\,m^{-2}$. By day in temperate latitudes Q^* is usually positive, whereas on a cloudless night it is negative.

By day the energy due to net radiation is used in three processes: downward flux of heat into the cooler soil (ground heat flux Q_G); upward flux of heat into the air by convection (sensible heat flux, Q_H); and evaporation of surface moisture into vapour, which is carried away from the surface by turbulent motions (latent heat flux, Q_E). The fluxes are positive when energy is flowing away from the surface: by night, as Q^* becomes negative, the surface is cooler than both the soil and the air, and the fluxes reverse sign. The *surface energy balance* is given by Eq. (9.1) and is shown in Figure 9.1

$$Q^* = Q_H + Q_E + Q_G \tag{9.1}$$

The ratio between the sensible and latent heat flux can be used to characterize the amount of energy partitioned into each component, known as the *Bowen ratio* (β):

$$\beta = \frac{Q_H}{Q_E} \tag{9.2}$$

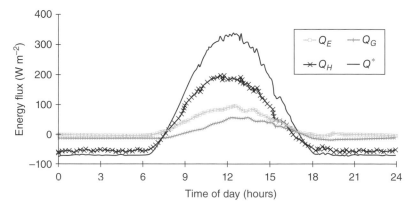

Figure 9.1 The surface energy balance measured at the University of Reading Atmospheric Observatory on 17 March 2006.

A higher ratio indicates less energy is partitioned into evaporation and more energy is used to heat the atmosphere. The Bowen ratio varies over different surfaces. Typical daytime values for different surfaces are: water, 0.1; irrigated crops, 0.2–0.4; grassland, 0.3–0.6; urban, 1–5; and desert, greater than 10.

9.1.2 Diurnal Cycle of the Boundary Layer

The diurnal cycle of heating and cooling of the surface causes distinct changes in boundary layer structure, as shown in Figure 9.2. Heating of the surface leads to *convection*, whereby buoyant air rises, causing turbulence, which mixes heat upwards through the depth of the boundary layer. This is called *convective instability* and is associated with a decrease in potential temperature, θ, with height. (The potential temperature is defined as the temperature that would be reached by an air parcel if it is moved adiabatically down from its original height in the atmosphere to 1000 mbar. Simply put, the temperature lapse due to decrease in atmospheric pressure with height is taken into account by using potential temperature, so that changes due to heating and cooling may be more clearly observed.) Heat can rise in localized plumes called *thermals* – a cumulus cloud may form on top of a thermal, as water vapour is lifted to a height where its dewpoint temperature is reached. Where there are no clouds, birds may be seen circling to stay within the warm updraught, which has vertical speeds of $1-2\,\mathrm{m\,s^{-1}}$.

Some thermals are buoyant enough to overshoot the temperature inversion marking the top of the boundary layer – that is, they keep rising a short distance, even though the air around them is warmer. This turbulent motion mixes warmer air downward and into the boundary layer, and is called the *entrainment heat flux*. Observations have shown this flux to be proportional to the *sensible heat flux* at the surface, and approximately 25% of its magnitude. For example, if the sensible heat flux is $100\,\mathrm{W\,m^{-2}}$ (upwards), then the entrainment heat flux is $-25\,\mathrm{W\,m^{-2}}$ (downwards). The fluxes of heat from the ground and the free atmosphere into the boundary layer together cause it to grow in depth throughout the day.

Following sunset, the surface heat flux reverses sign and convection ceases. This transition takes us to the *nocturnal boundary layer* in which the surface cools by infra-red radiation. As the surface cools much more rapidly than the air above, there is an increase of potential temperature with height, creating a *stable boundary layer* at the surface that grows in depth throughout the night as cooling continues. Turbulent

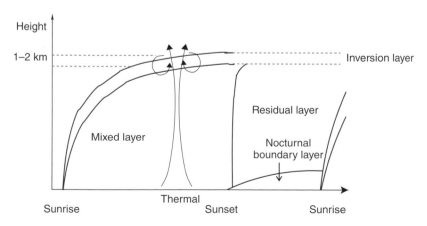

Figure 9.2 Diurnal cycle of the boundary layer.

overturning of the air is suppressed as the air is negatively buoyant, causing layers of air to form that do not mix with each other. If the air is saturated with water vapour then a layer of fog may form with a distinct top, due to the lack of turbulent mixing. Above the stable boundary layer, a *residual layer* forms in which remaining turbulence from the day before gradually dies out due to lack of convective motion. The temperature inversion marking the top of the previous day's convective boundary layer still remains. At sunrise, the surface warms, the sensible heat flux becomes positive and convection recommences.

Figure 9.3 shows schematic profiles of potential temperature, wind and specific humidity within the boundary layer by day and night. By day it can be seen that there is a *surface layer* occupying the lowest

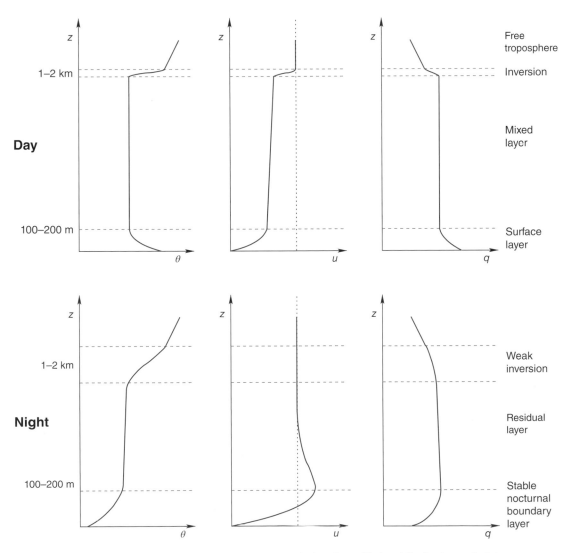

Figure 9.3 Schematic profiles of potential temperature, wind, and specific humidity by day and night.

10% of the boundary layer, which is *unstable* because the potential temperature decreases with height. Thermals are driven by surface heating but their action is to cause the rest of the boundary layer to be a *well-mixed layer*, with *neutral stratification* (i.e. potential temperature is constant with height). In the surface layer there is an increase of wind speed with height, and within the well-mixed layer there is little or no wind shear. *Specific humidity* is the mass of water vapour per mass of air, and is a passive scalar that is carried around by turbulent motions: it shows a similar profile to temperature, except that air above the boundary layer is much drier and hence a sharp negative gradient exists.

By night, the stable boundary layer occupies a typical depth of 100–200 m. At the top of this layer is a maximum in the wind profile that is called the *nocturnal jet*. This forms because there is negligible momentum exchange down across the stable layer towards the ground due to the lack of turbulence. (Momentum per unit volume of air is defined by multiplying air density and wind velocity.) In the residual layer, the flow is thus effectively decoupled from the surface – it 'feels' no friction – and hence is free to accelerate. The flow is influenced by the Coriolis force and the direction of the jet changes slowly throughout the night.

9.1.3   Surface Layer

This section focuses on the lowest 10% of the boundary layer – the *surface layer*. We have already seen that the wind speed decreases towards the surface in response to friction. The gradient in wind with height causes *shear instability*, where turbulent overturning is created, causing a flux of momentum down to the surface. Figure 9.4 shows a short time series of turbulence data measured at a height of 5 m. Wind velocity and temperature were measured using a sonic anemometer, which responds rapidly enough to capture turbulent variations. Components of the velocity vector U are usually defined as $U = (u,v,w)$, where u is in line with the downstream wind direction, v is laterally perpendicular to the

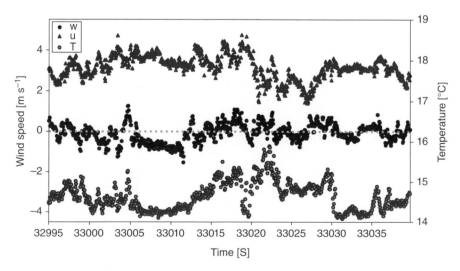

Figure 9.4   Turbulence data measured at the Chilbolton Observatory (UK) on 21 May 2011. Streamwise wind component *u*, vertical wind component *w*, and temperature, *T*.

wind direction, and w is vertically perpendicular. Each wind-speed component, such as u, can be separated into mean ($\bar{u}$) and fluctuating (u') parts over a given time period (t):

$$u(t) = \bar{u} + u'(t) \tag{9.3}$$

This is called *Reynolds averaging* and is a consequence of the observation that turbulent fluctuations happen on time scales much shorter (seconds or less) than the mean variations due to diurnal heating or mesoscale weather systems (hours to days). Fluctuations around the mean usually show a near-Gaussian distribution, meaning that the standard deviation (σ_u) is given by

$$\sigma_u = \sqrt{\overline{u'^2}} \tag{9.4}$$

This term is proportional to the amount of turbulence in the flow, in this case in the downstream, or 'streamwise' component of flow, u. A measure of wind gustiness is given by the *turbulence intensity*, defined as σ/U, where U is the mean wind speed. Turbulence intensity is typically 0.1–0.3 in the surface layer. The square of turbulence intensity depends on the fraction of total kinetic energy of the flow associated with turbulent fluctuations in each wind component. We can define the *turbulent kinetic energy*, e (per unit mass), as follows:

$$e = \frac{1}{2}\left(\overline{u'^2} + \overline{v'^2} + \overline{w'^2}\right) \tag{9.5}$$

Careful examination of Figure 9.4 shows that peaks in u are sometimes correlated with troughs in w. This shows that downward gusts transport horizontal momentum towards the surface. The covariance between u and w over a given time period, $\overline{u'w'}$, is a measure of the *momentum flux* – a negative value implies a downward flux. The momentum flux, or *shear stress*, τ, is thus given by Eq. (9.6), where ρ is the air density and u_* is the *friction velocity*, which is a scaling velocity to represent the strength of turbulence due to friction:

$$\tau = -\rho\overline{u'w'} = \rho u_*^2 \tag{9.6}$$

The rate at which mean wind speed U increases with height z above the ground is proportional to the friction velocity u_*. The vertical wind profile $U(z)$ in the surface layer under neutral conditions (i.e. when turbulence is due to shear instability, not buoyancy) is defined by the *log law*:

$$U(z) = \frac{u_*}{k} \ln\left(\frac{z}{z_0}\right) \tag{9.7}$$

where z_0 is the *roughness length* and is defined as the height where $U = 0$. It is not a physical length, although it can be thought of as a length scale representing the roughness of the surface. For example, over short grass, $z_0 \sim 0.001$ m and a rule of thumb states that $z_0 \sim 0.1\,h$, where h is the mean height of surface roughness elements. Through experimentation, the value of k has been established to be ~0.4. This is called *von Kármán's constant*.

A final look at Figure 9.4 shows that peaks in vertical velocity w are sometimes correlated with peaks in temperature, T. This shows that upward gusts carry warm air upwards – as these data were measured during the day, this is consistent with a positive sensible heat flux Q_H. In fact, we can write the sensible heat flux in terms of the covariance:

$$Q_H = \rho c_p \overline{w'T'} \tag{9.8}$$

where ρ is the air density and c_p is the specific heat capacity of dry air at constant pressure. There are occasional 'ramps' in the temperature trace – increases in T by several degrees followed by sudden drops. Ramps can be associated with passing thermals embedded in the mean flow.

9.1.4 Atmospheric Stability

Qualitatively, an unstable boundary layer is associated with a positive heat flux, and a stable boundary layer with a negative heat flux. The amount of turbulent kinetic energy is hence clearly dependent on production or suppression of turbulence by buoyancy. But as outlined in the previous section, it also depends on shear production. So, despite suppression of turbulence near the surface at night, if there is significant wind shear, then turbulence will be created. To quantify the ratio between buoyant production or suppression, and shear production of turbulence, the *Monin–Obukhov stability parameter* can be defined as:

$$\frac{z}{L} = \frac{-\left(g/\theta_0\right)\left(Q_H/\rho c_p\right)}{u_*^3/kz} \tag{9.9}$$

where z is the height above ground, θ_0 is the potential temperature at the surface, and L is the *Obukhov length*. For convective conditions, with large positive Q_H, low wind speed (and thus low u_*), z/L is large and negative. For windy, overcast days, Q_H is small and positive, but u_* is large, and thus z/L is negative and near zero. Clear, cloudless nights with low wind speed lead to the most stable conditions, where z/L is positive. For $z/L > 1$, turbulence 'switches off' almost completely in the highly stable conditions. There are only occasional bursts of turbulence, possibly triggered by the nocturnal jet high above the ground. L is, again, a length scale rather than a physical length, but one physical interpretation is that for $z > |L|$, convective motions dominate the turbulence, whereas for $z < |L|$, shear instability due to the surface dominates.

The beauty of this stability parameter is that it tells us something fundamental about the atmosphere near the surface: measurements might be made at different heights, with different Q_H and u_*, in different locations, but if the value of z/L is the same, then the turbulence should have similar properties. This is the *Monin–Obukhov similarity theorem*, which allows us to generalize profiles of mean and turbulent characteristics, and hence use the resulting equations to model the atmosphere for numerical weather prediction, dispersion models, etc. For instance, the wind profile equation can be modified to allow for non-neutral conditions:

$$U(z) = \frac{u_*}{k}\left\{ ln\left(\frac{z}{z_0}\right) - \psi_m\left(\frac{z}{L}\right) \right\} \tag{9.10}$$

where ψ_m is a function of stability parameter z/L, derived from measurements. A famous set of experiments made in Kansas in 1968 (Kaimal and Wyngaard 1990) provided one of the first, major datasets from which such stability functions could be derived.

Worked Example 9.1

A student measured fluxes on a sunny but breezy day. She obtained the following values at a height of $z = 2\,\text{m}$:

> sensible heat flux $Q_H = 200\,\text{W}\,\text{m}^{-2}$
> momentum flux $\tau = 0.1\,\text{kg}\,\text{m}^{-1}\,\text{s}^{-2}$
> surface potential temperature $\theta_0 = 18\,^\circ\text{C}$

What was the stability parameter z/L, and what did this tell her about the convective state of the atmosphere? Assume that the air density $\rho = 1.2\,\text{kg}\,\text{m}^{-3}$, and the specific heat capacity of dry air at constant pressure $c_p = 1004\,\text{J}\,\text{kg}^{-1}\,\text{K}^{-1}$.

z/L is given by Eq. (9.9), for which we need surface potential temperature in degrees Kelvin:

$$\theta_0 = 18 + 273 = 291K$$

and we can calculate the friction velocity using Eq. (9.6):

$$u_* = \sqrt{\frac{\tau}{\rho}} = \sqrt{\frac{0.1}{1.2}} = 0.29\,ms^{-1}$$

From Eq. (9.9):

$$\frac{z}{L} = \frac{-\left(g/\theta_0\right)\left(Q_H / \rho c_p\right)}{u_*^3 / kz} = \frac{-\left(9.81/291\right)\times\left(200/\left(1.2\times1004\right)\right)}{0.29^3/\left(0.4\times2\right)} = -0.18$$

z/L is negative, showing that the atmosphere is unstable. Its magnitude is small, but not near-zero (which we usually define as being approximately $-0.03 < 0 < 0.03$). Therefore, we describe the conditions as *slightly unstable*.

9.1.5 Turbulent Time and Length Scales

Figure 9.4 showed that variability in u exists at many different scales: rapid fluctuations are superimposed on slower changes; slower changes tend to cause larger changes in wind speed. This indicates that turbulence consists of larger, stronger eddies, generated by wind shear and convection, which generate smaller and smaller eddies, until the eddies are small enough to dissipate their energy through friction due to molecular viscosity. L.F. Richardson (1922), inspired by R.L. Stevenson, was so struck by this fact that he penned a poem:

> 'Big whorls have little whorls that feed on their velocity,
> and little whorls have smaller whorls and so on to viscosity.'

One consequence of such structure within the seemingly random mess of turbulence is this: if we take measurements of velocity at two points located near to each other, fluctuations will be correlated with each other to some degree. As we separate the measurements, the correlation will be weaker, until at a certain separation, there is no correlation. This length scale is known variously as the *decorrelation length scale* or the *integral length scale*, L_x. Its size is similar to the largest, most dominant eddies in the atmosphere: in a convective boundary layer these are limited only by the boundary layer depth (of order 1000 m), whereas in stable conditions, they may be tens to hundreds of metres.

It is not convenient to make measurements at two locations up to a kilometre apart to test this idea! An important principle used for interpreting turbulence measurements made at a single point in space is *Taylor's frozen turbulence hypothesis*. This assumes that most of the time variation at an observation point in space is associated with the advection of eddies past that point by the mean wind speed U. The relationship between the spatial structure of turbulence and the temporal variations recorded by the instrument is thus given by $x = Ut$: we can deduce an integral time scale T_L from a time series of the wind speed, and deduce the integral length scale if we know the mean wind speed by $L_x = UT_L$. For more information about the statistics used to do this, Ibbetson (1981) and Kaimal and Finnigan (1994) provide clear explanations. The importance of the integral time scale will become clear when we consider dispersion of pollutants in Section 9.4.

9.2 Flow over Vegetation

It has been seen so far that the boundary layer responds to surface roughness and heat fluxes. A large fraction of the Earth's surface is covered with some kind of vegetation – forest, grass, crops, etc. Vegetation forms an interface between the physical and the biological spheres: its annual cycle of growth has an impact on the exchanges of carbon dioxide and water vapour with the atmosphere, and also on the radiative characteristics of the Earth's surface. We now consider vegetation canopies and the impact they have on the surface energy balance and flow dynamics.

9.2.1 Canopy Characteristics

Air flowing over a plant encounters many obstacles: leaves, twigs, stems, branches, and trunks. In terms of gas exchange, a leaf is covered in many, tiny holes called stomata that allow exchange of carbon dioxide and water vapour with the air. The plant can close or open stomata to regulate gas exchange – for instance, if the plant is water-stressed, stomata will close to prevent water loss.

A single leaf affects the radiation balance: it reflects shortwave radiation; it absorbs different wavelengths of the visible spectrum (VIS) (think: Why do most leaves look green?), and shades other leaves beneath it. It exchanges heat with the air through turbulent fluxes, like any other surface. The flow over the leaf depends on its orientation, and how smooth it is. Also, unlike other obstacles on the ground, leaves can flutter and move, which generates turbulence.

All these physical processes are important, but we cannot hope to model them for all the plants on the planet. Instead we consider the bulk effects of a collection of plants, i.e. a *vegetation canopy*. Each canopy has a different structure, depending on species, that can be quantified by using morphometric variables, e.g. mean height and spacing. As leaves are a crucial interface with the physical environment, we define the *leaf area index* (LAI) as the plan area of leaves (one side of leaf only) per unit area of ground. Some typical LAI values are grass, 2; winter wheat, 5–8; desert plants, 1; tundra, 3; tropical forest, 5.

9.2.2 Radiative Characteristics

The distribution of leaves within the canopy will affect the radiative characteristics: sunlight is absorbed more by higher leaves, and scattered throughout the canopy. Typical albedo values for canopies range from 0.1 to 0.3, which means that canopies look quite 'dark' in the visible part of the spectrum. Albedo decreases for taller canopies: the deeper the canopy, the more scattering that occurs and the lower the albedo. In terms of longwave characteristics, vegetation behaves almost like a blackbody: typical values of emissivity are 0.93 for French beans and 0.99 for geraniums.

As large areas of the planet are covered in vegetation, satellites have become increasingly important for monitoring the amount of vegetation cover. The spectrum of radiation reflected or emitted by foliage is distinct from that of soil or water. We can characterize this spectral 'signature' by measuring the emitted radiation in the near infrared (NIR) and the red part of the VIS. If we know what the incident radiation is (in practice, this is determined from reflection from a white surface such as snow), then we can define the following ratios:

ρ_{NIR}: emitted NIR/incident NIR
ρ_{VIS}: emitted VIS/incident VIS

These ratios will vary according to plant type and stage of growth. The chlorophyll in a plant strongly absorbs visible radiation for photosynthesis, meaning that ρ_{VIS} is small for a healthy, green plant. In contrast, the cell structure of a leaf will reflect much of the incident NIR radiation, making ρ_{NIR} large. The *normalized difference vegetation index* (NDVI) can indicate the type of vegetation:

$$NDVI = \frac{\rho_{NIR} - \rho_{VIS}}{\rho_{NIR} + \rho_{VIS}} \tag{9.11}$$

Typical values for vegetation of increasing greenness range from 0.1 to 0.8. Values between 0 and 0.1 can represent rocks, bare soil, and wintertime deciduous forests. Negative values can indicate rain, cloud, or snow.

Worked Example 9.2
A deciduous forest absorbs 90% of incident visible light, and reflects 50% of NIR radiation. What is the value of the NDVI?

Answer
For visible light, 90% is absorbed, so 10% is emitted. Expressed as a fraction, the ratio of emitted to incident visible radiation is:

$$\rho_{VIS} = 0.1$$

For NIR radiation, 50% is reflected, which is the same as saying that it is emitted. So:

$$\rho_{NIR} = 0.5$$

NDVI is given by Eq. (9.11):

$$NDVI = \frac{\rho_{NIR} - \rho_{VIS}}{\rho_{NIR} + \rho_{VIS}} = \frac{0.5 - 0.1}{0.5 + 0.1} = \frac{0.4}{0.6} = 0.67$$

This is a relatively high value, showing that the deciduous forest is actively growing with good leaf cover. If the forest experienced water stress, or was about to drop its leaves in autumn, NDVI might reduce to 0.1–0.2.

9.2.3 Gas Exchange – Carbon Dioxide

One of the key gases that vegetation exchanges with the atmosphere is carbon dioxide (CO_2). Vegetation absorbs CO_2 during photosynthesis (i.e. there is a downward flux of CO_2 to the canopy) and emits CO_2 (i.e. upward flux) as it respires at night. Water vapour is also lost from the plants, and the sign of the moisture flux depends on the specific humidity gradient – positive (upward) fluxes during the day, negative (downward) fluxes at night.

First, let us consider typical values of CO_2 concentrations measured in the Amazon rainforest, one of the 'green lungs' of the planet. There is a large diurnal range of concentrations, between approximately 550 ppm (night) and 375 ppm (day); the range reduces to a few tens of ppm for cooler days due to a reduction in photosynthesis. CO_2 flux changes in sign from negative during the day to positive at night. The small net difference in flux, if negative, is the net uptake by the canopy, which means that carbon is locked away in solid form in the vegetation rather than in gaseous form in the atmosphere.

Figure 9.5 shows a schematic of the CO_2 profile in the rainforest. Figure 9.5a shows the distribution LAI as a function of height – the crowns of the trees contain more leaves, and thus near the top of the

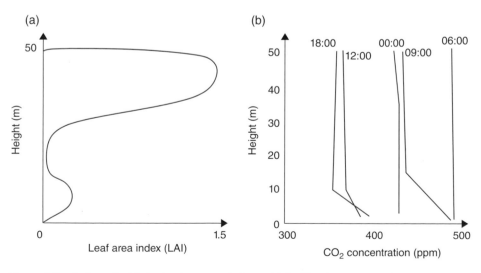

Figure 9.5 Carbon dioxide in the Amazon rainforest. (a) Profile of leaf area index (LAI) with height. (b) Profiles of carbon dioxide concentration by day and night.

canopy there is a large 'sink' of CO_2 during the day, and 'source' at night. Organisms in the soil also cause CO_2 emissions. Figure 9.5b shows the CO_2 profiles: note increased values near the ground source during the day; generally well-mixed profile (i.e. small gradients) within the canopy; diurnal cycle between low values during the day, and large values at night.

Despite the nonuniform distribution of carbon dioxide sources and sinks, there is an approximately uniform concentration profile. This suggests that the flow is good at mixing CO_2 throughout the canopy. The next section considers how a canopy changes the wind speed and turbulence characteristics.

9.2.4 Dynamics of Canopy Flows

9.2.4.1 Above Canopy

Over a canopy of objects (e.g. forest of trees, crop of wheat, city of buildings) we have to introduce an offset in height to allow for upward displacement of the flow. Thus, the neutral stability log law (Eq. 9.7) becomes

$$U\left(z\right) = \frac{u_*}{k} ln\left(\frac{z-d}{z_0}\right)$$

(9.12)

where d is the *displacement height*, or the plane above which height is effectively determined. Another rule of thumb states that $d \sim 2/3\,h$, where h is the mean canopy height. Figure 9.6 shows mean wind speed as a function of height $U(z)$ over and within a canopy of mean height h. Note that the y-axis has a logarithmic scale, and so the relationship given in Eq. (9.12) holds above the canopy and is extrapolated as a dotted line to intersect with the y-axis at $z = d + z_0$. The actual wind profile deviates away from the logarithmic relationship down into the canopy, but first we consider the logarithmic part of the profile.

Rules of thumb for how z_0 and d depend on the height of the canopy have already been given. However, as we now know more about the structure of the canopy, we can be more precise about the exact relationship. For example, Shaw and Pereira (1982) found that z_0 and d depend on the density of the canopy, i.e. the number of plants per unit area. Figure 9.7a and b show schematically how z_0/h and d/h depend on

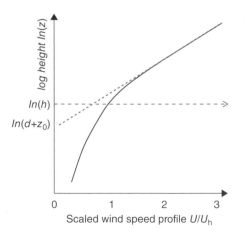

Figure 9.6 Wind speed profile above and within a canopy. Note the logarithmic scale on the y-axis.

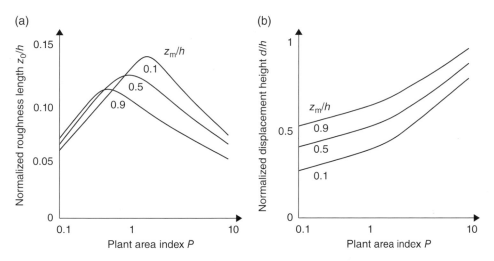

Figure 9.7 Roughness parameters (a) z_0/h and (b) d/h as a function of plant packing density P, and relative height of foliage maximum z_m/h.

the plant area index (P), which is the total area of plants occupying an area of ground. Figure 9.5a showed that the distribution of leaves and branches in a canopy is not uniform with height. As the drag exerted on the flow will be proportional to the amount of foliage, this is taken into account by assuming that P is a simple function of height: P is zero at canopy top and bottom, and maximum at some height $z = z_m$, varying linearly with height in between.

Note that z_0/h has a peak, the location of which depends on where maximum foliage is, whilst d increases monotonically. So the most dense canopy is not the most rough – as plants become more densely packed (i.e. P increases), flow is displaced further upwards (d increases) and the flow loses less momentum (z_0 decreases).

It should be noted that as the flow within a canopy is influenced by every obstacle it encounters, it is highly *inhomogeneous*. A profile measured at a location is different from one measured 10 cm away. In order to deal with this, we take spatial averages of the flow within the canopy, remembering not to include the volume taken up by the plants (as they usually do not move very fast!). So the mean wind speed at a given height is the mean over space and time. Above the canopy, the wind profile is not sensitive to the location where it is measured, and the turbulent quantities are said to be *homogeneous*: that is, they are the same everywhere (see also Figure 9.9).

Similar to wind speed, temperature and humidity profiles are also logarithmic:

$$\theta(z) - \theta_0 = \frac{\theta_*}{k}\left\{ \ln\left(\frac{z-d}{z_h}\right) - \psi_h\left(\frac{z-d}{L}\right)\right\} \tag{9.13}$$

$$q(z) - q_0 = \frac{q_*}{k}\left\{ \ln\left(\frac{z-d}{z_q}\right) - \psi_q\left(\frac{z-d}{L}\right)\right\} \tag{9.14}$$

where $\theta_* = -\overline{w'\theta'}/u_*$ and $q_* = -\overline{w'q'}/u_*$ are the scaling parameters for the potential temperature and specific humidity profiles respectively and are analogous to the friction velocity, u_*. The values θ_0 and q_0 are the surface values of potential temperature and humidity. The surface is defined at the height $z = d + z_h$ and $z = d + z_q$, respectively. Variables z_h and z_q are the roughness lengths for heat and moisture: just as z_0 is proportional to how much momentum is absorbed by the canopy, z_h and z_q are proportional to how much heat and moisture, respectively, are emitted or absorbed by the canopy. ψ_h and ψ_q are the stability functions for heat and moisture, respectively, analogous to the stability function for momentum defined in Eq. (9.10).

9.2.4.2 Within Canopy

Within vegetation canopies the wind speed is small and turbulence intensity is very high. Flow measurements have only relatively recently been accurate enough to cope with this challenging environment. Research has shown (Kaimal and Finnigan 1994) that measured flow profiles within various canopies, e.g. pine forests, corn fields, artificial wind-tunnel models, are similar despite differing canopy size. This means that there are common flow characteristics if profiles are scaled correctly, i.e. height is divided by mean canopy height, z/h, mean wind speed is divided by mean wind speed at the top of the canopy, U/U_h, and turbulent velocities are divided by friction velocity, σ/u_*. The following flow characteristics emerge:

- *Mean wind speed, U.* Figure 9.6 shows a decrease of wind speed with height within the canopy which is not logarithmic. This behaviour can be described mathematically by an exponential decay function

$$\frac{U}{U_h} = exp\left[-v_e\left(1 - \frac{z}{h}\right)\right] \tag{9.15}$$

where v_e is an extinction coefficient, which is approximately proportional to LAI.
- *Momentum flux, τ.* Within the canopy, τ decreases rapidly with height, showing that the ground itself absorbs very little momentum from the flow. This also shows that turbulence is strongly inhomogeneous in the canopy, i.e. its properties vary with height, in contrast to the surface layer above.
- *Standard deviations, σ_u, σ_w.* Turbulent fluctuations also decrease in size towards the ground. Note that as mean wind speed also decreases, it can be seen that high turbulence intensity is maintained right down to the ground, i.e. there is considerable mixing throughout the canopy. This explains why the profiles of carbon dioxide concentration shown in Figure 9.5b were almost constant with height.

9.3 The Urban Boundary Layer

The world's population is becoming increasingly urbanized, with the proportion living in cities expected to reach 6 billion people, or two-thirds of the global population, by the year 2050. In industrialized nations, this figure is already higher, with 70–80% of the population living in urban areas in 2003. Urban areas cover only 1–3% of Earth's surface, and their impact on global climate is small. However, their impact on local weather is large, and designing cities able to withstand future climates requires more sophisticated parameterizations of urban areas in global climate models (WMO 2014). Urban areas are also dominant sources of many pollutants, and therefore quantifying concentrations and emission rates of gases such as carbon dioxide requires an understanding of the urban boundary layer (Barlow 2014).

From the point of view of a city planner or architect, knowledge of the physics of heat exchange within a city can help to design cool buildings to withstand a warming climate (Hacker et al. 2005). The European heat wave in summer 2003 led to many deaths, particularly in Paris, and may become a more common event. Dispersion of pollutants, or even a catastrophic release of toxic gas, depends on the nature of urban flow, and is considered in the final section of this chapter.

9.3.1 Why Are Urban Areas Different from Rural Areas?

Built environment materials are generally impervious, which means that the surface is less moist and therefore latent heat fluxes are smaller. The materials have an impact on the balance of shortwave and longwave radiation at the surface. In addition, human activity leads to heat release into the atmosphere through combustion, the anthropogenic heat flux. Urban areas consist of groups of relatively large obstacles, which make them rougher aerodynamically. We first consider their impact on flow at three horizontal length scales: street, neighbourhood, and city scale (Britter and Hanna 2003).

9.3.1.1 Street Scale

This is the scale of a single building or street, up to approximately 100 m. Flow around a building can be broken down into several features: the flow impacts on a building, slowing considerably, thus forming a *stagnation point*. The flow then diverges around and over it, and separates (overshoots) at the corners, which causes vortices, particularly in the building wake. The wake extends several building heights downstream and is highly turbulent.

Two lines of buildings form a *street canyon* (see Figure 9.8). The *aspect ratio* is a key parameter: the ratio of the mean building height, H, to the street width, W. The flow forms three different *flow regimes*.

- *Isolated roughness* ($H/W < 0.3$) – each building and the wake it creates is isolated from the next one.
- *Wake interference* ($0.3 < H/W < 0.65$) – the wake of a building interferes with the flow over the next building. The flow in the street is highly turbulent, causing relatively good ventilation of pollution into the air above.
- *Skimming flow* ($H/W > 0.65$) – there is a single vortex within the street and little exchange with the air above.

Figure 9.8 Sideview of street canyon flow for isolated roughness (top), wake interference (middle), and skimming flow regimes (bottom).

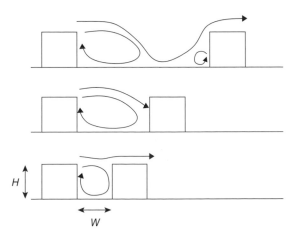

Figure 9.8 provides sketches of these regimes; note that these pictures are relevant if the flow is perpendicular to the orientation of the streets. If the mean flow is oriented parallel to the streets then there can be a *channelling* effect, where flow is accelerated along the street due to being confined by it.

9.3.1.2 Neighbourhood Scale

The neighbourhood scale is approximately 1–2 km, where the groups of streets and buildings can be treated collectively as a very rough surface, similar to a canopy of trees. To define a neighbourhood, buildings, trees, and other objects do not vary too much in size and form across it, i.e. it is relatively homogeneous. Some of the flow characteristics are very similar to those of vegetation canopies already discussed. The *urban canopy* encompasses the region from the ground up to the mean building height H. We can define two other layers in terms of height z – the *roughness sublayer* of depth z_*, and the *inertial sublayer* above it – which have the following characteristics.

- *Roughness sublayer* $(z < z_*)$ – flow is three-dimensional around the buildings, as shown in Figure 9.9a up to $z_* = 3H$. Also shown are vertical dashed lines showing the positions of measured wind profiles which are shown in Figure 9.9b. Within the roughness sublayer (i.e. $z < 3H$), each wind profile is spatially dependent, i.e. different values upwind and downwind of a building. Also shown is the spatially averaged profile, which is smooth and exponential below H.
- *Inertial sublayer* $(z > z_*)$ – flow is horizontally homogeneous. Figure 9.9b shows that local wind profiles are identical to the spatially averaged wind profile above $3H$.

The surface layer described in Section 9.1 encompasses both the inertial sublayer and roughness sublayer. The depth of the roughness sublayer, z_*, depends on how closely packed the buildings are and is at least $2H$ over densely packed buildings, and up to $5H$ for more sparse urban canopies (Rotach 1999). A roughness sublayer will develop over any rough surface – even short grass! In the case of grass, it is so

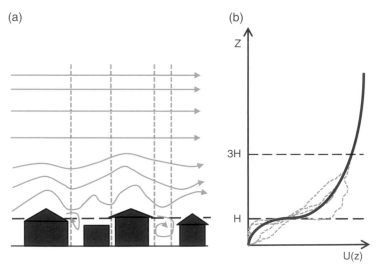

Figure 9.9 Roughness and inertial sublayers. (a) Flow over buildings – vertical dashed lines show positions of measured wind profiles. (b) Measured (dashed) and spatially averaged (bold, solid) wind profiles.

shallow that all measurements will be located in the inertial sublayer. In cities, we have the opposite situation: the roughness sublayer may be tens of metres in depth, and it is difficult to measure in the inertial sublayer above.

Hence, the mean wind profile over an urban canopy is very similar to a vegetation canopy. Figure 9.9b shows an urban canopy of height H, with a log law wind profile in the inertial sublayer above $3H$; the roughness sublayer extending from just above the buildings down to the ground; and an exponential profile within the urban canopy layer. Also similar to vegetation canopies, it has been found that the roughness length z_0 and displacement height d depend on how densely the buildings are packed. One measure of this is the *frontal area index*, λ_F. This is defined as the ratio of total frontal area of buildings to the total ground area they occupy. Note that λ_F varies between 0 (no buildings) and 1 (buildings only), which is different to vegetation canopies, where LAI can be more than 1. In addition, the maximum amount of foliage – and hence, drag on the wind – may be at a certain height z_m, whereas in an urban canopy the building 'mass' is generally more uniform with height. One final difference is that buildings are large, 'bluff bodies' taking up a large volume within the canopy, whereas plants are porous – the wind flows through them – and take up a smaller volume.

Despite these differences in canopy structure, z_0 and d behave in a similar way: z_0 reaches a maximum for intermediate values of λ_F, d increases monotonically with λ_F. Some of the turbulent characteristics within the urban canopy are similar: the momentum flux τ reaches a peak near the top of the urban canopy, although some studies have found a maximum just above the mean building height (Kastner-Klein and Rotach 2001). The standard deviations, σ_u, σ_v, and σ_w, decrease in magnitude towards the ground. However, in the presence of moving traffic, there can be an increase in turbulent kinetic energy up to a height of a few metres (Kastner-Klein et al. 2003). This is important to consider when modelling pollutant dispersion.

9.3.1.3 City Scale

At the scale of a whole city (up to 10–20 km), the whole boundary layer adjusts as air flows from the rural surface upstream over the urban surface. The change in roughness will modify the wind profile. Consider the simplest case of a sudden change in surface roughness from smooth (rural) to rough (urban) in neutral conditions, shown in Figure 9.10. The upstream flow is in equilibrium with the rural surface, of

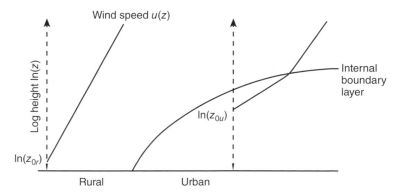

Figure 9.10 Roughness change from rural to urban surface. Internal boundary layer (IBL) grows with distance downstream. Wind profiles with logarithmic height axis are marked in upstream and downstream of roughness change.

roughness length z_{0r}, and the surface layer wind profile is logarithmic (see Eq. 9.7). Upon encountering the change in roughness at the edge of the city, an *internal boundary layer* (IBL) forms of depth $h_b(x)$, where x is the distance from the edge of the city. The IBL is the region where there is greater momentum exchange with the new, urban surface, and thus the wind profile has a different value of u_*. Usually we define h_b as being the height at which the wind (or momentum flux) returns to a value within a few percent of that at the same height over the upstream surface. Often, the top of the IBL is marked by a 'kink' in the vertical mean wind profile, where the shear changes rapidly. The urban surface gradually affects a greater depth of the atmosphere; hence, the IBL 'grows' at a rate determined by the turbulent diffusion of momentum – it will grow in depth more quickly over a rougher surface. Urban areas may have roughness lengths of order 1–2 m. About 10 km downstream of the roughness change, the IBL will reach a depth of 1 km, thus occupying the full depth of the boundary layer. As the rural boundary layer re-establishes itself downstream of the urban area, an *urban plume* remains aloft and carries pollutants, etc., downstream of the urban area. Urban plumes can extend tens or even hundreds of kilometres downstream and can affect cloud formation, as particulate pollution interacts with water droplets.

When air encounters an urban surface, it is not only rougher but warmer. This can cause buoyant uplift, which sucks air into the city at the surface from all directions, as shown in Figure 9.11. Here, we see an example produced with a turbulence-resolving model (large eddy simulation). The area in the middle of the domain has a sensible heat flux twice the size of the surrounding surfaces. This higher sensible heat flux causes larger updrafts than the surroundings and a *thermally induced circulation* develops. The horizontal wind close to the surface is inward towards the area of enhanced heating, and higher up in the atmosphere the flow is outwards. Over urban areas, the flow induced at the surface is known as the *urban breeze* and is only present for strong heating and low wind speeds. Observational and modelling studies have found that the urban breeze is not observed for background wind speeds exceeding 3–4 m s^{-1}. Some studies have suggested that increased uplift over an urban area can enhance convective

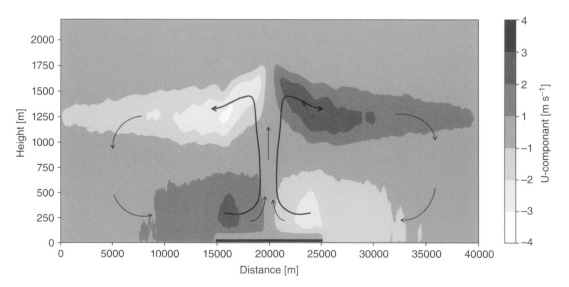

Figure 9.11 Large eddy simulation of a thermally induced circulation over a city.

rainfall. This has been observed over Atlanta, Georgia, a large US city in the middle of flat terrain, at a latitude with strong solar input. The boundary layer is often deeper over urban areas, having implications for pollution dispersion: as the effective mixing height of the pollutants is increased, the concentration decreases.

9.3.1.4 Surface Energy Balance

One of the first difficulties in describing the urban surface energy balance is defining what the surface is. Given the complexity of the environment, one of the easiest definitions is to define the urban surface as the top of the urban canopy. Then the incoming energy interacts with not only individual building and ground surfaces, but also with the layer of air between the buildings. Second, the *anthropogenic heat flux* is due to sources mostly within the urban canopy layer. Although it can be explicitly included in the surface energy balance, it is difficult to quantify. Some of it will be partitioned into the sensible heat flux and some into the latent heat flux, as energy is used in evaporation. Hence, if we measure radiation and turbulent fluxes above the urban canopy layer, the surface energy balance equation is given by:

$$Q^* + Q_F = Q_H + Q_E + \Delta Q_S + \Delta Q_A \tag{9.16}$$

By comparison with Eq. (9.1), there are significant differences. There are two additional terms, namely the anthropogenic heat flux Q_F, which is the energy related to human activities (e.g. waste heat from buildings and fuel combustion), and the advective flux, ΔQ_A. This term represents the horizontal transport of heat by the wind, which can be large where there is a distinct change of surface (e.g. from warm buildings to a cool park). Over homogeneous surfaces, this term is neglected as all heat transport is assumed to be vertical. The ground heat flux, Q_G, has been neglected and replaced by the storage term, ΔQ_S. Heat is still conducted down into the relatively small amount of exposed ground, but is also conducted into buildings and stored for slow release later in the day. A small amount is used in warming the air between the buildings: ΔQ_S is the sum of heat stored in the urban fabric and air. Practically, the storage term is hard to measure and is often determined as the residual of the other terms, where measurements have been placed well downstream of any changes in roughness and surface type to minimize heat advection.

So how does an urban surface affect each of the terms in Eq. (9.16)? Consider first the net radiation, Q^*. Due to pollutants such as sulphate particles, shortwave radiation input is reduced by scattering compared with that over a rural area – this reduction may be as large as 30%. The urban albedo is generally lower than most rural areas, with a typical value of 0.15. The reason for this is partly due to the urban geometry, as shortwave radiation is reflected from vertical building surfaces and scattered repeatedly from other walls. Hence, the outgoing shortwave radiation is smaller, but as the incoming shortwave is also reduced by pollution effects, the shortwave contribution to the net radiation is similar over urban areas when compared to rural areas. Incoming longwave radiation can be greater because of pollution. Carbon-based pollutants, such as soot, are particularly good absorbers and emitters of longwave radiation. Nocturnal longwave cooling to space can be reduced over the urban area compared to a flat surface. This is because the street canyon geometry reduces the *sky-view factor*. A flat emitting surface has a sky-view factor of unity. However, for a street, radiation emitted from the canyon at angles of less than about 60° to the horizontal are absorbed by the buildings. Hence the sky-view factor is reduced to approximately 0.5. Given all these effects, it is perhaps surprising that Q^* can be similar over urban and rural areas.

When comparing the diurnal variation in surface energy balance terms over a suburban area and a rural area, certain characteristic differences emerge. The suburban latent heat flux is considerably smaller during the day, due to the lack of moisture sources. The sensible heat flux is considerably larger, some of which is due to anthropogenic sources, and peaks slightly later in the day. The storage term (equivalent to the ground heat flux for the rural surface) is large and positive during the day, and becomes negative at sunset, as the buildings start to cool. Sensible heat flux remains positive after sunset: it is thought that the heat stored in the buildings is released slowly due to the building material's large heat capacity. As the surface is warmer than the air, the sensible heat flux must be positive. This means that daytime unstable or neutral conditions persist over urban areas for longer than over rural areas, which has consequences for the dispersion of pollutants.

It is observed, mostly at night, that the urban area can be several degrees warmer than the surrounding rural area, forming an *urban heat island* (UHI). A schematic of the daily temperature variation and the resulting UHI is shown in Figure 9.12. As a direct consequence of the differences in surface energy balance in urban and rural areas, there is a difference in net heating and cooling rates: the urban area

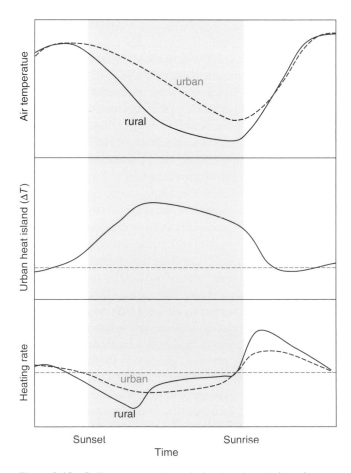

Figure 9.12 Daily temperature variation in urban and rural areas.

cools off more slowly. The resulting temperature difference between urban and rural areas, ΔT, quantifies the UHI intensity. The smaller cooling rates over the urban area are caused by a combination of factors: (i) slow release of heat stored in buildings (ΔQ_S), (ii) anthropogenic heat release (Q_F), and (iii) the sensible heat flux remaining positive after sunset. The UHI peaks in intensity about two to three hours after sunset, a typical maximum intensity being 6 °C for a moderately sized city during clear and calm conditions.

When calculating UHI intensities it is important to remember that two environments are compared, the urban and the rural environment. The cooling rate in the rural environment is equally important. For example, we have a network of temperature sensors set up at four different locations that have all taken measurements at 22:00 local time. One is located in the centre of the city, measuring 24 °C, one at the edge of the city in a suburban neighbourhood, measuring 22.5 °C, one at a forest site outside the city, measuring 20 °C, and the last site is located in a grass field, measuring 18.5 °C. The temperature difference will be smaller if the central urban temperature is compared to the forest site ($\Delta T = 4$ °C) compared to with the grassland site ($\Delta T = 5.5$ °C). Forests also tend to trap heat and have a lower cooling rate, so ΔT will evolve differently through the night. Lowry (1977) created a framework for the effects included in taking a single measurement (M) during weather type (i), time period (t) at station (x):

$$M_{itx} = C_{itx} + L_{itx} + E_{itx} \tag{9.17}$$

The measurement depends on the background climate (C), the local climate (L) – which could include effects of orography, proximity to a coast, etc. – and the effect of local urbanization (E). In order to analyse only the effect of local urbanization all the other factors must be kept constant. Going back to the example of the four measurement sites, it is important to keep in mind how the background climate and local climate effects will be different for each of the sites. Imagine that the grass field site is located at the bottom of sloped terrain and during clear, calm nights katabatic flows bring cool air down from further up the slope. The night-time temperatures will be much cooler at this site compared to if the grass field was located in flat terrain. If these katabatic winds do not influence the rest of the measurement locations, L_{itx} will be different for i = clear, calm weather, t = night time at x = grassland station. In order to better compare the measurements over different types of urban surfaces and attribute them to local climatic effects, Stewart and Oke (2012) designed a classification of urban and rural surfaces called local climate zones. These include 10 different urban zones based on building height, spacing, and function and 7 rural zones ranging from dense trees to bare soil and water.

9.4 Dispersion of Pollutants

Gases and particulate pollutants emitted into the atmosphere are dispersed by air movements that carry them away from the source and diffuse them into larger volumes of air by turbulent eddies. Such dispersion and dilution is affected by density differences between the pollutants and atmosphere, by deposition to the surface, and by flow around obstacles, such as the canopies we have been discussing thus far. In this section, we will concentrate on the principles of dispersion, and discuss the effect of atmospheric motions on pollutant concentrations for rural and urban terrain.

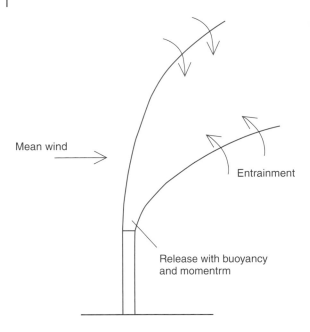

Figure 9.13 Dispersion processes near a buoyant release. *Source:* from Carruthers (2003), with permission from Blackwell Publishing.

9.4.1 Motions of Gases and Particles near a Source

A release of gas from a fixed point, such as a chimney, may consist of polluting gases mixed with carbon dioxide and water vapour. It may be released at a higher temperature than the ambient environment, and so the emission is positively buoyant. Thus, the gas release has some momentum that drives it upwards.

At first, when material is released into the atmosphere, it does not immediately travel with the velocity of the airflow: it is an *active scalar*. For a positively buoyant release, such as water vapour from a cooling tower, this is known as plume rise (Briggs 1984). Figure 9.13 shows the processes acting on warm gases released from a chimney. Differences between the velocities of the release gases and the airflow are confined within a small layer immediately surrounding the emitted plume of gases. The wind shear across this layer leads to small-scale turbulent mixing, which causes entrainment of cleaner, colder, slower air into the plume. Eventually, this mixing causes the emitted gases to lose their initial buoyancy. At that point, the gases are *passive scalars* that no longer generate motion and just 'go with the flow'. The distance over which the active mixing process occurs depends on the buoyancy and momentum of the release, its height and direction. For large plumes from tall chimneys it may last several kilometres from the source; for weakly buoyant releases near the ground, the distance may be less than 100 m.

9.4.2 Dispersion Processes

If gas is released from a point source, the mean flow advects it downwind, whereas turbulence diffuses the pollution laterally and vertically over time to form a *plume*. The combination of these two processes (*advection* and *diffusion*) is called *dispersion*. The shape of the plume therefore depends on the strength of mixing due to turbulence, which is proportional to σ_u, σ_v, and σ_w. In highly unstable conditions, large eddies of buoyant origin can cause the entire plume to meander. In stable conditions the plume is confined in width, particularly in the vertical direction.

We can estimate gas concentrations within the plume using a *Gaussian plume model*. Although it is impossible to predict the exact position of a polluted air parcel at any moment (due to the unpredictability of turbulence), it is possible to calculate the probability distribution of concentrations within the plume over a given time. We assume a Gaussian probability distribution, p, for the turbulent fluctuations around the mean for each component of the wind, i.e. for streamwise component u:

$$p(u) = \frac{1}{\sqrt{2\pi}\sigma_u} exp\left[\frac{-(u-\bar{u})^2}{2\sigma_u^2}\right]$$ (9.18)

We can thus derive (not shown here) an expression for the expected concentration of pollution as a function of time and space:

$$\langle C(x,y,z,t)\rangle = \frac{q}{2\pi\bar{U}\sigma_y\sigma_z} exp\left[\frac{-y^2}{2\sigma_y^2} - \frac{-z^2}{2\sigma_z^2}\right]$$ (9.19)

where angle brackets denote expected value. The source emission rate is represented by q. Variable σ_y represents mixing in the lateral direction, and is related to σ_v: similarly, σ_z is related to σ_w. Both mixing variables are a function of x, i.e. they increase with distance from the source, allowing the plume to spread. Mixing in the streamwise direction (σ_x) has been neglected, as transport by the mean wind speed U dominates. Finally, co-ordinates (x, y, z) are defined relative to the source location, and time is defined with respect to the time of release. A plume is depicted schematically in Figure 9.14.

As discussed in Section 9.1, the strength of turbulence varies according to stability and height within the boundary layer, but by using Eq. (9.19), we can take account of such variations by suitable parameterization of σ_y and σ_z. The next section describes how plume spread varies with distance downstream of the source, based on ideas originally introduced by Taylor (1921).

9.4.3 Turbulent Diffusion near the Source

If the flow is turbulent, parcels of gas from the source are randomly displaced downstream of the source, each one having a location (x, y, z). If we consider many separate parcels of gas, each following different trajectories in the turbulent windfield, then the lateral plume spread σ_y can be thought of as the root mean

Figure 9.14 Schematic showing Gaussian plume from an elevated source. *Source:* from Carruthers (2003), with permission from Blackwell Publishing.

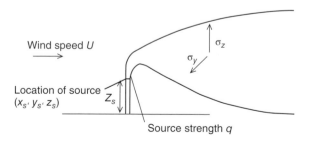

square of the lateral displacement of all parcels, $\sqrt{\left(\overline{y^2}\right)}$, and similarly for the vertical displacement $\sqrt{\left(\overline{z^2}\right)}$. Close to the source, the lateral and vertical plume spread parameters, σ_y and σ_z, are proportional to the standard deviations σ_v and σ_w respectively. After a time of travel t from the source:

$$\sigma_y = \sigma_v t \text{ and } \sigma_z = \sigma_w t \tag{9.20}$$

where t is greater than the integral time scale of the turbulence, but smaller than the time scale for changes in background meteorology. (Remember that the time of travel downstream t is related to the distance x by $x = Ut$, where U is the mean wind speed. So when we talk about time of travel from the source, we might as well be talking about distance downstream of the source.) When the turbulence is weak compared with the mean wind speed (i.e. turbulence intensity is low, perhaps in neutral or stable conditions), then the parameterizations in Eq. (9.20) can be used in Eq. (9.19) to estimate concentrations near the source. However, when the turbulence intensity is large, then mean concentrations may not reach a steady-state distribution, such as is given in Eq. (9.19): the paths of air parcels within the overall plume become too unpredictable.

In a convective boundary layer ($z/L < -0.3$) the vertical turbulence component can be significantly non-Gaussian due to strong thermal updraughts embedded in the flow, surrounded by weak down-draughts. Consequently, the height of maximum concentration within the plume decreases with distance downstream (instead of being constant) and the vertical profile of concentration is non-Gaussian. A modified non-Gaussian concentration profile is used instead of Eq. (9.19) in this case.

9.4.3.1 Turbulent Diffusion Far from the Source

As parcels of gas become dispersed further away from the source, the plume spreads in a different way. Recall that turbulence consists of many, different-sized eddies, meaning that the flow at a point is correlated with flow at other points up to the decorrelation length scale, L_x. Beyond this separation distance, the parcels of gas start to move off in completely different directions. As a consequence, $\sqrt{\left(\overline{y^2}\right)}$ and $\sqrt{\left(\overline{z^2}\right)}$ do not increase linearly with time of travel, but with the square root of time of travel, that is:

$$\sigma_y \propto \sigma_v \sqrt{t} \text{ and } \sigma_z \propto \sigma_w \sqrt{t} \tag{9.21}$$

Equation (9.21) shows that the plume spreads out less rapidly with time of travel, and hence plume width increases more slowly further downstream of the source. The transition between the two regimes of plume spread occurs at a time that is dependent on the integral time scale T_L. For time far less than the integral time scale of the turbulence ($t \ll T_L$, typically less than a few hundred seconds), the plume spreads linearly with time, and for $t \gg T_L$, the plume spreads in proportion to the square root of time. Another factor in plume spread is stability: vertical spread is enhanced in convective conditions due to large *thermals* transporting pollutant quickly, and suppressed in stable conditions where air parcels are negatively buoyant and cannot rise.

9.4.3.2 Dispersion in Nonuniform Flow

The simple model for plume dispersion that has been described so far captures a lot of the basic processes that control pollutant concentrations. However, real flows are more complicated. The plume spread described by Eqs. (9.20) and (9.21), and depicted in Figure 9.14, assumes that the atmosphere remains exactly the same over many hours. In reality, large mesoscale eddies, fronts and synoptic-scale variations cause the mean wind speed, stability, and turbulent characteristics to vary over time. To account for this, it is necessary to use meteorological measurements averaged over suitable periods (e.g. 30 minutes), or local climatological estimates of wind direction and speed. In addition, mean wind speed and turbulent characteristics vary as a function of height, which was shown in Figure 9.3. Ultimately, as a plume spreads vertically, it will be limited by the inversion layer at the top of the boundary layer, where the size of turbulent eddies and thus vertical mixing reduce significantly. Hence, the vertical growth of the plume in Figure 9.14 is not entirely realistic.

9.4.3.3 Averaging Times

The expressions for plume spread given in Eqs. (9.20) and (9.21) correspond to the expected concentration distribution across the plume when averaged over a sufficiently long time. An analogy for this is if we were to take a photograph of a dispersing plume of smoke with a very long exposure time. 'Very long' means more than the integral time scale – that is, the time scale associated with the largest energy producing eddies in the boundary layer ($T_L \sim 1000$ s for boundary layer of depth 1 km). However, at times shorter than this – we take instantaneous snapshots – the plume will be much narrower, and it will meander randomly as large eddies transport the plume without mixing and diluting it.

For measurements made at a point, this results in large instantaneous changes in concentration, as the plume meanders across the sensor in time. Such large changes in concentrations are smoothed out as the averaging time is increased above the integral time scale. So is the concentration predicted by a simple model such as Eq. (9.19) really representative of the range of concentrations actually measured at a point? It is more useful to describe the concentration probability distribution, that is, to quantify the range of concentrations that will be experienced, rather than just the mean. The form of such a distribution will depend on both in-plume structure and plume meander (Mylne and Mason 1991).

If the point measurement is one of the many pollution monitoring stations, such as are installed across the UK (see https://uk-air.defra.gov.uk/interactive-map), then measured data will exhibit large fluctuations in concentrations from second to second. Similarly, human beings walking around in the plume will breathe air with highly fluctuating concentrations of pollutants. Clearly, when it comes to managing air quality across a whole country, monitoring short-term fluctuations in concentrations is impractical. As a workable solution for monitoring pollution levels, the UK government published Air Quality Objectives in January 2000. They stipulate acceptable concentrations averaged over a given time period that is never shorter than the integral time scale for several key pollutants, given the effect on health and the environment, and serve as benchmark values to monitor whether air pollution is improving.

9.4.3.4 Mean Concentrations over Flat Terrain

Given our discussion of the processes of dispersion, here we consider the concentrations predicted by the dispersion model ADMS 3 (Carruthers et al. 1997), based on the simple Gaussian plume ideas presented in the previous sections. Figure 9.15 shows how the ground level concentration (g.l.c.) due to a pollutant emitted from a point source varies as a function of downstream distance. For Figure 9.15a, the source is at the ground, and for Figure 9.15b it is an elevated source, such as a chimney. The averaging time is one hour, and three stability conditions are considered: convective, neutral, and stable.

(a)

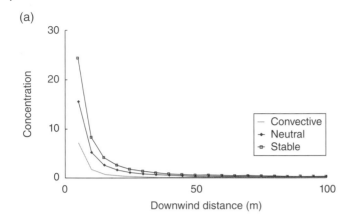

Downwind distance (m)

Figure 9.15 Variation of ground-level concentration with distance for (a) ground level and (b) elevated source. *Source:* from Carruthers (2003), with permission from Blackwell Publishing.

(b)

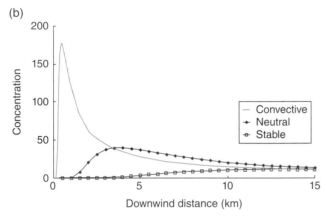

Downwind distance (km)

When the source is at ground level, the highest concentrations are experienced near the source, and decrease quite rapidly over a distance of approximately 20 m. The rate of decrease clearly depends on the stability – in convective conditions, increased vertical mixing results in low concentrations near the ground compared to the stable case. It can thus be seen that stable conditions can lead to high pollutant concentrations near the ground, and winter temperature inversions are often responsible for severe pollution episodes in the UK. If the source is a line of traffic travelling along a road, the result in Figure 9.15a shows that pollution concentrations are highest next to the road. It has been found that pollution exposure is significantly reduced by choosing walking or cycling routes away from major roads (Kaur et al. 2005).

For the elevated source in Figure 9.15b it might seem surprising that the highest concentrations at ground level occur in convective conditions. However, recall that vertical mixing is most rapid in this case; and that the skewed distribution of turbulent fluctuations results in the plume maximum concentration being closer to the ground. The peak concentrations thus occur much further downstream for neutral and stable conditions, by which time lateral mixing has had time to act, resulting in lower concentrations. Note the typical distances in this case: peak concentrations at ground level can occur several kilometres downstream of an elevated source. Hence, using a tall chimney to emit pollutants can be a good strategy to reduce exposure of the population living nearby.

9.4.3.5 Dispersion around Buildings

As most pollution is released in urban areas, we now focus on the effect of buildings on pollutant dispersion. So far, our discussion of dispersion has focused on mixing by turbulence, and downstream transport by the mean wind, with the stability of the atmosphere governing the strength of the turbulence. However, obstacles can have a profound effect on both turbulence and mean flow patterns. We now describe qualitatively the effects of buildings at the street, neighbourhood and city scales considered previously. Belcher (2005) provides more detailed discussion of the flow processes described here and Belcher et al. (2013) give a state-of-the-art review of urban dispersion modelling.

For an isolated building, the flow obviously has to deflect around and over it. For pollution released upstream of it, this can mean that lateral dispersion is wider as the flow diverges around the building, compared to plume spread by turbulence alone, represented by σ_y. Similarly, in front of the building the flow lifts the pollution rapidly upwards – but also rapidly downwards in the recirculation behind the building as shown in Figure 9.8 (sometimes called 'downwash'). Strong turbulence in the wake rapidly mixes gases, but turbulent diffusion out of the wake is relatively slow. Observations of swirling litter or leaves show that such vortex structures are ubiquitous in urban areas, and can lead to persistent, localized areas of high pollution concentrations near buildings.

For a street canyon, Figure 9.8 showed that the flow at street level is in the opposite direction to that above. For pollutants released at street level (i.e. traffic fumes) concentrations are much higher on the 'upwind' side of the street than the other side. Turbulence disperses the pollution, which reduces its concentration, and intermittent gusts vent it vertically out of the street (Salmond et al. 2005).

At the neighbourhood scale, the interaction of wakes from multiple buildings produces a highly turbulent environment. This produces rapid diffusion of gas released near the surface up to the top of the urban canopy, particularly in the wake interference regime (Barlow and Belcher 2002). This effect can be parameterized in simple dispersion models by enhancing the vertical plume spread parameter σ_z for dispersion near the source (Hanna and Chang 1992). Laterally, plume spread is enhanced as the pollution moves through the network of streets, which can be approximated by increasing σ_y. So-called 'street network models' (Belcher et al. 2013) capture this process by assuming that pollution is well-mixed by turbulence and split between streets at each intersection. This simple approach is promising but requires maps of street layout as an input.

The unmodified plume model, using the mean wind and direction, gives inaccurate predictions within the urban canopy. However, once the pollution rises above the buildings, dispersion can be described adequately by a Gaussian plume model – transport is in the downstream mean wind direction and mixing occurs laterally and vertically. Above the roughness sublayer, surface layer turbulence parameterizations may apply as long as there are no significant changes in surface roughness. If the plume is dispersing over an urban canopy and then encounters an extensive park, then its form will change as the flow adjusts to the smoother surface by accelerating. At city scale, recall that as flow adjusts from the urban to the rural surface, the turbulence characteristics near the ground will differ from those at height: a plume that has been lifted high above the urban surface may well not be mixed down to the ground in the rural area downstream of a city, causing the urban plume mentioned earlier.

Finally, the topography and type of terrain surrounding a city can have a very strong effect on the dispersion of pollutants within it (Turco 1997). In Los Angeles, with sea to the west and hills to the east, pollution that is released within the city by day can be blown inland by a sea breeze until it meets the 'barrier' of the hills. By night, flow is reversed due to the rapid cooling of the land compared with the ocean, and a land breeze can transport pollution out to sea. Due to stable conditions over the sea at night, pollution is not dispersed, and the next day it is transported back into Los Angeles as the sea breeze develops again. Mexico City is surrounded by steep hills – at night, due to enhanced radiative cooling at the hill tops, cold air flows down into the city, also known as a *density current*. This can cause a stable layer to form over the city in the valley, despite the UHI. Clearly, pollution levels would be enhanced in this situation.

9.5 Summary

The atmospheric boundary layer is worthy of study as it is the part of the atmosphere most influenced by the Earth and its inhabitants. The interaction between the biosphere and the atmosphere is mediated by micrometeorological processes that vary for different surface types. In this chapter we have focused on vegetation canopies, which exchange carbon dioxide with the air, and urban areas, where anthropogenic emissions of carbon dioxide and pollutants influence the whole atmosphere.

Our starting point was to consider boundary layer structure over flat terrain, as the most coherent concepts and theories have been developed for this case. The structure of the boundary layer is determined mostly by heating and cooling of the surface over a diurnal time scale. Heat input into the atmosphere is controlled by the surface energy balance, which is determined by the physical characteristics of the surface: whether it is reflective, whether it absorbs and stores heat, etc. Clearly, when the surface is covered with vegetation or buildings, the boundary layer structure and its development in time are also changed.

No discussion of the boundary layer can proceed without exploring the nature of turbulence. Turbulence is responsible for vertical exchange of gases in the atmosphere, as well as coupling the atmosphere to the surface through exchange of momentum and heat. A statistical description of turbulence helps us to quantify its intensity and to recognize some of the structures within it. The Monin–Obukhov similarity theorem identifies the key processes acting near the surface, and provides a framework whereby changes in the vertical profiles of wind, temperature, and turbulent characteristics can be related in a quantitative way to the stability of the atmosphere.

Canopies, whether of vegetation or buildings, alter the exchange of momentum with the surface, and thus also alter the fluxes of heat and gases into the atmosphere, when compared to a flat surface. The complex, intermittent turbulence of the roughness sublayer contrasts with the more homogeneous inertial sublayer above. Although research is ongoing into vegetation canopies, certain characteristics have emerged, such as gas exchange: in a vegetation canopy, the plants themselves regulate the flow of gases through physiological means. However, this is dependent on the evapotranspiration of water, which is partly a physical effect, demonstrating the interplay between biology and meteorology.

Urban areas have a large impact on the surface energy balance by being generally hotter and drier than rural areas. Heat stored in the urban fabric and released at night can lead to an UHI, where the city cools more slowly than the rural surroundings. Buildings are relatively large obstacles, in contrast to plants in a vegetation canopy. Thus, there are distinctive changes in flow at the scale of a single building or street. Flow at the scale of a neighbourhood is similar to flow over a vegetation canopy. At the scale of a whole city, the flow must adjust to the change from a warm, rough urban to cool, smooth rural surface.

This changes the boundary layer structure by causing an IBL to develop, with a distinct difference in turbulence characteristics near the surface compare to those aloft.

Finally, dispersion of gases released within the boundary layer depends on the flow developed over the surface, and the structure of the boundary layer. A simple model, the Gaussian plume model, captures the main features of dispersion from a source, namely transport of gases downstream by the wind, and dilution with background air through turbulent mixing. Although it can work well above the urban canopy, research is ongoing to find suitable mathematical models that capture simply the effect of urban canopies on dispersion, the street network model being a recent development.

Questions

1 One day, the latent heat flux Q_E is measured to be $150 \, \text{W} \, \text{m}^{-2}$. Given that the latent heat of vaporization of water, λ, is $2.5 \times 10^6 \, \text{J} \, \text{kg}^{-1}$, calculate the moisture flux, E, and its units.

2 We usually measure daily rainfall in terms of millimetres per day. Given that the density of water is $1000 \, \text{kg} \, \text{m}^{-3}$, from your answer to *1* above, what is the evaporation rate in millimetres per day?

3 Find a tree to stare at. Estimate its LAI.

4 Find a forest to stare at. Using Figure 9.7a, estimate the ratio of roughness length to height, z_0/h. Find an orchard and do the same. Is there much of a difference between the answers?

5 Which flow regime might be present when the wind is blowing perpendicular to your street? Use Figure 9.8 to help you decide.

Hints and Answers

1 $E = 6 \times 10^{-5} \, \text{kg} \, \text{m}^{-2} \, \text{s}^{-1}$. This is the rate of moisture loss from the surface.

2 $5.2 \, \text{mm} \, \text{day}^{-1}$. The key is to envisage that 1 kg of water spread over $1 \, \text{m}^2$ has a depth of 1 mm.

3 Make some reasonable assumptions about the area of a single leaf; the number of leaves on a branch; the number of branches; the plan area of the tree. Or, estimate the volume that the tree occupies, and how many leaves there might be per cubic metre.

4 The orchard probably has a lower value due to being less densely planted, so $P \sim 0.1$. For a forest, $P \sim 1$ or more if trees have overlapping branches. Your answer might depend on what you estimate the maximum height of foliage to be, z_m/h, for either case.

5 Try to estimate the mean height of buildings, H, and mean street width, W, to calculate aspect ratio H/W.

References

Barlow, J.F. (2014). Progress in understanding and modelling the urban boundary layer. *Urban Climate* 10: 216–240.

Barlow, J.F. and Belcher, S.E. (2002). A wind tunnel model for quantifying fluxes in the urban boundary layer. *Boundary-Layer Meteorology* 104: 131–150.

Belcher, S.E. (2005). Mixing and transport in urban areas. *Philosophical Transactions of the Royal Society* 363: 2947–2968.

Belcher, S.E., Coceal, O., Hunt, J.C.R. et al. (2013). A review of urban dispersion modelling. Report for Atmospheric Dispersion Modelling Liaison Committee, ADMLC-R7 (January 2013). https://admlc.com/publications (accessed 2 October 2018).

Briggs, G.A. (1984). Plume rise and buoyancy effects. In: *Atmospheric Science and Power Production*, Report DOE/TIC-27601 (ed. D. Randerson). Washington, DC: Technological Information Center, Office of Science and Technology Information, US Department of Energy.

Britter, R.E. and Hanna, S.R. (2003). Flow and dispersion in urban areas. *Annual Review of Fluid Mechanics* 35: 469–496.

Carruthers, D.J. (2003). Atmospheric dispersion and air pollution meteorology. In: *Handbook of Atmospheric Science* (eds. C.N. Hewitt and A. Jackson), 255–274. Oxford, UK: Wiley Blackwell.

Carruthers, D.J., Edmunds, H.A., Bennet, M. et al. (1997). Validation of the ADMS dispersion model and assessment of its performance relative to R-91 and ISC using archived LIDAR data. *International Journal of Environment and Pollution* 8: 264–278.

Garratt, J.R. (1992). *The Atmospheric Boundary Layer*. Cambridge University Press.

Hacker, J.N., Belcher, S.E., and Connell, R.K. (2005). Beating the Heat: Keeping UK Buildings Cool in a Warming Climate. UKCIP briefing report, UK Climate Impacts Programme, Oxford.

Hanna, S.R. and Chang, J.C. (1992). Boundary layer parameterizations for applied dispersion modelling over urban areas. *Boundary Layer Meteorology* 58: 229–259.

Ibbetson, A. (1981). Some aspects of the description of atmospheric turbulence. In: *Dynamical Meteorology: An Introductory Selection* (ed. B.W. Atkinson), 138–152. London, UK: Methuen.

Kaimal, J.C. and Finnigan, J.J. (1994). *Atmospheric Boundary Layer Flows: Their Structure and Measurement*. Oxford, UK: Oxford University Press.

Kaimal, J.C. and Wyngaard, J.C. (1990). The Kansas and Minnesota experiments. *Boundary Layer Meteorology* 50 (1–4): 31–47.

Kastner-Klein, P. and Rotach, M.W. (2001). A wind tunnel study of organised and turbulent air motions in urban street canyons. *Journal of Wind Engineering and Industrial Aerodynamics* 89 (9): 849–861.

Kastner-Klein, P., Fedorovich, E., Ketzel, M., and Berkowicz and Britter, R. (2003). The modelling of turbulence from traffic in urban dispersion models – part II: evaluation against laboratory and full-scale concentration measurements in street canyons. *Environmental Fluid Mechanics* 3 (2): 145–172.

Kaur, S., Nieuwenhuijsen, M., Colvile, R. et al. (2005). Personal exposure of street canyon intersection users to $PM_{2.5}$, ultrafine particle counts and carbon monoxide in Central London, UK. *Atmospheric Environment* 39 (20): 3629 3641.

Lowry, W.P. (1977). Empirical estimation of urban effects on climate: a problem analysis. *Journal of Applied Meteorology* 16 (2): 129–135.

Mylne, K.R. and Mason, P.J. (1991). Concentration fluctuation measurements in a dispersion plume at a range of up to 1000 m. *Quarterly Journal of the Royal Meteorological Society* 117: 177–206.

Rotach, M.W. (1999). On the influence of the urban roughness sublayer on turbulence and dispersion. *Atmospheric Environment* 33: 4001–4008.

Salmond, J.A., Oke, T.R., Grimmond, C.S.B. et al. (2005). Venting of heat and carbon dioxide from urban canyons at night. *Journal of Applied Meteorology* 44 (8): 1180–1194.

Shaw, R.H. and Pereira, A.R. (1982). Aerodynamic roughness of a plant canopy: a numerical experiment. *Agricultural Meteorology* 26: 51–65.

Stewart, I.D. and Oke, T.R. (2012). Local climate zones for urban temperature studies. *Bulletin of the American Meteorological Society* 93 (12): 1879–1900.

Taylor, G.I. (1921). Diffusion by continuous movement. *Proceedings of the London Mathematical Society* 20: 196–212.

Turco, R.P. (1997). *Earth under Siege: From Air Pollution to Global Change*. Oxford, UK: Oxford University Press.

WMO (2014). Towards integrated urban weather, environment and climate services. *WMO Bulletin* 63 (1) https://public.wmo.int/en/resources/bulletin/towards-integrated-urban-weather-environment-and-climate-services (accessed 4 October 2018).

Further Reading

Arya, S.P. (1999). *Air Pollution Meteorology and Dispersion*. Oxford University Press, 320 pp.

Monteith, J.L. and Unsworth, M.H. (1990). *Principles of Environmental Physics* (2). Butterworth-Heinemann, London, UK.

Oke, T.R., Mills, G., Christen, A. and Voogt, J.A. (2017). *Urban Climates*. Cambridge University Press, Cambridge, UK.

10

Urban Air Pollution

Zongbo Shi

School of Geography, Earth and Environmental Science, The University of Birmingham, Birmingham, United Kingdom

10.1 Introduction

Urban regions are by their nature concentrations of humans, materials and activities. They therefore exhibit both the highest levels of pollution and the largest targets of direct impacts.

Urban air pollutants come from two distinct sources: local emissions and regional transport. Local sources, such as transport, domestic combustion, cooking, and industry, emit primary pollutants, including sulphur dioxide (SO_2), particulate matter (PM), nitrogen oxides (NO_x), and volatile organic compounds (VOCs) into the atmosphere. Regional transport refers to air pollutants from outside the urban areas, which can be from anthropogenic or natural sources (e.g. mineral dust from desert, sea-salt particles from the ocean, and ozone formed from primary pollutants). These local and regional pollutants are mixed together and undergo complex physical and chemical processes, leading to enhanced levels of air pollutants such as ozone, nitrogen dioxide (NO_2), and PM that haunt inhabitants in cities.

In contrast to regional and global pollution, local efforts to mitigate urban-scale air pollution may generally have direct and observable effects. One of the exceptions is ozone pollution, which has a much more complex response to air pollution mitigation measures.

10.2 Urban Air Pollution – A Brief History

Initially, air pollution was an indoor phenomenon, caused by open fires without controlled venting. This is still observed in present-day dwellings in the developing world (Smith 1993). Ambient urban air pollution, however, is as old as cities, and literature as well as historical records testify that the problems were extensive. Whilst the use of wood and charcoal as fuels would have been important in early times, it is the large-scale use of coal in cities that was associated with periodic episodes of very poor air quality. There was recognition that air pollution episodes were associated with adverse effects on human health as far back as 1661. Attempts at regulation have been documented for at least 2000 years. Some problems may even have been underestimated, because generally people were less critical about their living conditions and they had no means

Atmospheric Science for Environmental Scientists, Second Edition. Edited by C.N. Hewitt and Andrea V. Jackson.
© 2020 John Wiley & Sons Ltd. Published 2020 by John Wiley & Sons Ltd.

of evaluating long-term impacts of, for example, carcinogens. Further, many of the records concern aesthetic impacts in the form of smell and soiling, which are not deleterious to health in themselves. It is interesting to note that up to the Second World War, there was an ambivalent attitude towards pollution, which to some extent was perceived as a symbol of wealth and growth. Thus, advertisements showed pictures of fuming chimney stacks and cars with visible exhaust – images hardly anyone would accept today!

Human population increased steadily from 1 to 3 billion from 1804 to 1960 and the rate of increase is accelerating. Along with this increase, the percentage of people living cities rose from about 7% in 1800 to 16% in 1900 and 29% in 1950 (Ritchie and Roser 2018). These two factors together contribute to a worsening of urban air pollution since the Industrial Revolution. In the first half of the twentieth century, there were a number of notorious episodes of air pollution in developed countries involving coal burning as the major source, such as in London and New York, or a combination of coal burning and industrial emissions as in the Meuse Valley episode in Belgium in 1930 and Donora, Pennsylvania, in 1948 (Harrison et al. 2014). The serious health effects following the December 1952 London smog, causing 4000 premature deaths (Figure 10.1), led to the introduction of Clean Air Act in 1956. These pollution events marked the beginning of extensive air pollution control.

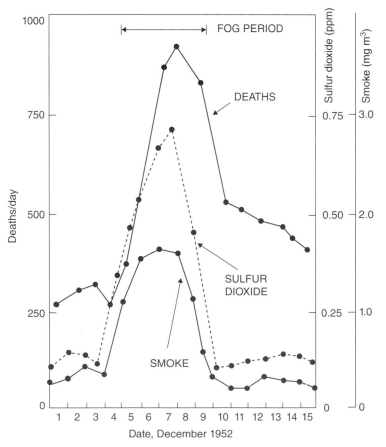

Figure 10.1 Daily concentration of smoke and sulphur dioxide and daily mortality during the December 1952 London smog. *Source:* reprinted from Harrison et al. 2014 with copyright permission.

Air pollution control measures from the 1950s onwards led to a steady reduction in the airborne concentrations of smoke (i.e. combustion particles) and SO_2 in Western countries (Harrison et al. 2014). Concurrently, however, there were marked increases in the use of road vehicles and the air pollution climate in developed countries progressively changed from one dominated by coal combustion to an atmosphere dominated by vehicular emissions, particularly oxides of nitrogen and hydrocarbons, and their chemical reaction products such as ozone. In less developed countries, there is a continuing trend of migration of people from the countryside to the cities. Today, around 40% of the population now lives in cities in these countries (Ritchie and Roser 2018). The air pollution climatology in these areas is unprecedentedly more complex with characteristics of traditional London and Los Angeles smogs as well as road traffic and industrial pollution. Recently notorious pollution episodes in developing megacities such as Beijing and Delhi have prompted national governments to initiate a fresh battle against air pollution, leading to continued improvement in air quality (Vu et al. 2019).

10.3   Scale of Urban Air Pollution

The scale of the processes within the urban atmosphere is fundamentally different from those of the regional and global atmospheres (Figure 10.2). There is always a regional background pollution, that can either be from natural sources, such as biomass burning, wind-blown dust and sea salt or from pollutants transported from long distances. Studies in the Mediterranean region and southern Europe have indicated that in certain periods the urban areas may be significantly affected by sources located hundreds of kilometres away. Primary emissions of air pollutants in the city, in particular from anthropogenic activities such as road traffic, lead to a city increment. However, the magnitude of the increment is highly dependent on the time and locations within the city as well as the size of the city. For example, within heavily trafficked streets set between continuous rows of high buildings, there exists a possibility

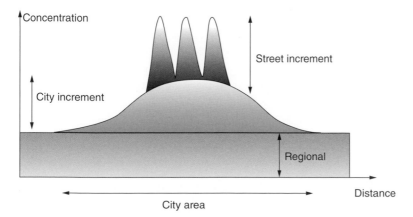

Figure 10.2   The scale of air pollution in the urban atmosphere. In addition to regional background emissions, there is a city-scale emission. Further pollutant emissions are also seen from within the street canyon. *Source:* courtesy Professor Roy Harrison in the opening presentation for the Faraday Discussions – Chemistry in the Urban Atmosphere (Shi et al. 2016).

to build up high levels of traffic-generated air pollutants due to very restricted atmospheric dispersion. This leads to local pollution hotspots. This air pollution problem in the street canyon is perhaps the smallest of the scales of air pollution problems. Away from the street canyon, a number of physical and chemical processes take place, leading to the mixing with urban background pollutants (Shi et al. 2016). The result is that the concentrations of traffic-generated pollutants tend to be rather uniform and significantly higher than those outside the urban area (Harrison et al. 2014). Megacities pose a different scale of air pollution problem, where pollution level is usually significantly higher than in surrounding areas due to greater emission strength associated with anthropogenic activities. Even the most advanced megacities in the world, such as London, still face serious air pollution problems, not to say those in developing countries, such as Beijing and Delhi.

10.4 Air Pollutants and Their Sources in the Urban Atmosphere

Air pollutants are often classified as primary or secondary in origin. *Primary pollutants* refer to pollutants that are emitted directly to the atmosphere from sources such as road traffic and industry. This term is in contrast to *secondary pollutants,* which are formed in the atmosphere from primary pollutants.

A huge number of primary anthropogenic air pollutants have been identified – most of them organic compounds. The pollutants that are of particular regulatory interests are PM, SO_2, nitrogen oxides ($NO_x = NO + NO_2$) and carbon monoxide (CO).

10.4.1 Characteristics of Pollutants

Particulate matter was originally determined as soot or 'black smoke' for which there is a European Union (EU) air quality limit value. Later, the concept of total suspended particulate (TSP) matter was introduced, but since 1990 size fractionating has been attempted by measurements of PM_{10} and $PM_{2.5}$ (particles with aerodynamic diameter <10 or 2.5 μm). $PM_{2.5}$, often termed as fine particles, has been of particular interest recently because it can enter into human lungs, which directly cause damage to human health. In urban atmospheres, the actual size spectra usually show a more pronounced number peak at nanoparticle (<100 nm) size range but mass peaks at 100–1000 nm and 1000–2500 nm size range (Figure 10.3). Mass concentration of $PM_{2.5}$ is typically in the range of a few to $1000 \, \mu g \, m^{-3}$ in the urban atmosphere (Table 10.1). The number concentration of particles is typically from a few thousands to tens of thousands per cm^3, which are dominated by nanoparticles. PM is a complex mixture of tens of thousands of compounds, many of which are not yet identified. The major chemical groups in typical urban PM include water-soluble ions (e.g. sulphate, nitrate, chlorine, and ammonium), organic matter, elemental carbon, crustal elements (e.g. silicon, aluminium, calcium, and iron), and trace elements (such as chromium, vanadium, arsenic, and mercury).

Nitrogen oxides ($NO_x = NO + NO_2$) are mainly formed by oxidation of atmospheric nitrogen or nitrogen compounds in the fuel during combustion. The main part, especially from petrol cars, is emitted in the form of the nontoxic nitric oxide (NO), which is subsequently oxidized in the atmosphere to the toxic pollutant NO_2. A fraction of NO_x from diesel vehicles is in the form of NO_2, a major factor leading to the breach of the air-quality limit value for NO_2 in many European cities (Grange et al. 2017). NO_x mass concentration is typically in the range of a few to hundreds of $\mu g \, m^{-3}$ in the urban atmosphere.

Figure 10.3 Typical urban aerosol number, surface, and volume (proportional to mass) distributions. *Source:* reprinted from Seinfeld and Pandis 2016 with copyright permission.

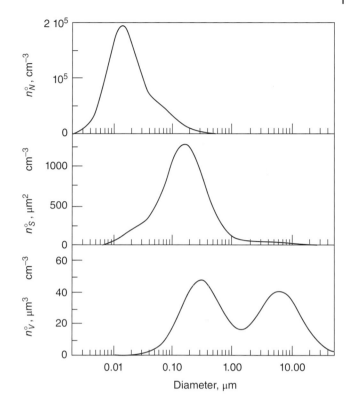

Sulphur dioxide (SO$_2$) is the classic air pollutant associated with sulphur in fossil fuels. Emissions can be successfully reduced using fuels with low sulphur content (e.g. natural gas or oil instead of coal). On larger plants in industrialized and some developing countries desulphurization of the flue gas is an established technique. Typical mass concentration of SO$_2$ in the urban atmosphere ranges from a few to hundreds of µg m^{-3} in the urban atmosphere.

Carbon monoxide (CO) is the result of incomplete combustion. The emissions can be reduced by increasing the air/fuel ratio, but with the risk of increasing the formation of nitrogen oxides. In cars the most effective reductions are carried out with catalytic converters. CO mass concentration is typically in the range of a few hundred to 3000 ppb (sometimes also shown as ppbv) (parts per billion by volume of air) in the urban atmosphere.

VOCs as air pollutants are released from combustion of fuels or emitted from plants. Some industrial processes and the use of solvents also result in the emission of VOC. In the urban air, the most important compounds are benzene, 1,3 butadiene, ethene, propene, and a series of aldehydes (see Case study 10.1). Biogenic VOCs, emitted from vegetation, do not pose a health risk in themselves, and the sources in cities are modest, but they must be taken into account in relation to regional photochemical air pollution, which, in turn, may influence urban air quality. The concentration of specific VOCs vary greatly in the urban atmosphere, but are typical in the ppt (parts per trillion by volume of air) to ppb ranges.

Table 10.1 Pollutant concentrations from various urban areas in 2017 (Unit: $\mu g\,m^{-3}$).

Site	Pollutant	Measurement	Concentrations
Marylebone Road, London, UK	Carbon monoxide	Annual mean	350
		Highest 8-hour running mean	1900
	Nitrogen dioxide	Annual mean	84
	Sulphur dioxide	Annual mean	7
	Ozone	Annual mean	16
		Highest daily max 8-hour running mean	71
	PM_{10}	Annual mean	27
		Highest 24-hour running mean	103
	$PM_{2.5}$	Annual mean	15
		Highest 24-hour running mean	100
Los Angeles, USA (Central LA)	Carbon monoxide	Highest 1-hour average	1541
		Annual mean	692
	Nitrogen dioxide	Highest 1-hour average	85
	Sulphur dioxide	Annual mean	9
	Ozone	Highest 1-hour average	139
	PM_{10}	Annual mean	36
	$PM_{2.5}$	Annual mean	23
Delhi, India (Mandir Marg station)	PM_{10}	Annual mean	186
	$PM_{2.5}$	Annual mean	102
	NO_2	Annual mean	50
	SO_2	Annual mean	12
	CO	Annual mean	1290
	O_3	Annual mean	24
Beijing, China	PM_{10}	Annual mean	84
	$PM_{2.5}$	Annual mean	58
	NO_2	Annual mean	46
	SO_2	Annual mean	8
	CO	Daily max 95 percentile	2100
	O_3	8-hourly max 90 percentile	193

Note: all data from open sources.

Case Study 10.1   Toxic Organic Pollutants

Dioxins

Dioxins are 210 closely related polychlorinated dibenzodioxine (PCDD) and furane (C_4H_4O) compounds. Their main source is combustion, typically incineration of material containing chlorine.

Although dioxins are air pollutants, it is not direct inhalation that is crucial for their impact on health. They are very stable, are deposited on the soil and water surface, and as they are fat-soluble, they move easily through the foodchain and are then ingested by humans.

Dioxins are some of the most toxic compounds known. They are carcinogenic, may have endocrine-disrupting effects, and may inflict damages to the unborn child. Special attention has been given to the content in mother's milk.

Polycyclic Organic Material

Combustion of material – typically wood and fossil-fuel containing cyclic hydrocarbons – generates a series of mutagenic and carcinogenic compounds. The polycyclic aromatic hydrocarbons (PAH) are the most important.

Environmental concern for these compounds dates back to the nineteenth century, when many incidences of skin cancer were observed in workers in the tar industry. The carcinogenic compound, benzo(a)pyrene, was first identified in tar in 1933, but its importance as an air pollutant was not recognized until the 1970s.

Lighter and more volatile PAHs exist in the gaseous phase, but the heavier and more carcinogenic PAHs have a tendency to attach to particles. In the urban air, PAHs are mainly from fossil fuel and biomass burning. Inhalation of such particles has been shown to cause cancer.

Benzene and Methyl-tert-butyl Ether

Changes in the composition of petrol to increase its combustion characteristics (octane number) may increase the emission of aromatic hydrocarbons, including benzene. However, the introduction of catalytic converters and controls on the benzene content of fuel may reduce ambient concentrations.

An alternative fuel additive methyl-tert-butyl ether (MTBE) causes both immediate eye and respiratory irritation and long-term risk of cancer. It has caused contamination of soil and groundwater, especially around fuel-filling stations.

Compounds that increase the greenhouse effect (carbon dioxide, methane, nitrous oxide) or deplete the ozone layer (e.g. chlorofluorocarbons, CFCs) are not urban pollutants per se but control of their emissions is related to emission of other pollutants, and their impacts have a bearing on urban quality of life.

Airborne concentrations typical of a range of urban locations are shown in Table 10.1. It shows that PM and SO_2 pollution in London and Los Angeles is relatively low but NO_2 is still relatively high in London. $PM_{2.5}$ level in developing megacities such as Delhi is very high, sometimes reaching over $1000\,\mu g\,m^{-3}$. $PM_{2.5}$ level in Beijing is much lower than in 2017 that 10 years ago (He et al. 2001), but it is still approximately six times higher than the World Health Organization (WHO) guideline.

10.4.2 Main Sources of Air Pollutants

Different sources emit by and large similar compounds, but in varying proportions, under different conditions and with different temporal patterns. Major anthropogenic sources of primary pollutants include:

- *Traffic*, and especially individual motorized traffic in the form of cars and motorbikes, is a major source of air pollution in most urban areas, not only in terms of local emissions, but also in terms of resulting pollution levels, because the emissions take place at low height and often in street canyons. In Europe, vehicle emission is the single-most important source of CO and NO_x and contributes significantly (>30%) to ambient VOCs. Traffic can also generate non-tailpipe pollutants such as road dust, and brake and tyre wear.
- *Residential source* includes domestic energy use for cooking and heating. This is particularly important in the winter when heating is needed. Biomass and coal are still the main domestic energy source in developing countries, including China, India, and major African and South American countries. Biomass fuel combustion for heating remains an important source of air pollutants in developed countries, including the United Kingdom. The key air pollutants emitted from this source include PM, SO_2, and CO.
- *Industry* continues to be an important source of urban air pollutants, particularly in developing countries. Industrial processes, such as iron- and steel-making directly produce PM and VOCs. Combustion processes associated with industrial processes, including machinery and energy use, also contribute to emission of air pollutants, including PM, NO_x, CO, and VOCs. In most of the modern megacities, industrial plants are moved outside of the city but their emissions can still indirectly affect urban air quality via long-range transport.
- *Power generation* with fossil fuels, in some cases combined with district heating, is, despite use of cleaner fuels and flue gas cleaning, still an important source of SO_2 and NO_x. In general, however, the impacts on air quality in developed megacities are modest, because many plants are located outside the cities and equipped with pollutant removal equipment.
- *Incineration of waste* in larger plants has some similarities to power production, and is in many cases used for district heating. Due to the mixed fuels, the emission of heavy metals and toxic organic compounds (e.g. dioxin) must be considered. Depending on the production, emission of organic compounds and heavy metals may be significant. If the emissions arise from diffuse sources, they may be difficult to control and have relatively large local impacts.
- *Solvent use* in households, painting, and minor industries gives rise to evaporation of VOCs. A more responsible use, including a switch to water-based coatings, can by and large solve the problem.

Important natural sources of primary pollutants include:

- *Windblown dust* is particularly important in cities located in arid and semi-arid regions. Sporadic dust storms from deserts can lead to a rapid deterioration of air quality, for example, in Beijing and Delhi.
- *Vegetation* emits VOCs as well as primary biological particles. Vegetation is unlikely to be the dominant source of VOCs in the urban atmosphere but it is non-negligible. Primary biological particles usually contribute little to total PM mass but they are key allergens that have important health impacts.
- *Oceans* not only generate bubbles, which become sea-salt aerosols once entered into the atmosphere, as well as VOCs. The impact is usually on coastal cities.

10.4.3   Emission Inventories

Emissions from individual sources can be measured by monitoring instruments or sampling followed by offline analyses, but because most emissions arise from a series of similar sources a different approach is normally used. Experiments determine how much pollution is emitted in a given activity, which results in so-called emission factors. They are often expressed in terms of pollution connected with use of a particular fuel, and can have the form '0.013 kg CO emitted per gigajoule or tons of fuels, e.g. when natural gas or coal is used in district heating plants'. They can also be related, for example, to emission from a car or a domestic animal. The emission is then calculated as the product of activity and emission factor.

Emission inventories are carried out at different scales and degrees of resolution in time and space. National inventories of emissions – as they are carried out for example in the form of the EMEP/EEA (Environmental Monitoring, Evaluation and Protection/European Environment Agency) database – are used in international negotiations of emission reductions and may for lack of better information suggest general trends in air pollution levels. For studies of transboundary air pollution a spatial resolution of 50 km is normally sufficient. For urban areas, however, more detailed investigations are necessary with time and space resolutions relevant to the applied scale. Proxy data for larger areas can be generated on the basis of information on traffic patterns, but detailed investigations of individual streets must be based on actual traffic counts.

10.5   From Emissions to Airborne Concentrations

Emission of air pollutants from urban sources determines to a large extent the urban air quality. However, physical and chemical transformation are also important. Therefore, the impacts of different sources and their relative importance cannot be evaluated on the basis of emission inventories alone. It must be realized that the relations between emissions and resulting concentrations are by no means simple. Measurements are still the foundation of the understanding, but application of mathematical modelling, and also of physical modelling in wind tunnels, is of increasing importance in urban air pollution management.

10.5.1   The Urban Climate

An urban area differs from the surrounding rural region in important ways, which influence its climate and the possibilities of dispersion of air pollutants, and their resultant impacts: a city is generally darker than its surroundings and thus has higher energy absorption (lower albedo). This effect is amplified by the rougher surface created by buildings, which trap solar radiation. Finally, various energy-consuming activities add to the heating. All in all, it may result in temperatures a few degrees above the surrounding rural areas. At low wind speeds, this so-called *urban heat island* can give rise to a circulation where the pollution is trapped. This situation can be aggravated by the topography, e.g. if the city is situated in a valley, which further restricts horizontal mixing. At higher wind speeds, the urban area acts as the source of a plume with elevated pollution levels downstream. The relative humidity in the city centre is generally lower – partly because of the elevated temperature and partly because of enhanced run-off of precipitation and the lack of evapotranspiration from vegetation in urban areas.

10.5.2 Physical Processes Determining Air Pollutant Concentrations

Dispersion mechanisms have received special interest with the increasing traffic in built-up areas, where street canyons exhibit special flow patterns (Case study 10.2, Figure 10.4). Wind speed is also of importance. Thus, air pollution with NO_2 at rush hours in a street canyon can be substantially lower when it is windy.

Concentrations of urban air pollutants tend to be highest in more stable atmospheric conditions, typically with little vertical mixing and low wind speeds. Atmospheric stability also plays a key role in the diurnal variation of air pollutant concentrations. For example, air pollutants accumulate at night when the atmosphere becomes increasingly stable at night, especially under clear skies; during daytime, air pollutants mix more vigorously in the atmosphere as heating of the ground leads to turbulence. In particular, the formation of a temperature inversion (where a layer of cool air at the surface is overlain by warmer air) causes accumulation of pollutants in the boundary layer. The phenomenon has contributed to almost all of the serious air pollution events in cities.

Case Study 10.2   Dispersion in a Street Canyon

In street canyons, the dispersion of air pollution from traffic is complex. If the wind blows perpendicular to the street, a vortex may be created (Figure 10.4), causing much more pollution at the leeside than in the windward side. Both measurements and dispersion calculations have shown that the wind direction is important in determining long- and short-term pollution levels.

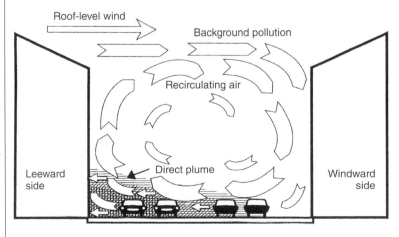

Figure 10.4   Flow pattern of air pollutants in a street canyon.

Urban parks provide a surface for deposition of pollutants, such as black carbon and NO_2, which may also affect air pollutant concentrations within the cities.

10.5.3 Chemical Transformation

During their physical mixing in the air, also undergo chemical transformations. For example, black carbon particles may be coated with inorganic and organic compounds (Liu et al. 2017). This can affect the size, toxicity, and optical properties of the primary particles.

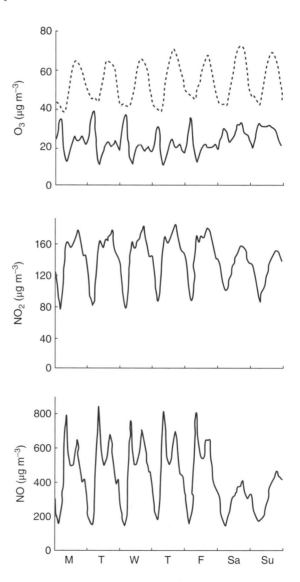

Figure 10.5 Variations of NO, NO_2, and O_3 at different times of day in a week.

Many of the gaseous pollutants, such as NO_x and VOCs, also react with each other under the influence of sunlight, which creates secondary pollutants such as ozone and other oxidants. This is the key process in the photochemical smog, which was first recognized in Los Angeles in the mid-1940s as an urban phenomenon related to car exhaust emissions in a subtropical, topographically confined region. Photochemical smog is now observed in many parts of the world, but with distinctly different patterns. In the south of Europe, cities such as Athens and Rome may experience 'summer smog' of the Los Angeles type, but in many cases it is a large-scale phenomenon. The formation of photochemical smog is highly complex and thus a very limited introduction is given here on the key atmospheric chemical processes.

The driving force of the photochemical smog is sunlight, which controls the two most influential photochemical processes in air pollution chemistry. The first one is the photolysis of ozone:

$$O_3 + hv \rightarrow O_2 + O(^1D) \tag{10.1}$$

Only light with wavelength less than 315 nm have enough energy to photolyse ozone molecules. The product includes singlet state (excited) oxygen atom, which is highly reactive. It reacts with water vapour to form hydroxyl radicals (OH):

$$O(^1D) + H_2O \rightarrow 2OH \tag{10.2}$$

Phytolysis of HONO is another key source of OH in the urban atmosphere (Crilley et al. 2016). OH is of key importance in air pollution chemistry, as it is responsible for much of the chemical reactions in the atmosphere.

The second important photochemical process is the photolysis of nitrogen dioxide, which produces triplet (ground state) oxygen atoms which further react with oxygen to form ozone:

$$NO_2 + hv \rightarrow NO + O(^3P) \tag{10.3}$$

$$O(^3P) + O_2 \rightarrow O_3 \tag{10.4}$$

In cities, ozone can be destroyed by nitric oxide to form oxygen and nitrogen dioxide:

$$NO + O_3 \rightarrow NO_2 + O_2 \tag{10.5}$$

Reactions (10.3) to (10.5) suggest that photolysis of NO_2 will give rise to a small steady-state concentration of ozone. Reaction (10.5) also indicates that ozone levels are generally lower at ground level in the streets than at roof level or in the surrounding countryside due to higher emissions of NO from traffic. Urban ozone levels are *higher* during weekends with *low* traffic and may be practically nil during some pollution episodes (Figure 10.5). Note also in Figure 10.5 that the concentration of NO follows the traffic intensity with rush hours and weekends much more closely than NO_2, the concentration of which is largely determined by the available O_3, supplied from outside the city or formed in-situ. Assuming the rate of loss (Reaction (10.3)) equals the rate of formation (Eq. (10.5)), we can see that

$$J_{NO2}[NO_2] = k[NO][O_3]$$

where J_{NO2} refers to sunlight dependent rate of NO_2 photolysis and k is the rate coefficient for the reaction of NO with O_3. So,

$$[O_3] = \frac{J_{NO2}}{k} \frac{[NO_2]}{[NO]}$$

But the atmosphere is not as simple as this. Other species, particularly VOCs, could react with hydroxyl radical (Reaction (10.2)) to build up higher concentration of ozone. Using methane and carbon monoxide as examples, the series of reactions include:

$$CH_4 + OH \rightarrow CH_3 + H_2O \tag{10.6}$$

$$CH_3 + O_2 \rightarrow CH_3O_2 \tag{10.7}$$

$$CH_3O_2 + NO \rightarrow CH_3O + NO_2 \tag{10.8}$$

$$CO + OH \rightarrow CO_2 + H \tag{10.9}$$

$$H + O_2 \rightarrow HO_2 \tag{10.10}$$

$$NO + HO_2 \rightarrow NO_2 + OH \tag{10.11}$$

Thus, in the sequence of reactions, peroxy radicals, CH_3O and HO_2 are formed, which then convert NO to NO_2 in a catalytic cycle involving the regeneration of HO_2. In this way, NO is converted to NO_2 without involvement of O_3 (Reaction (10.5)). As a result, O_3 concentration can build up from reactions (10.3) and (10.4). Other VOCs will also contribute to the formation of peroxy radicals. The series of reactions above are usually responsible for the formation of photochemical smog in different parts of the world.

Since ozone is a secondary pollutant, it can be regulated only via the primary pollutants. Chemical-transport models demonstrate the effects of changes in the emissions and ozone concentrations, which are non-linear.

Many other important air-pollution processes in the urban atmosphere can also affect the concentration of primary and secondary pollutants, which is not practical to elaborate further upon in this chapter. Readers are referred to Seinfeld and Pandis (2016) for further reading.

10.6 Urban-Scale Impacts

Urban air pollution has a series of impacts including on human health and well-being, materials, vegetation (including urban agriculture), and visibility. These impacts depend in the first instance on the relevant pollution levels, but also on other factors such as climate and lifestyle and the possibility of interaction between different components. For short reviews with references, see, for example, chapters 18–21 in Fenger et al. (1998).

10.6.1   Human Health and Well-Being

Impacts of air pollution on human morbidity and mortality have been unambiguously documented during acute episodes in the past, for example, during the 1952 London smog (Figure 10.1). Air pollution is considered one of the greatest environmental risks by the WHO. It claims 7 million lives every year and contributes to one in nine deaths (WHO 2014). Even in developed countries where air quality is substantially improved over the years, it still contributes significantly to premature deaths. For example, air pollution in the United Kingdom is linked to 40 000 premature deaths per year (Holgate et al. 2016). Human health impacts depend on the exposure and nature of the air pollutants, as well as sensitivity of individuals, with elderly, children, and asthmatics being especially sensitive to acute respiratory impacts.

PM is the most important air pollutant in terms of health effects. There is no safe threshold below which no adverse effect would be anticipated (Figure 10.6). Increasing PM pollution is associated with increased health risks, including pulmonary and cardiovascular diseases and mortality although there is some evidence the dose-response relationship becomes less steep at higher concentrations, e.g. over $200\,\mu g\,m^{-3}$. The penetration of airborne particles is highly dependent upon their size. In particular, fine particles can directly enter into human lungs (Figure 10.7), and this has received increasing attention in the past three decades (Pope et al. 1995; Maynard and Howard 1999; Burnett et al. 2014). The impact goes beyond the respiratory system, however. There is growing evidence that PM exposure is associated with cardiovascular hospitalizations and mortality, although the mechanisms are complicated (Pope and Dockery 2006). Some of the possible mechanisms include an increase in viscosity and coagulability of the blood, increasing the risk of heart attacks. Nanoparticles can also enter into the human brain, which might contribute to dementia (Maher et al. 2016).

In an often-cited investigation of the association between air pollution and mortality in six US cities, Dockery (1993) reported a strong correlation between the concentrations of small particles ($PM_{2.5}$) and mortality rate.

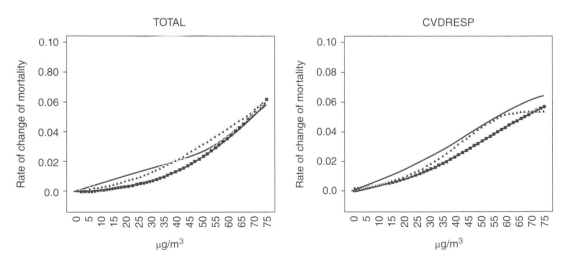

Figure 10.6   Mortality-PM_{10} dose–response curves for total (TOTAL) mortality and cardiovascular and respiratory (CVDRESP) mortality, 20 largest US cities, 1987–1994. The dose–response curves for the mean lag, current day, and previous day PM_{10} are denoted by solid lines, squared points, and triangle points, respectively. *Source:* reprinted from Daniels et al. 2000 with permission from Oxford University Press.

In a larger study, Pope et al. (1995) linked ambient air pollution data from 151 US metropolitan areas to the survival or death of 0.55 million persons. After correction for a series of confounders, a significant association between fine particulate exposure and survival emerged. Adjusted relative risk ratios showed a value of 1.17 between the highest and lowest polluted area. In view of this surprisingly high ratio, the results of these investigations should not be transferred uncritically to other countries with different age distribution, lifestyle, climate, and pollution mix. Nevertheless, taking together the two studies, the WHO estimates a relative risk of 1.1 per long-term exposure to $10\,\mu g\,PM_{2.5}\,m^{-3}$. Numerous later epidemiological studies of short-term and long-term effects of air pollution have shown that $PM_{2.5}$ at the present levels are responsible for significant pulmonary impacts, especially on people already suffering from respiratory and cardiopulmonary diseases.

Gaseous compounds (ozone, sulphur and nitrogen dioxide, and aldehydes) act directly as eye irritants, but the dominant route is via the respiratory system. The rate of uptake of gaseous pollutants depends on their solubility. Sulphur dioxide is soluble, and normally more than 95% is deposited in the upper airways, whereas NO_2 and O_3 penetrate deeper into the lungs.

Short-term exposure to high ambient concentrations of O_3 can cause inflammation of the respiratory tract and irritation of the eyes, nose, and throat. High levels may exacerbate asthma or trigger asthma attacks, and some non-asthmatic individuals may also experience chest discomfort whilst breathing. O_3 exposure is also linked to reduced lung capacity. Evidence is also emerging of health effects due to long-term exposure (DEFRA 2017).

SO_2 is a respiratory irritant that irritates the nose, throat, and airways to cause coughing, wheezing, shortness of breath, or a tight feeling around the chest. People with asthma are considered to be particularly sensitive. Health effects can occur very rapidly, making short-term exposure to peak concentrations important. Long-term exposure to SO_2 may harm the respiratory system.

Short-term exposure to concentrations of NO_2 higher than $200\,\mu g\,m^{-3}$ can cause inflammation of the airways. NO_2 can also increase susceptibility to respiratory infections and to allergens. It has been difficult to identify the direct health effects of NO_2 at ambient concentrations because it is emitted from the same sources as other pollutants such as PM. Studies have found that both day-to-day variations and long-term exposure to NO_2 are associated with higher mortality and morbidity (COMEAP 2018). Evidence from studies that have corrected for the effects of PM is suggestive of a causal relationship, particularly for respiratory outcomes. The range of estimates of the annual mortality burden of human-made air pollution in the UK is estimated as an effect equivalent to 28 000–36 000 deaths (COMEAP 2018).

Figure 10.7 Deposition of particles in the human respiratory system. Note that coarse particles are predominantly caught in the upper airways, whereas fine and especially nanoparticles penetrate deep into the lungs.

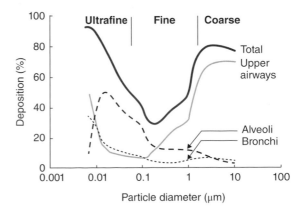

CO affects the ability of the blood to take up oxygen from the lungs as it binds with haemoglobin in the blood to form carboxyhaemoglobin that limits the ability of the blood to transport oxygen. At high concentrations, CO can kill human beings. Long-term exposure to CO may cause adverse health effects (Townsend and Maynard 2002).

Some important problems remain to be satisfactorily elucidated. First of all the statistical procedures are debated. Thus, it is a question of how to relate daily mortality with longer-term mortality effects and especially how to filter out a possible 'harvesting effect' on weak citizens. Second, it must be borne in mind that epidemiological studies provide little (or no) information on the underlying impact mechanisms. Finally, it will always be a difficult job to ascertain a particular case of mortality to air pollution.

10.6.2 Material Damage

Urban air pollution degrades materials, which leads to direct economic loss, failure of equipment, and deterioration of irreplaceable cultural values. The local urban climate is important because temperature and relative humidity control the moisture layer on surfaces in the absence of precipitation. This effect partially counteracts the higher pollution levels in the centre of a city (Kucera and Fitz 1995). This can also explain the noticeable impacts in fairly clean Nordic cities compared with cities in the south.

a) *Soiling*. Particulate pollution has different effects: the soiling itself represents an aesthetic loss, and cleaning can be expensive and may further impose mechanical wear. Soiling, especially with hygroscopic compounds (e.g. sulphates), facilitates formation of moisture. Finally, some compounds act as catalysts for various reactions – notably oxidation of SO_2 and NO_x.

b) *Corrosion of metals*. In dry air and at normal temperatures, oxygen reacts with most metals and in some cases forms protective layers of oxides, which inhibit further corrosion; thus, aluminium is covered with Al_2O_3 and iron with a mixture of oxides. In a polluted atmosphere, a less-protecting layer of sulphates is formed. In humid air, the corrosion is generally a more rapid electrochemical process in a moist layer on the metal surface, e.g. where iron is converted to rust, which is peeled or washed off.

c) *Stone materials*. Stone used as building material and for monuments has a wide range of composition, texture, and structure. Most important is the porosity, ranging from 0.5% for the dense granites and marbles to 25% for some limestones and sandstones. The classic pollutant causing degradation is SO_2, which is dissolved in the moisture layer to form sulphite and eventually is oxidized to sulphate. This attacks calcium carbonate ($CaCO_3$) and forms gypsum. The process initially results in a weight increase, but gradually the gypsum is washed off by rain.

d) *Organic materials*. Degradation of organic materials is mainly related to ozone, which attacks double bonds in unsaturated polymers. Thus, accelerated wear of tyres in California was an early sign of photochemical air pollution. The exposure to ozone results in fading and cracking of outdoor paints and other coatings, but the impacts may be difficult to distinguish from direct action from sunlight. However, indoor effects have also been observed. Thus, accelerated laboratory exposure to a mixture of photochemical oxidants leads to extensive fading of colourants, and humidity enhances the effect. It appears that in some museums without proper air conditioning, especially in the tropics, serious fading may result within a few months.

10.6.3   Urban Ecosystems Including Vegetation

The natural environment in urban areas is firstly of utilitarian value. Many city dwellers grow food for their consumption – mainly in Asia, Africa, and Latin America. Commercial production of high-value crops are normally found on the outskirts, but in densely populated regions such as The Netherlands more intensive agriculture activity is also incorporated into urban planning. Exposure of vegetation to air pollutants may produce a range of symptoms in plants, including leaf blemishes, inhibition of growth, and increased susceptibility to other stresses such as disease or insect attack (Harrison et al. 2014). In ecologically sensitive areas they may lead to a reduction in biodiversity through the loss of sensitive species.

Air pollutants, including SO_2, NO_2, and O_3, can be harmful to plants directly. Direct absorption of NO_2 can lead to accumulation of toxic nitrite ions. SO_2 may impair the plant's metabolism. At high concentrations, SO_2 can inhibit photosynthesis, whilst at low concentrations, it can interfere with the ability to translocate sugars from the leaves that are produced to the rest of the plant. The effects of O_3 on plants have been extensively studied, and is probably the gaseous pollutant with greatest impact globally. Exposure to O_3 produces a series of symptoms associated with oxidative stress. One common plant response to the oxidative stress is to release the gaseous plant hormone ethylene, and this may itself react with gaseous O_3 in the stomatal cavity to yield highly reactive hydroperoxides that can cause further damage to the plant. Symptoms include reduction of growth at lower concentrations, with visible damage to leaves at higher concentrations.

Acid rain, which is a direct outcome of acidic air pollutant (e.g. SO_2 and NO_x) emissions, can also indirectly affect vegetation. For example, acid deposition to forests results in yellowing (chlorosis) of leaves, decreasing frost tolerance, increasing susceptibility to pest and disease attack, and eventually the death of trees. These symptoms may be partly due to the leaching of nutrients such as such as phosphorus, potassium, and calcium from the soil by the acids, causing nutrient deficiency. Acid rain may also render the plants more susceptible to attack by pests and pathogens.

10.6.4   Visibility

The most visible impact of air pollution is visibility degradation. An early example is the London 'pea soup' in 1952, when the visibility was as low as a few metres. In the modern context, 'haze' – marked by low visibility is one of the most widely used words in China and other developing countries due to their frequent occurrence. It is pleasant when the air is clear and it is advantageous in many respects, including road safety. A reduction of visibility is therefore not only unsavoury, but also dangerous. We can see an object because it reflects light from the sun to reach us. When there are more particles in the air, some of the reflected light will not be able to reach us as they are scattered (e.g. by sulphate aerosol) or absorbed (e.g. by black carbon particles). The more particles in the air, the lower the visibility. Light scattering and absorption by the particles depend on the size of the particles. Thus, humidity also plays an important role, as higher humidity will make fine particles larger, causing more light scattering. On the other hand, for relatively large particles such as fog droplets the scattering is largely independent of the wavelength and results in a diffuse white light, which may reduce the visibility to nearly zero.

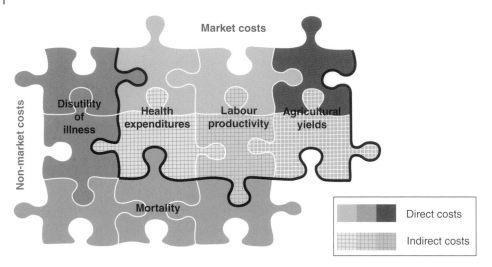

Figure 10.8 Direct and indirect costs of air pollution. Reprinted from OECD 2016 under CC-BY licence. See colour plate section for the colour representation of this figure.

10.6.5 Economy

Air pollution has a number of direct and indirect impacts on the economy (Figure 10.8). The direct costs include increased health expenditure, loss of labour productivity and decreases in agricultural yields, and the indirect costs disutility of illness and mortality (OECD 2016). OECD (2016) estimated that global air pollution-related health-care costs are as high as US$21 billion in 2015, which is projected to increase to US$176 billion in 2060 without stringent pollution control. UK air pollution reduces productivity, leading to 0.11% of GDP loss (Ricardo-AEA 2014). OECD (2016) estimated that annual number of lost working days was 1.2 billion at the global level. Air pollution reduces average life expectancy (e.g. six months for the UK population as a whole and as high as several years in the most polluted cities). The annual global welfare costs associated with the premature deaths from outdoor air pollution, calculated using estimates of the individual willingness-to pay to reduce the risk of premature death, are projected to rise from US$3 trillion in 2015 to US$18–25 trillion in 2060 without stringent pollution control (OECD 2016); and the annual global welfare costs associated with pain and suffering from illness are projected to be around US$2.2 trillion by 2060 without stringent pollution control, up from around US$300 billion in 2015, based on results from studies valuating the willingness-to-pay to reduce health risks (World Bank 2016). Furthermore, global crop production losses totalled 79–121 million metric tons, worth US$11–18 billion annually in year 2000 due to O_3 damage only (Avnery et al. 2011).

Material damage is better documented than health impacts, and experiments are less controversial. Various attempts based on estimates of maintenance costs (Tidblad and Kucera 1998) suggest total European damage in the order of several billion euro per year. But a special problem is posed by damage of irreplaceable cultural monuments and works of art, the reason being partly that the value is often related to a thin layer of decoration and not to mechanical strength. Attempts to evaluate the benefit derived in the form of pleasure from such objects (contingent valuation) suggest that the economic loss of damage is in the same order of magnitude as damage to trivial materials.

Overall, the cost to the wider UK economy of outdoor air pollution, mostly in the cities, is more than £20 billion per year, in terms of health-care costs, premature illness, and the impact on business (Holgate et al. 2016). Globally, the direct costs of air pollution is about 0.3% of world GDP in 2015 (OECD 2016) and the total welfare loss costs about 7% of world GDP (World Bank 2016). A large fraction of those estimated costs are associated with urban air pollution.

10.7 Means of Mitigation

The impacts of urban air pollution can be mitigated at all links of the *pollution chain*. The classic solution with dispersion from high stacks tends only to transfer the problem to other regions. With the growing and eventually merging of urban areas this option is therefore only acceptable in combination with advanced flue gas cleaning. The problem with increasing traffic and its emissions can in principle be attacked in two ways: on the technical level by reductions or elimination of the individual emissions; and on the economic and planning level by reduction of the activities – either by reducing the need for transport as such or by promoting public transport. Some solutions based on traffic planning (e.g. clean air zones), however, are of debatable value. They improve the air quality in city centres, but transfer problems to the outskirts and may even increase the total activity.

10.7.1 Mitigation of Urban Air Pollutant Emissions

10.7.1.1 Legislation
Past experiences have with depressing clarity shown that existing technical possibilities and recommended management practises will not be used unless legally or economically enforced. Air quality expressed as pollutant concentrations is controlled by limit values. With the possible exception of ozone limit values, they are only relevant for pollution levels in urban or industrial areas. The scientific foundation is experiments on humans or animals and epidemiological investigations.

Most countries have established limit values for the major air pollutants (MAP). Some of the important examples are the legislation in the European Union and the United States, which in many cases have served as models for other regions. In the following, the EU legislation will be briefly introduced as an example.

10.7.1.2 Air Quality Legislation in the European Union
In the EU, the setting of limit values (Table 10.2) is a multistep process with a system of EU directives, the first being adapted in 1980. These directives, and the related more detailed protocols, are subsequently ratified in the individual member states in the form of national legislation. So far, this system comprises SO_2, PM_{10}, $PM_{2.5}$, NO_2, CO, benzene, O_3, As, Cd, Ni, Pb, and polycyclic aromatic hydrocarbons. Threshold values for each pollutant were regulated on various timescales and concentration levels considering their acute and long-term health impacts. EU air pollutant limit values are legally binding in all member states, although they can set lower values.

EU directives also set up emission standards and ceilings. The EU legislation on vehicle emissions and fuel-quality standards has evolved greatly since the first directive in 1970. The early legislation had the dual purpose of reducing pollution and avoiding barriers to trade due to different standards in different member states. It is now giving way to designs aimed at meeting air-quality targets. Figure 10.9 compares

Table 10.2 EU air quality standards.

Pollutant	Concentration	Averaging period	Legal nature	Permitted exceedences each year
Fine particles (PM2.5)	25 µg/m^3	1 year	Target value to be met as of 1.1.2010	n/a
			Limit value to be met as of 1.1.2015	
Sulphur dioxide (SO2)	350 µg/m^3	1 hour	Limit value to be met as of 1.1.2005	24
	125 µg/m^3	24 hours	Limit value to be met as of 1.1.2005	3
Nitrogen dioxide (NO2)	200 µg/m^3	1 hour	Limit value to be met as of 1.1.2010	18
	40 µg/m^3	1 year		n/a
PM10	50 µg/m^3	24 hours	Limit value to be met as of 1.1.2010	35
	40 µg/m^3	1 year	Limit value to be met as of 1.1.2005	n/a
Lead (Pb)	0.5 µg/m^3	1 year	Limit value to be met as of 1.1.2005 (or 1.1.2010 in the immediate vicinity of specific, notified industrial sources; and a 1.0 µg/m^3 limit value applied from 1.1.2005 to 31.12.2009)	n/a
Carbon monoxide (CO)	10 mg/m^3	Maximum daily 8 hour mean	Limit value to be met as of 1.1.2005	n/a
Benzene	5 µg/m^3	1 year	Limit value to be met as of 1.1.2010	n/a
Ozone	120 µg/m^3	Maximum daily 8 hour mean	Target value to be met as of 1.1.2010	25 days averaged over 3 years
Arsenic (As)	6 ng/m^3	1 year	Target value to be met as of 31.12.2012	n/a
Cadmium (Cd)	5 ng/m^3	1 year	Target value to be met as of 31.12.2012	n/a
Nickel (Ni)	20 ng/m^3	1 year	Target value to be met as of 31.12.2012	n/a
Polycyclic Aromatic Hydrocarbons	1 ng/m^3 (expressed as concentration of Benzo(a)pyrene)	1 year	Target value to be met as of 31.12.2012	n/a

Comparison of NO$_x$ emission standards for different Euro classes

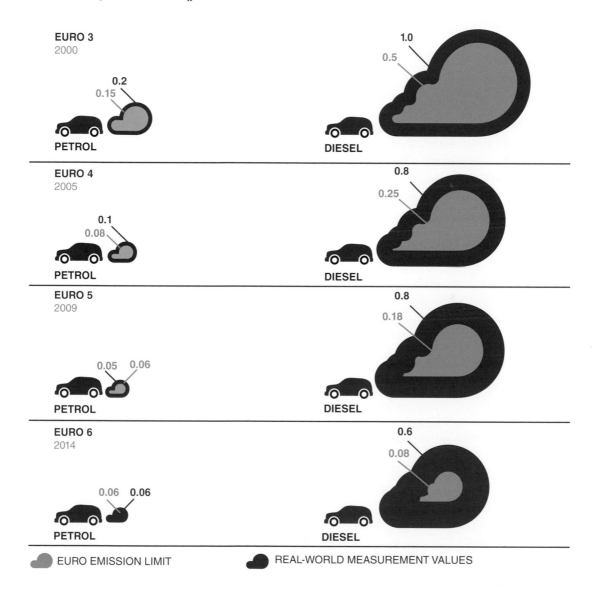

Figure 10.9 Comparison of NO$_x$ emission from different euro class vehicles. *Source:* reprinted from EEA 2016 with permission.

the EU limit on NO_x emissions for petro and diesel vehicles. Measures are being taken to tackle the higher real-world emissions than that based on EU test cycles to avoid the next Volkswagen scandal.

EU legislation on industrial air pollution has always taken air-quality limit values into account and required the operators to use 'Best available technology, not entailing excessive costs'. EU directive 2010/75/EU on industrial emissions is the main EU instrument regulating pollutant emissions from industrial installations. It is based on a Commission proposal recasting seven previously existing directives (including in particular the integrated pollution prevention and control [IPPC] Directive) following an extensive review of the policy. The directive entered into force on 6 January 2011 and had to be transposed by member states by 7 January 2013. Around 50 000 installations undertaking the industrial activities listed in Annex I of the directive are required to operate in accordance with a permit (granted by the authorities in the member states). This permit should contain conditions set in accordance with the principles and provisions of the directive. The Industrial Emissions Directive is based on several pillars, in particular (i) an integrated approach, (ii) use of best available techniques, (iii) flexibility, (iv) inspections, and (v) public participation.

10.7.1.3 Economic Incentives

Attempts to reduce urban driving by various types of economic incentives have had some success, but they are often opposed by trade. It must also be considered that driving restrictions *in* cities may promote the growth of large shopping centres, hotels, and office buildings *outside* the cities, where they can offer free parking spaces and other facilities, thus often resulting in an increase in total traffic. In such cases, improvement in urban air quality is paid for with more pollution on a larger geographical scale.

10.7.1.4 Taxes

Taxes can be imposed on the purchase of a vehicle or as an annual tax related to the vehicle weight, energy consumption, or emissions. Fuel taxes will in the long run lead to more effective driving and will increase the demand for smaller and more efficient cars. The total effect on energy consumption and emissions is much higher (and more logical) than any other form of taxation.

The basic problem with taxes is that the real political purpose often appears to be a means to provide additional revenue rather than protection of the environment. In addition, the attempts to reduce transport are in contrast to the general belief that mobility in society is a prerequisite for economic growth.

10.7.1.5 Traffic Management and Clean-Air Zones (or Low-Emission Zones)

More rational types of economic incentive aim at reducing traffic in sensitive areas. Driving in city centres can thus be reduced by restricting the number of parking places or by increasing the fees for their use. Pricing the infrastructure in the form of toll roads is used in many areas. Typically, a fee is paid for the use of individual stretches of highways, or for driving in a designated urban area (e.g. the London *congestion charge*). These zones are often called *clean-air zones* or *low-emission zones*. Oxford recently proposed a *zero-emission zone* where no petro or diesel vehicles are allowed to enter. Such measures usually result in an improvement in air quality in the designated zones, but the pollutant levels in surrounding areas go up.

A more advanced, in principle more fair, but not yet fully developed system is road pricing. Here the position of each car is determined by remote sensing and the information on movements is stored electronically in the car or elsewhere. This allows detailed taxing according to type of vehicle, different zones, and time of the day, etc., and thus encourages economic driving. A general objection to the system is the vision of a 'big brother society' where all movements of all people are registered.

10.7.1.6   City Planning

The result of a choice between public and individual transportation depends on the proximity of access to the public system. In an industrialized country, a distance of more than 1 km often means the use of a car. The impacts of urban air pollution can therefore be mitigated by constructive city planning. The complete separation of industry and habitation, originally envisaged as an environmental improvement and a reasonable solution in a society with heavily polluting industries, is now outdated and often leads only to increased commuting traffic and congestion.

The ideal now is integrated land use, which minimizes transport and thus total urban emissions. Open spaces and parks can be used to improve the environmental quality. In existing cities, the possibilities of restructuring are limited, but the construction of ring-roads, which lead part of the traffic around the city centre is one of the options. In this planning, which to a large extent is planning of traffic, it must be realized that air pollution is not its only environmental impact, and probably not even the most important.

10.7.2   Mitigation of Long-Range Transport of Air Pollutants

Air pollutants have no respect for national boundaries. The long-range transport of acidic pollutants from the United Kingdom to Scandinavian countries in the mid-twentieth century was a major environmental problem – acid rain – in these countries, which had important impact on terrestrial ecosystems. Therefore, international conventions on long-range transport of air pollutants have been established in a number of parts of the world that limit pollutant emissions. This includes the Geneva Convention on long-range, transboundary air pollution, which was established and signed in 1979. The related protocols are all aimed at protecting natural systems, and a new multipollutant–multieffect protocol will comprehensively address acidification, eutrophication, and photochemical air pollution. However, since the main part of the relevant

---

Case Study 10.3   Urban Pollution Patterns

The complex interplay between human activities, technical and legislative development, and natural parameters give rise to completely different pollution patterns in different cities of the world. However, there are some common features for cities in the industrialized world (mainly exemplified by Western Europe) and the developing world (mainly East Asia), respectively.

Comprehensive Records of Urban Air Quality

In most of the industrialized world urban air pollution is now monitored routinely. Since 1974, WHO and UNEP have, within the 'Global Environment Monitoring System', collaborated on a project to monitor urban air quality, the so-called GEMS/AIR. These data and similar data give an indication of trends in ambient air quality at the national level and in selected cities. Often, however, the data are based on only a few monitoring stations, placed at critical sites and thus representing microenvironments. It should also be taken into account that the coverage of stations varies from country to country, and that average values therefore can be biased in different ways.

Air pollution in the developing countries and in some countries with economies in transition is not documented in detail and longer time-series are very rare. In most cases, a general trend in air quality can be estimated only on the basis of dubious emission inventories and to a certain degree, visibility, which is more likely to be available from airports. Data presented in the open literature are seldom up to date and normally concern specific cities, which may not be fully representative.

General Pollution Development

Seen over longer periods, air pollution in major cities tends to: increase during the expansion phase; pass through a maximum level; and then reduce as abatement strategies are developed (Figure 10.10). Depending on the time of initiation of emission control the stabilization and subsequent improvement of the air quality may occur sooner or later in the development.

As Bertholt Brecht has put it, 'Erst kommt das Fressen, dann kommt die Moral' – or translated into modern English: 'First development and only later pollution control'. This was first seen in Western countries, and is now occurring in China and India. Other developing countries, including those in Africa and Southeast Asia, look set to follow this route (Figure 10.10) unless major efforts are being made to control air pollution before it worsens during the economic development. In the industrialized western world urban air pollution is in some respects in the last stage with effectively reduced levels of SO_2 and PM.

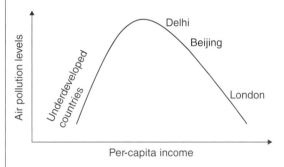

Figure 10.10 Hypothetical air pollution Kuznets curve. Delhi, Beijing, London, and underdeveloped African countries are sitting in different parts of the curve. (Drawn by the author.)

emissions takes place in urban areas, the necessary reductions have direct impacts on urban air quality. Thus, the later decreases in levels of SO_2 in the 1980s are related to the use of natural gas, low-sulphur fuel oil and desulphurization (or alternative sources of energy), necessary to comply with the international agreements. Also, attempts to reduce ecological impacts of large-scale photochemical air pollution will directly influence urban ozone levels – especially in the north of Europe. International agreements allocate the emissions to different industrial sectors or even to specific companies or operating sites in signature countries. In the EU, emissions standards for road vehicles are also set on an international level.

In other parts of the world, notably in East Asia, increasing regional and transboundary impacts have led to widespread haze events that span thousands of square kilometres.

10.7.3 Industrialized (OECD) Countries

10.7.3.1 Europe

Europe is a highly urbanized continent with more than 70% of the population living in cities. The large resources of coal was the primary source of energy during the industrial revolution, but in recent decades oil, gas and renewable energy have gradually replaced coal as the main energy source in the more developed western Europe, resulting in reduced emissions of SO_2 and particles. The total emission of SO_2 steadily increased from about 5 million tons in 1880 to a maximum of nearly 60 million in the 1970s, interrupted only by the Second World War. It peaked in the mid-1970s, but has now been reduced to less than half. For the traffic-related pollutants NO_x, CO, and VOCs, an increase has been reversed only recently by the introduction of three-way catalytic converters.

European road traffic currently accounts for the majority of the total CO emissions, more than half the NO_x-emissions and a third of the VOCs, but only a minor part of SO_2. European cities differ in various ways, which influence the relations between emissions and resulting pollution levels: western Europe is influenced by the predominant westerly wind bringing moist air from the sea, a climate that also favours long-range transport. In the northern part of Europe the relatively small amount of sunlight favours persistent inversions with poor dispersion conditions. In central and eastern Europe, synoptic high-pressure situations with air stagnation and accumulation of local pollution are frequent. During the summer, the climate in the Mediterranean region likewise favours accumulation of local emissions, whereas during the winter large-scale wind systems are more frequent. Formation of photochemical oxidants depends on sunlight, which in combination with poor dispersion conditions result in frequent pollution episodes during summer.

The general air quality in European cities has improved in recent decades – often in spite of an increase in population density and standard of living – but air pollution is still considered a top priority environmental problem with both urban and large-scale impacts.

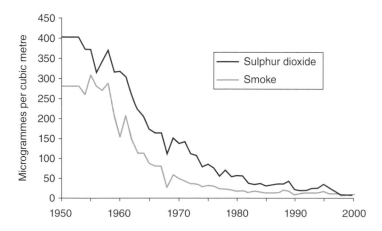

Figure 10.11 Annual average smoke and sulphur dioxide concentrations in London 1950–2000. Before 1954, data was only published as five-year averages (after 2000, the concentrations are too low to plot on this graph). *Source:* Greater London Authority (2002). See colour plate section for the colour representation of this figure.

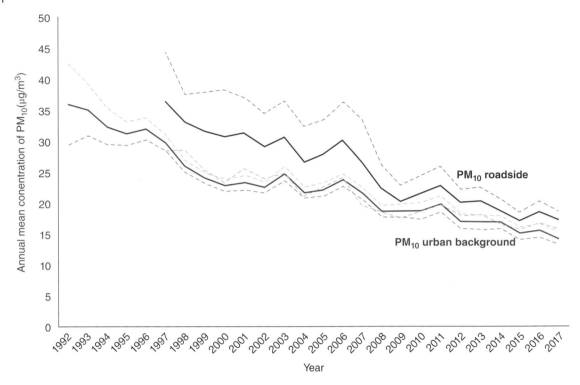

Figure 10.12 Annual concentrations of PM_{10} at urban background and roadside sites in the UK, 1992–2017. The PM_{10} index shows the annual mean, averaged over all included sites that had data capture greater than or equal to 75%. The dotted lines represent upper and lower bounds of the 95% confidence interval for the annual mean concentration for roadside sites and urban background sites. These intervals narrow over time because of an increase in the number of monitoring sites for both roadside and urban background sites and a reduction in the variation between annual means for PM_{10} measured at roadside sites. *Source:* reprinted from DEFRA (2018).

The drastic reduction in ambient 'black smoke' in cities in the 1960–1970s (e.g. in the London, Figure 10.11), partly brought about by a change from coal to less-polluting fuels for domestic heating and partly by the closing down of polluting industry, has also resulted in a marked reduction in incidence of haze events and fogs. Since 1990s, more widespread monitoring of PM_{10} became available. Generally speaking, urban background and roadside particulate pollution (PM_{10}) has shown long-term improvement. For example, the PM_{10} at urban background sites in the United Kingdom reduced by 50% from 1992 to 2017 (Figure 10.12). A similar decreasing trend in $PM_{2.5}$ was also observed, for example, in the UK since 2009 when its monitoring became operational (DEFRA 2018).

In many Western European cities, severe pollution by SO_2 is a thing of the past (e.g. Figure 10.11). This is partly because of the replacement of coal with cleaner fuels such as natural gases and partly by the adoption of SO_2 removal technologies in power plants and industries. The SO_2 levels in the provincial cities are not much higher than at rural sites, indicating that here most of the SO_2 is due to long-range transport. This development is the result of several causes. An important aspect is the Geneva Convention on Trans-boundary Air Pollution, which resulted in reduction of total emissions,

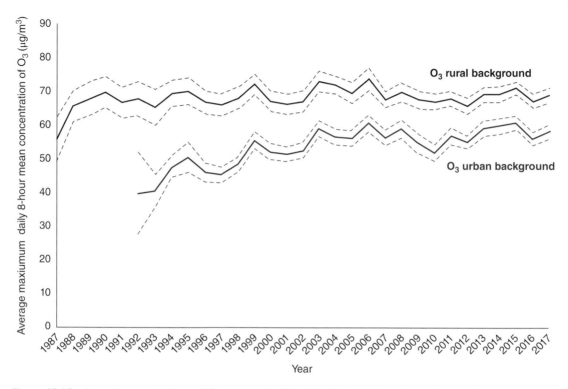

Figure 10.13 Annual concentrations of O_3 in the UK, 1987–2017. The O_3 index shows the annual mean, averaged over all included sites that had data capture greater than or equal to 75%. The dotted lines represent upper and lower bounds of the 95% confidence interval for the annual mean concentration for urban background sites and rural background sites. *Source:* reprinted from DEFRA (2018).

but also a widespread transition from individual to district space heating produced by large units with high stacks (often as combined heat and power production) has played a role.

In countries that have eliminated lead in petrol, the lead levels have substantially declined, especially in countries with few or no lead-emitting industries. In Denmark, where lead from petrol in 1977 accounted for 90% of the national lead emissions, the problem has virtually disappeared now. Low lead concentrations are now essentially due to long-range transport. In the early 1980s, 5% of Europe's urban population in cities with reported lead levels were exposed to more than the WHO guideline of $0.5\,\mu g\,m^{-3}$. At the end of the decade, levels above the guideline value were no longer reported from the Western countries.

European ozone levels appear to have increased from about $20\,\mu g\,m^{-3}$ around 1900 to about double that in 2009, with the most rapid rise between 1950 and 1970 concurrent with the rise in emissions of primary precursors. More recently, urban background ozone pollution in the United Kingdom has remained fairly stable between 2003 and 2017, although concentrations have shown a long-term increase since monitoring began (Figure 10.13). Summer smog with high ozone concentrations occurs in many European countries. As an urban phenomenon, it is most serious in Athens and

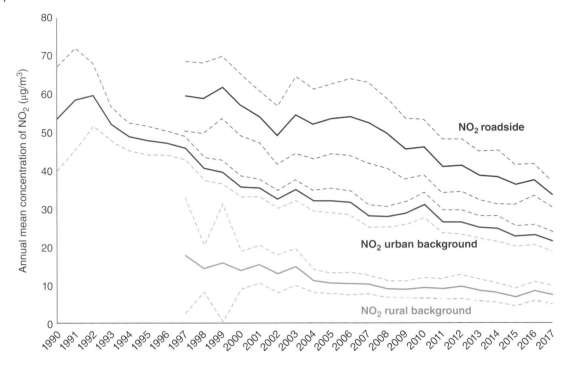

Figure 10.14   Annual mean concentrations of NO_2 at roadside, urban background, and rural background sites in the UK, 1990–2017. The NO_2 index shows the annual mean, averaged over all included sites that had data capture greater than or equal to 75%. The dotted lines represent upper and lower bounds of the 95% confidence interval for the annual mean concentration for roadside sites, urban background sites and rural background sites . The intervals narrow over time because of an increase in the number of monitoring sites and a reduction in the variation between annual means for NO_2. *Source:* reprinted from DEFRA (2018) with permission.

Barcelona with concentrations up to $200 \, \mu g \, m^{-3}$, but Frankfurt, Krakow, Milan, Prague, and Stuttgart are also affected. Generally, the present European ozone concentrations decrease from southeast to northwest. The latest EU air quality directive contains a target eight hourly mean value of $120 \, \mu g \, m^{-3}$ not to be exceeded more than 25 times a year averaged over three years (Table 10.2). The EU also defines an ozone threshold of $180 \, \mu g \, m^{-3}$ beyond which its member state must inform the public. The threshold reflects 'a level beyond which there is a risk to human health from brief exposure for particularly sensitive sections of the population'. In 2017, a number of sites in the United Kingdom exceeded the EU ozone public information threshold, which triggered air pollution alerts by the government.

In recent years, NO_x emission have decreased substantially despite the huge increase in vehicle population on the road. For example, in the UK, NO_x emission reduced by over 60% from 1992 to 2010 (Harrison et al. 2015). However, NO_2 concentration showed a much slower decreasing trend, particularly after 2000 (Figure 10.14). Indeed, many cities still break NO_2 limits: an annual average concentration of 40

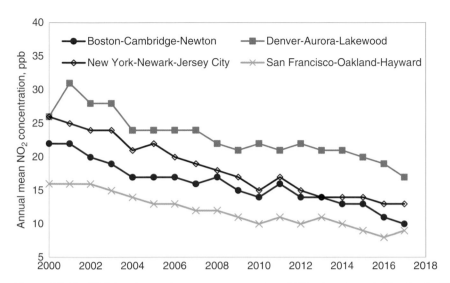

Figure 10.15 Weighted annual mean concentration of $PM_{2.5}$ in selected cities in the USA. Each of the annual average value were averaged over 2–12 sites monitoring sites. *Source:* data were obtained from https://www.epa.gov/air-trends/air-quality-cities-and-counties.

and no more than 18 hourly concentration to be over $200\,\mu g\,m^{-3}$ (Table 10.2). It was reported in earlier 2017 that a London area breached the annual NO_2 limits in just five days. This is why the European Commission has referred France, Germany, and the United Kingdom to the Court of Justice of the EU. NO_2 pollution is the biggest air quality policy challenge in many European cities now. Much of the problem comes from the extensive use of diesel cars, which emit much more NO_2 than petrol ones of the same Euro regulation class (Figure 10.9), the higher than expected emissions of NO_2 under real driving conditions (Figure 10.9) and the defeat devices installed by the car manufacturers to cheat on the emission tests, as illustrated by the Volkswagen scandal.

10.7.3.2 North America

The United States and Canada are amongst the wealthiest countries in the world, both with respect to natural resources and production. This has previously led to serious urban pollution, especially in the United States.

Many cities in the early industrialized United States were, like those in Europe, characterized by heavy smoke and subsequently gone through the typical development phases. Generally, the overall emissions and atmospheric concentrations of SO_2, CO, VOCs, and lead drastically reduced from the 1970s (Harrison et al. 2015). Reductions in PM_{10} and $PM_{2.5}$ were also observed in most but not all of the cities since 1990–2015 (Figure 10.15). A majority of the cities met the US EPA's National Ambient

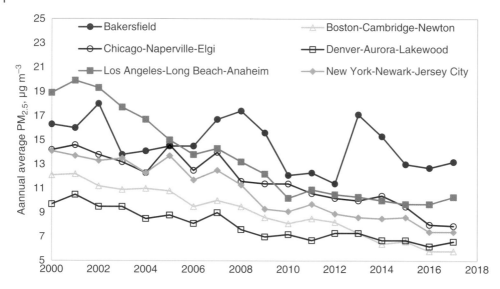

Figure 10.16 Annual arithmetic mean concentration of hourly concentrations of NO_2 in selected cities in the USA. Each of the annual average value were averaged over 2–12 sites monitoring sites. 1 ppb of NO_2 is approximately $1.91\,\mu g\,m^{-3}$ under 20 °C and 1013 mb. *Source:* data were obtained from https://www.epa.gov/air-trends/air-quality-cities-and-counties.

Air Quality Standards for $PM_{2.5}$ of $12\,\mu g\,m^{-3}$. But some cities still break the EPA standard. For example, $PM_{2.5}$ in Bakersfield, California, was over $12\,\mu g\,m^{-3}$ in 2017 (Figure 10.15).

In the USA, ozone concentrations showed a significant decrease of 30% from 1988 to 1993 in urban residential areas, both as an average and in the most polluted cities. Ozone concentration in general decreased in cities but the rate is very slow from 2000 to 2017. Many counties find it hard to comply with the 2015 ozone limit of 70 ppb (approximately $140\,\mu g\,m^{-3}$), which is the annual fourth-highest daily maximum eight-hour concentration, averaged over three years.

Unlike European cities, NO_2 pollution is less serious, due to less prevalent use of diesel engines in cars. Figure 10.16 shows the trend of annual average NO_2 concentrations in some US cities, which halved from 1992 to 2017. None of these cities have NO_2 levels higher than $40\,\mu g\,m^{-3}$.

In Canada, emissions of SO_2 and PM have been reduced significantly since the early 1970s, and lead has virtually disappeared. However, some central Canadian cities experience unacceptable air quality with high levels of ozone and PM, especially during the summer.

10.7.4 Developing Countries – East Asia as an Example

East Asia contains three of the world's largest countries (China, India, and Indonesia), several small land locked states, and a series of island states (including the highly industrialized Japan). The region contains about half the largest cities in the world. Air pollution, particularly the 'haze' pollution in developing countries, has become a concern for the general public and the government for years. In January

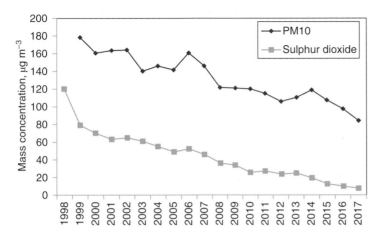

Figure 10.17 Annual average PM_{10} and SO_2 mass concentration in Beijing. *Source:* data from publicly available open resources collected from Beijing Environmental Protection Bureau.

2013, the widespread haze events covered huge areas of northern China, when the $PM_{2.5}$ levels reached over 500 $\mu g\,m^{-3}$. During the Diwali week in Delhi in 2017, the $PM_{2.5}$ level sometimes reached over 1000 $\mu g\,m^{-3}$. These events have led to major efforts to control air pollution, which contributed to improvement in quality in some of the megacities in East Asia.

PM_{10} levels reduced from about 180 $\mu g\,m^{-3}$ in 1999 to less than 90 $\mu g\,m^{-3}$ in 2017 in Beijing, and SO_2 reduced from 120 $\mu g\,m^{-3}$ in 1998 to only 8 $\mu g\,m^{-3}$ in 2017 (Figure 10.17). $PM_{2.5}$ pollution level has reduced dramatically in Beijing from over 120 $\mu g\,m^{-3}$ in 2001 (He et al. 2011) to <60 $\mu g\,m^{-3}$ in 2017 (publicly available data from Beijing Environmental Protection Bureau). However, PM pollution remains the most significant air pollution problems in developing countries.

In a series of major cities, where energy production is based on gas or low-sulphur coal (e.g. Mumbai, Kolkata, Bangkok) or is in the transition from coal to natural gas (e.g. Beijing), SO_2 pollution is not a serious problem.

But urban transport is an increasing problem in many cities. This contributed to the PM pollution but more seriously the NO_x pollution and associated photochemical smog. Taking Beijing as an example, O_3 annual average concentration has been increasing since the late 1990s.

10.8   Summary

Urban air pollution and its impact on urban air quality is a worldwide problem. It manifests itself differently in different regions, depending on economic, political, and technological development, upon the climate and topography, and last – but not least – on the nature and quality of the available energy sources. Nevertheless, a series of general characteristics emerges.

10.8.1   From Space Heating to Traffic

Originally, urban air pollution was a strictly local problem mainly connected with space-heating and primitive industry. In the earlier stages of modern industrialization, it was considered unavoidable or even a symbol of growth and prosperity. The situation in the industrialized Western world has in most respects proved this viewpoint outdated. Emissions from industry and space heating are by and large controllable, but the urban atmosphere in most developed cities is now dominated by traffic emissions and in developing cities by both traffic and traditional pollutant sources (coal combustion and industry) with documented impacts on human health. The attention has thereby been shifted from sulphur dioxide and soot to nitrogen oxides, the whole spectrum of organic compounds, and particles of various sizes and composition, which are reported to be carcinogenic and/or cause a significant reduction of life through respiratory and cardiovascular diseases. These pollutants require much more detailed investigations, both in the form of chemical analysis and computer modelling.

In principle, the control of emissions of sulphur and nitrogen oxides is relatively straightforward when they are related to power production by large plants, which can be compelled to use clean fuels and equipped with proper cleaning technology. Traffic emissions are more difficult to control, because they arise from small units. Meeting the increasing stringent air pollution targets is therefore not an easy task. According to the conclusion of the 'Auto Oil Programme', even with the maximum technical package introduced in the EU not all cities will be able to comply.

The situation in developing countries is mixed. In some major cities in Asia, sulphur emissions have been brought under control and $PM_{2.5}$ has also decreased, e.g. via a transition from coal to natural gas. However, many of the developing cities, particularly megacities, face complex pollution with characteristics of both traditional London smog $PM_{2.5}$ and/or ozone pollution remaining the biggest challenge in developing megacities such as Beijing and Delhi.

10.8.2   Regional Impacts on Urban Air Pollution

The interactions between the cities and their surroundings are becoming increasingly important. With the expanding and often merging urban areas and diminishing emissions in the cities, the pollution levels are to a large extent determined by long-range transport. The same applies to lead pollution in countries where lead has been removed from petrol. Another example is photochemical air pollution, which in many cases is a large-scale phenomenon, where emissions and atmospheric chemical reactions in one country may influence urban air quality in another.

10.8.3   Cities as Sources of Pollution

A more far-reaching problem is the city as a source of pollution. In the past, local problems were attacked by dispersing pollutants from high stacks, but this only resulted in a transfer to a larger geographical scale in the form of acidification and other transboundary phenomena. Now long-lived greenhouse gases, especially carbon dioxide, threaten the global climate – irrespective of their origin. This problem can be solved only by a general reduction in net emissions.

In the industrialized countries the development in technology and legislation to protect air and water quality has in many cases resulted in improved energy efficiency and emission reductions, although some means of improving urban air quality, such as catalytic converters, which consume energy and emit nitrous oxide (another greenhouse gas) are contrary to this goal. On a global basis, the growing population and its demand for a higher material standard of living have so far counteracted any reductions in total emissions. Therefore, the responsibility for the future is both national and global, comprising many actors such as national environmental agencies and international organizations.

10.8.4   A Systematic Approach

The realization that traffic is rapidly becoming *the* urban air-quality problem, both in industrialized and developing countries, calls for comprehensive solutions, where traffic-related air pollution is seen in connection with other impacts of traffic such as noise, accidents, congestion, and general mental stress. As a consequence, technological improvements in the form of less-polluting vehicles are not sufficient. Also, support of infrastructures, where the need of transport is minimized, and where use of public means of transport dominates over individual private cars and motorbikes should be encouraged. Unfortunately, most attempts at control will be perceived as a restriction of individual freedom, and they are frequently met with outspoken opposition. Obviously, a change in attitude is called for.

Acknowledgement

I thank Jes Fenger from National Environmental Research Institute, Denmark, for agreeing to let me use some of the material in the previous edition of this book for this chapter. I also thank doctoral students Ms. Jinxiu Han and Ms. Dewi Komalasari from the University of Birmingham for collecting some of the data from open resources. Research under funding from Natural Environment Research Council (NE/R005281/1 and NE/N007190/1) contributed to this chapter.

Questions

1   Why are ozone levels *higher* in the weekends in northern European cities than during the working week, and why might urban ozone concentrations increase in the future?

2   Why can material damage be *higher* in northern than in southern cities, although pollution levels are generally *lower*?

3   Why does a city such as Copenhagen generally have lower pollution levels than other cities with similar emission densities?

4   What are the differences between *primary* and *secondary* air pollutants?

5   What are the arguments *against* restricting driving in cities?

6   What are the differences between MAP and hazardous air pollutants (HAP)?

References

Avnery, S., Mauzerall, D.L., Lu, J., and Horowitzc, L.W. (2011). Global crop yield reductions due to surface ozone exposure: 1. Year 2000 crop production losses and economic damage. *Atmospheric Environment* 45: 2284–2296.

Burnett, R.T., Pope, C.A. III, Ezzati, M. et al. (2014). An integrated risk function for estimating the global burden of disease attributable to ambient fine particulate matter exposure. *Environmental Health Perspectives* 122: 397–403.

COMEAP (2018). Understanding the links between nitrogen dioxide (NO2) and our health. Report by the Committee on Medical Effects of Air Pollutants. https://assets.publishing.service.gov.uk/government/uploads/system/uploads/attachment_data/file/734799/COMEAP_NO2_Report.pdf (accessed 1 October 2018).

Crilley, L. R., Kramer, L., Pope, F. D., Whalley, L. K., Cryer, D. R., Heard, D. E., ...Bloss, W.J. (2016). On the interpretation of in situ HONO observations via photochemical steady state. *Faraday Discussions*, 189, 191-212. doi:10.1039/c5fd00224a

Daniels, M.J., Dominici, F., Samet, J.M., and Zeger, S.L. (2000). Estimating particulate matter-mortality dose-response curves and threshold levels: an analysis of daily time-series for the 20 largest US cities. *American Journal of Epidemiology* 152: 397–406.

DEFRA (2018). Defra National Statistics Release: Air quality statistics in the UK 1987 to 2017. https://assets.publishing.service.gov.uk/government/uploads/system/uploads/attachment_data/file/702712/Air_Quality_National_Statistic_-_FINALv3.pdf (accessed 2 October 2018).

DEFRA (Department of Environment, Food & Rural Affairs) (2017). Air pollution in the UK 2016. DEFRA annual report. https://uk-air.defra.gov.uk/library/annualreport (accessed August 2019).

Dockery, D.W. (1993). An association between air pollution and mortality in six U.S. cities. *The New England Journal of Medicine* 329: 1753–1759.

EEA (2016). Comparison of NOx emission standards for different Euro classes. https://www.eea.europa.eu/media/infographics/comparison-of-nox-emission-standards/view (accessed on 1 October 2018).

Fenger, J., Hertel, O., and Palmgren, F. (eds.) (1998). *Urban Air Pollution. European Aspects.* Dordrecht, the Netherlands: Kluwer Academic Publishers.

Grange, S.K., Lewis, A.C., Moller, S.J., and Carslaw, D.C. (2017). Lower vehicular primary emissions of NO_2 in Europe than assumed in policy projections. *Nature Geoscience* 10: 914–918.

Greater London Authority (2002). 50 years on: The struggle for air quality in London since the great smog of December 1952. London.

Harrison, R.M., Pope, F.D., and Shi, Z. (2014). Air pollution. In: *Reference Module in Earth Systems and Environmental Sciences* (ed. S.A. Elias). Elsevier Sciences.

Harrison, R.M., Pope, F.D., and Shi, Z. (2015). Trends in local air quality 1970–2014. In: *Still Only One Earth: Progress in the 40 Years Since the First UN Conference on the Environment*, Environmental Science and Technology No. 40 (eds. R.E. Hester and R.M. Harrison), 58–106. Royal Society of Chemistr.

He, K., Yang, F., Ma, Y. et al. (2011). The characteristics of PM2.5 in Beijing, China. *Atmospheric Environment* 35: 4959–4970.

Holgate, S., Grigg, J., Agius, R. et al. (2016). *Every Breath We Take: The Lifelong Impact of Air Pollution, Report of a Working Party*. London: Royal College of Physicians.

Kucera, V. and Fitz, S. (1995). Direct and indirect air pollution effects on materials including cultural monuments. *Water, Air, and Soil Pollution* 85: 153–165.

Liu, D., Whitehead, J., Alfarra, M.R. et al. (2017). Black-carbon absorption enhancement in the atmosphere determined by particle mixing state. *Nature Geoscience* 10: 184–188.

Maher, B.A., Ahmed, I.A.M., Karloukovski, V. et al. (2016). Magnetite pollution nanoparticles in the human brain. *PNAS* 113: 10797–10801.

Maynard, R.L. and Howard, C.V. (1999). *Particulate Matter. Properties and Effects Upon Health*. Oxford, UK: Bios Scientific Publishers.

OECD (2016). The economic consequences of outdoor air pollution – policy highlights. https://www.oecd. org/env/the-economic-consequences-of-outdoor-air-pollution-9789264257474-en.htm (accessed August 2019).

Pope, C.A. III and Dockery, D.W. (2006). Health effects of fine particulate air pollution: lines that connect. *Journal of the Air & Waste Management Association* 56: 709–742.

Pope, C.A. III, Dockery, D.W., and Schwartz, J. (1995). Review of epidemiological evidence of health effects of particulate air pollution. *Inhalation Toxicology* 7: 1–18.

Ricardo-AEA (2014). Valuing the impacts of air quality on productivity. Report for the Department for Environment, Food and Rural Affairs. Ricardo-AEA/R/3417.

Ritchie H. and Roser, M. (2018). Urbanization. OurWorldInData.org (accessed 27 September 2018).

Seinfeld, J. and Pandis, S. (2016). *Atmospheric Chemistry and Physics: From Air Pollution to Global Change*, 3e. New York, NY: Wiley.

Shi, Z. (2016). Highlights from Faraday discussion: chemistry in the urban atmosphere, United Kingdom. *Chemical Communications* 52: 9162–9172.

Smith, K.R. (1993). Fuel combustion, air pollution exposure, and health: situation in the developing countries. *Annual Review of Energy and the Environment* 18: 529–566.

Tidblad, J. and Kucera, V. (1998). Materials damage. In: *Urban Air Pollution, European Aspects* (eds. J. Fenger, O. Hertel and F. Palmgren), 343–361. Dordrecht, the Netherlands: Kluwer Academic Publishers.

Townsend, C.L. and Maynard, R.L. (2002). Effects on health of prolonged exposure to low concentrations of carbon monoxide. *Occupational and Environmental Medicine* 59: 708–711.

Vu, T. V., Shi, Z., Cheng, J., Zhang, Q., He, K., Wang, S., & Harrison, R. M. (2019). Assessing the impact of clean air action on air quality trends in Beijing using a machine learning technique. *Atmospheric Chemistry and Physics*, 19(17), 11303-11314. doi:10.5194/acp-19-11303-2019

WHO (2014). Burden of disease from Household Air Pollution for 2012. http://www.who.int/phe/health_topics/outdoorair/databases/FINAL_HAP_AAP_BoD_24March2014.pdf?ua=1 (accessed on 27 September 2018).

World Bank (2016). *The Cost of Air Pollution: Strengthening the Economic Case for Action*. Washington, DC: World Bank Group http://documents.worldbank.org/curated/en/781521473177013155/The-cost-of-air-pollution-strengthening-the-economic-case-for-action.

Further Reading

Finlayson-Pitts, B.J. and Pitts, J.N. Jr. (2000). *Chemistry of the Upper and Lower Atmosphere: Theory, Experiments and Applications*. San Diego, CA: Academic Press.

Greadel, T.E. and Crutzen, P.J. (1995). *Atmosphere, Climate and Change*. New York: Scientific American Library.

Mage, D., Ozolins, G., Peterson, P. et al. (1996). Urban air pollution in megacities of the world. *Atmospheric Environment* 30: 681–686.

The European Environmental Agency (2005). *The European Environment. State and Outlook 2005.* Copenhagen, Denmark: European Communities.

World Health Organization (WHO)/United Nations Environment Programme (UNEP) (1992). *Urban Air Pollution in Megacities of the World.* Oxford, UK: Blackwell.

11

Global Warming and Climate Change Science

Atul Jain[1], Xiaoming Xu[1], and Nick Hewitt[2]

[1] *Department of Atmospheric Sciences, University of Illinois, Urbana, IL, United States of America*
[2] *Lancaster Environment Centre, Lancaster University, Lancaster, United Kingdom*

Prior to the Industrial Revolution, the climate of the Earth was entirely governed by natural influences. Records preserved in ice cores and ancient sediments reveal large fluctuations in global climate. The Milankovitch theory holds that long-term natural variations in temperature change were primarily governed by 26 000–100 000-year cycles in solar flux due to the Earth's orbital variations. Comparison of historical records of temperature and greenhouse gas concentrations (Figure 11.1) reveals that: variations in temperature are well-correlated with variations in major *greenhouse gases* (*GHGs*) from biogenic sources, such as carbon dioxide (CO_2) and methane (CH_4); and historic temperature variations were also affected by periodic episodes of high dust levels related to increases in volcanic activity or widespread arid conditions.

Until just a few decades ago, it was believed that human presence on this planet was insufficient to affect the global natural environment (e.g. Keller 1998). The truth is that, commencing in the 1600s and rapidly accelerating after the Industrial Revolution of the 1700s and 1800s, human activities began to intervene in the natural world, and the scale of emissions from human activities is now sufficient to potentially jeopardize life on Earth. Worldwide monitoring has provided concrete evidence that human activities such as fossil-fuel combustion, agriculture development, waste generation, synthetic chemical production, biomass burning, and changes in land use are significantly altering levels of radiatively and chemically active GHGs and aerosols in the atmosphere. These changes have resulted in greenhouse-gas-induced temperature rise, known as *global warming* or climate change.

Greenhouse warming is a natural phenomenon that is caused by naturally occurring GHGs in the Earth's atmosphere. These GHGs absorb much of the Earth's infrared (IR) radiation that would otherwise escape to space. The trapped radiation warms the lower atmosphere and the Earth, keeping the Earth's surface over 30 °C warmer than it would otherwise be. This phenomenon, known as the *greenhouse effect,* enables life to survive on Earth. The major GHGs contributing to this natural effect are those with significant natural sources: water vapour (H_2O), carbon dioxide (CO_2), methane (CH_4), nitrous oxide (N_2O), and tropospheric ozone (O_3).

Atmospheric Science for Environmental Scientists, Second Edition. Edited by C.N. Hewitt and Andrea V. Jackson.
© 2020 John Wiley & Sons Ltd. Published 2020 by John Wiley & Sons Ltd.

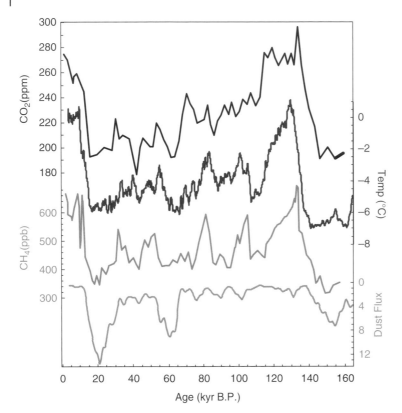

Figure 11.1 Climate and atmospheric composition records from the Vostok ice-core (East Antarctica) covering the past 160 000 years. These include CO_2 and CH_4 greenhouse gas records, which are closely tied to Antarctic temperature variations over the last full glacial–interglacial climate cycle (Lorius et al. 1985; Barnola et al. 1987; Jouzel et al. 1987; Chappellaz et al. 1990). Temperature data are plotted as deviations from the present-day mean annual temperature. Also included is the record of the flux of dust to the area (shown on an inverted scale for comparison purposes) (Petit et al. 1990). See colour plate section for the colour representation of this figure.

The problem of global warming is what is known as the *enhanced* greenhouse effect. Since the beginning of industrial times, the greenhouse effect has been enhanced by steadily increasing GHG concentrations in the atmosphere. These concentrations have been increasing due to emissions from human-related activities. Increasing concentrations alter the radiative balance of the Earth and result in a warming of the Earth's surface. The main GHGs affected by human activities are CO_2, CH_4, N_2O, tropospheric O_3, and CFCs (chlorofluorocarbons; mainly CFC-11 and CFC-12), which are also responsible for stratospheric O_3 loss. Aerosols resulting from human activities, particularly from fossil-fuel combustion, can scatter or absorb solar and IR radiation, both cooling and/or heating the Earth's surface. Emissions, concentrations, and radiative effects of these gases and aerosols are the topics of this chapter.

Measures imposing substantial reductions in CO_2 and other GHG emissions would severely limit global fossil fuel energy use. If GHG emissions are not limited, there are strong scientific grounds to believe they will cause significant global warming in the next few decades. Such warming would be global, but with regional impacts of varying severity. Concomitant changes are expected to occur in regional patterns of temperature, precipitation and sea level, with resultant impacts on most societal, economic, and environmental aspects of existence on this planet. With increasing world population, impacts can be expected not only in the areas of agriculture and water resources, affecting food security, but also by sea-level rise affecting coastal settlements, including a number of the world's major cities. These impacts are expected to impinge on the economy and the welfare of every nation.

This chapter provides the latest scientific understanding of global warming – that has been created by the growing influence of human activities on the Earth–atmosphere system. We begin by providing the historical and current greenhouse gas concentrations and presenting their impacts on climate. We then describe their conceivable future trends. We conclude this chapter with the discussion of the potential impacts of climate and current policy considerations.

11.1 Historical Evidence of the Impact of Human Activities on Climate

11.1.1 Increased Greenhouse Gas and Aerosol Concentrations

Analysis of air bubbles trapped in ice cores reveals that prior to the Industrial Revolution, atmospheric concentrations of heat-trapping GHGs were relatively constant. Since the mid-1700s, however, the world has become increasingly industrialized. Dramatic increases in the use of carbon-containing fossil fuels such as coal, oil, and natural gas have occurred to meet the demand for energy. Synthetic chemicals have been manufactured to supply important products. At the same time, population has grown exponentially and agricultural activities have developed accordingly. This has resulted in additional energy use and deforestation, particularly in tropical regions. Taken together, increased fossil-fuel burning, chemical production, and agriculture have caused emissions of GHGs and aerosol precursors to rise rapidly, resulting in a significant increase in atmospheric levels of these same GHGs and aerosols. The most conclusive evidence for such a connection between human activities and atmospheric change comes from a comparison between historical emissions from energy use and agricultural activities, and the observed atmospheric concentrations of various GHGs. For example, Figure 11.2 shows fossil-fuel-related carbon emissions, whilst Figure 11.3 shows the observed increasing concentrations for CO_2.

Following is a brief discussion of the observed trends in atmospheric concentrations and the estimated current emissions for the most important GHGs – CO_2, CH_4, N_2O, tropospheric O_3, CFC-11, and CFC-12 – and aerosols. There are also several gases, such as carbon monoxide (CO), non-methane hydrocarbons (NMHC), and nitric oxides (NO_x), that are not important GHGs in themselves but are important precursors of tropospheric O_3 and will be discussed in that section.

11.1.1.1 Water Vapour

Water vapour (H_2O) is the most important GHG in the atmosphere, responsible for much of the naturally occurring greenhouse effect. Water vapour absorbs infrared (IR) radiation over a large range of wavelengths, leaving only an atmospheric window region between 9 and 12 μm through which IR radiation emitted by the Earth can escape to space. (The definition of a non-H_2O greenhouse gas is that it has at least one strong absorption band in the atmospheric window region where H_2O is not absorbing.)

Water vapour concentration is highly variable, ranging from over 20 000 ppmv in the lower tropical atmosphere to only a few ppmv in the stratosphere (Peixoto and Oort 1992). On a global scale, tropospheric

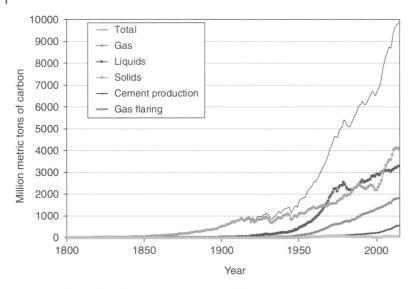

Figure 11.2 Global CO_2 emissions from fossil-fuel burning, cement manufacture, and gas flaring: 1751–2014. *Source:* Boden and Andres (2017). See colour plate section for the colour representation of this figure.

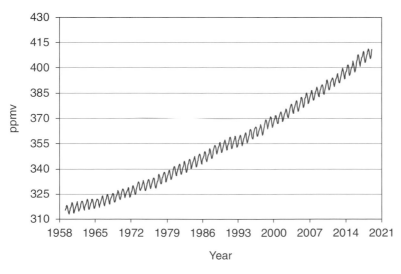

Figure 11.3 Observed monthly average CO_2 concentration (ppmv) from Mauna Loa, Hawaii (Keeling et al. 2001). Seasonal variations are primarily due to the uptake and production of CO_2 by the terrestrial biosphere.

concentrations of water vapour are determined by climate parameters such as evaporation, condensation, and precipitation rates, which are in turn affected by temperature and atmospheric dynamics. As such, tropospheric water vapour is not directly affected by human emissions, but rather is determined internally within the climate system. However, a human-induced climate warming will affect water vapour in the lower atmosphere. As temperatures rise, more water vapour will evaporate into the atmosphere.

In this way, water vapour is projected to be a significant *positive feedback* to global warming. (A positive feedback acts to enhance the original forcing. In the case of water vapour, increased temperature due to global warming will enhance evaporation, raising water vapour levels in the lower atmosphere. Since water vapour is a greenhouse gas, more water vapour in the atmosphere means more absorption of IR radiation, adding to the initial warming effect.)

Water vapour concentrations in the upper troposphere and lower stratosphere are very low. However, due to the lower temperatures at that altitude, the effectiveness of water vapour as a greenhouse gas is enhanced. The main sources of stratospheric water vapour are transport upwards from the troposphere and the oxidation of methane.

11.1.1.2 Carbon Dioxide

Carbon dioxide is the most important GHG produced by human activities. Based on measurements from air bubbles trapped in ice cores, preindustrial CO_2 concentration was about 280 ppmv. Accurate, real-time measurements of CO_2 began in 1958 (Figure 11.3), and show that the annually averaged atmospheric CO_2 concentration had risen from 316 ppmv in 1959 to 411 ppmv in 2019 (Keeling et al. 2001; https://www.esrl.noaa.gov/gmd/ccgg/trends). Concentrations of CO_2 exhibit a seasonal cycle due to the uptake and release of atmospheric CO_2 by terrestrial ecosystems.

Atmospheric CO_2 is primarily affected by three main human activities:

1) the burning of fossil fuels, which results in the oxidation of the carbon contained in the fuels by atmospheric oxygen to form CO_2;
2) the manufacturing of cement, which releases the carbon contained in limestone as CO_2 into the atmosphere;
3) changes in land use, mainly due to deforestation, that release the carbon contained in biomass into the atmosphere as CO_2.

Annual global emissions from fossil-fuel burning and cement production in 2015 were estimated to be 9.9 Pg C year^{-1}, with 0.6 Pg C year^{-1} from cement production (Boden and Andres, 2017). Net land-use flux, made up of a balance of CO_2 emissions due to deforestation versus CO_2 uptake due to regrowth on abandoned agricultural land, can be estimated based on land-use statistics and simple models of rates of decomposition and regrowth. The uncertainty in CO_2 emissions due to land-use changes is quite large (Jain and Yang 2005). The annual flux of carbon from land-use changes for the 1990s and 2000s have been estimated to be 1.4 (0.7–2.1) Pg C year^{-1} and 1.3 (0.6–2.0) Pg C year^{-1} (Le Quéré et al. 2018).

According to the Global Carbon Budget 2018 (Le Quéré et al. 2018), during the last decade (2008–2017), fossil CO_2 emissions were 9.4 ± 0.5 Pg C year^{-1}, emissions from land use and land-use change (mainly deforestation) were 1.5 ± 0.7 Pg C year^{-1}, the growth rate of atmospheric CO_2 concentration was 4.7 ± 0.02 Pg C year^{-1}, the oceanic CO_2 sink was 2.4 ± 0.5 Pg C year^{-1}, and the terrestrial CO_2 sink was 3.2 ± 0.8 Pg C year^{-1}. There was a small budget imbalance of 0.5 Pg C year^{-1}, indicating overestimated emissions and/or underestimated sinks.

For the year 2017 alone, the growth in fossil CO_2 emissions was about 1.6% and emissions increased to 9.9 ± 0.5 Pg C year^{-1}, emissions from land use and land-use change were 1.4 ± 0.7 Pg C year^{-1}, the growth rate of atmospheric CO_2 concentration was 4.6 ± 0.2 Pg C year^{-1}, the oceanic CO_2 sink was 2.5 ± 0.5 Pg C year^{-1}, and the terrestrial CO_2 sink was 3.8 ± 0.8 Pg C year^{-1}, with a budget imbalance of 0.3 Pg C.

In contrast to other GHGs, CO_2 is not removed from the atmosphere by chemical reactions with other atmospheric gases. For this reason, atmospheric CO_2 does not have a specific lifetime. Instead, it is part of a cycle whereby carbon is transferred between terrestrial, oceanic, and atmospheric reservoirs over time scales ranging from tens to thousands of years.

11.1.1.3 Methane

Per molecule, CH_4 is approximately 34 times more effective as a greenhouse gas than CO_2 (IPCC 2013). When its effectiveness as a GHG is combined with the large increase in its atmospheric concentration, methane becomes the second-most important GHG contributing to global warming. Based on ice-core measurements and direct measurements, which began in 1978, atmospheric CH_4 concentration increased steadily from its preindustrial concentration of about 0.7–1.8 ppmv by 2011, a 250% increase (IPCC 2013). The CH_4 concentrations in the atmosphere have been variable in recent decades. They were relatively stable for about a decade in the 1990s, but then started growing again in 2007 (Figure 11.4a). The exact drivers of the renewed growth since 2007 are still debated. Climate-driven fluctuations of CH_4 emissions from natural wetlands are the main drivers of the global inter-annual variability of CH_4 emissions (IPCC 2013). The short-term variations in 1990s are thought to be linked to chemical- and temperature-related impacts due to the Pinatubo eruption in 1991, whilst the longer-term decline in the growth rate from late 1990s to mid-2000s may signal a decoupling between traditional CH_4 sources and actual emissions. The renewed increase in global CH_4 levels since 2007 may be due to the rise in natural wetland emissions and fossil-fuel emissions (Kirschke et al. 2013).

The source strength for methane is about 678 Tg CH_4 per year based on bottom-up estimates in the 2000s, with 50–65% of CH_4 emissions contributed by anthropogenic sources (IPCC 2013). Human-related sources are mainly biogenic in origin and related to agriculture and waste disposal, including cows and other ruminants, rice paddies, biomass burning, animal and human waste, and landfills. Methane is emitted naturally by wetlands, termites, other wild ruminants, ocean, and hydrates.

Over 90% of CH_4 is removed from the atmosphere through reactions with the hydroxyl radical (OH; see Chapter 5). Reactions with OH oxidize CH_4 to carbon monoxide, which, in turn, reacts with OH, lowering OH levels and slowing the removal of the original CH_4. For this reason, although the chemical lifetime of CH_4 is 8.7 years, due to the CH_4–OH–CO feedback cycle, emissions of CH_4 from human activities are removed from the atmosphere over an adjustment time of approximately 12 years. The remaining 10% of CH_4 not oxidized by OH is removed through uptake by soil or transport into the stratosphere.

11.1.1.4 Nitrous Oxide

Nitrous oxide, also known as laughing gas, is the third-most important GHG produced by human activities. On a molecule basis, N_2O is about 298 times more efficient than CO_2 in absorbing IR radiation (IPCC 2013). Levels of N_2O have continued to grow in the global atmosphere since preindustrial times. Ice-core measurements show that prior to industrialization, concentrations of N_2O were relatively stable at about 270 ppbv. By 2018, the mean global atmospheric concentrations of N_2O had increased 22% to 330 ppbv (NOAA 2018). Direct measurements of the global mean N_2O concentrations for the period 1978–2018 show a continuous increase in the N_2O concentration at a rate of 0.75 ppbv year^{-1} (Figure 11.4b).

Nitrous oxide is produced from a wide variety of microbial sources in soils and water. Based on recent estimates, total global N_2O emissions are about 17.9 Tg N year^{-1}, of which 39% of emissions arise from human-related activities whilst 61% are of natural origin (IPCC 2013). Major anthropogenic sources of

Figure 11.4 Observed global annual average (a) CH$_4$, (b) N$_2$O, and (c) CFC-11 and CFC-12 concentrations since 1978. The data are based on measurements from NOAA Climate Monitoring and Diagnostics Laboratory (http://www.cmdl.noaa.gov/hats/graphs/graphs.html).

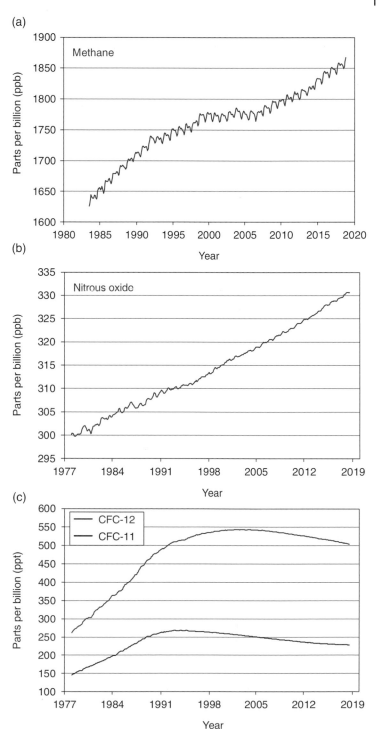

N_2O include fertilized cultivated soils, biomass burning, industrial sources, and cattle and feedlots. The main natural sources for N_2O are the ocean, tropical soils such as wet forests and dry savannas, and temperate soils such as forests and grasslands.

Atmospheric N_2O is primarily removed in the stratosphere by photolysis and reaction with electronically excited oxygen atoms.

11.1.1.5 Halocarbons

Halocarbons are unique in that they contribute to both global warming and stratospheric ozone depletion (see Chapter 8). These gases are important GHGs because they have strong IR absorption lines in the atmospheric window region. Due to their relatively high concentrations, the most potent halocarbons in the current atmosphere are the industrially produced chlorofluorocarbons $CFCl_3$ (CFC-11) and CF_2Cl_2 (CFC-12). A molecule of CFC-11 and CFC-12 is about 12400 and 15800 times more effective, respectively, at absorbing IR radiation than an additional molecule of CO_2 (Jain et al. 2000).

In addition to their radiative properties, halocarbons are extremely long-lived. There are no significant removal mechanisms for these gases in the troposphere, so most halocarbons are transported directly to the stratosphere. In the stratosphere, halocarbons are broken down by high-energy radiation from the Sun, releasing the chlorine and bromine they contain. These highly reactive forms of Cl and Br can then catalyse thousands of O_3-destroying reactions before being converted to less-reactive compounds.

Since the beginning of the twentieth century, halocarbons have been manufactured for use in many industrial and commercial products. Their inertness and long lifetimes have made them attractive chemicals for use as foam-blowing agents, coolants, propellants, fire retardants, and other industrial applications. With the exception of the naturally occurring and short-lived emissions of CH_3Cl and CH_3Br from volcanic and oceanic activity, essentially all the halocarbons in the atmosphere are man-made. The ozone-depleting halocarbons with the largest potential to influence climate are chlorofluorocarbons (CFCs), mainly CFC-11 ($CFCl_3$), CFC-12 (CF_2Cl_2), and CFC-113 ($CF_2ClCFCl_2$). Less-effective ozone-depleting substances include hydrochlorofluorocarbons (HCFCs) and the hydrofluorocarbons (HFCs) that contain fluorine instead of chlorine. The chemical link between these compounds and stratospheric O_3 destruction was first suggested by Molina and Rowland (1974) and has since been confirmed by numerous observational and modelling studies (WMO 2019a). In addition, it is now clear from measurements in polar firn air that there are no natural sources of CFCs, HCFCs, or HFCs (Butler et al. 1999).

Chlorofluorocarbons CFC-11 and CFC-12 have the largest atmospheric concentrations, at 0.26 and 0.54 ppbv, respectively. Tropospheric concentrations of both of these gases were increasing at about 4% per year through the 1980s and in the early 1990s. However, their growth rates have now slowed appreciably, and have started to decrease, as shown in Figure 11.4c. This decrease is due to the prohibition of the use of CFCs and a number of other ozone-depleting compounds since 1996 under the Montreal Protocol and its amendments. Atmospheric concentrations of several other halocarbons have until recently been growing at even a faster rate than CFC-11 and CFC-12. For example, the concentration of CFC-113 ($C_2F_3Cl_3$), was increasing about 10% per year in the early 1990s but has also slowed greatly, with a current concentration of about 0.07 ppbv (WMO 2019a).

Recent measurements indicate CH_3CCl_3 concentration continued to decline between 2012 and 2016, decreasing to 2.6 ± 0.7 ppt in 2016, 2% of its maximum value of 133 ± 4 ppt, which was reached in 1992 (WMO 2019a). The abundance of HCFC-22 (CHF_2Cl) continued to increase and was around 237 ppt in 2016, but its growth rate has declined to the level in the early 2000s (WMO 2019a, and the references therein).

The halocarbons whose production has been banned due to environmental concerns had many important uses to humanity. Therefore, alternatives to these chemicals were required to meet expanding worldwide needs for refrigeration, air conditioning, energy-efficient insulation, plastic foams, solvents, and aerosol propellants, such as in medical products. The proposed replacements, particularly HFCs, retain many of the desirable properties of CFCs; however, as a result of adding one hydrogen into their molecular structure, they have a much shorter lifetime in the atmosphere, reducing their contribution to global warming. In addition, HFCs do not contain chlorine; therefore, these chemicals have no potential for ozone depletion. Because of their increasing use in recent years, HFCs are increasing at accelerating rates (WMO 2019a).

Other perfluorinated (PFCs) species such as CF_4, C_2F_6, and sulphur hexafluoride (SF_6) have very long lifetimes (greater than 1000 years) and significant IR absorption lines in the atmospheric window region. Species CF_4 and C_2F_6 are byproducts released to the atmosphere during the production of aluminium, and SF_6 is mainly used as a dielectric fluid in heavy electrical equipment. Most of the current emissions of C_2F_6 and SF_6 arise from human activities. Harnisch and Eisenhauer (1998) have shown that CF_4 and SF_6 are naturally present in fluorites, and outgassing from these materials leads to natural background concentrations of 40 pptv for CF_4, and 0.01 pptv for SF_6. However, at present, human emissions of CF_4 greatly exceed natural emissions by a factor of 1000 or more.

Atmospheric concentrations of CF_4 and SF_6 are increasing. Global average concentrations of SF_6 and CF_4 in year 2016 were about 8.9 and 82.7 ppt, respectively (WMO 2019a). Because of their long lifetime, comparatively small emissions of these species will accumulate and lead to a significant impact on climate over the next several hundred years. For this reason, these species are included as part of the 'basket' of GHGs to be reduced under the Kyoto Protocol (UN 1997).

11.1.1.6 Stratospheric Ozone

Over 90% of the ozone in the atmosphere is located in the stratosphere (see Chapter 8), where it has two important effects. First, stratospheric ozone absorbs the Sun's short-wavelength radiation, heating the stratosphere and causing stratospheric temperature to increase with height, which has a cooling effect on the Earth's surface. Second, stratospheric ozone protects the Earth's surface from harmful ultraviolet (UV) B radiation from the Sun that would otherwise have adverse effects on human, animal, and plant life.

Stratospheric ozone is produced by the photodissociation of oxygen by UV radiation, a process unaffected by human emissions. Ozone is then removed in catalytic mechanisms involving free radicals such as NO_x, HO_x, and ClO_x. In the past, stratospheric O_3 production and destruction were in balance, resulting in a constant protective ozone layer. However, human emissions of CFCs and other halocarbons have altered the balance by increasing concentrations of reactive Cl and Br. Both substances are extremely effective catalysts in the O_3 removal cycles that can turn over thousands of times before being converted to less-reactive forms.

The increase in stratospheric Cl and Br loading has resulted in about a 5% decrease in global total lower stratosphere ozone since the 'ozone hole' was first observed over Antarctica in the 1970s, with the largest decreases occurring in both hemispheres at mid- to high-latitudes.

11.1.1.7 Tropospheric Ozone

Approximately 10% of the atmosphere's O_3 is located in the troposphere (see Chapter 5). In contrast to the beneficial effects of stratospheric ozone, lower-level ozone is an important greenhouse gas, contributing to global warming. At high levels, such as occur downwind of many urban areas, O_3 also has a toxic effect on human, animal, and plant life. However, as O_3 photolysis is the main source of OH, the atmosphere's principal cleansing agent, O_3, is also the primary driver of photochemical processes that remove pollutants from the troposphere (see Chapter 5).

Ozone is not emitted directly from human activities; however, we are responsible for emissions of a number of gases that are *ozone precursors* – i.e. gases that are involved in chemical reactions that produce ozone. These gases include NO_x, CO, CH_4, and other NMHCs. Increased emissions of these gases are thought to have significantly raised tropospheric O_3 levels since preindustrial times. Levels of NO_x are particularly crucial, as approximately 50% of tropospheric O_3 is produced by the photolysis of NO_2. The other 50% of tropospheric O_3 results from downward transport from the stratosphere. Due to its short lifetime and inhomogeneous distribution, NO_x is often the limiting species on O_3 production. It is estimated that only 10% of potential tropospheric O_3 production is currently being realized, with a net production of O_3 in high-NO_x areas and a net destruction of O_3 in low-NO_x regions.

Over the past 100 years, ozone production in photochemical smog, such as occurs in the atmospheric boundary layer over and downwind of many urban areas, has increased steadily. Although smog-induced ozone has major health impacts, it is not a significant contributor to climate. Rising levels of CH_4, NO_x, CO, and NMHCs as well as stratospheric O_3 loss will enhance O_3 production, whilst increasing temperature and water vapour from global warming could increase or decrease O_3 levels. Due to the complexity of the factors influencing tropospheric O_3 abundances, the amount and even the sign of future tropospheric O_3 change is uncertain – although recent modelling studies project a net increase over the next few decades (Royal Society 2008).

11.1.1.8 Aerosols

Aerosols differ from GHGs in two primary aspects: (i) they can have either a cooling or a warming effect on climate, depending on aerosol composition, and (ii) they have short lifetimes of only a few days to weeks. Due to this short lifetime, aerosols are inhomogeneously concentrated over areas such as the Northern Hemisphere continents. This patchy distribution leads to a regionally dependent climate impact, in contrast to the longer-lived, well-mixed GHGs.

A number of major types of aerosols are recognized to currently affect climate to a significant degree (see Chapter 7). The most important type, sulphate aerosol, has a net cooling effect on climate. Sulphate aerosols are not directly produced from human activities. Instead, combustion of fossil fuels containing sulphur results in emissions of sulphur dioxide (SO_2), the precursor for sulphate aerosols. In the atmosphere, SO_2 is quickly oxidized to sulphuric acid, which in turn condenses onto cloud-droplet and aerosol-particle surfaces to form sulphate aerosols. Coal-fired electricity generation is one of the most important sources of anthropogenic aerosols, generating a major portion of global emissions. Other human-related emissions arise from combustion of other fossil fuels and biomass burning. Ocean sea salt also contributes approximately 15% of global atmospheric sulphur in the form of dimethyl sulphide (DMS). Finally, sulphate and other aerosol types are episodically mass-produced by volcanic eruptions. Under sufficiently violent eruptions, large amounts of aerosols of mixed composition can be injected directly into the stratosphere. In the stratosphere, aerosol lifetime is extended from days to months and even years. As bands of volcanic aerosols circle the Earth, they exert a significant impact on the Earth's climate. Their radiative effects also have an impact on natural emissions and uptake of GHGs including CO_2 and CH_4. The Mount Pinatubo eruption, for example, is thought to have been responsible for large fluctuations in the growth rate of CO_2, CH_4, CO, and other gases, enhanced stratospheric O_3 destruction, and a short-term cooling at northern latitudes.

Carbonaceous aerosols can have either a warming or cooling effect on climate, depending on whether they contain black carbon (BC) or organic carbon (OC). BC contained in soot forms the nuclei of IR-absorbing carbonaceous aerosols that have recently been identified as significant contributors to positive radiative forcing. Major anthropogenic sources of BC are split evenly, with coal and diesel combustion accounting for approximately 50% of emissions, whilst biomass burning makes up the remainder. Combustion also results in emissions of OC, with the ratio between black and organic emissions being primarily dependent on the combustion temperature. Since most large-scale combustion facilities operate at very high temperatures, industry is a major source of BC, whilst domestic fuel use and biomass burning are responsible for the majority of OC emissions. Organic aerosols exhibit great variety, which makes it difficult to fully characterize their impacts on atmospheric chemistry and climate.

Other human and natural aerosols, including nitrate and mineral dust aerosols, for example, also affect the radiative balance of the atmosphere. In the past, aerosol emissions were relatively constant, punctuated by occasional volcanic eruptions that have been detected by dust deposits in glacier and polar ice-cores and correlated with historical records dating back thousands of years. However, the beginning of the Industrial Revolution and large-scale fossil-fuel use signalled a significant increase in sulphate, BC, and OC aerosols. Estimates of historical energy use and fuel sulphur content show that sulphur emissions have increased from only a few megatons of sulphur (Mt S) in 1850 to approximately 75 Mt S by 1990 (Lefohn et al. 1999), with recent reductions due to power-plant sulphur controls in developed countries. Since emissions of BC and OC are tied to fossil-fuel use and biomass burning, concentrations of these aerosols are also estimated to have increased over the past 100 years (Tegen et al. 2000).

11.1.2 Changes in Radiative Forcing

Clear evidence has been presented in the previous section that the atmospheric concentrations of a number of radiatively active GHGs have increased over the past 100 years or more as a result of human activity. By absorbing outgoing infrared radiation, they increase the heat-trapping ability of the atmosphere, driving climate change. As discussed above, human-related activities have also been responsible for increases in aerosol particles in the atmosphere, which can have either heating or cooling effects, depending on the aerosol type. A change in the concentrations of GHGs and aerosols will change the balance between the incoming solar radiation and the outgoing IR radiation from the Earth. This change in the planetary radiation budget is termed the *radiative forcing* of the Earth's climate system (see Chapter 3). Changes in Earth's surface temperature are approximately linearly proportional to the radiative forcing inducing those changes, although there is some nonlinearity induced by the sensitivity of climate response to height, latitude, and the nature of the forcing (e.g. Jain et al. 2000).

11.1.2.1 Greenhouse Gases

Recent assessments of the direct radiative forcing due to the changes in well-mixed GHG concentrations are generally in good agreement, determining an increase in radiative forcing of about $3.00 \, \mathrm{W \, m^{-2}}$ relative to 1750 to the present time (Figure 7.10, IPCC 2013). The overall growth rate in radiative forcing from all well-mixed GHGs is smaller over the last decade than in the 1970s and 1980s, owing to a reduced rate of increase in the combined non-CO_2 radiative forcing (IPCC 2013). By far the largest effect on radiative forcing has been the increasing concentration of carbon dioxide, accounting for about 56% ($1.68\,\mathrm{W\,m^{-2}}$) of the total forcing. The radiative forcing due to changes in concentrations of CH_4, N_2O, and CFCs and other halocarbons relative to 1750 is 0.97, 0.17, and $0.18\,\mathrm{W\,m^{-2}}$, respectively (Figure 7.10).

11.1.2.2 Stratospheric Ozone

New detection and attribution studies of lower stratospheric temperature changes support an assessment that anthropogenic forcing, dominated by stratospheric ozone depletion due to ozone-depleting substances, has led to a detectable cooling of the lower stratosphere since 1979 (IPCC 2013). Recent estimates (IPCC 2013) place the forcing due to the stratospheric ozone changes at $-0.05 \pm 0.1\,\mathrm{W\,m^{-2}}$.

11.1.2.3 Tropospheric Ozone

Observational evidence for tropospheric ozone increases over the past 100 years at the global scale is weak. However, as discussed previously, recent modelling studies that account for emissions and concentrations of ozone precursors estimate a net increase in tropospheric O_3 since preindustrial times. Estimates of the change in radiative forcing relative to 1750 for tropospheric ozone are positive, at $0.40 \pm 0.20\,\mathrm{W\,m^{-2}}$ (IPCC 2013).

11.1.2.4 Aerosols

Tropospheric aerosols that are thought to have a substantial anthropogenic component include sulphate, BC, OC, mineral dust, and nitrate aerosols. Due to their short lifetime, the geographical distribution plus the diurnal and seasonal patterns of the radiative forcing from aerosols are quite different from that of GHGs.

Anthropogenic aerosols influence the radiative budget of the Earth–atmosphere system in two different ways. The first is the direct effect, whereby aerosols scatter and absorb solar and thermal IR radiation, thereby altering the radiative balance of the Earth–atmosphere system. The second is the indirect effect, whereby aerosols modify the microphysical and hence the radiative properties and lifetimes of clouds.

The radiative forcing of the total aerosol (primarily sulphate, OC, BC, nitrate, and dust) effect in the atmosphere, including cloud adjustments due to aerosols, is $-0.9\,\mathrm{W\,m^{-2}}$ (IPCC 2013).

11.1.2.5 Solar Variability

The Sun's energy output is known to vary by small amounts over the 11-year cycle associated with sunspots. There are also indications that the solar output may vary by larger amounts over longer time periods. The IPCC (2013) estimated the recent forcing due to solar variability at about $0.05\,\mathrm{W\,m^{-2}}$.

11.1.2.6 Combined Effects

Based on the above discussion, Figure 7.10 shows the current global, annual-mean averaged radiative forcing estimates from 1750 to present for the GHGs, aerosols, and the forcings discussed above. This is 2.29 [1.13–3.33] $\mathrm{W\,m^{-2}}$ (in 2011) (Figure 7.10). This figure clearly shows a large uncertainty in the exact amount of forcing. This is particularly so for aerosols because of the difficulty in quantifying the indirect effects of aerosols, and the overall changes in aerosol and ozone concentrations. Because of the hemispheric and other inhomogeneous variations in concentrations of aerosols, the overall change in radiative forcing could be much greater or much smaller at specific locations than the globe, with the largest increase in radiative forcing expected in the Southern Hemisphere (where aerosol content is smallest).

11.1.3 Past Changes in Temperature

11.1.3.1 Observed Land and Ocean Surface Temperature

One of the main pieces of evidence of global climate change is an increase in historical surface temperatures over land and ocean. Figure 11.5a shows the global average surface temperature increase (i.e. the area-weighted average in the increase of the land-surface air temperature and the sea-surface temperature) of 0.85 [0.65–1.06] °C since 1880 (IPCC 2013). As indicated in this figure, the increase in temperature is by no means a uniform one. Instead, the increase is seen to have occurred in two distinct periods: from 1910 to 1945, and from 1976 to the present. The past four years (2015–2018) are the top four warmest on record, according to the World Meteorological Organization (WMO 2019b). The 20 warmest years on record since 1850 have all occurred in the past 22 years (1997–2018).

There are, of course, uncertainties in the temperature records. For example, during this time period recording stations have moved, and techniques for measuring temperature have varied. Also, marine observing stations are scarce. In spite of these uncertainties, confidence that the observed variations are real is increased by noticing that the trend and the shape of the changes are similar when different selections of the total observations are made. For instance, the separate records from the land- and sea-surface and from the Northern and Southern Hemispheres are in close accord (Brohan et al. 2006).

11.2 Future Outlook of Climate Change

11.2.1 Modelling Tools for Climate Science

Assessment of the future climate and stratospheric ozone responses to the GHG emissions requires mathematical models based on a set of fundamental physical and chemical principles governing the Earth–atmosphere system. A hierarchy of models is available for such studies. These models begin with one-dimensional models where altitude is the only dimension considered, such as the very early surface energy balance model of Arrhenius (1896) or the 1-D radiative-convective model of Manabe and Wetherald (1967), which was used to explore possible changes in the vertical distribution of temperature in radiative-convective equilibrium for various values of CO_2 concentrations. Two-dimensional climate models consider both altitude and latitude, and solve the primitive equations in finite difference form for mass continuity, the continuity of momentum, and the thermal energy balance for a number of variables (e.g. zonal mean temperature) on a grid with nodes in latitude and in the vertical dimension (e.g. MacKay and Khalil 1994).

More recently, advances in computing power have allowed large, three-dimensional general circulation models (GCMs) of the Earth's climate system to be developed, and these now form the basis of weather forecasting, understanding the climate system and exploring possible future climate change. Although usually described as being 3-D, these models are more properly four-dimensional, as they apply discrete equations for fluid motion on a rotating sphere integrated forward in time. Over the past 20 years or so, more complete GCMs that couple the atmosphere and oceans have become the norm (e.g. Sun and Hansen 2003). However, it is obvious that the Earth's climate system is modulated by several external factors, such as changes to the carbon cycle, the formation of atmospheric aerosols from pollution emissions, albedo changes caused by changes in land use, changes to uptake and storage of carbon by the biosphere, etc. Bringing these components into a 3-D GCM allows human influences and feedback effects on the climate system to be better explored. Such models are referred to as *Earth System Models* (*ESMs*) or *global*

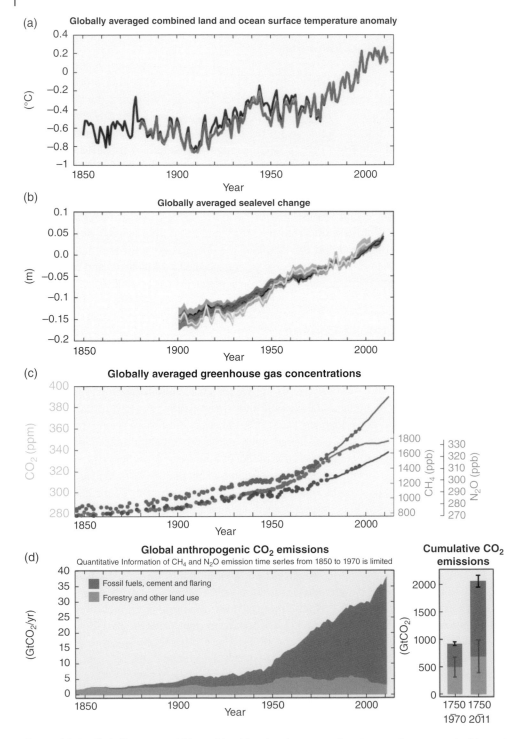

Figure 11.5 Globally averaged (a) combined land and ocean surface temperature anomaly, (b) sea-level change, (c) greenhouse gas concentrations, and (d) anthropogenic CO_2 emissions, since 1850, as presented in the IPCC AR5 report (IPCC 2013). See colour plate section for the colour representation of this figure.

climate models, and what mainly distinguished them from climate models is their ability to simulate the carbon cycle. Examples are the National Center for Atmospheric Research's Community Earth System Model (CESM), the European Earth System Model (EC-Earth), and the UK's Earth System Model (UKESM). All of these are under continuing development but have been used for climate change research purposes.

A disadvantage of GCMs and ESMs is that they are very computationally intensive. For this reason, an alternative approach known as *integrated assessment* has been developed, in which a relatively simple (usually 2-D) climate model is linked with models of the social and economic factors that drive the emissions of GHGs to give a climate change-integrated assessment model (IAM) (e.g. Jain 2019). Whilst these bring together very different types of information (e.g. knowledge about climate, economics, ecology) in a coherent framework, they are not predictive models. They can provide a framework for understanding how global climate might change and for informing judgements about the relative value of different options for dealing with climate change, but they cannot provide definitive answers about how to respond to a particular outcome. The limitations of IAMs have been discussed by Ackerman et al. (2009).

Climate models are mathematical representations or descriptions of the hugely complex interactions occurring between the Sun, the atmosphere, the Earth's surface (in all its physical forms), and the biosphere, together with the artificial human influences now perturbing the natural system. These models are built to make predictions about the future state of the Earth's climate, yet can only be tested by hindcasting or backtesting. However, hindcasting now shows that climate models are able to reliably simulate the observed temperature and sea-level rises of the last century, giving confidence in their ability to predict future climate change. In fact, the degree of confidence in them is sufficient for major international policy actions concerning climate change to be taken in the last two decades.

11.2.2 International Climate Policy

International policy action on climate change began with the adoption of the United Nations Framework Convention on Climate Change (UNFCC) in 1992. This has the broad objective to 'stabilize greenhouse gas concentrations in the atmosphere at a level that would prevent dangerous anthropogenic interference with the climate system' and initially set nonbinding greenhouse gas emission limits for individual countries with no enforcement or sanctions mechanisms. There are currently nearly 200 signatories to the Convention. In 1997, the Kyoto Protocol was agreed, establishing legally binding emissions reduction targets for developed countries in the period 2008–2012. In 2010, agreement was reached at the UN Climate Change Conference in Cancun, Mexico, stating that future global warming should be limited to below 2.0 °C relative to the preindustrial level. This was later amended in the Doha Agreement to cover the period 2013–2020. In 2015, a further agreement (the Paris Agreement) was adopted, lowering the global temperature target to 1.5 °C relative to the preindustrial level and setting out emissions reductions and other climate-related actions from 2020 in the form of 'Nationally Determined Contributions'.

The Paris Agreement requires each 'Party' (i.e. signatory state) to set out and adopt its own 'Nationally Determined Contribution' by specifying post-2020 emissions reductions and other domestic mitigation measures. For example, the European Union's Nationally Determined Contribution, submitted in 2016 on behalf of all the EU's 28 member states, committed to at least a 40% domestic reduction in greenhouse gas emissions by 2030 relative to a base year of 1990. The Contribution covered all GHGs, including both the most abundant ones (carbon dioxide, methane, and nitrous oxide) and the more exotic synthetic compounds (hydrofluorocarbons, perfluorocarbons, sulphur hexafluoride, and nitrogen trifluoride). It covers all sectors, including energy, industry and agriculture. In the case of agricultural emissions, it states that a policy on how to include land use, land-use change, and forestry into the 2030 greenhouse gas mitigation framework will be established before 2020.

As an example of the Nationally Determined Contribution of a less-developed nation, Afghanistan has committed to a 13.6% reduction in greenhouse gas emissions (carbon dioxide, methane, and nitrous oxide) by 2030 compared to a business-as-usual scenario for 2030, conditional on receiving external financial support of US\$17 billion. In practise, this means that Afghanistan is committed to growing its greenhouse gas emissions from $19.3 \, \text{Mt. year}^{-1}$ (CO_2 equivalent) in 2005 to $35.5 \, \text{Mt year}^{-1}$ in 2020 and to $42.7 \, \text{Mt year}^{-1}$ in 2030, instead of the business-as-usual projected emissions of $48.9 \, \text{Mt year}^{-1}$ in 2030.

Whether or not the UNFCC and its constituent agreements, including the Paris Agreement, have any impact on greenhouse gas emissions in the future remains to be seen. Current indications as of 2019 are not positive, as global CO_2 emissions continue to rise (2.7% in 2018), despite a brief period (in 2015) when a slight reduction in emissions gave rise to a more optimistic outlook (Le Quéré et al. 2018).

11.2.3 The Intergovernmental Panel on Climate Change

In 1988, the World Meteorological Organization and the United Nations Environment Programme established the Intergovernmental Panel on Climate Change (IPCC). The IPCC is tasked with producing comprehensive and definitive *assessment reports* (*ARs*) that can inform policy makers, and in particular can contribute to the UNFCC. The IPCC does not carry out original research or monitoring itself but rather assesses the available published evidence to produce an objective, scientific view of climate change, its natural, political, and economic impacts and risks, and possible policy and technological response options. This is work done by thousands of scientists working voluntarily, aided by a secretariat and technical support unit. Its main outputs are the ARs, published in 1990, 1995, 2001, 2007, and 2014. AR6 is due to be completed in 2022, at which point the data and assessments contained within it will supersede those summarized in this present chapter.

In AR1, the IPCC concluded that it was certain that anthropogenic CO_2 emissions were causing, on average, an additional warming of the Earth's surface, but that unequivocal detection of enhanced global warming was not yet possible. This report served as the basis of the UNFCC. By the time AR5 was published in 2013, the IPCC were able to say, 'Warming of the climate system is unequivocal, and since the 1950s, many of the observed changes are unprecedented over decades to millennia'. AR5 also said 'the overall risks of climate change impacts can be reduced by limiting the rate and magnitude of climate change' and 'without new policies to mitigate climate change, projections suggest an increase in global mean temperature in 2100 of 3.7–4.8 °C, relative to pre-industrial levels (median values; the range is 2.5–7.8 °C including climate uncertainty)'. Importantly, AR5 also says 'the current (i.e. 2013) trajectory of global greenhouse gas emissions is not consistent with limiting global warming to below 1.5 or 2 °C, relative to pre-industrial levels'. Depressingly, as of 2019, the same conclusion would have to be drawn.

11.2.4 Recent Observed Changes in the Climate System

As noted in Chapter 1 and in this chapter, above, the Earth's surface temperature is higher now that it was prior to the Industrial Revolution (by about 1 °C; see Figure 1.7) as a result of sustained and continuous (if not uniform) warming. This warming appears to be continuing, with each of the last three decades warmer than any preceding decade since 1850. The 20 warmest years on record have all occurred since 1995, with the five warmest years occurring since 2010. The warmest year on record of all was 2016, and 2018 was the fourth warmest on record. As the atmosphere has warmed, sea level has risen. Figure 11.5 shows the globally averaged (i) combined land and ocean surface temperature anomaly, (ii) sea-level change, (iii) greenhouse gas concentrations, and (iv) anthropogenic CO_2 emissions, since 1850, as presented in the IPCC AR5 report (IPCC 2013). As well as temperature and sea-level changes, the amounts of snow and ice on the Earth's surface have decreased, precipitation has increased, and the oceans have become more acidic. Importantly, AR5 ascribes degrees of confidence to all these conclusions, so, for example, it was able to say *with a high degree of confidence* that the Greenland and Antarctic ice sheets lost mass over the period 1992–2011.

On the basis of all this observed evidence, it is now clear that human activities, particularly the emissions of CO_2 and other GHGs, have caused, and likely will continue to cause, changes in the Earth's climate. More precisely, IPCC AR5 says,

> ... Anthropogenic GHG emissions have increased since the pre-industrial era, driven largely by economic and population growth, and are now higher than ever. This has led to atmospheric concentrations of carbon dioxide, methane and nitrous oxide that are unprecedented in at least the last 800,000 years. Their effects, together with those of other anthropogenic drivers, have been detected throughout the climate system and are extremely likely to have been the dominant cause of the observed warming since the mid-20th century ... (IPCC 2013).

By *extremely likely*, the IPCC means that there is a 95–100% probability of this being correct.

The character of extreme weather events has also been observed to have changed since about 1950, with a decrease in cold weather extremes, an increase in warm weather extremes, and an increase in extreme rainfall events in some regions of the world. Attributing specific changes in the weather to anthropogenic climate change is challenging but becoming increasingly possible (e.g. IPCC 2013; National Academies of Sciences, Engineering and Medicine report 2016). For example, Pall et al. (2011) showed that it is very likely climate change is increasing the occurrence of autumn flooding in the United Kingdom.

11.2.5 Future Climate Change

11.2.5.1 Representative Concentration Pathways

Climate projections in IPCC AR5 are based on *representative concentration pathways* (*RCPs*). These supersede the emission scenarios (called SRES projections) previously used by the IPCC. Which of these RCPs turns out to be closest to the actual future trajectory in GHG concentrations over time depends, obviously, on how much GHGs are emitted in the future. Four RCPs have been selected for investigation, in an attempt to describe a range of possible climate futures. These are known as RCP2.6, RCP4.5, RCP6, and RCP8.5, and correspond to radiative forcings of +2.6, +4.5, +6, and +8.5 W m^{-2} in 2100 relative to preindustrial values.

RCP2.6 assumes that global GHG emissions (as CO_2-equivalents) peak before 2020, with emissions declining rapidly thereafter. RCP4.5 and RCP6 have emissions peaking around 2040 and 2080, respectively, with declines thereafter, and in RCP8.5 emissions continue to rise, only tailing off but still rising towards the end of the twenty-first century. On the basis of current evidence, it is far too early to say which RCP, if any, is the most likely to represent future reality, although it is increasingly unlikely that the more optimistic RCP2.6 can be realized. Equally, it is possible that global reserves and supplies of fossil fuels will be insufficient for the increasing emissions of RCP8.5 to be sustained. The detailed description for RCPs can be found in Moss et al. (2010) and van Vuuren et al. (2011).

11.2.5.2 Climate Projections Based on the RCPs

IPCC AR5 presented projections of global warming and mean sea level rise for the mid and late twenty-first century (2046–2065 and 2081–2100 averages, respectively) for the four RCPs. These are shown in Table 11.1 (the Integrated Science Assessment Model [ISAM] simulation results are also presented as comparison). Global mean temperature is projected to rise by 0.3–4.8 °C and mean sea level is projected to rise by 0.26–0.82 m by 2100 in all four RCPs. Even in the moderately optimistic RCP4.5 it is likely that global mean temperature rise will exceed the Paris Agreement target of warming less than 1.5 °C relative to the preindustrial level towards the end of the twenty-first century. RCP 6.0 and RCP 8.5 would both cause the Paris Agreement goal of limiting warming to much less than 2 °C to be breached before the end of the century.

AR5 also projects climate change beyond 2100. In the extended RCP2.6 it is assumed that net anthropogenic emissions become negative after 2070 (i.e. more GHGs are taken out of the atmosphere by human activities than are emitted) with atmospheric CO_2 concentrations being around 360 ppmv in 2300 (cf. current concentrations of greater than 410 ppmv). This gives a projected global warming of 0–1.2 °C by the late twenty-third century, relative to 1986–2005. On the other hand, the extended RCP8.5 assumes continuing rises in CO_2 emissions, with concentrations reaching 2000 ppmv in 2250, and a projected warming of 3.0–12.6 °C by 2300.

Table 11.1 Projections of global mean temperature increases and global mean sea level rises to the mid and late twenty-first century for four Representative Concentration Pathways (IPCC 2013 and the Integrated Science Assessment Model [ISAM] results).

Scenario	2046–2065	2081–2100	2046–2065	2081–2100
	Global temperature increase (°C) (mean and likely range)		Global mean sea-level rise (m) (mean and likely range)	
RCP2.6 (IPCC)	1.0 (0.4–1.6)	1.0 (0.3–1.7)	0.24 (0.17–0.32)	0.40 (0.26–0.55)
RCP2.6 (ISAM)	1.4	1.4	0.20	0.25
RCP4.5 (IPCC)	1.4 (0.9–2.0)	1.8 (1.1–2.6)	0.26 (0.19–0.33)	0.47 (0.32–0.63)
RCP4.5 (ISAM)	1.7	2.1	0.22	0.32
RCP6.0 (IPCC)	1.3 (0.8–1.8)	2.2 (1.4–3.1)	0.25 (0.18–0.32)	0.48 (0.33–0.63)
RCP6.0 (ISAM)	1.6	2.7	0.21	0.36
RCP8.5 (IPCC)	2.0 (1.4–2.6)	3.7 (2.6–4.8)	0.30 (0.22–0.38)	0.63 (0.45–0.82)
RCP8.5 (ISAM)	2.4	4.0	0.27	0.48

In a Special Report on global warming of 1.5 °C published in 2018, the IPCC concludes that a temperature rise of 1.5 °C is very likely to occur between 2030 and 2052. For this not to happen, global CO_2 emissions would have to decrease by about 45% (range 40–60%) from their 2010 levels by 2030, reaching net zero by around 2050. Such an emissions trajectory is more stringent than RCP2.6 and would require very rapid (starting as soon as possible after 2018), unprecedented and far-reaching transitions in energy, land use, transport, buildings, and industry (IPCC 2018).

This Special Report also concludes that in order to limit global warming to 1.5 °C it will be necessary to remove CO_2 from the atmosphere over the course of the twenty-first century, i.e. net emissions will have to become negative. This CO_2 removal (CDR) would have to be on the scale of 100–1000 Gt CO_2 over the century in order to compensate for residual emissions. The more rapidly emissions decrease from now (2019), the less CDR would be required to meet the 1.5 °C target (IPCC 2018).

It is worth noting at this point that climate models project that 2 °C warming will cause significantly greater climate-risks to health, livelihoods, food security, water supply, human security, and economic growth than will warming of 1.5 °C and that adaptation to 2 °C warming would be more difficult, more disruptive and more costly than to 1.5 °C warming. For example, the proportion of the global population subjected to water stress (floods and shortages) under 1.5 °C warming might be half that at 2 °C. IPCC (2018) predicts, with high confidence, that limiting warming to 1.5 °C reduces the risks to both natural and human systems compared with 2 °C and that 2 °C warming will cause the most venerable populations to be subjected to multiple and compound risks. As of the time of writing, there is a growing political consensus that 2 °C warming will be dangerous to global society, but there is less agreement on how serious 1.5 °C warming (which actually means an additional ~0.5 °C warming from 2019 over and above the ~1 °C warming that has already occurred since the Industrial Revolution) will be.

11.2.5.3 Climate Risks at 1.5 and 2 °C Warming

At the present time, it is not possible to say with certainty what a '1.5 °C or 2 °C warmer world' will be like as humans are more sensitive to local and short-term phenomena ('weather') than they are to global or regional longer-term means ('climate'). In addition, the rate of change in state will influence how effective or indeed possible will be adaptation. How large any overshoot is before the climate reverts to a new steady-state will also be significant. Notwithstanding this, the IPCC Special Report on 1.5 °C (IPCC 2018) does go into great detail about the possible or likely regional-scale climate risks of 1.5 and 2 °C worlds.

The strongest warming is expected to occur at mid-latitudes in the warm season (with increases of up to 3 °C for 1.5 °C of global warming) and at high latitudes in the cold season (up to 4.5 °C for a 1.5 °C world). Extreme heatwaves are likely to become more frequent and more extreme in the tropics. Limiting warming to 1.5 °C would result in about 420 million fewer people being frequently exposed to extreme heatwaves compared with a 2 °C world.

Risks of local species losses and, consequently, risks of extinction are much less in a 1.5 °C compared with a 2 °C warmer world, but significant damage to biodiversity is predicted under both scenarios. Changes to sea level rise and coastal flooding, ocean water acidity, productivity and biodiversity, increasing magnitude and frequency of both floods and water shortages, thawing of the tundra and dramatic declines in sea ice extent and duration are all predicted in both 1.5 and 2 °C worlds. Crop yield reductions in many regions of the world, leading to reductions in food availability, are likely to be much more significant at 2 °C and other risks to human health are similarly likely to be greater at 2 °C than at 1.5 °C.

The largest reductions in economic growth at 2 °C compared to 1.5 °C of warming are projected for low- and middle-income countries and regions (Africa, SE Asia, India, Brazil, and Mexico), although this is said with low to medium confidence (IPCC 2018).

11.3 The Integrated Science Assessment Modelling (ISAM)

Integrated assessment modelling provides an integrated view of human interaction with the physical world. Rather than attempting to use a range of multidimensional and complicated expert models, IAMs build on the knowledge achieved by each individual scientific discipline. As shown in Figure 11.6, there are four key components to an IAM. The *Human Activities* and *Ecosystems* modules cover population, technology and economic changes that contribute to energy use; shifts in land use prompted by the evolution of human societies; and the elements leading to emissions of GHGs. The *Atmospheric Composition* and *Climate* sections deal with the modification of climate induced by these emissions. The impact of climate change is dealt with through interaction between the Human Activities and Ecosystems components.

We have developed an (ISAM) in order to better represent the spatial variations and processes relevant to evaluating biogeochemical cycles, determining atmospheric composition, and projecting the resulting global and regional climatic effects. Integrating a process-level understanding of regional impacts into an IAM will help to improve the understanding of climate change impacts and extend the range of issues that can be addressed in an IAM framework.

The current (2019) version of ISAM consists of several sub-models, including a terrestrial-biosphere–ocean carbon-cycle model and two-dimensional (latitude and height) models for chemical transport, radiative forcing and climate. In the next sections, we use the reduced-form ISAM model (Harvey et al.

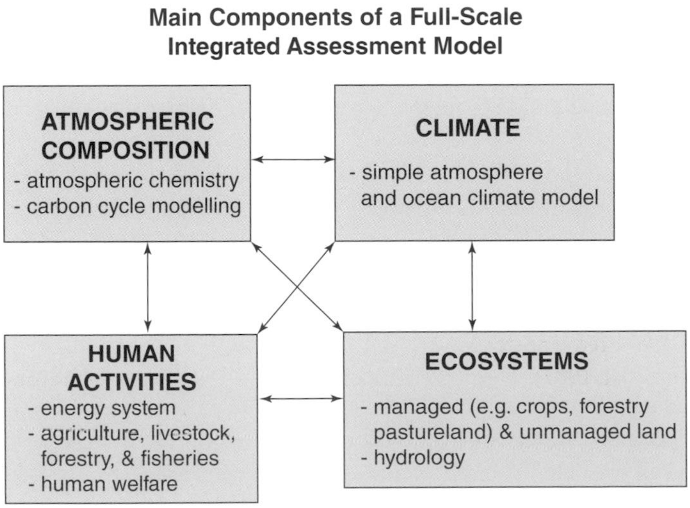

Main Components of a Full-Scale Integrated Assessment Model

ATMOSPHERIC COMPOSITION
- atmospheric chemistry
- carbon cycle modelling

CLIMATE
- simple atmosphere and ocean climate model

HUMAN ACTIVITIES
- energy system
- agriculture, livestock, forestry, & fisheries
- human welfare

ECOSYSTEMS
- managed (e.g. crops, forestry pastureland) & unmanaged land
- hydrology

Figure 11.6 Main components of a full-scale integrated assessment model (IAM).

1997; Jain 2019) to estimate the relationship between the time-dependent rate of GHG and aerosol emissions and quantitative features of climate including GHG concentrations, equivalent effective stratospheric chlorine (EESC) and ozone, global temperature, and sea level.

11.3.1 Carbon Cycle

The ISAM's global carbon cycle component simulates CO_2 exchange between the atmosphere, carbon reservoirs in the terrestrial biosphere, and the ocean column and mixed layer (Jain et al. 1994, 1995, 1996; Kheshgi et al. 1999). The model consists of a homogeneous atmosphere, an ocean mixed layer and land biosphere boxes, and a vertically resolved upwelling–diffusion deep ocean. Ocean and land-biosphere components of this model are described by Jain et al. (1995, 1996). The model takes into account important feedbacks, such as the fertilization and climatic feedback processes. The ocean component of the carbon cycle model takes into account the effects of temperature on CO_2 solubility and carbonate chemistry (Jain et al. 1995), whereas the biospheric component takes into account the temperature effect through the temperature-dependent exchange coefficient between terrestrial boxes.

11.3.2 Methane Cycle

In its reduced form as used in this study, atmospheric CH_4 concentrations are calculated by simulating the main atmospheric chemical processes influencing the global concentrations of CH_4, CO, and OH, using a global CH_4–CO–OH cycle model (Jain and Bach 1994; Kheshgi et al. 1999). The removal rates of CH_4 and CO take into account oxidation by OH, soil uptake, and stratospheric transport (Jain and Bach 1994). Reaction with the OH radical is responsible for the removal of up to 90% of tropospheric CH_4. In ISAM, the concentration of OH is determined by the photochemical balance between the total tropospheric production of OH and the loss rate due to reaction with CH_4, CO, and NMHC (Jain and Bach 1994). The production rate of OH is based on the NO_x emissions and CH_4 concentrations as discussed in Kheshgi et al. (1999).

11.3.3 Other Greenhouse Gases

In its reduced form, ISAM past and future atmospheric concentrations for N_2O and halocarbons are calculated by a mass balance model as described by Bach and Jain (1990).

11.3.4 Climate Model

The ISAM calculates temperature based on a reduced-form energy-balance climate model of the type used in the 1990 IPCC assessment (Harvey et al. 1997). Thermohaline circulation is schematically represented by polar bottom-water formation, with the return flow upwelling through the one-dimensional water column to the surface ocean from where it is returned to the bottom of the ocean column as bottom water through the polar sea. The climate component of ISAM calculates the perturbations in radiative forcings from CO_2 and other GHGs based on updated seasonal and latitudinal GHG radiative forcing analyses (Jain et al. 2000).

The response of the climate system to the changes in radiative forcing is principally determined by the climate sensitivity, ΔT_{2x}, defined as the equilibrium surface temperature increase for doubling of atmospheric CO_2 concentration. This parameter is intended to account for all climate feedback processes not explicitly dealt with in the model, particularly those related to clouds and water vapour and related processes. Analysis of an atmosphere–ocean general circulation model (AOGCM) with constraints from observations suggest that ΔT_{2x} range from 2.1 to 4.4 °C (IPCC 2013). The IPCC (2013) analysis also suggests that it is very unlikely to be less than 1.5 °C.

11.3.5 Importance of Greenhouse Feedbacks in Estimating Future Changes in Climate

There are number of important feedback processes occurring in the Earth–ocean–atmosphere system that could significantly modify future CO_2 and other GHG concentrations in the warmer world. Some of these are discussed briefly here. These feedbacks have the potential to be either positive (amplifying the initial change) or negative (dampening them).

As increasing GHG concentrations alter the Earth's climate, changing climate, and environmental conditions in turn react back on the carbon cycle and atmospheric CO_2. For example, temperature change affects the growth, disturbance and respiration of plants and soils in the terrestrial component of the carbon cycle model. Therefore, net emissions of CO_2 from terrestrial ecosystems will be elevated if higher temperature (resulting from CO_2 and other GHG increases) increases respiration at a faster rate than photosynthesis. This would be a positive feedback, resulting in more CO_2 emissions and a further increase in temperature. Climate can also affect the ocean's role in the global carbon cycle. If the ocean became warmer, its net uptake of CO_2 could decrease because of changes in the chemistry of CO_2 in the seawater, biological activities in the surface water and the rate of exchange of CO_2 between the surface layers and the deep ocean. This would also be a positive feedback, as increasing temperatures would lead to more CO_2 build-up in the atmosphere, again enhancing the initial temperature rise. In the reduced form of ISAM, the carbon cycle's influence on climate feedback due to the radiative effects of CO_2, other GHGs, and aerosols is accounted for using a climate sensitivity parameter as discussed above.

As discussed earlier, observed CH_4 concentrations vary seasonally (Figure 11.4a), mainly because CH_4 emissions from natural wetlands and rice paddies are particularly sensitive to temperature and soil moisture. Biogenic emissions from these sources are significantly larger at higher temperatures and increased soil moisture; conversely, a decrease in soil moisture would result in smaller emissions. Note that the CH_4 component of reduced form ISAM does not take these climate feedbacks into account due to its coarse horizontal resolution.

11.3.6 Chemical Feedbacks on Atmospheric Lifetimes for Non-CO_2 Greenhouse Gases

The atmospheric lifetime of a GHG is another important factor when estimating future GHG concentrations and hence the impact of GHG emissions on the Earth's climate. Because of their very long lifetimes (Table 11.2), CO_2, CFCs and HCFCs, and N_2O are removed slowly from the atmosphere. Hence, their atmospheric concentrations take decades to centuries to adjust fully to a change in emissions. In contrast, some of the halocarbons replacement compounds (HFCs) and CH_4 have relatively short atmospheric lifetimes. This enables their atmospheric concentrations to respond fully to emission changes within a few decades.

Table 11.2 Lifetime estimates for various greenhouse gases (GHGs).

Species	Lifetime (years)
CO_2	Variable (50–100)
CH_4	12.4
N_2O	121
CFC-11,CFC-12	45–100
Other CFCs, HCFCs, HFCs	0.01–1000
SF_6 and PFCs	2000–4100

CFC, chlorofluorocarbons; HCFC, hydrochlorofluorocarbons; HFC, hydrofluorocarbons; PFC, perfluorogenated carbon.

For all GHGs except CO_2 and H_2O, lifetime is calculated based on the balance between emissions and the chemical reactions in the atmosphere. In the case of CO_2, the lifetime is determined mainly by the slow exchange of carbon between the surface and deep ocean, and the atmosphere and terrestrial biosphere. GHGs containing one or more H-atoms (e.g. CH_4 and HFC) are removed primarily by reactions with hydroxyl radicals (OH), which takes place mainly in the troposphere. The GHGs N_2O, PFCs, CF_6, CFCs, and halons do not react with OH in the troposphere. These gases are destroyed in the stratosphere or above, mainly by solar UV radiation at short wavelengths (<240 nm).

How do these chemical feedbacks affect the future concentrations of a gas? An example is the early IPCC estimate for future CH_4 concentrations (Houghton et al. 1996), which chose to ignore the changing emissions of short-lived gases, such as CO, VOC (volatile organic compounds) and NO_x that affect OH and hence CH_4 removal. As discussed above, the concentration of OH and hence CH_4 depends critically on these compounds.

11.3.7 Future Emission Projections for Major Greenhouse Gases and Aerosols

In order to study the potential implications on climate from further changes in human-related emissions and atmospheric composition, a range of scenarios for future emissions of GHGs and aerosol precursors has been produced by the IPCC RCP (RCP, IPCC 2013). These scenarios are for use in modelling studies to assess potential changes in climate over the twenty-first century. As described above, four different narrative story lines, RCP2.6, RCP4.5, RCP6.0, and RCP8.5, were developed to describe the relationships between emissions driving forces and their evolution. All the scenarios based on the same storyline constitute a scenario family. None of these scenarios should be considered as a prediction of the future, but they do illustrate the effects of various assumptions about economics, demography, and policy on future emissions. In this study we investigate these four RCPs as examples of the possible effect of GHGs on climate. Each scenario is based on a narrative, describing alternative future developments in economics, technical, environmental, and social dimensions. These scenarios are no more or less likely than any other scenarios but they have received the closest scrutiny. Details of these storylines and the RCP process can be found elsewhere (IPCC 2013).

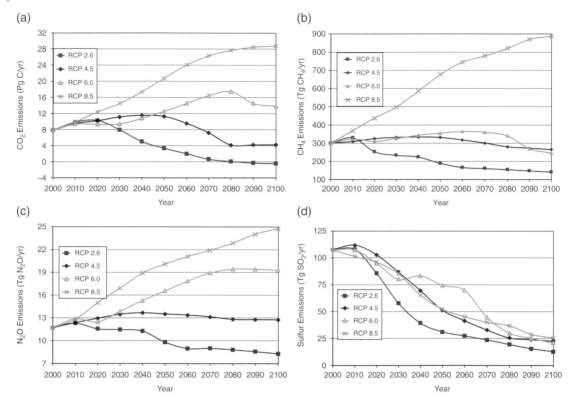

Figure 11.7 Anthropogenic emissions in the four scenarios of the Representative Concentration Pathway (RCP) for (a) CO_2, (b) CH_4, (c) N_2O, and (d) SO_2. Note that the RCP emission values are standardized such that emissions in 2000 are identical in all scenarios. See colour plate section for the colour representation of this figure.

Figure 11.7a–d shows the anthropogenic emissions for four of the most important gases of concern to climate change: CO_2, CH_4, N_2O, and SO_2. Carbon dioxide emissions span a wide range, from nearly four times the 2000 value by 2100 to emissions that rise and then fall to below their 2000 value. The N_2O and CH_4 emission scenarios reflect these variations and have similar trends. However, global sulphur dioxide emissions decline by 2100 to below their 2000 levels in all RCP scenarios, because rising affluence increases the demand for emissions reductions to improve local air quality. Note that sulphur emissions, particularly to mid-twenty-first century, differ fairly substantially between the scenarios. Emissions for halocarbons and related GHGs controlled under the Montreal Protocol and its amendments are derived from the RCP Database Version 2.0 (available at: www.iiasa.ac.at/web-apps/tnt/RcpDb).

11.3.7.1 Projection of Future Concentrations

Figure 11.8a–c shows the ISAM-calculated global mean GHG concentrations from 2000 to 2100 for the RCP scenarios. The concentrations of the major GHGs CO_2, CH_4, and N_2O all follow the same order: RCP8.5 > RCP6.0 > RCP4.5 > RCP2.6. The various scenarios lead to substantial differences in projected greenhouse gas concentration trajectories. The RCP emission scenarios imply the following changes in greenhouse gas concentrations from 2000 to 2100: CO_2 changes range from 13 to 155% (Figure 11.8a), CH_4 from −36 to +111% (Figure 11.8b),

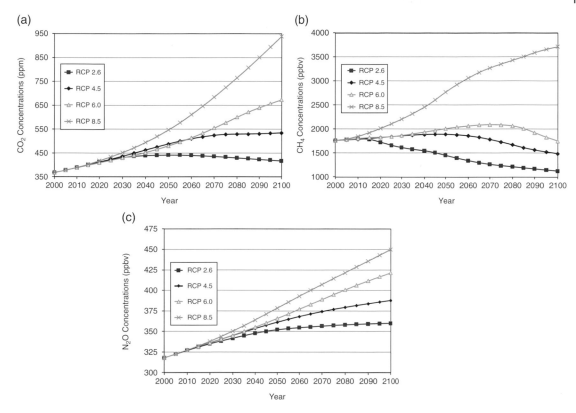

Figure 11.8 Integrated science assessment model (ISAM) estimated atmospheric (a) CO_2, (b) CH_4, and (c) N_2O concentrations from 2000 to 2100 for the RCP scenarios. See colour plate section for the colour representation of this figure.

and N_2O from 13 to 42% (Figure 11.8c). In the case of halocarbons (not shown here), the concentrations first increase before the trend reverses, due to the projected complete phase-out of halocarbon emissions.

11.3.7.2 Projection of Future Changes in Radiative Forcing

Figure 11.9 shows the derived globally averaged radiative forcing as a function of time for these scenarios, relative to the radiative forcing for the preindustrial background atmosphere. For sulphate aerosols, the direct and indirect radiative forcings were calculated on the basis of sulphur emissions, as discussed in Houghton et al. (2001), except for the aerosol direct forcing due to biomass burning. The radiative forcing due to aerosols from biomass burning, OC and BC direct aerosol forcings, the radiative forcings for all GHGs, and direct and indirect aerosol forcing are estimated based on the method described in Houghton et al. (2001). The direct radiative forcing for biomass burning emissions of OC and BC over the period 2000–2100 was assumed to remain constant at 2000 levels. The contribution from aerosols is probably the most uncertain part of any future radiative forcing projections. Figure 11.9 shows that for the RCP scenarios the 2100 value ranges between $2.85\,\mathrm{W\,m^{-2}}$ and $8.53\,\mathrm{W\,m^{-2}}$. The range of values represents a sizeable increase over the $1.63\,\mathrm{W\,m^{-2}}$ derived for 2000, implying a significant warming tendency. The negative forcing due to tropospheric aerosols offsets some of the GHG positive forcing in all scenarios.

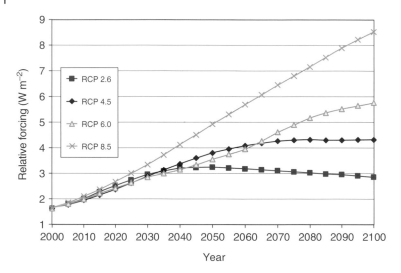

Figure 11.9 Integrated science assessment model (ISAM) estimated radiative forcing from 2000 to 2100 for the Representative Concentration Pathway (RCP) scenarios. See colour plate section for the colour representation of this figure.

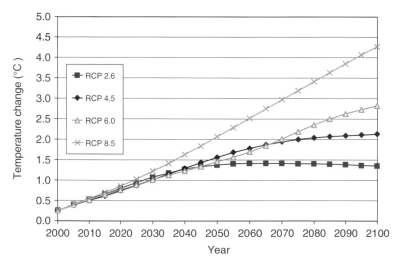

Figure 11.10 Integrated science assessment model (ISAM) estimated global mean temperature change since 2000 for the RCP scenarios. These calculations use a climate sensitivity of 2.8 °C for a doubling of CO_2, which appears to best represent the climate sensitivity of current climate models. See colour plate section for the colour representation of this figure.

11.3.7.3 Future Temperature Changes

The reduced form ISAM has been used in this analysis to estimate the global mean temperature changes for various RCP scenarios described earlier. The temperature change results for all scenarios are shown in Figure 11.10. For the RCP scenarios, the 2000–2100 temperature change ranges between 1.1 and 4.0 °C.

11.4   Potential Impacts of Climate Change

In the previous sections, we briefly discussed the climate change that we might expect in the future due to current and future human activities. There are many uncertainties in our predictions, particularly with regard to the timing, magnitude, and regional patterns of climate change. Nevertheless, scientific studies have shown that human health, ecological systems, and socioeconomic sectors (e.g. hydrology and water resources, food and fibre production, and coastal systems, all of which are vital to sustainable development) are sensitive to changes in climate, as well as to changes in climate variability. A great deal of work has been done to assess the potential consequences of climate change (from early work by Houghton et al. 2001; McCarthy et al. 2001; Metz et al. 2001 through to IPCC AR5 and more recent publications). These, as well as many other studies, have assessed how systems would respond to climate change resulting from an arbitrary doubling of equivalent atmospheric CO_2 concentrations. Here we restrict our discussion to only a brief overview of this important but evolving topic.

11.4.1   Ecosystems

Ecosystems both affect and are affected by climate. As carbon dioxide levels increase, the productivity and efficiency of water use by vegetation may also increase. Warming temperatures will cause the composition and geographical distribution of many ecosystems to shift as individual species respond to changes in climate. As vegetation shifts, this will, in turn, affect climate. Vegetation and other land cover determine the amount of radiation absorbed and emitted by the Earth's surface. As the Earth's radiation balance changes, the temperature of the atmosphere will be affected, resulting in further climate change. Other likely climate change impacts from ecosystems include reductions in biological diversity and in the goods and services that ecosystems provide society.

11.4.2   Water Resources

Climate change may lead to an intensification of the global hydrological cycle and can have major impacts on regional water resources. Reduced rainfall and increased evaporation in a warmer world could dramatically reduce runoff in some areas, significantly decreasing the availability of water resources for crop irrigation, hydroelectric power production, and industrial/commercial and transport uses. In light of the increase in artificial fertilizers, pesticides, feedlot excrement, and hazardous waste dumps, the provision of good-quality drinking water is anticipated to be difficult.

11.4.3   Agriculture

Crop yields and productivity are projected to increase in some areas and decrease in others, especially in the tropics and subtropics, which contain the majority of the world's population. The decrease may be so severe as to cause increased risk of hunger and famine in some locations that already contain many of the world's poorest people. These regions are particularly vulnerable, as industrialized countries may be able to counteract climate change impacts by technological developments, genetic diversity, and maintaining food reserves.

Livestock production may also be affected by changes in grain prices due to pasture productivity. Supplies of forest products such as wood during the next 100 years may also become increasingly inadequate to meet projected consumption due to both climatic and nonclimatic factors. Boreal forests are likely to undergo irregular and large-scale losses of living trees because of the impact of projected climate change. Marine fisheries production is also expected to be affected by climate change. The principal impacts will be felt at the national and local levels.

11.4.4 Sea-Level Rise

In a warmer climate, sea level will rise due to two factors: (i) the thermal expansion of ocean water as it warms, and (ii) the melting of snow and ice from mountain glaciers and polar ice-caps. Over the past 100 years, the global-mean sea level has risen about 25 cm. Over the next 100 years, ISAM projects a further increase of 25–53 cm in global-mean sea level for the RCP scenarios of greenhouse gas emissions discussed above (Figure 11.11). These calculations are based on the 'best guess' climate sensitivity of 2.8 °C. A sea-level rise in the upper part of this range could have very detrimental effects on low-lying coastal areas. In addition to direct flooding and property damage or loss, other impacts may include coastal erosion, increased frequency of storm-surge flooding, saltwater infiltration, and hence pollution of irrigation and drinking water, destruction of estuarine habitats, damage to coral reefs, etc.

11.4.5 Health and Human Infrastructure

Climate change can have an impact on human health through changes in weather, sea level and water supplies, and through changes in ecosystems that affect food security or the geography of vector-borne diseases.

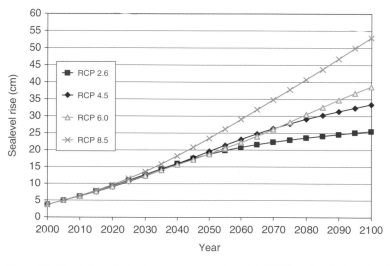

Figure 11.11 Integrated science assessment model (ISAM) estimated sea-level rise (cm) for the RCP scenarios using a 2.8 °C climate sensitivity. See colour plate section for the colour representation of this figure.

In terms of direct effects on human health, increased frequency of heat waves would increase rates of cardiorespiratory illness and death. High temperatures would also exacerbate the health effects of primary pollutants generated in fossil-fuel combustion processes and increase the formation of secondary pollutants such as tropospheric ozone. Changes in the geographical distribution of disease vectors such as mosquitoes (malaria) and snails (schistosomiasis) and changes in life-cycle dynamics of both vectors and infective parasites would increase the potential for transmission of disease. Non-vector-borne diseases such as cholera might increase in the tropics and subtropics because of climatic-change effects on water supplies, temperature, and microorganism proliferation. Concerns for climate-change effects on human health are legitimate; however, research on the impact of these effects is sparse and the conclusions reached by various studies are still highly speculative.

Indirect effects would also result from climatic changes that decrease food production, which would reduce overall global food security and lead to malnutrition and hunger. Shortages of food and fresh water and the disruptive effects of sea-level rise may lead to psychological stresses and the disruption of economic and social systems.

11.5   Summary

There is clear evidence that human activities are already having an impact on Earth's radiation budget and hence on global climate. Potential impacts are projected to be widespread, affecting human health, ecological systems, and socioeconomic sectors. As discussed in this chapter, it is now clear that a certain degree of climatic and ozone changes is now inevitable due to humankind's past and present activities. However, this conclusion has not led to inaction but, rather, to the growing consensus that only a concerted action can help solve the major environmental challenges that still lie ahead.

In its first attempt to confront a global environmental problem, the international community took action to protect the ozone layer through the Montreal Protocol. All nations contributing to the depletion of stratospheric ozone have agreed to control their production of ozone-destroying compounds. As a consequence, abundances of chlorine and other ozone-depleting substances in the atmosphere are decreasing, and stratospheric ozone is expected to complete its recovery within the next few decades.

In contrast, the problem of global warming is being caused by GHGs emitted from activities and energy resources that lie at the core of human society. Moreover, in tackling the problem of global warming, many of the challenges that lie ahead grow out of existing uncertainties. Measurements and modelling results are plagued with uncertainties, which is not surprising when one considers how difficult it is to measure and simulate the many interactions of the Earth's complicated climate system. We know from experience that further research will reduce the uncertainties in some areas. Due to inherent complexity, in other areas the uncertainties may increase and even new ones may be created. Because of these complexities and uncertainties, it is extremely difficult to develop a global policy to control global warming. However, a precautionary approach to the problem of global warming is warranted on the basis of its potential impact and the scale of the response that is necessary if serious impacts are to be avoided. In order to preserve its future, the world needs concrete action *now* to avert future changes in climate and associated consequences. In that sense, the Kyoto agreement (UN 1997) was an important first step in the right direction. Kyoto initiated a policy of small steps taken over 10 years based on a balanced portfolio of policy options. This portfolio began with 'no–regret' strategies that make economic and environmental sense whether or not there is a climate-change problem. It is unavoidable, however, that as a matter of precaution and prudence, global action will eventually have to go far beyond such a 'no-regret' policy. Current consensus is that the global community must follow a two-pronged strategy: namely, to conduct research to narrow down uncertainties in our knowledge, whilst at the same time, taking precautionary measures in response to current knowledge. The Paris Agreement of 2015 does this, but whether global society is able and/or willing to act sufficiently remains to be seen.

Acknowledgements

This work is supported in part by the US Department of Energy (DE-SC0016323) and the US National Science Foundation (NSF-AGS-12-43071).

Questions

1   What has caused climate to change in the past?

2   What is required to limit climate change/sea-level rise to 'acceptable' levels in the face of increasing population and per capita wealth?

3   How uncertain are our projections? What range of values for the climate sensitivity is consistent with the historical record? What do these different sensitivities imply for the future climate?

4   Can we mitigate our greenhouse emissions? What is the time scale of human impacts? If we stopped emitting CO_2 today, how long would it take for atmospheric CO_2 concentrations to return to preindustrial levels?

References

Ackerman, F., DeCanio, S.J., Howarth, R.B., and Sheeran, K. (2009). Limitations of integrated assessment models of climate change. *Climatic Change* 95: 297–315.

Arrhenius, S. (1896). On the influence of carbonic acid in the air upon the temperature of the ground. *Philosophical Magazine and Journal of Science* 5 (41): 238–276.

Bach, W. and Jain, A.K. (1990). CFC greenhouse potential of scenarios possible under the Montreal Protocol. *International Journal of Climatology* 10: 439–450.

Barnola, J.M., Raynaud, D., Korotkevich, Y.S., and Lorius, C. (1987). Vostok ice core provides 160 000-year record of atmospheric CO_2. *Nature* 329: 408–414.

Boden, T. and Andres, B. (2017). *Global CO_2 Emissions from Fossil-Fuel Burning, Cement Manufacture, and Gas Flaring: 1751–2014*. Oak Ridge, TN: Carbon Dioxide Information Analysis Center, Oak Ridge National Laboratory, U.S. Department of Energy.

Brohan, P., Kennedy, J.J., Harris, I. et al. (2006). Uncertainty estimates in regional and global observed temperature changes: a new data set from 1850. *Journal of Geophysical Research* 111: D12106. https://doi.org/10.1029/2005JD006548.

Butler, J.H., Battle, M., Bender, M.L. et al. (1999). A record of atmospheric halocarbons during the twentieth century from polar firn air. *Nature* 399: 749–755.

Chappellaz, J., Barnola, J.M., Raynaud, D. et al. (1990). Atmospheric CH_4 record over the last climatic cycle revealed by the Vostok ice core. *Nature* 345: 127–131.

Harnisch, J. and Eisenhauer, A. (1998). Natural CF_4 and SF_6 on Earth. *Geophysical Research Letters* 25: 2401–2404.

Harvey, D., Gregory, J., Hoffert, M. et al. (1997). *An Introduction to Simple Climate Models Used in the IPCC Second Assessment Report*, IPCC Technical Paper 2, 50 pp. (eds. J.T. Houghton, L.G.M. Filho, K. Griggs and D.J. Maskell). Cambridge/Geneva: Cambridge University Press/IPCC.

Houghton, J.T., Meira Filho, L.G., Callander, B.A. et al. (eds.) (1996). *Climate Change 1995. The Science of Climate Change. Intergovernmental Panel on Climate Change*, 572 pp. Cambridge University Press.

Houghton, J.T., Ding, Y., Griggs, D.J. et al. (eds.) (2001). *Climate Change 2001, the Scientific Basis*. Cambridge, UK: Cambridge University Press.

IPCC (2013). *Climate Change 2013: The Physical Science Basis. Contribution of Working Group 1 to the Fifth Assessment Report of the Intergovernmental Panel on Climate Change* (eds. T.F. Stocker, D. Qin, G.-K. Plattner, et al.), 1535. Cambridge, UK: Cambridge University Press.

IPCC (2018). Special Report 15, Global Warming of 1.5 °C. https://www.ipcc.ch/sr15.

Jain, A.K. (2019). *Web-Interface of the Reduced Form Version of Integrated Science Assessment Model (ISAM)*. Urbana, IL: University of Illinois http://climate.atmos.uiuc.edu/isam2/index.html.

Jain, A.K. and Bach, W. (1994). The effectiveness of measures to reduce the man-made greenhouse effect: the application of a climate-policy-model. *Theoretical and Applied Climatology* 49: 103–118.

Jain, A.K. and Yang, X. (2005). Modelling the effects of two different land cover change data sets on the carbon stocks of plants and soils in concert with CO_2 and climate change. *Global Biogeochemical Cycles* 19 https://doi.org/10.1029/2004GB002349.

Jain, A.K., Kheshgi, H.S., and Wuebbles, D.J. (1994). Integrated science model for assessment of climate change model. In: *Proceedings of Air and Waste Management Association's 87th Annual Meeting*. Cincinnati, OH: US Department of Energy 19–24 June.

Jain, A.K., Kheshgi, H.S., Hoffert, M.I., and Wuebbles, D.J. (1995). Distribution of radiocarbon as a test of global carbon cycle models. *Global Biogeochemical Cycles* 9: 153–166.

Jain, A.K., Kheshgi, H.S., and Wuebbles, D.J. (1996). A globally aggregated reconstruction of cycles of carbon and its isotopes. *Tellus* 48B: 583–600.

Jain, A.K., Briegleb, B.P., Minschwaner, K., and Wuebbles, D.J. (2000). Radiative forcings and global warming potentials of thirty-nine greenhouse gases. *Journal of Geophysical Research* 105: 20773–20790.

Jouzel, J., Lorius, C., Petit, J.R. et al. (1987). Vostok ice core: a continuous isotope temperature record over the last climatic cycle (160 000 years). *Nature* 329: 403–407.

Keeling, C.D., Piper, S.C., Bacastow, R.B. et al. (2001). *Exchanges of atmospheric CO_2 and $13CO_2$ with the terrestrial biosphere and oceans from 1978 to 2000. I. Global Aspects*, SIO Reference Series, No. 01–06, 88 pages. San Diego, CA: Scripps Institution of Oceanography http://escholarship.org/uc/item/09v319r9.

Keller, C.F. (1998). *Global Warming: An Update*. Los Alamos, CA: Las Alamos National Laboratory.

Kheshgi, H.S., Jain, A.K., Kotamarthi, R., and Wuebbles, D.J. (1999). Future atmospheric methane concentrations in the context of the stabilization of greenhouse gas concentrations. *Journal of Geophysical Research* 104: 19183–19190.

Kirschke, S., Bousquet, P., Ciais, P. et al. (2013). Three decades of global methane sources and sinks. *Nature Geoscience* 6 (10): 813–823.

Le Quéré, C., Andrew, R.M., Friedlingstein, P. et al. (2018). Global carbon budget 2018. *Earth System Science Data* 10: 2141–2194.

Lefohn, A.S., Husar, J.D., and Husar, R.B. (1999). Estimating historical anthropogenic global sulfur emission patterns for the period 1850–1990. *Atmospheric Environment* 33: 3435–3444.

Lorius, C., Jouzel, J., Ritz, C. et al. (1985). 150 000-year climatic record from Antarctic ice. *Nature* 316: 591–595.

MacKay, R.M. and Khalil, M.A.K. (1994). Climate simulations using the GCRC 2-D zonally averaged statistical dynamical climate model. *Chemosphere* 29: 2651–2683.

Manabe, S. and Wetherald, R.T. (1967). Thermal equilibrium of the atmosphere with a given distribution of relative humidity. *Journal of the Atmospheric Sciences* 24: 241–259.

McCarthy, J.J., Canziani, O.F., Leary, N.A. et al. (eds.) (2001). *Climate Change 2001, Impacts, Adaptation and Vulnerability*. Cambridge, UK: Cambridge University Press.

Metz, B., Davidson, O., Swart, R., and Pan, J. (eds.) (2001). *Climate Change 2001, Mitigation*. Cambridge, UK: Cambridge University Press.

Molina, M.J. and Rowland, F.S. (1974). Stratospheric sink for chlorofluoromethanes: chlorine-atom catalyzed destruction of ozone. *Nature* 249: 810–814.

Moss, R.H., Edmonds, J.A., Hibbard, K.A. et al. (2010). The next generation of scenarios for climate change research and assessment. *Nature* 463: 747–756.

National Academies of Sciences, Engineering and Medicine report (2016). *Attribution of Extreme Weather Events in the Context of Climate Change*. Washington, DC: The National Academies Press.

NOAA (2018). Climate Monitoring and Diagnostics Laboratory Nitrous Oxide (N_2O). Combined Data Set. https://www.esrl.noaa.gov/gmd/hats/combined/N2O.html.

Pall, P., Aina, T., Stone, D.A. et al. (2011). Anthropogenic greenhouse gas contribution to flood risk in England and Wales in autumn 2000. *Nature* 470: 382–385.

Peixoto, J.P. and Oort, A.H. (1992). *Physics of Climate*, 520 pp. New York, NY: American Institute of Physics.

Petit, J.R., Mounier, L., Jouel, J. et al. (1990). Paleoclimatological and chronological implications of the Vostok core dust record. *Nature* 343: 56–58.

Royal Society (2008). Ground-level ozone in the 21st century: future trends, impacts and policy implications. Science policy report 15/08 ISBN 978–0–85403-713-1.

Sun, S. and Hansen, J.E. (2003). Climate simulations for 1951–2050 with a coupled atmosphere–ocean model. *Journal of Climate* 16: 2807–2826.

Tegen, I., Doch, D., Lacis, A.A., and Sato, M. (2000). Trends in tropospheric aerosol loads and corresponding impact on direct radiative forcing between 1950 and 1990: a model study. *Journal of Geophysical Research* 105: 26971–26989.

UN (1997). *Kyoto Protocol to the United Nations Framework Convention on Climate Change*. New York: United Nations.

Van Vuuren, D.P., Edmonds, J., Kainuma, M. et al. (2011). The representative concentration pathways: an overview. *Climatic Change* 109 (1–2): 5–31.

WMO (2019a). *Scientific Assessment of Ozone Depletion: 2018. Report No. 58, Global Ozone Research and Monitoring Project*. Geneva, Switzerland: World Meteorological Organization https://www.esrl.noaa.gov/csd/assessments/ozone/2018.

WMO (2019b). WMO provisional statement on the state of the global climate in 2018. http://ane4bf-datap1.s3-eu-west-1.amazonaws.com/wmocms/s3fs-public/ckeditor/files/Draft_Statement_7_February.pdf?5.6rzIGwBm5lwDSTPbgprB2_EgrjzRVY.

Appendix

Suggested Web Resources

Chapter 1

University of Reading, Climate Analysis Group: http://www.reading.ac.uk/research/themes/theme-environment/rd-climate.aspx

The University Corporation for Atmospheric Research (UCAR): www.ncar.ucar.edu

UK Met Office Climate information: https://www.metoffice.gov.uk

NOAA Climate Analysis Branch: www.noaa.gov

Chapter 2

The young faint Sun paradox and the age of the solar system: http://www.answersingenesis.org/tj/v15/i2/faintsun.asp

The atmosphere of early Earth: http://eesc.columbia.edu/courses/ees/climate/lectures/earth.html

Origin and evolution of the atmosphere: http://www.chem1.com/acad/webtext/geochem/09txt.html

Highly recommended, with many links: http://www.chem1.com/acad/webtext/geochem

Chapter 3

The Earth's energy budget: http://earthobservatory.nasa.gov/Library/RemoteSensingAtmosphere

The structure of the atmosphere: https://www.metoffice.gov.uk/learning/learn-about-the-weather

Aerosol effects on climate (see Chapter 7): https://www.ipcc.ch/report/ar5/wg1/#

Chapter 4

Site information for the UK Government's air quality monitoring networks: https://uk-air.defra.gov.uk/networks/network-info?view=automatic

Intergovernmental Panel on Climate Change: www.ipcc.ch

Carbon Dioxide Information Analysis Centre (CIDAC) reports available at: https://cdiac.ess-dive.lbl.gov

Chapter 5

Air pollution in the USA: www.airnow.gov

Air pollution in the UK: www.airquality.co.uk

Air quality in India: http://safar.tropmet.res.in/index.php

Air quality data worldwide (an amazing real-time resource): www.aqicn.org/here

Atmospheric Science for Environmental Scientists, Second Edition. Edited by C.N. Hewitt and Andrea V. Jackson.
© 2020 John Wiley & Sons Ltd. Published 2020 by John Wiley & Sons Ltd.

Chapter 6

On-line Aerosol Inorganics Model (AIM) Clegg, Brimblecombe and Wexler: http://www.aim.env.uea.ac.uk/aim/aim.htm

Henry's law constants: https://www.atmos-chem-phys.net/15/4399/2015/acp-15-4399-2015.pdf

Chapter 7

Global Atmosphere Watch Programme: https://public.wmo.int/en/programmes/global-atmosphere-watch-programme

US Environmental Protection Agency: www.epa.gov

Chapter 8

Scientific Assessment of Ozone Depletion: 2018: https://www.esrl.noaa.gov/csd/assessments/ozone/2018

Stratospheric Processes and Climate (SPARC) initiative of the World Climate Research Programme: http://www.atmosp.physics.utoronto.ca/SPARC/index.html

Information on reaction rate coefficients for reactions of importance in the stratosphere and troposphere: http://www.iupac-kinetic.ch.cam.ac.uk, http://jpldataeval.jpl.nasa.gov

Chapter 9

UN reports on urbanization of the world's population: http://www.un.org/en/development/desa/population/theme/urbanization/index.shtml

NASA's Earth Observatory data on normalized difference vegetation index: https://earthobservatory.nasa.gov/global-maps/MOD_NDVI_M

International Association for Urban Climate (free membership, informative newsletters): www.urban-climate.org

UK national air quality archive, with information about pollution, data, official reports, etc.: https://uk-air.defra.gov.uk

Chapter 10

European Environment Agency emissions inventory: https://www.eea.europa.eu/publications/emep-eea-guidebook-2016

USEPA (United States Environmental Protection Agency): https://www.epa.gov/environmental-topics/air-topics

EEA (European Environment Agency): www.eea.europa.eu

IEA (International Energy Agency): www.iea.org

GEMS/AIR (Global Environment Monitoring System): http://apps.who.int/iris/handle/10665/58355

OECD (Organization for Economic Cooperation and Development): www.oecd.org

EMEP (European Monitoring and Evaluation Programme): www.emep.int

Chapter 11

Intergovernmental Panel on Climate Change: www.ipcc.ch

US Environmental Protection Agency: www.epa.gov

UK Government guidance on climate change: https://www.gov.uk/guidance/climate-change-explained#climate-change-now

Index

Note: Page numbers in *italics* refer to Figures; those in **bold** to Tables.

Atmospheric Science for Environmental Scientists, Second Edition. Edited by C.N. Hewitt and Andrea V. Jackson.
© 2020 John Wiley & Sons Ltd. Published 2020 by John Wiley & Sons Ltd.

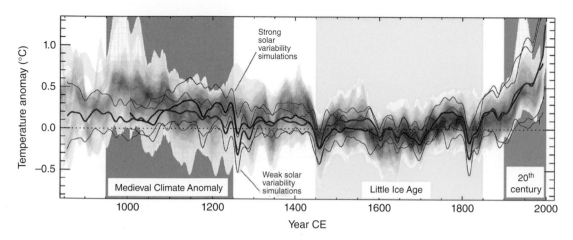

Figure 1.6 Simulated (lines) and reconstructed (grey shading) Northern Hemisphere average surface temperature changes since the Middle Ages. All data are expressed as anomalies from their 1500–1850 mean and smoothed with a 30-year filter. *Source:* data from, and more details available at, http://www.ipcc.ch/report/ar5/syr.

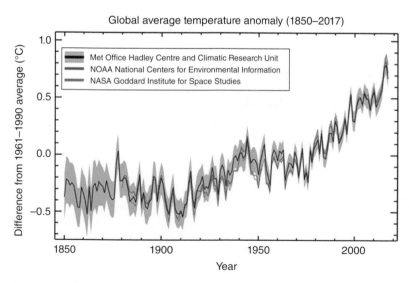

Figure 1.7 Global average surface temperature anomaly 1850–2017, relative to the 1981–2010 average. The grey shading shows the 95% uncertainty for the Met Office Hadley Centre/Climatic Research Unit HadCRUT4 data set and the solid lines show two other data sets. The figure is reproduced from the Hadley Centre for Climate Prediction and Research and more details are available at www.metoffice.gov.uk/research/news/2018/global-surface-temperatures-in-2017.

Atmospheric Science for Environmental Scientists, Second Edition. Edited by C.N. Hewitt and Andrea V. Jackson.
© 2020 John Wiley & Sons Ltd. Published 2020 by John Wiley & Sons Ltd.

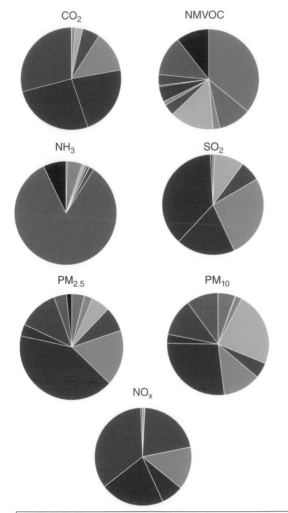

Figure 5.2 UK emission statistics by United Nations Economic Commission of Europe (UNECE) source category (1, combustion in energy production and transformation; 2, combustion in commercial, institutional, residential and agriculture; 3, combustion in industry; 4, production processes; 5, extraction and distribution of fossil fuels; 6, solvent use; 7, road transport; 8, other transport and mobile machinery; 9, waste treatment and disposal; 10, agriculture, forestry and land-use change; 11, nature).

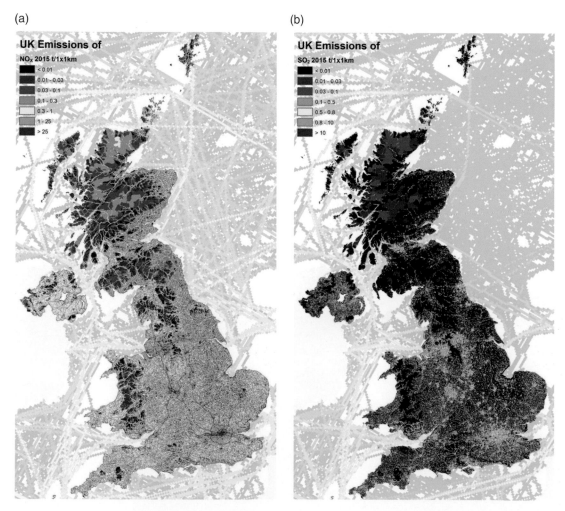

Figure 5.3 Emission maps (2002) for the UK on a 1 km × 1 km grid for (a) NO$_2$ and (b) SO$_2$. *Source:* data from UK National Atmospheric Emissions Inventory, www.naei.org.uk.

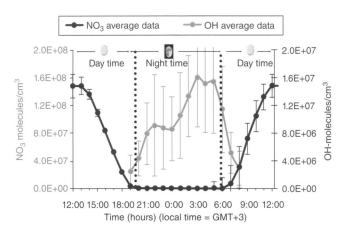

Figure 5.13 Concentrations of N_2O_5 and NO_3 from airborne measurements off the UK (Kennedy et al. 2011).

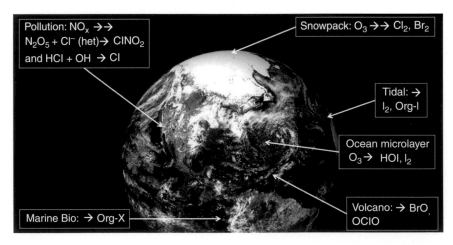

Figure 5.17 Primary sources of reactive halogen species or their precursor reservoir species (Simpson et al. 2015).

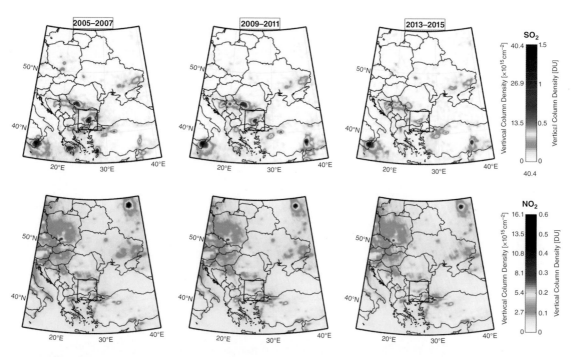

Figure 5.20 Three year average satellite instrument measured SO_2 over Eastern Europe. The largest source is Mt. Etna, Sicily, Italy. The blue box is around the coal mining and burning power stations in Bulgaria, showing a 50% reduction owing to the installation of flue gas desulphurization. (Krotkov et al. 2016).

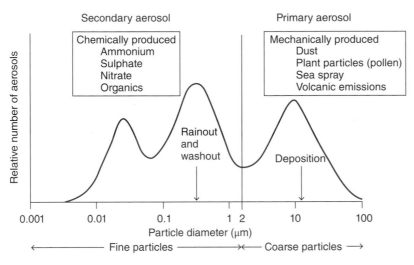

Figure 5.23 Size and relative number distribution of typical urban aerosols. *Source:* adapted from Whitby (1978).

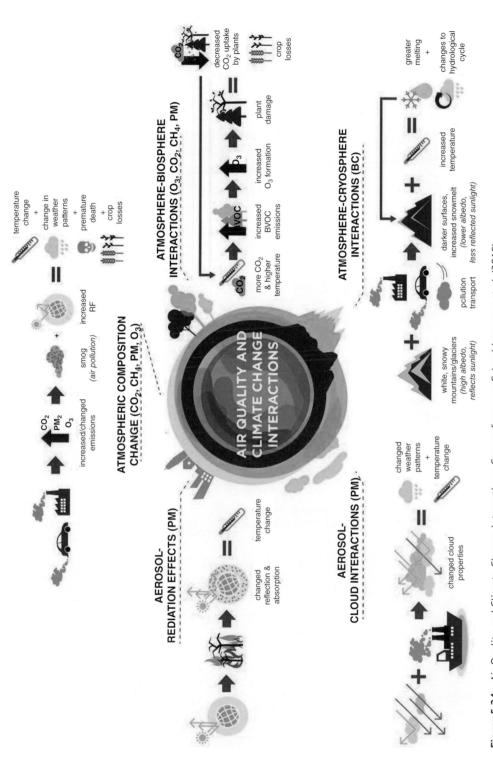

Figure 5.24 Air Quality and Climate Change Interactions. *Source:* from von Schneidemesser et al. (2015).

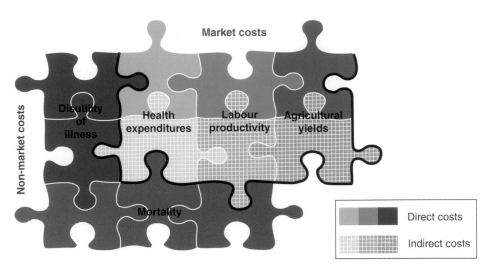

Figure 10.8 Direct and indirect costs of air pollution. Reprinted from OECD 2016 under CC-BY licence.

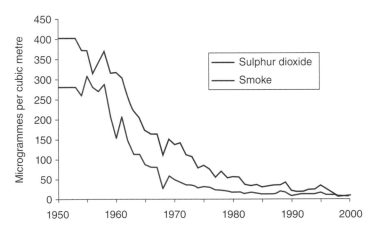

Figure 10.11 Annual average smoke and sulphur dioxide concentrations in London 1950–2000. Before 1954, data was only published as five-year averages (after 2000, the concentrations are too low to plot on this graph). *Source:* Greater London Authority (2002).

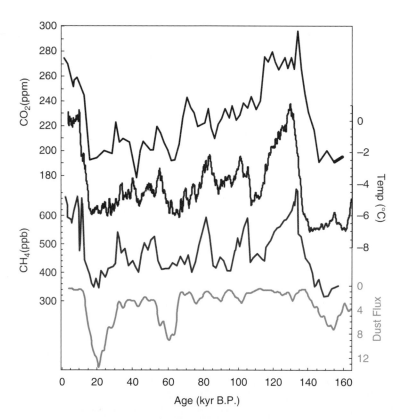

Figure 11.1 Climate and atmospheric composition records from the Vostok ice-core (East Antarctica) covering the past 160 000 years. These include CO_2 and CH_4 greenhouse gas records, which are closely tied to Antarctic temperature variations over the last full glacial–interglacial climate cycle (Lorius et al. 1985; Barnola et al. 1987; Jouzel et al. 1987; Chappellaz et al. 1990). Temperature data are plotted as deviations from the present-day mean annual temperature. Also included is the record of the flux of dust to the area (shown on an inverted scale for comparison purposes) (Petit et al. 1990).

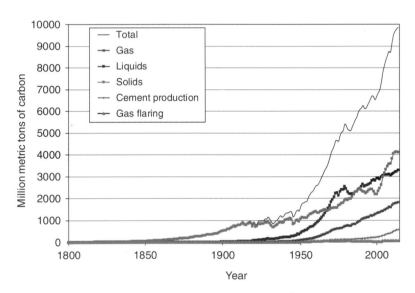

Figure 11.2 Global CO_2 emissions from fossil-fuel burning, cement manufacture, and gas flaring: 1751–2014. *Source:* Boden and Andres (2017).

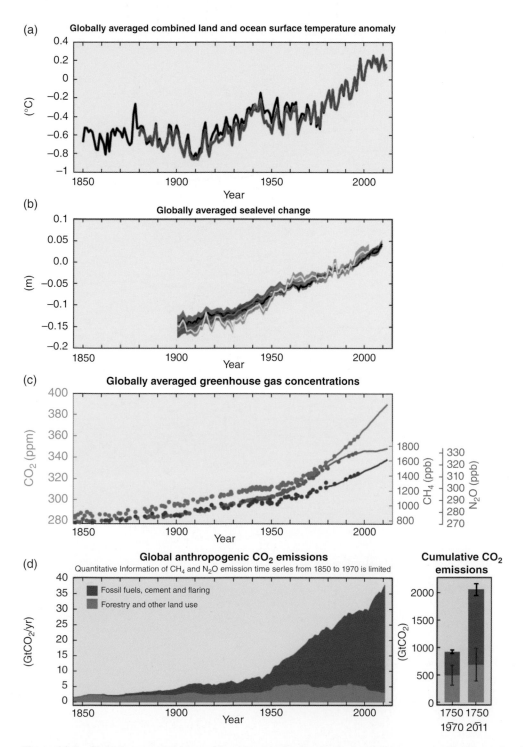

Figure 11.5 Globally averaged (a) combined land and ocean surface temperature anomaly, (b) sea-level change, (c) greenhouse gas concentrations, and (d) anthropogenic CO_2 emissions, since 1850, as presented in the IPCC AR5 report (IPCC 2013).

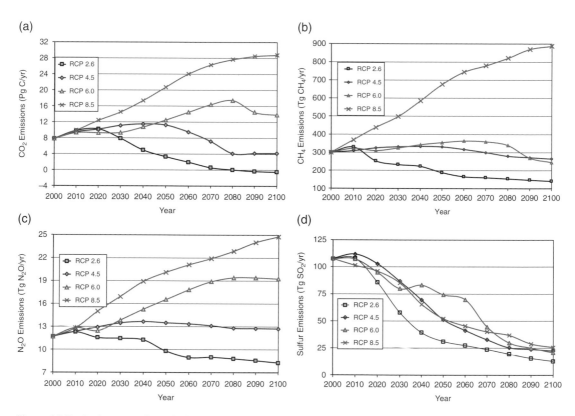

Figure 11.7 Anthropogenic emissions in the four scenarios of the Representative Concentration Pathway (RCP) for (a) CO_2, (b) CH_4, (c) N_2O, and (d) SO_2. Note that the RCP emission values are standardized such that emissions in 2000 are identical in all scenarios.

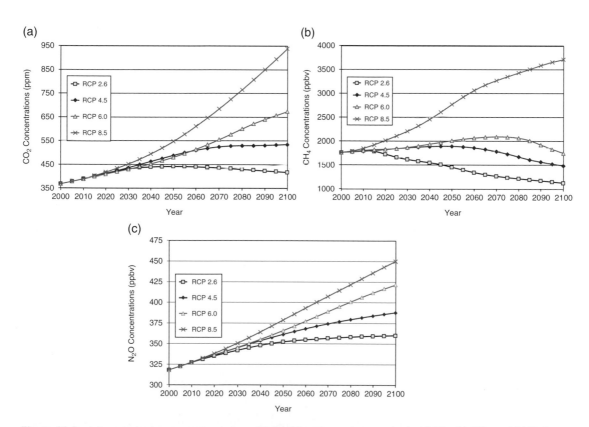

Figure 11.8 Integrated science assessment model (ISAM) estimated atmospheric (a) CO_2, (b) CH_4, and (c) N_2O concentrations from 2000 to 2100 for the RCP scenarios.

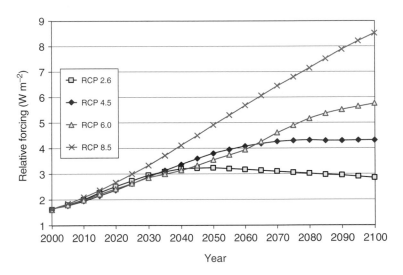

Figure 11.9 Integrated science assessment model (ISAM) estimated radiative forcing from 2000 to 2100 for the Representative Concentration Pathway (RCP) scenarios.

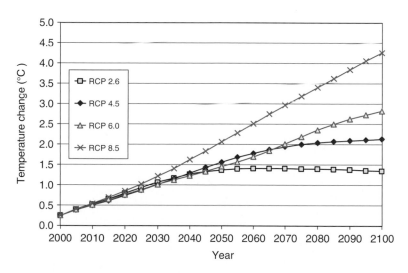

Figure 11.10 Integrated science assessment model (ISAM) estimated global mean temperature change since 2000 for the RCP scenarios. These calculations use a climate sensitivity of 2.8 °C for a doubling of CO_2, which appears to best represent the climate sensitivity of current climate models.

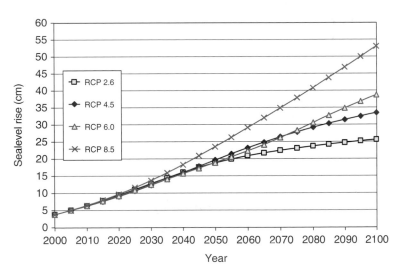

Figure 11.11 Integrated science assessment model (ISAM) estimated sea-level rise (cm) for the RCP scenarios using a 2.8 °C climate sensitivity.

Atmospheric Science for Environmental Scientists